Catesby's
Natural History

AF729643

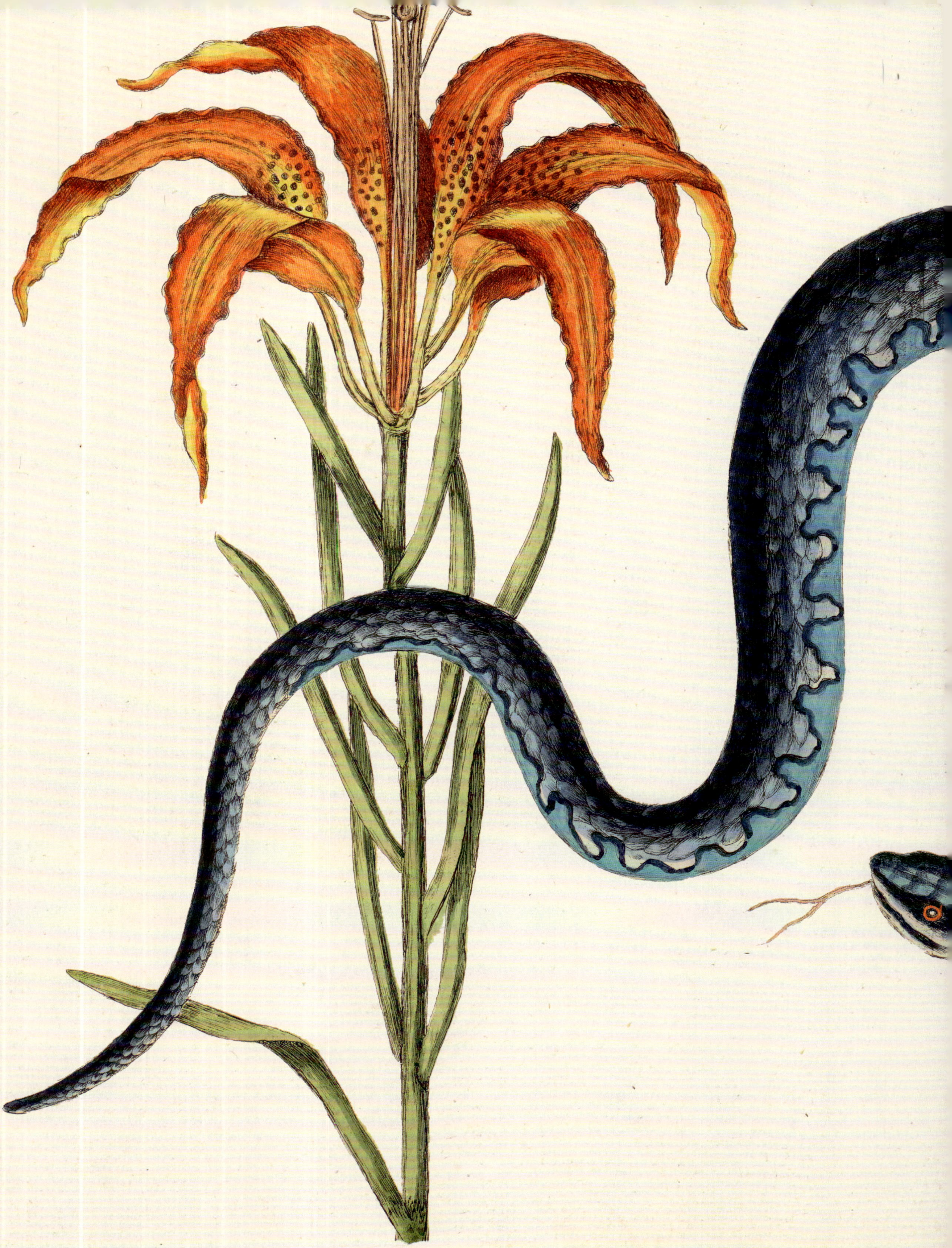

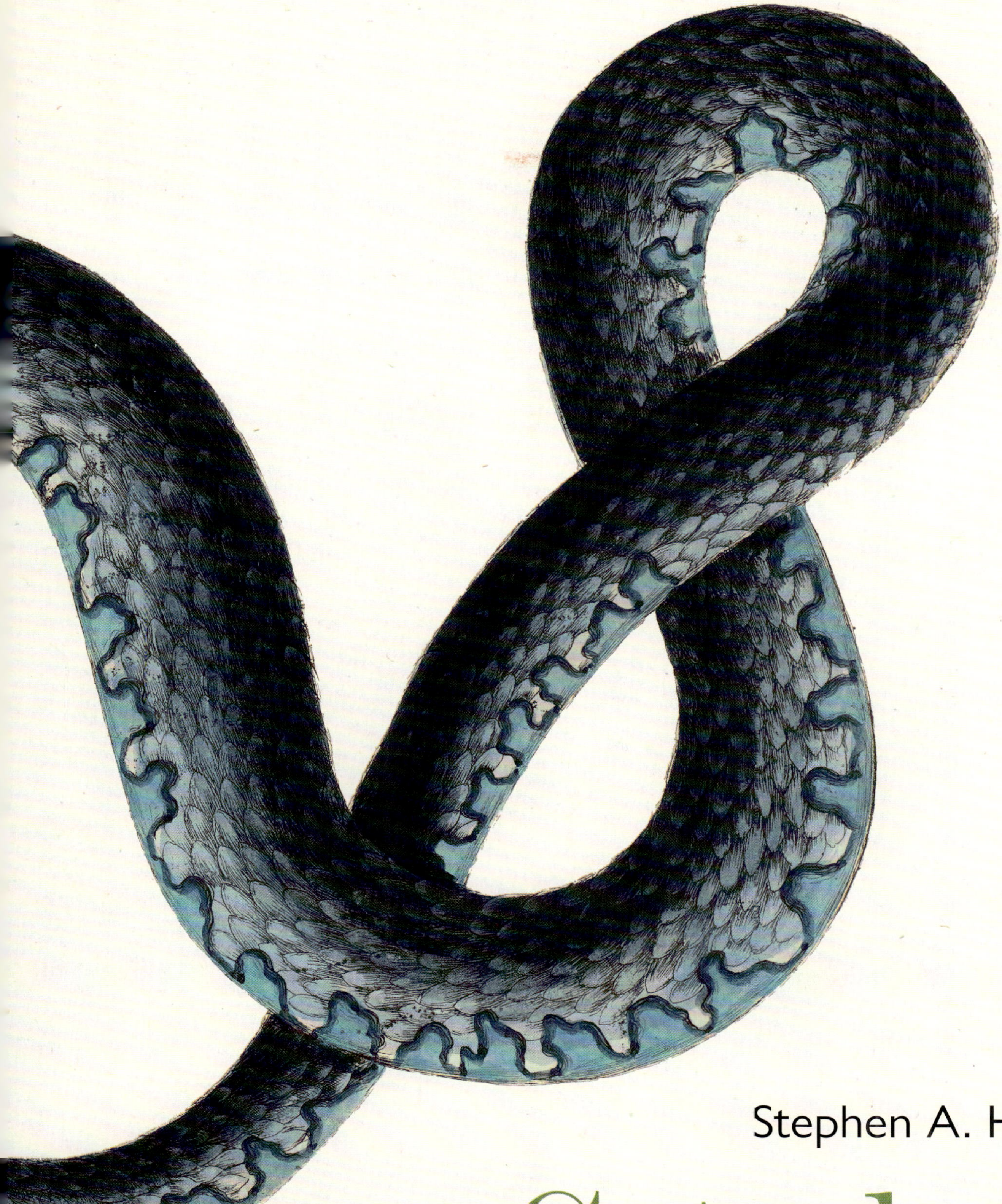

Stephen A. Harris

Catesby's Natural History

BODLEIAN
LIBRARY
PUBLISHING

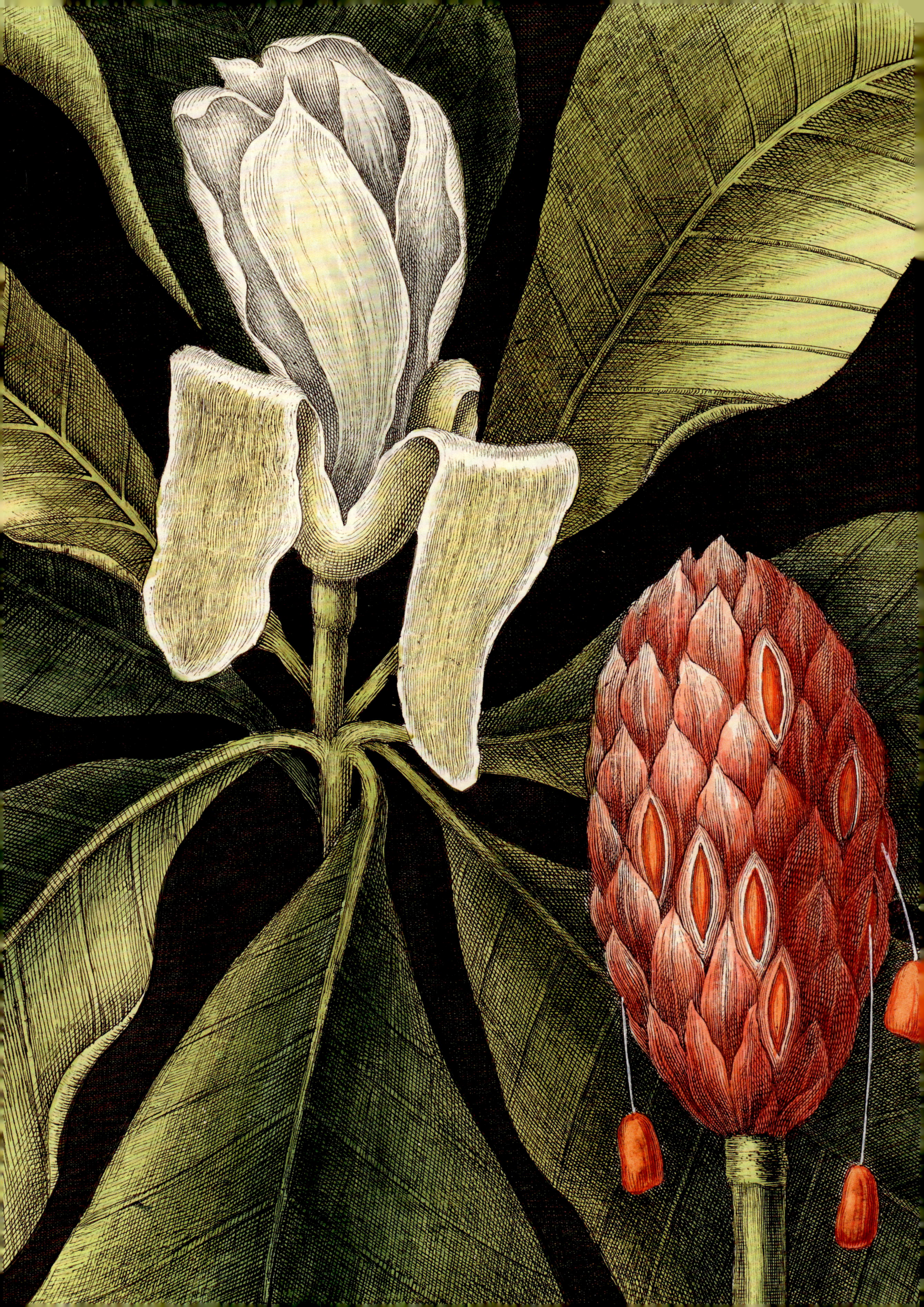

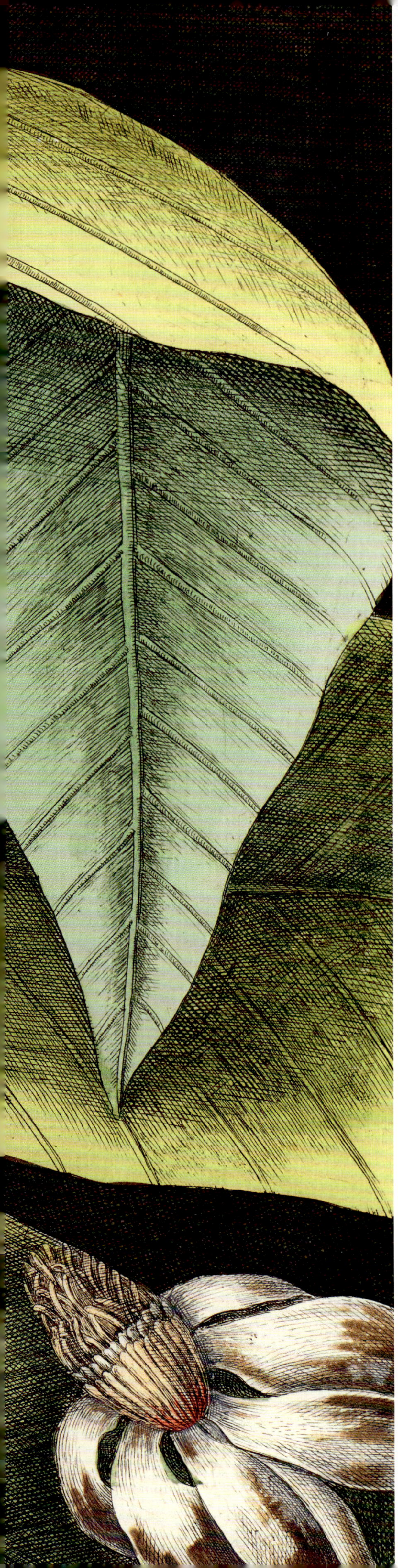

Contents

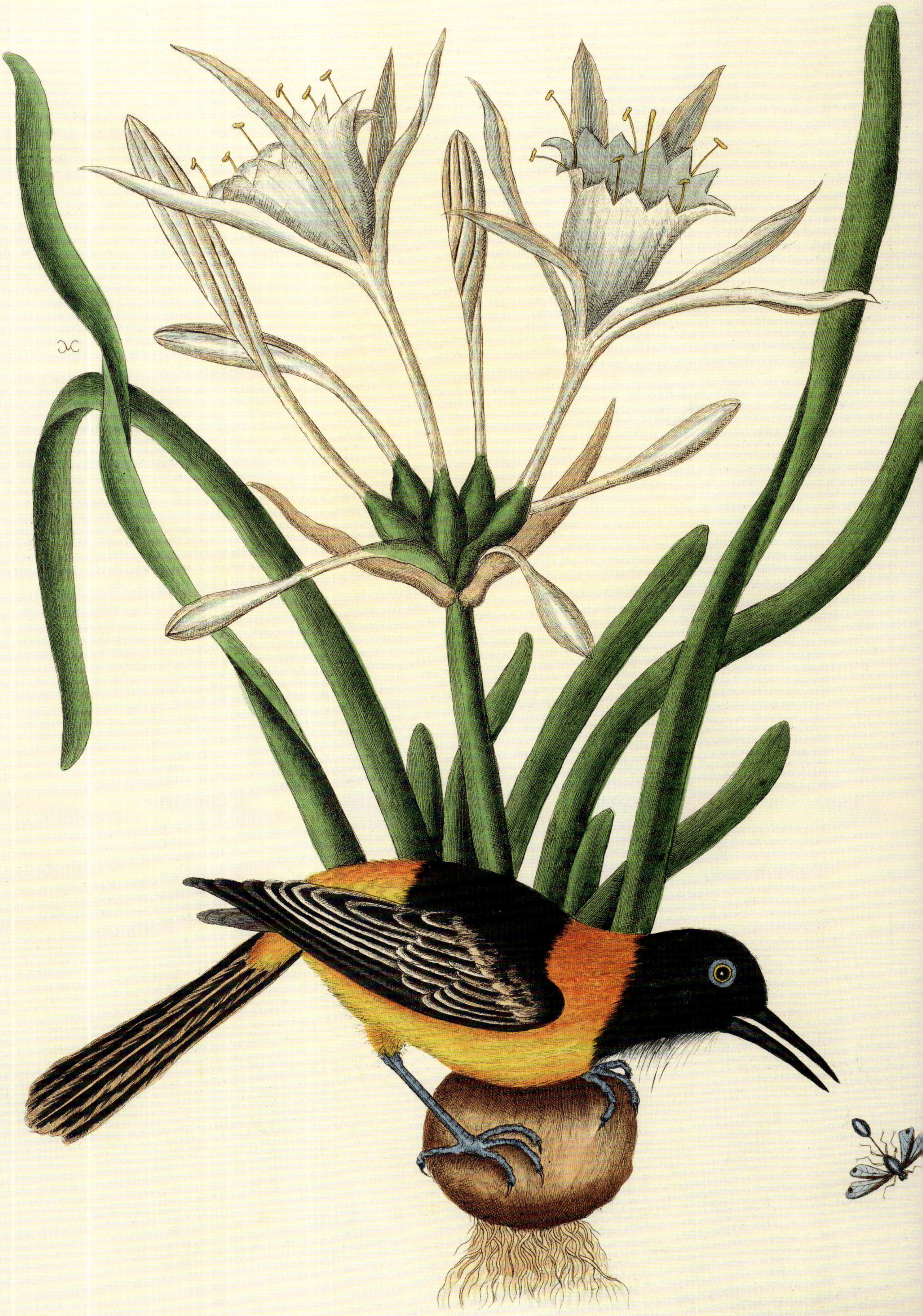

Preface

Mark Catesby, one of Philip Miller's so-called 'Procurers of Plants',[1] was a naturalist, sponsored by a group of English gentlemen to explore the southernmost Thirteen Colonies and the Bahamas in the early eighteenth century. His authorship and illustration of *The Natural History of Carolina, Florida and the Bahama Islands* (1729–47) made his name among the virtuosi of Georgian England. Moreover, the living plants he sent to England became incorporated into the gardens of the powerful and wealthy. Today, some of these plants remain commonplace horticultural staples.

In eighteenth-century Europe there was sufficient wealth and interest in natural history to support a market for high-quality illustrated books about areas that Europeans were only starting to explore. Systematic curiosity about the economic potential of these areas placed a high value on accurate information. Within this framework, Catesby's book was novel – there would be no equivalent for the plants and animals of North America until the nineteenth century. Books such as John Lawson's *A New Voyage to Carolina* (1709) were authoritative but had few illustrations, while illustrated volumes contained selections of images often based on knowledge gained from plants in European gardens or animals in European museums. In contrast, Catesby's *Natural History* combined the author's extensive field observations with his singular approach to the depiction of North American plants and animals across 220 hand-coloured plates. Catesby's audience, based primarily outside North America, were curious about the region's animals and plants.

The *Natural History* demonstrates the power of the right image, at the right time and in the right place, to capture people's imaginations. Had Catesby only collected specimens rather than prepared and illustrated the *Natural History*, it is likely that he would, like so many, have become a footnote in the history of our knowledge of the diversity of life in North America. His widely appreciated illustrations catalysed interest in the natural history of colonial America and, to modern audiences, also demonstrate the vulnerability and fragility of North American species and landscapes.[2]

Complex networks of patronage, obligation, commerce and precedent develop around the activities of naturalists, especially if their activities are to resonate beyond their lifetimes. Consequently, it was necessary for Catesby to foster other collaborations beyond the financial support of backers in London and the immediate practical help of people in the Carolinas and the Bahamas. Nurserymen and their clients, such as Thomas Fairchild, James Sherard, Charles Dubois, Peter Collinson and Thomas Knowlton, were needed to germinate and multiply the seeds he was returning, seeds that eventually helped establish North American plants in European horticulture. Moreover, if his observations were to have wide circulation, illustrators, printers, publishers and booksellers would also become necessary. Catesby's activities as a naturalist were cultural, influenced by his predecessors and peers and by the places, times and societies in which he lived.[3]

The illustrations reproduced in the present volume are all taken from a copy of the second edition of Catesby's *Natural History* (1754), housed in the Bodleian Library, University of Oxford.[4] It is appropriate that these reproduced images come from a copy of Catesby's work in the university where the botanical collections are founded upon the library and herbarium of William Sherard, one of Catesby's most ardent supporters but someone who never saw any of his protégé's work published. As the *Natural History* was sold unbound, the relative order of plates in different copies of the same edition may vary. In this book, the plates are presented in the order in which they appeared in

Plate 205 (detail)

volume 1 (**Plates 1–100**), volume 2 (**Plates 101–200**) and the appendix (**Plates 201–20**).[5]

The association of Catesby's images with Linnaean scientific names was first made by George Edwards, when the third edition of the *Natural History* was published in 1771.[6] Since then, numerous scholars, especially James Reveal, have critically evaluated Catesby's plates as to the identities of the organisms depicted.[7] The identifications presented here mostly follow Reveal's later views, with nomenclature modernized where appropriate. Each plate is accompanied by a caption that includes the most widely used English common name, with the scientific name in parentheses, followed by the volume and plate number, also in parentheses, for example, **Plate 189**: Northern needleleaf (*Tillandsia balbisiana*); Carolina grasshopper (*Dissosteira carolina*) (vol. 2, t. 89).

The plates are preceded by an extended essay that draws upon Catesby's published work, his letters and those of his patrons, together with his herbarium specimens and the work of Catesby scholars over the last fifty years. This informal essay is divided into eight chapters. Chapter 1 summarizes the state of natural history observations in the Thirteen Colonies before Catesby's work. Chapter 2 is a brief biography of Mark Catesby. His account of his North American activities in the *Natural History* is not a diary of a journey, where each location can be identified and the order in which places were visited determined. Rather it represents the considered reflections of someone publishing their field observations decades after the fact, perhaps wrestling with accumulating information and trying to appeal to an audience of book buyers. Catesby's struggle is reflected in subsequent chapters, where his observations on the landscapes through which he travelled are presented (Chapter 3), together with his reflections on Indigenous and enslaved peoples (Chapter 4) and his approaches to the collection of preserved and living specimens (Chapter 5). Catesby's illustrations are the focus of Chapter 6, whilst a brief history of the publication of the *Natural History* is presented in Chapter 7. The final chapter considers the legacies of Catesby's activities, which continue to have relevance in the modern world.

Note on language

To modern audiences some of the language and attitudes reflected in the quotes will be offensive, reflecting both Catesby's attitudes and those of his time. Catesby used the term 'Indian' to refer to Native Americans, rarely making more subtle distinctions between the multiple peoples who identify as Native Americans. The term 'negro' almost invariably refers to an enslaved black African person.

Acknowledgements

My introduction to Mark Catesby was not through his illustrations but via my interests in pre-Linnaean natural history collecting, botany and plant introductions to Britain. Oxford University Herbaria is fortunate to house many of the specimens collected by Catesby, which then led me to his *Natural History* and its illustrations. My understanding and appreciation of Catesby as an artist and illustrator have benefited greatly from conversations with Henrietta McBurney, whose book *Illuminating Natural History: The Art and Science of Mark Catesby* (2021) is a major resource for Catesby scholars.

Many thanks are due to Samuel Fanous, Head of Publishing, Bodleian Library, for initially suggesting that the illustrations in Mark Catesby's *Natural History* would be interesting to reproduce and for his willingness to include substantial introductory materials. The enthusiasm and professionalism of Janet Phillips, Leanda Shrimpton and Dot Little at Bodleian Library Publishing made the process of putting this book together a joy.

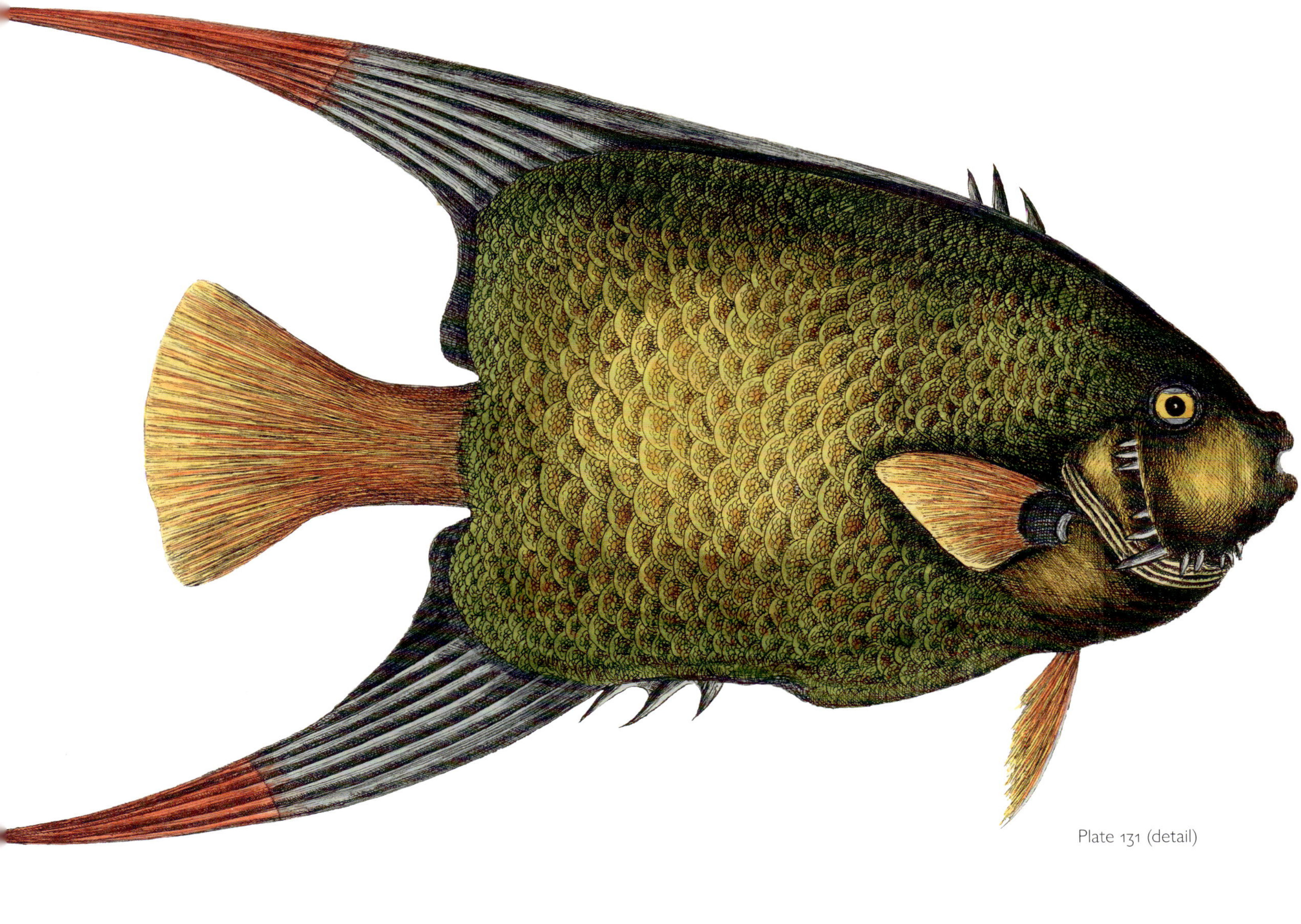

Plate 131 (detail)

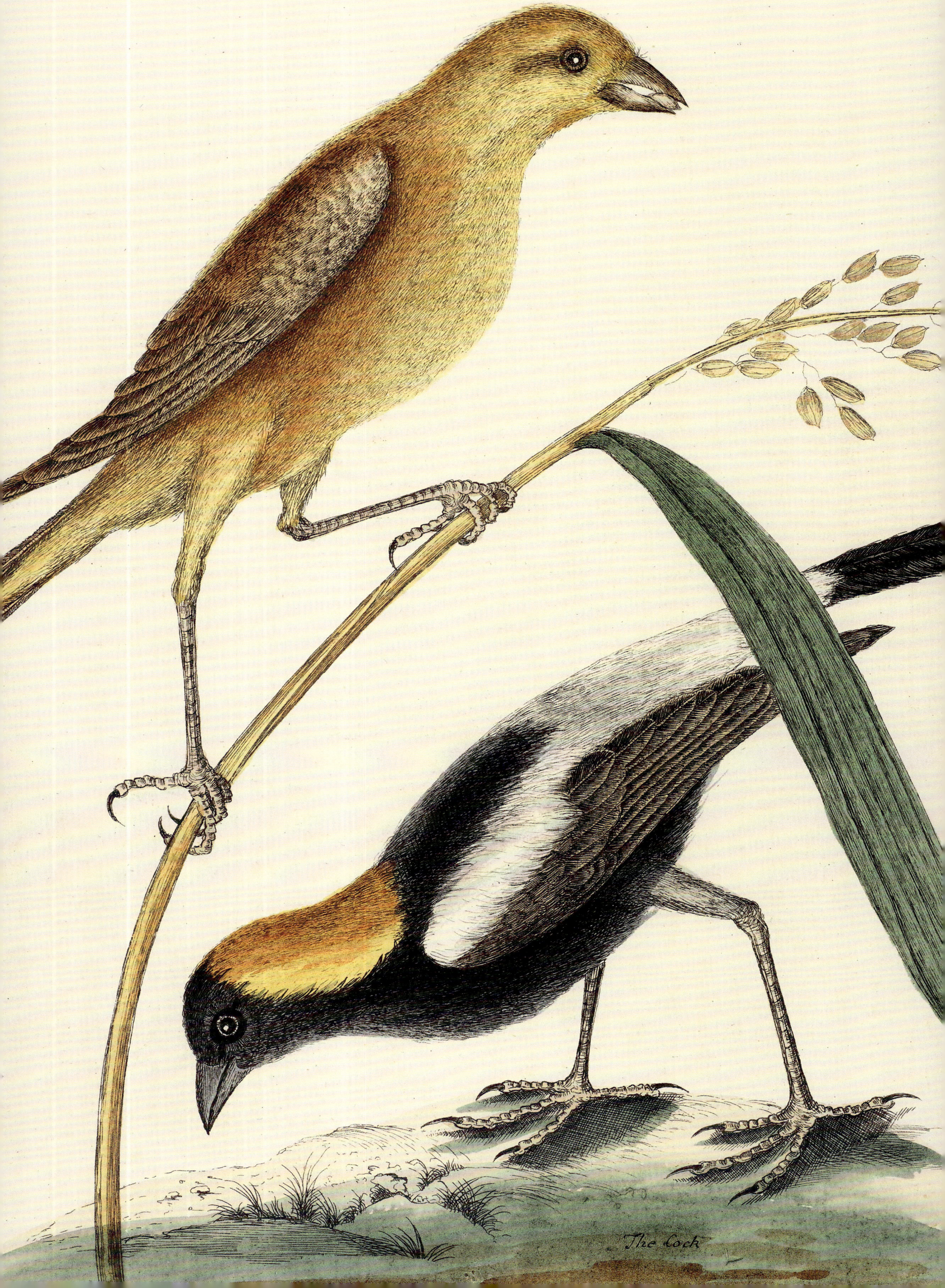
The Cock

Natural History in the Thirteen Colonies 1

Until … 1858, not a single work of any importance was published in this country [United States], either on anatomy or on the physiology of plants, not even a single one of the many [classification] systems ever proposed had its origin in America. And yet the labors of American and foreign scientists in America contributed their large share to the advancement of science.

FREDERICK BRENDEL[1]

Mark Catesby (1683–1749) arrived in North America fewer than 150 years after the first European colonies were set up there. The fringes of the coast and the larger rivers had been settled, forests stripped and agriculture established as the colonists tried to create economically viable settlements. In mainland North America, Catesby travelled through the southern colonies of Virginia, the Carolinas and Georgia and into the Spanish colony of Florida before venturing into the northern Caribbean. The legacies of earlier naturalists, who set the model for the exploration of the Thirteen Colonies in the early eighteenth century, are in evidence throughout Catesby's work, so it is necessary to delve back a little further into the early history of colonization.

Towards the end of the fifteenth century, Europeans stumbled upon lands beyond the western horizon of the Atlantic Ocean. They had found the Americas and the Caribbean, as they came to be known, and learnt that they were populated with diverse, sophisticated human societies with intimate, detailed and complex relationships with, and knowledge of, the natural world.[2] These people were the descendants of Palaeolithic hunter-gatherers who had crossed the Bering Strait from Asia more than 10,000 years earlier.[3] Besides people, Europeans found that the lands were replete with natural resources such as forests, wildlife and minerals. Amid the wonder of novelty, European churches and states saw the prospect of collecting souls, gathering wealth and wielding power by exploiting local people and their knowledge, and the lands' natural capital. Catesby interpreted this as little more than Europeans wanting to 'aggrandize their achievements in conquering a formidable people'.[4]

Plate 14 (detail)

Until the end of the sixteenth century, an envious British monarchy saw vast wealth extracted from the Americas filling continental Europe's coffers, especially those of Spain and France. Therefore, during the seventeenth century Britain became part of the rush to exploit this part of the world, especially the Caribbean, the Thirteen Colonies and parts of Canada. Before the American War of Independence in 1776, Britain's Thirteen Colonies were split between three regions along the east coast of North America. The northern colonies comprised New Hampshire, Massachusetts, Rhode Island and Connecticut, whilst the southern colonies were Maryland, Virginia, North Carolina, South Carolina and Georgia. Between these two blocks were New York, New Jersey, Pennsylvania and Delaware. The first successful English colony, Jamestown in Virginia (established 1607), was financed by the London Virginia Company in their search for gold. However, mineralogical treasure proved a short-term distraction. Long-term treasure was found in food, dyes and drugs extracted from plants, such as the cash crop tobacco (*Nicotiana tabacum*),[5] many of which were grown and harvested with the muscle of enslaved labour. In the Carolinas, during the eighteenth century, rice (*Oryza sativa*, **Plate 14**) became the major source of wealth, before 'King Cotton' took over in the antebellum United States.[6]

The hope offered by novel plants, especially drugs, from the Americas was both alluring and profitable. For example, in the 1600s the supposed pox-curing properties of tropical American holywood (*Guaiacum*

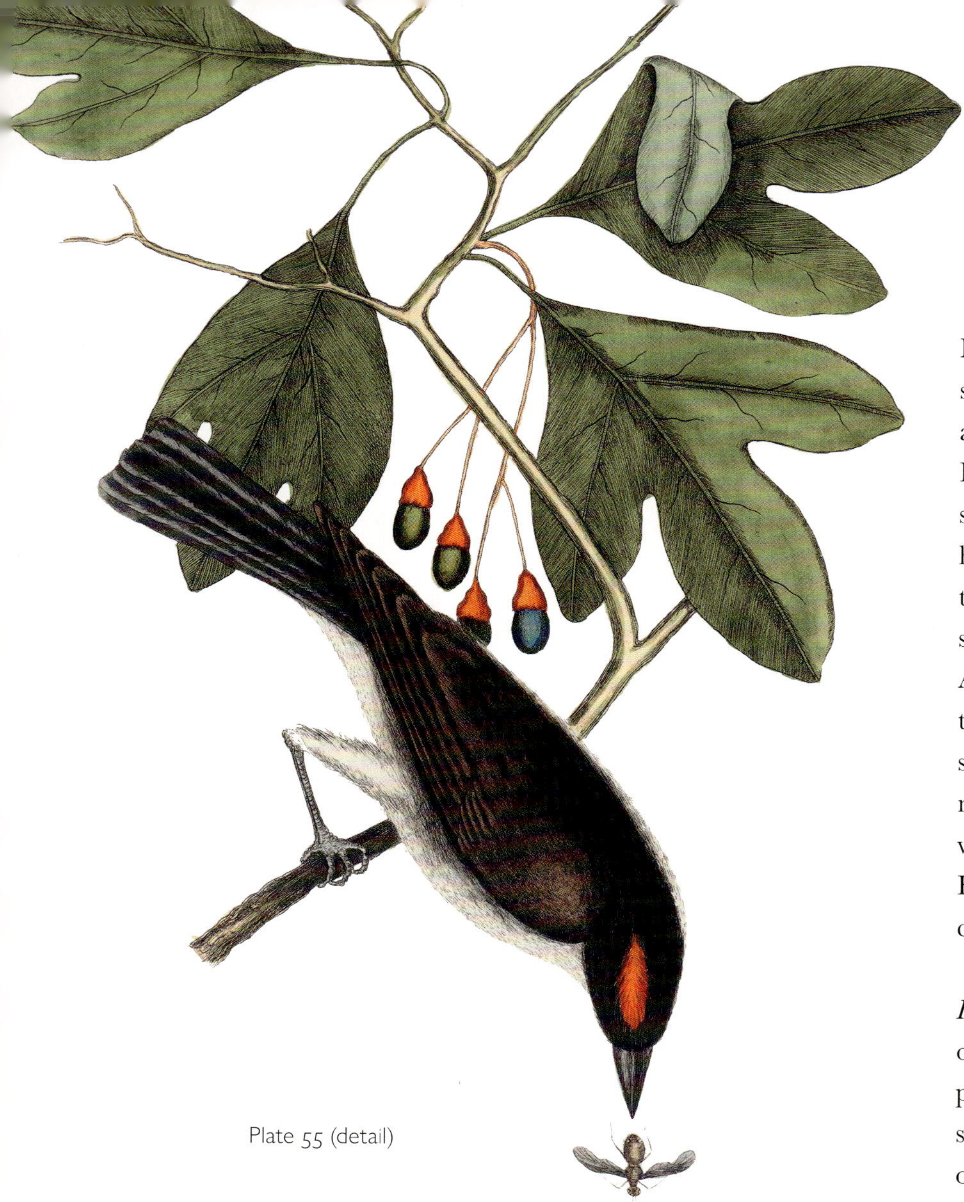

Plate 55 (detail)

santum) became a European obsession.[7] In the 1800s the anti-malarial quinine, extracted from Andean fever trees (*Cinchona*), facilitated European tropical colonization,[8] whilst in the late 1900s the rush for the anti-cancer drug taxol threatened natural populations of its source, the Pacific yew (*Taxus brevifolia*).[9] Moreover, such European encounters in the Americas, with their profound effects on the continents' Indigenous peoples, continue to resonate across the globe even today.[10]

In 1577 the English merchant John Frampton translated the Spanish physician Nicolás Bautista Monardes's *Historia medicinal de las cosas que se traen de nuestras Indias Occidentales* (Medical study of products imported from our West Indian possessions).[11] In this remarkable book, Monardes, who never left his home country, collated data about drug plants imported into Spain and first-hand reports of their use from travellers who had been to the Americas. Seven years later, Elizabeth I gave one of her favourite courtiers, Walter Raleigh (*c*.1552–1618), a time-limited royal patent in North America 'to discouer, search, finde out, and view such remote, heathen and barbarous lands, countries, and territories, not actually possessed of any Christian Prince, nor inhabited by Christian People'.[12] The sovereign may also have expected land bases to help her privateers harry Spanish treasure fleets returning to Europe. Between 1585 and 1587, expeditions supported by Raleigh made attempts to settle North America at Roanoke Island (North Carolina). Despite their lack of success, Raleigh's adventurers had started European exploration of North American natural history, particularly through the outstanding work of the astronomer and mathematician Thomas Harriot (*c*.1560–1621), a member of the first Roanoke colonization attempt.[13]

Harriot's *A Briefe and True Report of the New Found Land of Virginia* (1588) included short notes on 'marchantable commodities',[14] including two plants Frampton had introduced to English-speaking audiences. One – tobacco – became one of the economic mainstays of the Thirteen Colonies and a global currency and addiction.[15] Among the Indigenous people Harriot encountered, he observed that tobacco 'preserueth the body from obstructio[n]s ... whereby their bodies are notably preserued in health'.[16] Ironically, Harriot, a strong advocate for the use of tobacco in England, is thought to have died from smoking-induced cancer.

The other, equally marvellous, plant – based on properties reported by Monardes – was sassafras (*Sassafras albidum*, **Plate 55**), one of the most important drugs plants in early modern Europe.[17] Monardes credited the Indigenous people of Florida with showing French and Spanish soldiers the value of sassafras as a painkiller, a pox cure and a medicine to reduce fever. If such wonders were insufficient, sassafras was also claimed to make women fertile and to help those who were lame to walk. The tree, which the English dubbed the 'ague tree', is distributed through the forests and woodlands of eastern North America from Florida to Ontario. Throughout its range, Native Americans used sassafras as Harriot observed: 'Sassafras, called by the inhabitants *Winauk*, a kinde of wood of most pleasand and sweete smel, and of most rare vertues in phisick for the cure of many diseases. It is found by experience to bee farre better and of more uses then the wood which is called *Guaiacum*.'[18] By the end of the sixteenth century, the price of the spongy, reddish bark taken from the tree's woody roots reached extraordinary prices in Europe (e.g., in 1610 one ton was worth at least £50, or the equivalent[19] of around £8,500 today).[20] During the eighteenth century, sassafras tea was promoted as a panacea in the Thirteen Colonies: 'let him drink Sassafras Tea every Morning, live temperately, upon light and innocent Food and abstain intirely from strong Liquor. The Way to prevent this Calamity [cancer] is, to be very sparing in eating Pork, to forbear all salt, and high-season'd Meats, and live chiefly upon the Garden, the Orchard, and the Hen-house.'[21] However, in Britain, its use declined as safer, more effective drugs became available.

In 1597 the Elizabethan apothecary John Gerard precised Monardes's text and copied his illustration into his *Herball*. Thirty-six years later, when the much expanded and improved second edition, often called 'Gerard *emaculatus*', was published in 1633, little more information about sassafras was presented, although a shoot, taken from a sapling in the 'Garden of Master Wilmote at Bow',[22] was added to the original woodcut. By the mid-eighteenth century, sassafras was being grown in the gardens of Peter Collinson (1694–1768) at Peckham and of Mr Bacon at Hoxton, although Catesby cautioned that sassafras seed was fickle: 'the berries being somewhat succulent should be laid out to dry before they are put up, for fear of rotting in their passage, which should be short, for they will not endure being kept long out of the ground.'[23] Gerard and his contemporaries, who depicted other North American plants cultivated in Britain, heralded the flood that would make major contributions to our understanding of global plant diversity and remodel European gardens.[24]

Between 1628 and 1637, the English horticulturalist John Tradescant the Younger (1608–1662) sent seeds from North America for cultivation in British gardens.[25] Of course, North American botanical knowledge also reached Britain from continental European sources.[26] For example, the first book published on North American plants was *Canadensium plantarum* (1635) by the French physician Jacques-Philippe Cornut, an author who had never visited the Americas, while in 1655 the Dutch botanist Abraham Munting was sent seeds from Virginia via an Amsterdam-based merchant Johan van Wely.[27] The credulous English traveller John Jocelyn, one of the best documenters of the Thirteen Colonies' plants and animals, visited New England in the 1630s and 1660s.[28]

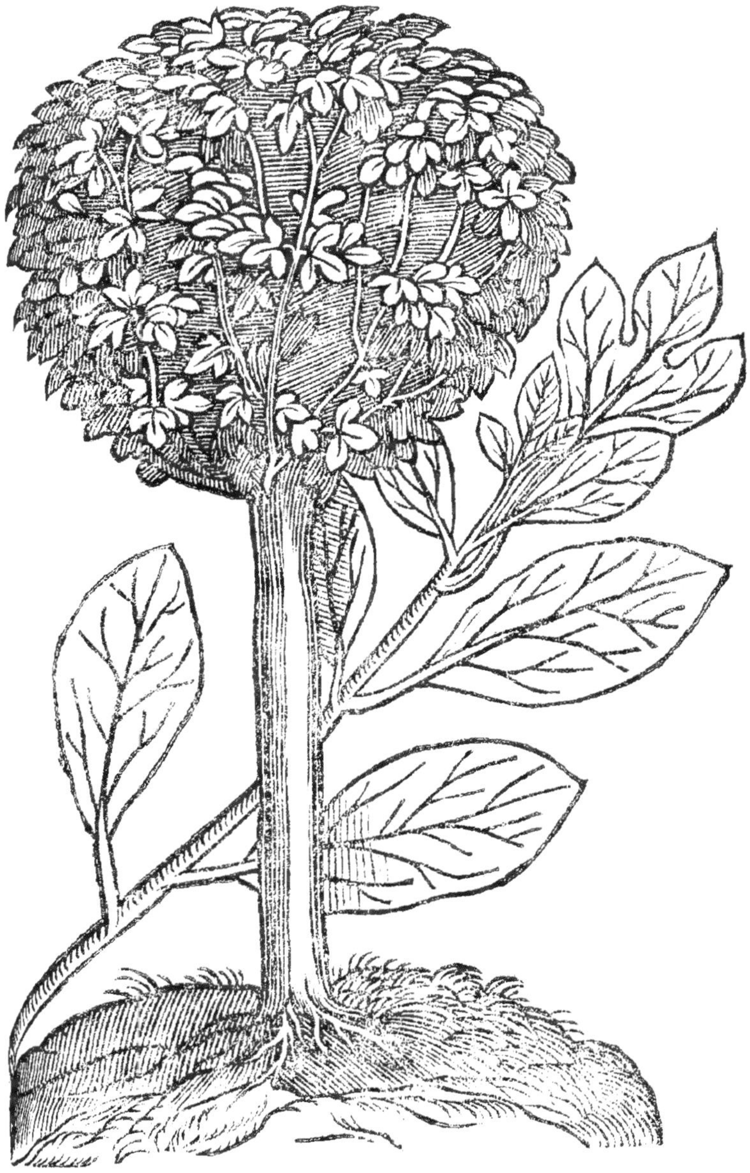

Fig. 1 Woodcut of sassafras (*Sassafras albidum*) from John Gerard's *Herball* (1633). The illustration of the tree was copied from the late sixteenth-century work of Nicolás Bautista Monardes, to which was added a sprig of a sapling cultivated in London.

The English missionary, clergyman and naturalist John Banister (1654–1692), 'the first martyr to science in the New World',[29] was among the earliest university-trained naturalists to work in North America.[30] From the late 1690s, Banister was sending living plants to fill Bishop Henry Compton's (1632–1713) episcopal garden at Fulham, and herbarium specimens and illustrations to the apothecary James Petiver (*c.*1665-*c.*1718) in London, John Ray (1627–1705) in Essex and Jacob Bobart (1641–1719) in Oxford. Cultivated plants sent by Banister to Compton included balsam fir (*Abies balsamea*), box elder (*Acer negundo*), honey locust (*Gleditsia triacanthos*), sweetgum (*Liquidambar styraciflua*, **Plate 165**), scarlet oak (*Quercus coccinea*)[31] and sweetbay magnolia (*Magnolia virginiana*; **Plate 215**).[32] At a similar time, the English naturalist William Vernon (*c.*1666–1711) and the German physician David Krieg (1667–1713) were collecting plant specimens in Maryland, and sending their discoveries to London-based collectors such as Hans Sloane (1660–1753), William Sherard (1659–1728) and Petiver.[33]

In mid-August 1700, a 25-year-old Englishman, John Lawson (1674–1711), arrived in Charles Town in the Carolinas – that vast hereditary fiefdom granted by Charles II to eight wealthy men (the Lords Proprietors). The Carolinas were the Lords Proprietors' reward for loyalty to the Crown during the English Commonwealth and their aid in restoring Charles to the English throne; they expected to profit from British colonization. About four months later, the day following his twenty-sixth birthday, Lawson set out on a fifty-seven-day journey through remote areas of Carolinas, where he eventually settled and established himself as a pelt trader, surveyor and land developer. In 1708 he was appointed to the lucrative post of surveyor general to the Lords Proprietors. Lawson briefly returned to England in 1709 to oversee publication of an account of his journeys, 'a thousand miles, travel'd through several nations of Indians'.[34] This book brought him to the attention of English naturalists, and he began sending preserved specimens of Carolinian plants to Petiver in London.[35] Besides its natural history, Lawson's book also promoted the Carolinas as a place for Europeans to settle, emphasizing the rewards to be had from colonization: 'with prudent Management, I can affirm, by Experience, not by Hear-say, That any Person, with a small Beginning, may live very comfortably, and not only provide for the Necessaries of Life, but likewise for those that are to succeed him.'[36] Lawson, however,

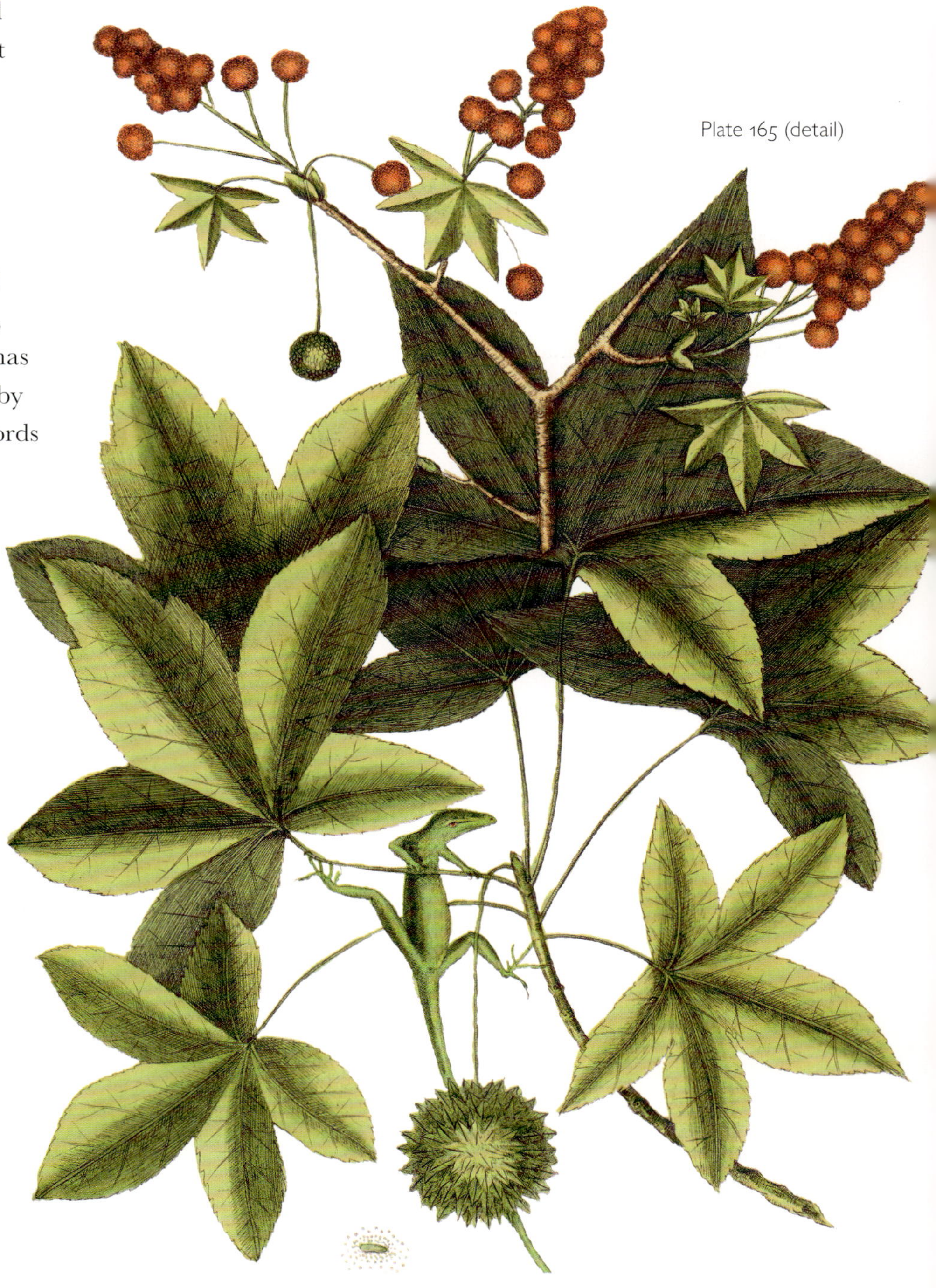
Plate 165 (detail)

considered his own work as inferior to that of Banister: 'had not the ingenious Mr. *Banister* (the greatest *Virtuoso* we ever had on the Continent) been unfortunately taken out of this World, he would have given the best Account of the Plants of *America*, of any that ever yet made such an Attempt in these Parts.'[37] Lawson also meet a violent end. By mid-September 1711, he was dead, killed by the Tuscarora at the start of the Tuscarora War, one the bloodiest eighteenth-century conflicts between European colonialists and Native Americans.[38]

European interests were not restricted to the mainland. Following their discovery by fifteenth-century Europeans, the islands of the Caribbean began to be exploited. One such example is the islands of the Bahamas, originally inhabited by a branch of the Taíno. The first European to make landfall on the Bahamas was Christopher Columbus in 1492. However, within twenty-five years the native people had died from disease or been shipped into slavery on Hispaniola, and the Bahamas were all but deserted.[39] By the middle of the seventeenth century, English colonists from Bermuda, seeking religious freedom, were trying to settle one of the Bahaman islands. However, in 1670 the islands were granted to the Lords Proprietors of the Carolinas, during a period when the islands became infamous as a pirates' republic. The Bahamas were made a Crown colony in 1718, under the governorship of Woodes Rogers (*c*.1679–1732), a privateer himself, who eventually brought his colleagues to heel.

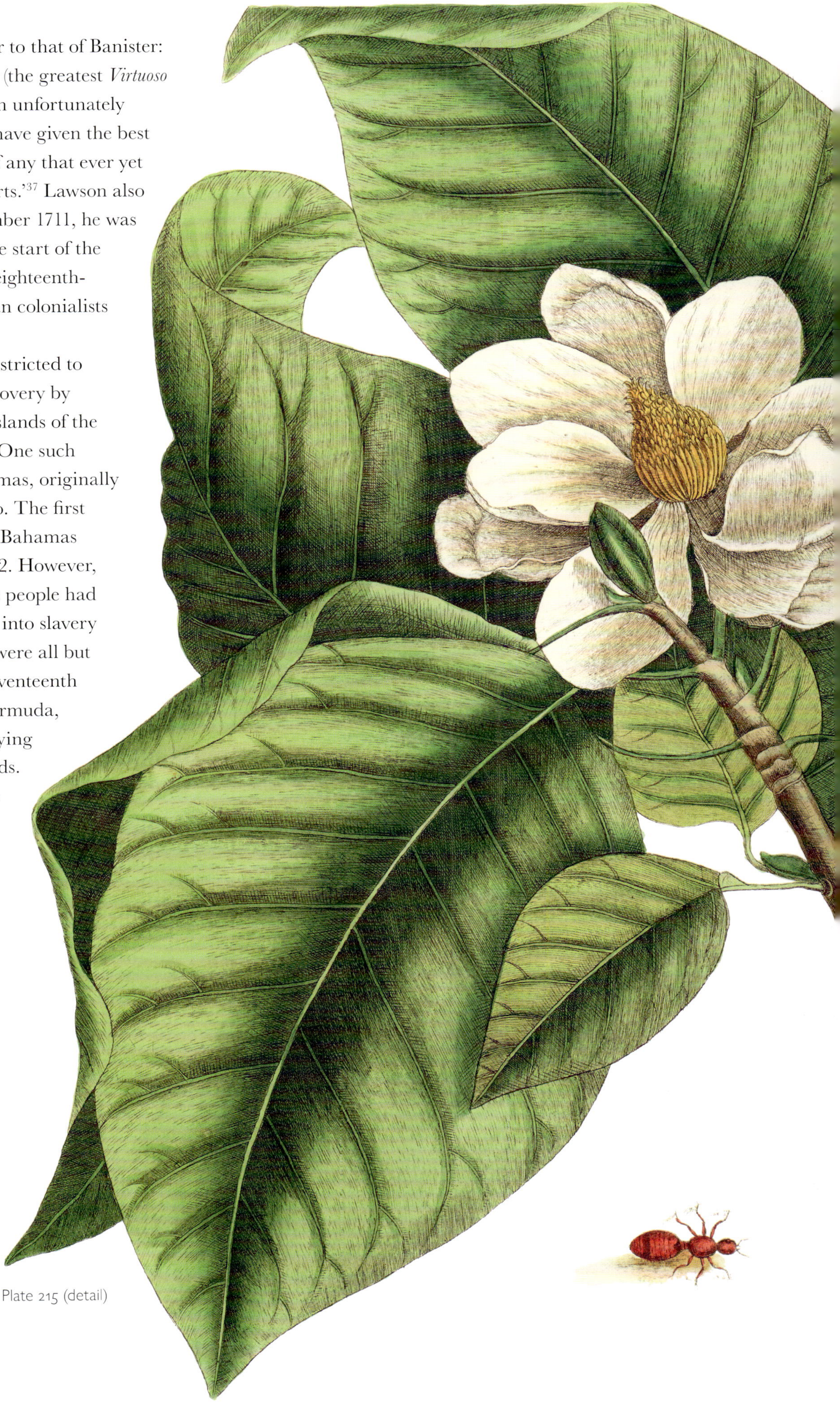

Plate 215 (detail)

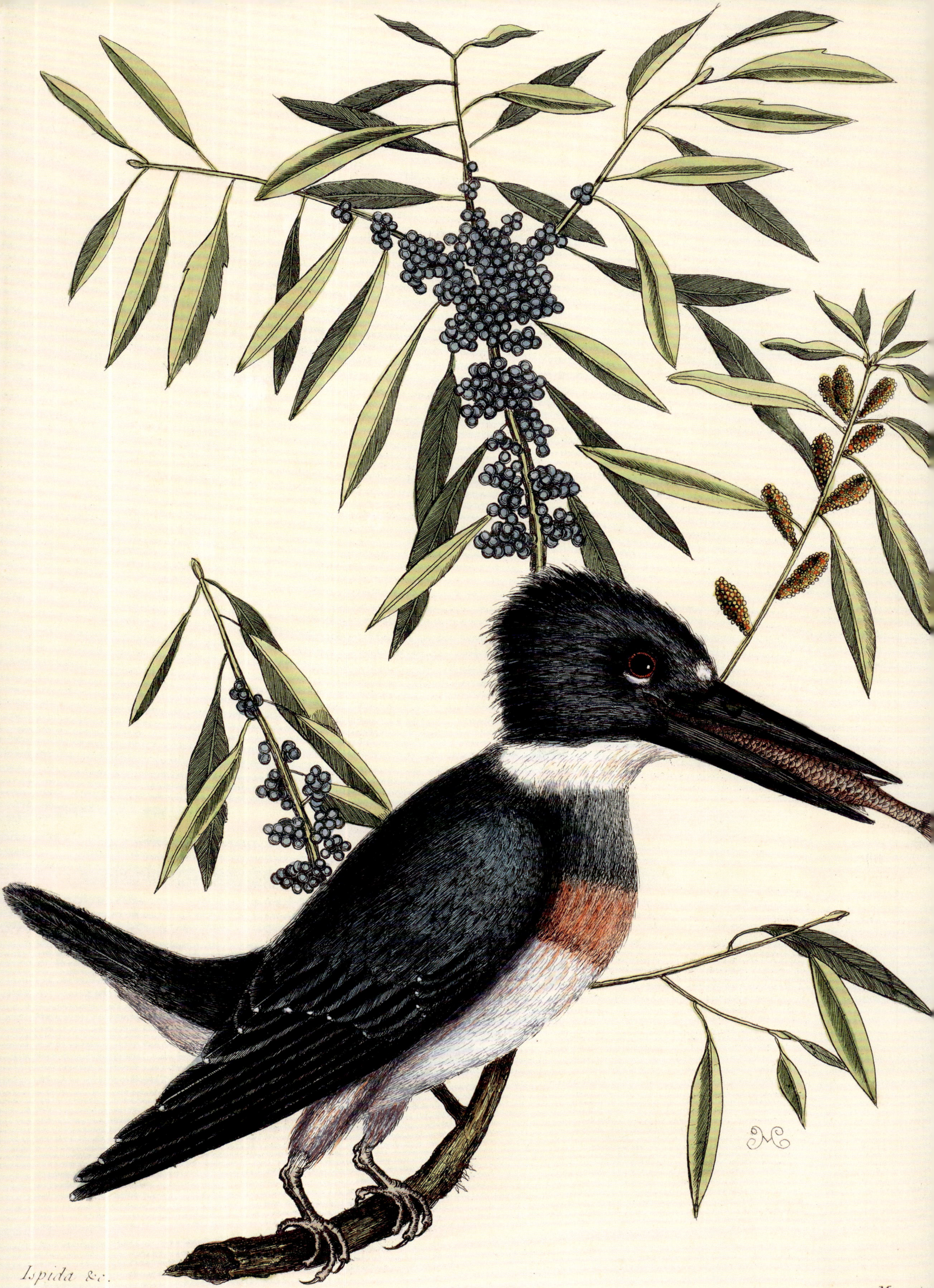

Ispida &c. *Myrtus*

Who was Mark Catesby? 2

At present the Curious may have a large Field to satisfy and divert themselves in, as Collections of strange Beasts, Birds, Insects, Reptiles, Shells, Fishes, Minerals, Herbs, Flowers, Plants, Shrubs, intricate Roots, Gums, Tears, Rosins, Dyes, and Stones, with several other that yield Satisfaction and Profit to those, whose Inclinations tend that Way.

JOHN LAWSON[1]

Mark Catesby was born in 1683 to John Catesby (*c.*1642–1703) and his wife Elizabeth (née Jekyll; *c.*1652–1708).[2] His father was town clerk and six times mayor of Sudbury. This wool-trading town on the Essex–Suffolk border was dismissed in the early 1720s by the writer Daniel Defoe: 'I know nothing for which this Town is remarkable, except for being very populous and very poor … [where] … the Number of the Poor is almost ready to eat up the Rich.'[3] John Catesby was evidently among the town's 'Rich'; when he died in 1703 he bequeathed substantial wealth to his wife and five surviving children, including 21-year-old Mark and 24-year-old Elizabeth (1680–1755); the latter became important for her brother's future career and reputation. Mark Catesby inherited properties and land in Sudbury, the surrounding villages and London.

We know nothing of his early life, although he probably spent these years on the Essex–Suffolk border. His interests in natural history may have been cultivated by his maternal uncle, Nicholas Jekyll, although he later considered these 'much suppressed by my residing too remote from *London*, the Centre of all Science' and being 'deprived of all Opportunities and Examples'.[4] However, among Jekyll's East Anglian acquaintances were John Ray, the most prominent English naturalist of the late seventeenth century, and the naturalist-apothecary Samuel Dale (1659–1739).[5]

Plate 69 (detail)

In 1712 the 29-year-old Catesby travelled to the Thirteen Colonies as chaperone to his sister and two of her children. Elizabeth's husband, William Cocke (1672–1720), had arrived in Virginia in 1710 and established himself as a colonial physician. The 'disobedient Daughter'[6] had caused consternation in her family by marrying Cocke in 1699, without her father's consent.

During the seven years he spent with his sister and her family, Catesby observed and admired 'the various Productions' of the colonies, rather than making any of the illustrations that would feature so prominently in his later work. He was learning to become an effective and efficient field naturalist. Moreover, he proved his worth as a collector to natural historians interested in American plants by returning 'some dried Specimens of Plants and some of the most Specious of them in Tubs of Earth, at the Request of some curious Friends'.[7] One of those friends was Dale, whose specimens are now part of the collection of the Natural History Museum, London. When Catesby returned to England, Dale ensured that his skills came to the attention of the influential naturalist brothers James (1666–1738) and William Sherard. James, an apothecary, owned a magnificent garden at Eltham, which would eventually grow many of Catesby's North American introductions, whilst William was at the centre of an extensive network of European botanists.

Catesby knew that patronage was essential if he was to return to North America, since 'expences were necessary'[8] for natural sciences, as with any cultural pursuit. By proving his worth to Dale and getting the attention of the Sherards, he was able to contemplate returning to North America as a naturalist, especially given the intellectual atmosphere of eighteenth-century London.

In 1660, approximately fifty years before Catesby first went to the Thirteen Colonies, the Royal Society of London had been founded by a group of men who argued that, through an interactive process of observation, hypothesis, experimentation, evaluation and retesting, the natural world could be understood without reference to supernatural interventions.[9] The Royal Society's methods demanded objective evidence – *Nullious in verba* (Take nobody's word for it) – from many different sources. Moreover, the society recognized that if data and knowledge were to be accepted, both had to be communicated so they could be criticized. If the Royal Society's methods were to progress, they needed to have people who would make accurate observations of the world that they could accept. Catesby became one of those observers.

Before a specimen is collected, an observation made, an illustration drawn or a foot is set upon unfamiliar ground, the naturalist must negotiate permissions, feed and clothe themselves, and pay for transport and assistants to get themselves to new places. Once collected, specimens must be preserved and transported to the collector's, or their sponsor's, base. Expenses do not stop there: specimens, notes and illustrations must be labelled, archived and catalogued in well-managed collections, preserved from the ravages of time, if they are to be permanently available for future generations. Funding for Catesby's return to North America, and his exploration of the southern American colonies, was negotiated by William Sherard. However, Sherard's activities would have come to naught had not Catesby's backers had faith in his ability to deliver on his promises.

When he finally published the *Natural History*, Catesby 'most gratefully acknowledge[d] the assistance and encouragement' of twelve 'Noble Persons and Gentleman'.[10] Hans Sloane, elected to the Royal Society in 1685, was England's most prominent collector of natural history objects. William Sherard, elected to the Royal Society in 1720, was one of the great European patrons of early eighteenth-century botany. Charles Dubois, treasurer of the East India Company and elected to the Royal Society in 1700, was a collector of both living plants and preserved specimens.[11] By the time Catesby left for the Carolinas in 1722, all three men were amassing the largest private collections of dried plants in western Europe. It is therefore not surprising that they, in the hope of getting novel plant specimens, should be prominent contributors to Catesby's enterprise. Sherard also named his plants,[12] but within two years of Catesby's return to England Sherard was dead. Instead, Sloane would become central to Catesby's efforts in publishing the *Natural History* in London.

How Sherard went about garnering support for Catesby is unclear but, notably, all but one of the remaining nine men were fellows of the Royal Society. The exception was Henry Goodricke (1677–1738), 4th Baronet of Ribstone,[13] a Justice of the Peace for the West Riding of Yorkshire. The supporter with the closest associations with the places Catesby would explore was the English soldier and colonial official Francis Nicholson, who was elected to the Royal Society in 1706.[14] He became the first royal governor of the colonial Province of South Carolina in 1721, following a rebellion by colonists against the Lord Proprietors, and relinquished the position in 1725. Another supporter, who eventually had ties with Catesby's areas of collection, was the Anglo-Irish politician John Perceval (1683–1748), 1st Earl of Egmont, who was elected to the Royal Society in 1701. In 1730 he became a Trustee for the Establishment of the Province of Georgia (established in 1733) and actively superintended the colonization of Georgia.[15] Three prominent English collectors who supported Catesby were the landowner and politician James Brydges (1674–1744; elected to the Royal Society in 1694), 1st Duke of Chandos; the politician Edward Harley (1689–1741; elected to the Royal Society in 1727), 2nd Earl of Oxford; and the royal physician Richard Mead (1673–1754; elected to the Royal Society in 1703), a collector of books, antiquities and natural history objects.[16] Notoriously, Brydges used his public offices to amass a fortune, which he lost in the South Sea Bubble – the collapse of the slaving-trading South Sea Company – although he continued to spend. Following his death, Brydges's son was forced to sell the grand country house Cannons (near Edgeware in Middlesex), with its magnificent gardens, to pay off some of the vast debts accumulated by his father.

Two other English politicians who supported Catesby were John Knight (*c.*1687–1733; elected to the Royal Society in 1706)[17] and the English politician George Markham (1666–1736; elected to the Royal Society in 1708), who is best known for the acrimonious arguments surrounding his will and substantial estates.[18] Another notorious backer of Catesby was the powerful English lawyer Thomas Parker (*c.*1666–1732; elected to the Royal Society in 1712), 1st Earl of Macclesfield. In 1725, he was convicted of corruption in public office by the House of Lords.[19]

Catesby's supporters were an eclectic mix of wealthy, powerful men who held positions in politics, medicine, law and commerce.[20] They are likely to have had multiple reasons for supporting his efforts. Perhaps they saw opportunities to make money, wanted new objects for their collections and gardens, wanted to be associated with a prestigious enterprise or were simply curious about North American plants and animals.

Reflecting on his journey years later in his life, Catesby reserved special praise for 'that great Naturalist and Promoter of Science Sir Hans Sloane, Bart to whose goodness I attribute much of the success I had in this undertaking'.[21] In 1687 Sloane had undertaken his own fieldwork in Jamaica, when he was personal physician to the island's governor Christopher Monck, 2nd Duke of Albemarle.[22] One year later, Sloane returned to London. He went on to establish himself as the heart of British scientific society in the first half of the eighteenth century, as president both of the Royal Society and of the College of Physicians. He published his own extensive Caribbean natural history observations, *A Voyage to the Islands Madera, Barbados, Nieves, S. Christophers and Jamaica* (1707, 1725), in two volumes illustrated with black-and-white metal engravings. The personal collection Sloane amassed during his lifetime became the foundation of the British Museum, the British Library and eventually the Natural History Museum. Catesby appears to have been able to interact easily with Hans Sloane and the primary figures of the scientific elite in early eighteenth-century London and the Royal Society, and through them with social elites.

Catesby's four-year exploration (1722–26) of the Thirteen Colonies saw him travel through the Carolinas into what would become Georgia, and Florida and finally the Bahamas. Catesby found collecting seed and botanical specimens in the field hard work, despite the experience he had acquired when living in Virginia. As he travelled long distances, through hostile landscapes in unpleasant weather, he quickly realized that 'to collect everything is impossible but with many years application'.[23] His collecting strategy – never collect twice in the same place or season – was driven by a combination of biology and the expediency of having to satisfy sponsors interested in novelty. As he travelled, he drew, but as he delivered on promises his work became more difficult as the number of sponsors increased.

Catesby returned to England in 1726. He had sent thousands of herbarium specimens and living plants, sometimes packed in precise and ingenious manners, to his English sponsors, and become an advocate of the economic potential of North American trees grown in British forests. His patience, attention to detail and capacity for hard work would be essential during the more than two-decade-long process of illustrating, writing and publishing the *Natural History*.[24] In April 1733 Catesby was elected a fellow of the Royal Society, sponsored by Peter Collinson, the antiquarian Roger Gale, Martyn (the recently appointed professor of botany at the University of Cambridge), Robert Paul and Hans Sloane.[25]

In his personal life, Catesby started to live with Elizabeth Rowland in about 1730, but they did not get married until October 1747, despite having at least four children together. During the final two years of his life, Catesby's health declined: his 'Legs swell & he looks badly, Drrs. Mead & Stack said there were little hopes for him long on this side of the Grave'.[26] He died on 23 December 1749, survived by his wife, eldest son, youngest daughter and a stepdaughter, and was buried at St Luke's Church, Old Street, London.

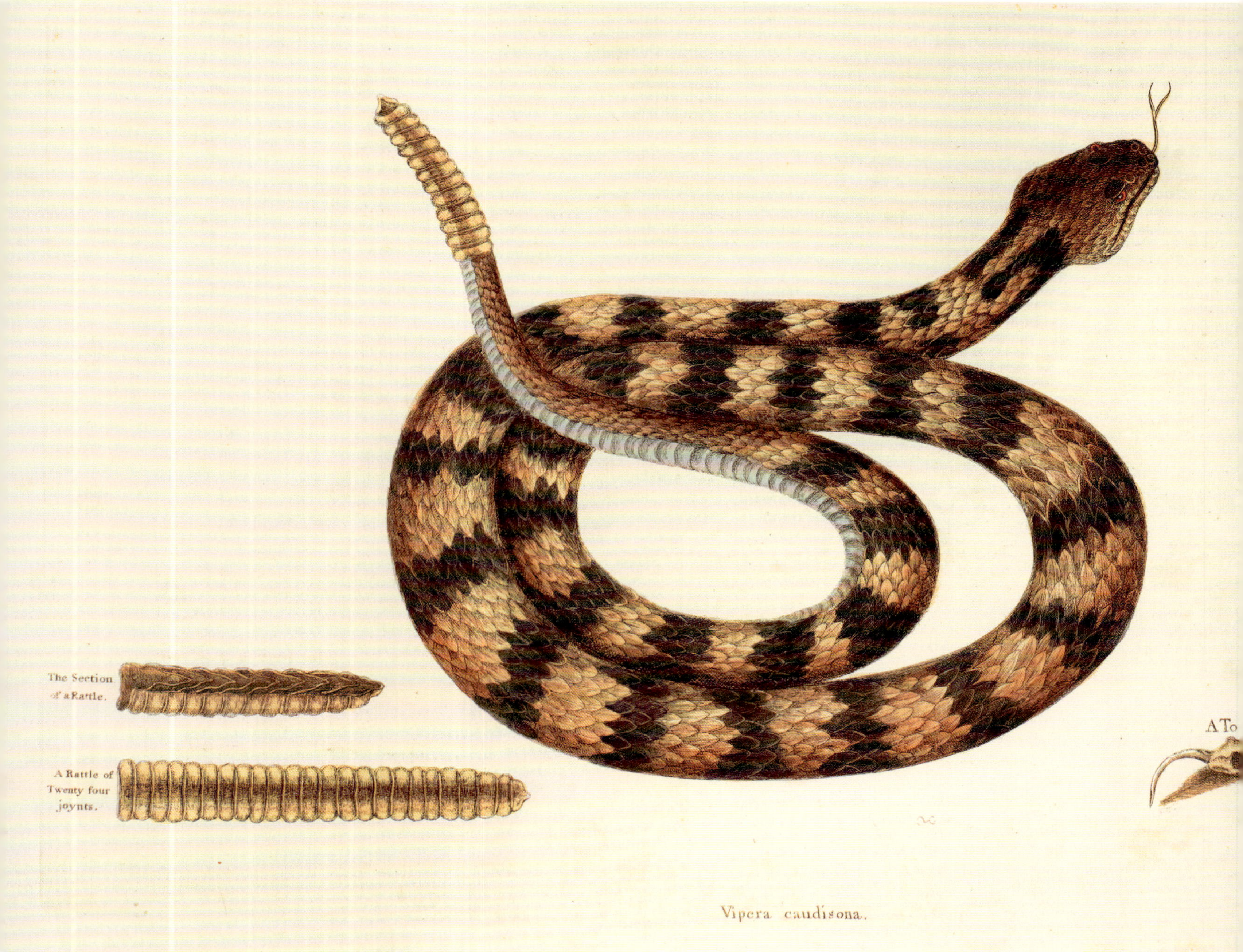

above Plate 141 (detail)

following pages Fig. 2 Map of the Carolinas and the Bahamas published by Mark Catesby in his *Natural History of Carolina, Florida, and the Bahama Islands* (1754), showing Georgia, which was found in 1732, six years after Catesby left North America for the final time.

On Habitats and their Uses 3

They [collectors] furnished the material for the work in all the other branches of botany, and particularly in the geography of plants. Most of them did a toilsome work, exposed in the wilderness to manifold fatigues and perils.

FREDERICK BRENDEL[1]

Natural history exploration is a serendipitous, seasonal activity that is dependant on the skills of the explorer and the places they choose to explore. Decisions made in the field determine the future value of any data that might be collected. Moreover, explorers rarely harvest everything they find; rather they are selective. They sample what they collect based on their knowledge and experience, and filter data associated with specimens based on their observational skills and attitudes, and the knowledge local peoples choose to impart. Modern understanding of the global diversity and distribution of the planet's species has come from the activities of early natural history explorers, people who travelled the world collecting and recording species' occurrences.[2] However, in recent years, the activities of such explorers, particularly those originating in Europe, have been re-evaluated according to modern perceptions influenced by who did the exploring; when, where and why they explored; the societies from which they came; and their interactions with the societies and landscapes through which they travelled. The actions and attitudes of explorers may also be judged by the manner in which they presented their findings and the purposes to which their finds were put – perhaps long after they were dead.

When Catesby arrived in the Carolinas, he relied on the kindness of strangers to help him through landscapes foreign to him, despite his Virginian experiences. His three-year stay in the Carolinas was facilitated by 'General [Francis] Nicholson, then Governor of that Province, who received me with much kindness' and 'hospitable and kind entertainment … amongst the Gentlemen of the country'. His sojourn in the Bahamas came at the invitation of his 'Excellency Charles Phinney, Esq; Governor of those Islands'.[3]

Natural history collectors establish bases within areas from which excursions of greater or lesser length are undertaken, whilst those who are itinerant regularly move from one area to another. Collectors who stay for short periods in areas can only collect opportunistically. Catesby was based in the Carolinas for years, and so was able to study animals and plants systematically across different seasons.[4] However, his activities were constrained by the demands of his patrons and the continual search for natural history novelties. Soon after his arrival, he assured his patron William Sherard that 'I am told up the rivers there are abundance of fossils and petrifactions and that those parts have not been searched for plants[,] which gives me hopes ~~(if God gives me health)~~ in some measure of effecting what you perticularly desire'.[5] Eventually, he was forced to remind his sponsors of the realities of his life in the American colonies; for example, 'I being at a great distance from where the St Andrews Cross [*Hypericum hypericoides*] grows at the time of ye ripe seed is the reason I have sent no more of it'.[6] In unfamiliar areas, Catesby was faced with getting lost, encountering wild animals and human hostility. In early 1723, he articulated such apprehensions: 'where a vein or Artery is pricked by the bite of a Rattle Snake [*Crotalus horridus*; **Plate 141**] no antidote will avail anything, but Death certainly and suddainly ensues sometimes in or minutes which I have more than once seen'[7] and 'the fear of … meeting with Indians as five of us hapned to doe as we were out a Buffello hunting tho' they hapned to be those we were at peace with[,] they were about 60'.[8]

In the Carolinas, Catesby found plants and animals familiar to him from Virginia, but also

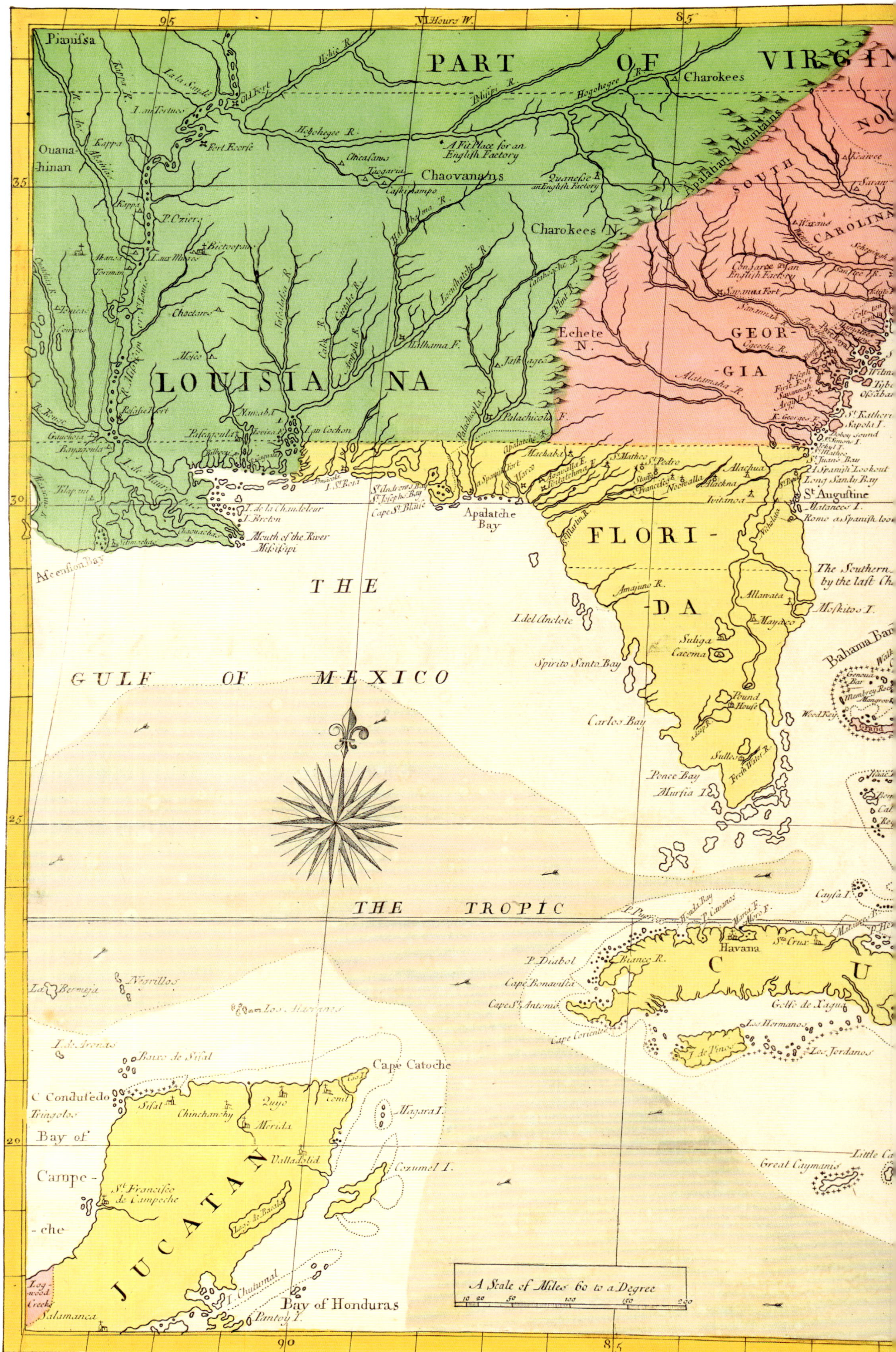

PART OF VIRGINIA
LOUISIANA
SOUTH CAROLINA
GEORGIA
FLORIDA
THE GULF OF MEXICO
THE TROPIC
JUCATAN
Bay of Campeche
Bay of Honduras
Apalatche Bay
Apalatian Mountains
Charokees
Chaovanans
Havana
Cape Catoche
Cape St. Antonio
Great Caymanis
St Augustine
A Scale of Miles 60 to a Degree

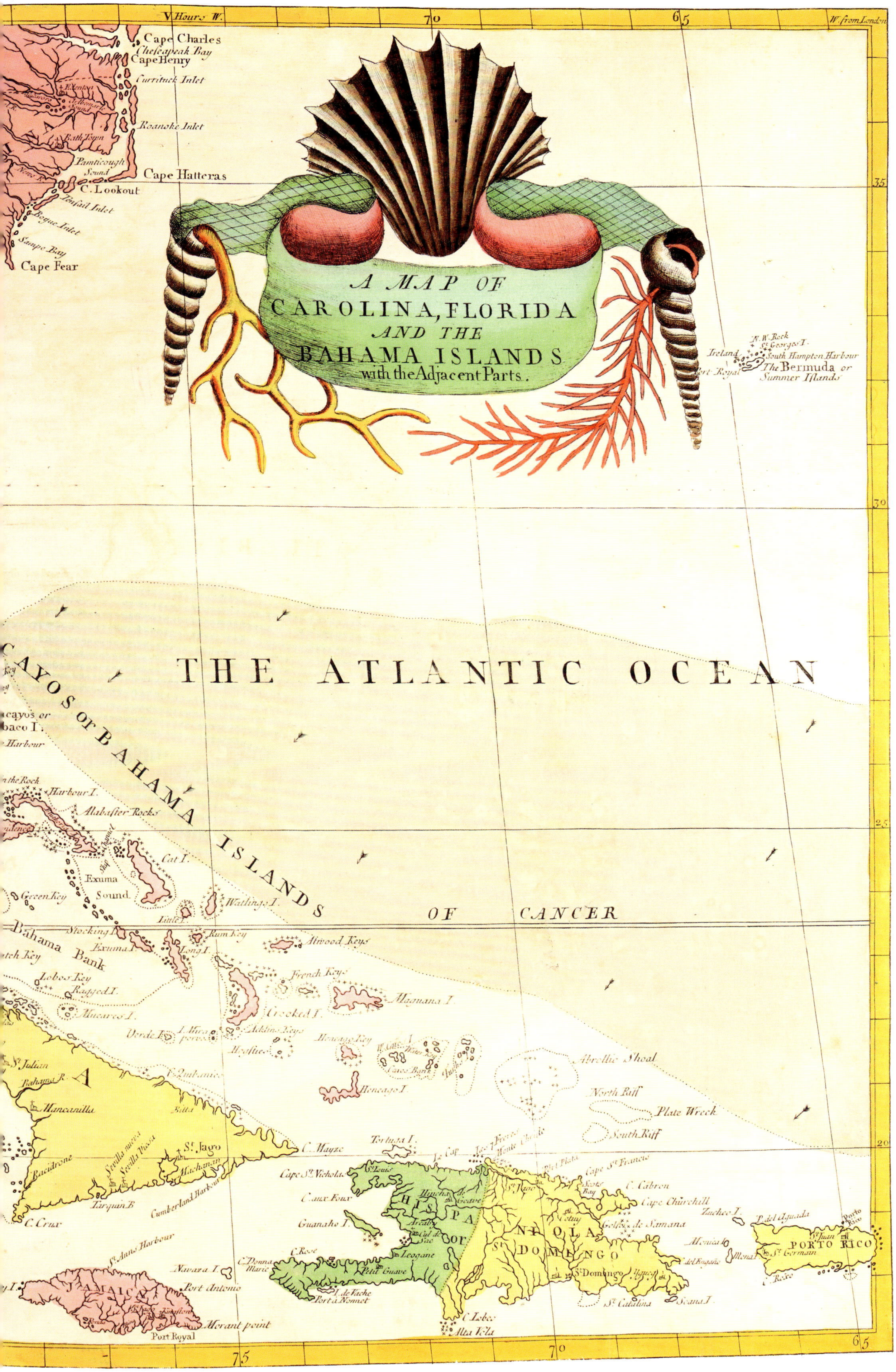

A MAP OF
CAROLINA, FLORIDA
AND THE
BAHAMA ISLANDS
with the Adjacent Parts.
THE ATLANTIC OCEAN
CAYOS or BAHAMA ISLANDS
OF CANCER
Cape Charles
Chesapeak Bay
Cape Henry
Currituck Inlet
Roanoke Inlet
Cape Hatteras
C. Lookout
Cape Fear
The Bermuda or Summer Islands
Exuma Sound
Bahama Bank
Cat I.
Watlings I.
Atwood Keys
French Keys
Maguana I.
Crooked I.
Heneaga I.
Abrollio Shoal
North Riff
Plate Wreck
South Riff
Tortuga I.
HISPANIOLA
St. DOMINGO
PORTO RICO
JAMAICA
Port Royal
Morant point
Port Antonio
Cumberland Harbour
St. Jago
C. Cruz
C. Mayze
Cape St. Nicholas
Cape Churchill
Cape St. Francis
C. Cabron
C. Lobos
Alta Vela
V Hours W.
W. from London
70
65
75
35
30
25
20

novelties that maintained his interest and, importantly, that of his patrons. He ranged where he safely could in the Carolinas from the sea to the inland mountains:

> I then went to the Upper uninhabited Parts of the Country, and continued at and about *Fort Moore* [near Savannah, Georgia], a small Fortress on the Banks of the River *Savanna*, which runs from thence a Course of 300 Miles down to the Sea, and is about the same distance from its source, in the Mountains. I was much delighted to see Nature differ in these Upper Parts, and to find here abundance of things not to be seen in the lower parts of the country. This encouraged me to take several Journeys with the *Indians* higher up the Rivers, towards the Mountains, which afforded not only a succession of new vegetable Appearances, but the most delightful Prospects imaginable, besides the Diversion of Hunting Buffaloes, Bears, Panthers, and other wild Beasts.[9]

Plate 49 (detail)

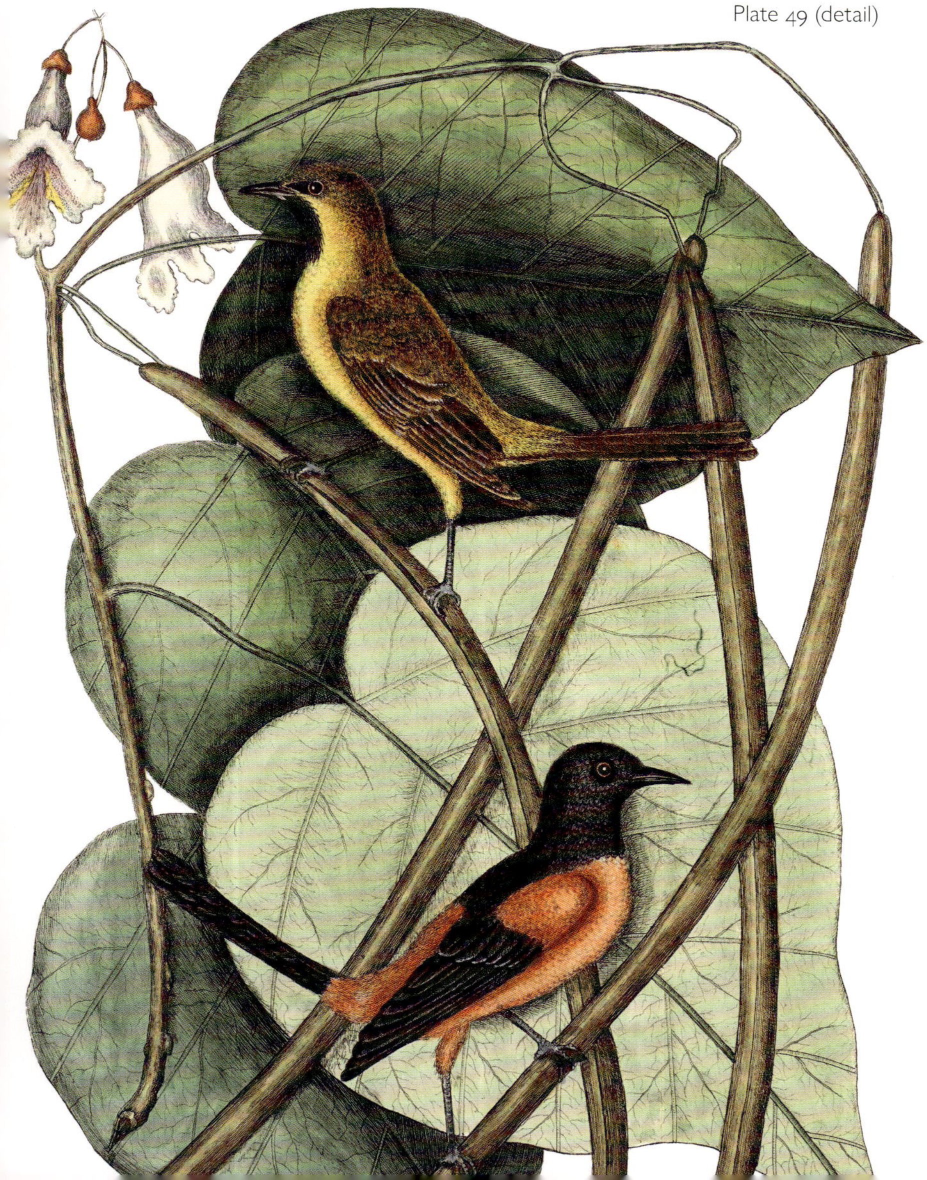

Arriving in the Carolinas, he established himself in Charles Town in the summer of 1722.[10] Many years later he wrote enthusiastically about the subtropical climate of the Carolinas, where summers are hot and winters mild: 'no Country enjoys a more temperate air. The Winter months are so moderate, and the air so serene, that it sufficiently compensates for the heats in Summer, in which it has the advantage of all our other Colonies on the Continent.'[11]

The broad landscape of Catesby's Carolinas is recognizable today. The Atlantic coast is protected by a chain of low-lying tidal and barrier islands, whilst inland there are more or less three geological zones and their associated vegetation: the Atlantic Coastal Plain, the Piedmont Plateau, and the Appalachian Mountains. The Coastal Plain is a low-lying matrix of dry, sandy and wet, silt- and clay-rich areas, which creates a pockmarked landscape of damp depressions (bays) dissected by streams, rivers, salt marshes and estuaries. Moreover, near the coast, habitats on the Coastal Plain are constantly modified by natural processes such as cyclones, floods, fire and drought: 'the Sea on these Coasts seldom makes any sudden or remarkable revolution, but gets and loses alternately and gradually.'[12] In contrast, the Piedmont is a hilly plateau that rises to 300 metres above sea level, separating the plain from the foothills of the Appalachian Mountains.

Two different forest types occur on the plain. On sandy soils, where fire is frequent, there are mixed forests of hickory, pine and oak, dominated by pines, oaks and southern catalpa (*Catalpa bignonioides*, **Plate 49**). Near the coast, there are evergreen broadleaved forests of trees such as southern magnolia (*Magnolia grandiflora*, **Plate 161**), sweetbay (**Plates 39 and 215**) and loblolly bay (*Gordonia lasianthus*, **Plate 44**), with understoreys of species such as wax myrtle (*Myrica cerifera*, **Plate 69**) and American devilwood (*Cartrema americana*, **Plate 61**). There are also diverse freshwater wetland communities dominated by swamp cypress (*Taxodium distichum*, **Plate 11**), water tupelo (*Nyssa aquatica*, **Plate 60**) and hickories, for example pignut hickory (*Carya glabra*, **Plate 38)**. Catesby found such swamps:

> filled with a Profusion of fragrant and beautiful Plants give a most pleasing Entertainment to the Senses, therein excelling other Parts of the Country, and by their Closeness and Warmth in Winter are a Recess to many of the Wading and Water-Fowls. This Soil is composed of a blackish sandy Loam, and proves good Rice-Land, but the Trouble of grubbing up and clearing it of the Trees and Underwood has been hitherto a Discouragement to the Culture.[13]

Denisons of these swamps included American pitcher plants, frogs and toads (**Plates 169–171**). In the Piedmont, the forests include many oaks species and hickories, such as pignut hickory (**Plate 38**) and mockernut hickory (*Carya tomentosa*, **Plate 38**), with understorey trees that include sassafras (**Plate 55**) and flowering dogwood (*Cornus florida*, **Plate 27**).

Wherever oaks are found, people have discovered their economic, cultural and strategic value. In eighteenth-century England, oaks were the pigs of the plant world – people used almost every part of them.[14] It was unsurprising that the English polymath and gardener John Evelyn opened his classic work on tree cultivation, *Sylva* (1664), with a chapter on the oaks. Despite stating that 'to enumerate now the incomparable *Uses* of this *Wood*, were needless',[15] he catalogued them. Oaks yielded timber for carpentry and shipbuilding; wood for fuel, barrels and food preservation; bark for tanning; galls for making ink; and acorns for pannage. Oak woodlands shaped landscapes and provided habitat for game. Moreover, they were associated with English identity. As Evelyn knew, the qualities of oak that are useful to humans are determined by species and growth conditions: 'a *Timber-tree* is a *Merchant Adventurer*, you shall never know what he is *worth*, till he is *dead*.'[16] In Britain, there are two sorts of native oak, English (*Quercus robur*) and sessile (*Quercus petraea*). In the *Natural History*, Catesby illustrated nine oaks from the Virginia and the Carolinas across eight plates (**Plates 16–23**). The immensity of American forests impressed the man who had grown up in East Anglia,[17] an area that had been cleared of its forests to raise sheep for their wool centuries before he was born.

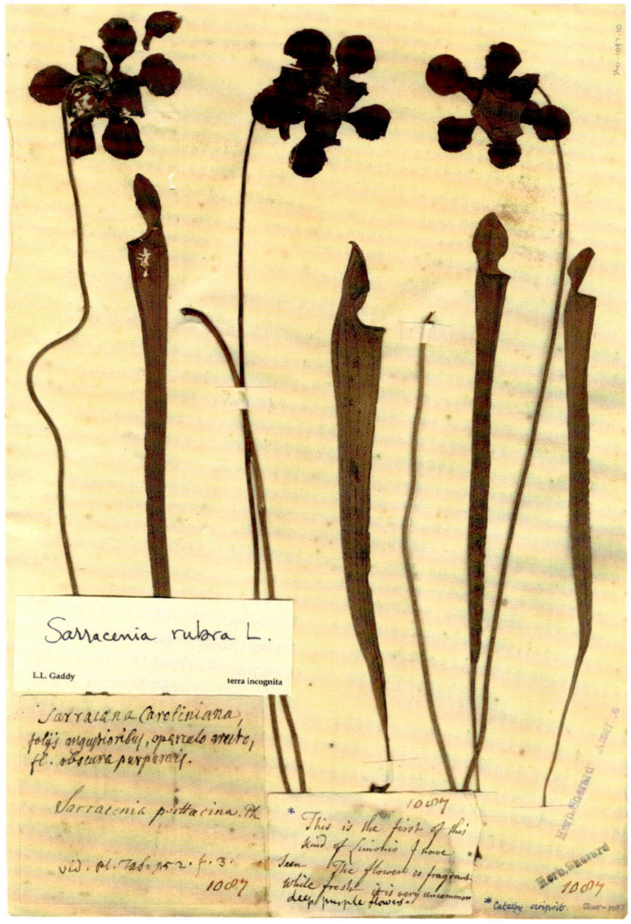

Fig. 3 Specimen of the sweet pitcher plant (*Sarracenia rubra*) collected by Mark Catesby. On his original label, bottom middle, Catesby notes that 'the flower is fragrant while fresh it is very uncommon'. On his label, bottom left, William Sherard's phrase name indicates the plant comes from the Carolinas and emphasizes that the flowers are dark purple. Other names have been added by nineteenth- and twentieth-century researchers.

Catesby was sensible to differences between the oaks he illustrated in terms of their appearance, habitat and uses. The information on the habitats of specific oaks is likely to have come from his own observations, whilst information on their properties probably came from his interactions with both colonists and Native Americans. In his opinion, the widespread white oak (*Quercus alba*, **Plate 21**) was the 'best oak in Virginia and Carolina'[18] and most closely resembled the English oak, although it did best on high barren ground associated with pines. Another oak with very useful wood identified by him was the swamp chestnut oak (*Quercus michauxii*, **Plate 18**), 'the tallest and largest of the Oaks in these parts of the world', which 'grows only in low and very good land'.[19] The least useful oak he illustrated was the soft, coarse-grained wood of willow oak (*Quercus phellos*, **Plate 16**), a lowland species of wet areas.[20] In the case of the water oak (*Quercus nigra*, **Plate 20**), a species of 'low waterish lands', the timber was also of little use, 'except for fencing in

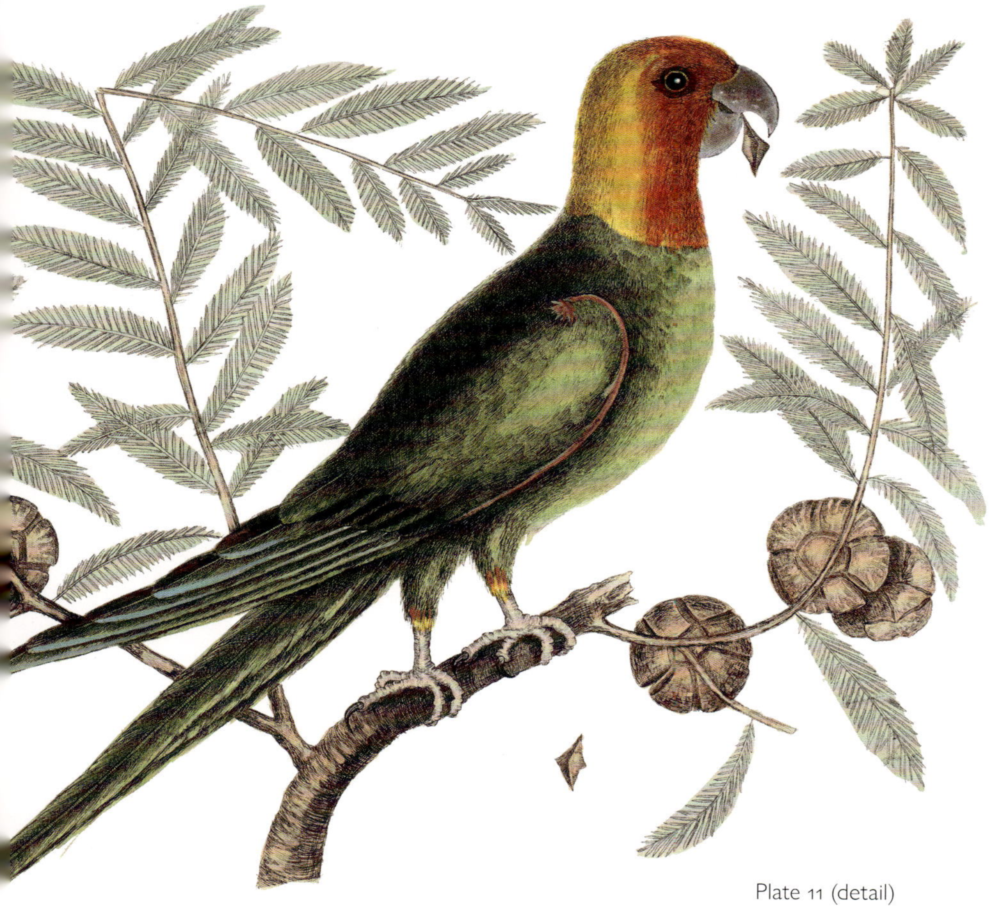

Plate 11 (detail)

fields';[21] even its acorns were rejected by pigs except when other food was unavailable. Another species with few uses, other than firewood, was the blackjack oak (*Quercus marilandica*, **Plate 19**).[22] In contrast, live oak (*Quercus virginiana*, **Plate 17**), a tall, coarse-grained species of salt marshes with a 'harder and tougher' wood than other American oaks he knew, had acorns that were 'the sweetest of all others; of which the *Indians* usually lay up store, to thicken their venison soop, and prepare them other ways. They likewise draw an oil, very pleasant and wholesome, little inferior to that of Almonds'.[23] The 'dark colour'd, very thick and strong' bark of the highland American turkey oak (*Quercus laevis*, **Plate 23**) was the best oak for tanning.[24] The bluejack oak (*Quercus incana*, **Plate 22**), found on poor, dry land, was successfully cultivated by Thomas Fairchild (1667–1729) at his nursery in Hoxton from seed supplied by Catesby.[25]

Through the association of species in his illustrations, Catesby made inferences about oaks as habitats for birds. Eight woodpeckers appear among the oak plates. The most famous of these illustrations is probably that of a male ivory-billed woodpecker (*Campephilus principalis*) clinging to a willow oak trunk (**Plate 16**). These birds ate 'Ants, Wood-worms, and other Insects, which they hew out of rotten trees. Nature having so formed their bills, that in an hour or two they will raise a bushel of chips.'[26] During Catesby's time the species' range probably extended from what is now eastern Texas to North Carolina and from Illinois to Florida and Cuba.[27] Today, one of the world's largest woodpeckers is probably extinct because of hunting and habitat loss.[28] Catesby recorded one of the bird's economic attractions:

> The bills … are much valued by the *Canada Indians*, who make coronets of them for their Princes and great warriors, by fixing them round a wreath, with their points outward. The Northern *Indians*, having none of these Birds in their cold country, purchase them of the Southern people at the price of two, and sometimes three buck-skins a bill.[29]

In his book, Catesby divided the soils of the Carolinas into '*Rice Land*, *Oak* and *Hiccory Land*, and *Pine barren Land*',[30] based on the quality of the land for agriculture, which broadly corresponded to the plain, Piedmont and mountains. The wet Rice Land soils were the most valuable but were useful only for cultivating rice. The most generally useful land was the well-drained, light sandy loams of the 'Oak and Hiccory Land', upon which a wide range of crops could be grown. The worst land was the light, sterile sands of the Pine Land, where the only useful trees were pines, although they produced important naval stores.[31]

In 1731 the chief exports from the Carolinas were 'Deer Skins, Pitch, Tarr, Turpentine, Beef, Pork, tann'd Leather, Cedar Wood, Deal boards, Pipe Staves, Timber of all Sorts, Masts, Yards, &c.'[32] However, one plant – rice (**Plate 14**) – became the primary source of the colony's wealth. When Catesby arrived in the Carolinas, approximately 4,000 tonnes of rice were exported annually; on the eve of the American Revolutionary War, this had risen to more than 34,000 tonnes.[33] Rewards from rice cultivation were so great that colonists became drawn to single-crop agriculture and all the risks associated with erratic year-to-year price swings produced by crop failures, harvest gluts and the collapse of market prices. Weathering these economic squalls contributed to the concentration of land in the hands of a few large families, the introduction of 'industrial' production techniques and the expansion of slavery. Between 1720 and 1776, approximately 600 slaves were imported annually to satisfy the demands of Carolinian rice production.[34]

Catesby's illustration of rice shows one of the risks faced by early eighteenth-century rice cultivators in the Carolinas – the bobolink, or rice bird (*Dolichonyx oryzivorus*, **Plate 14**). In his illustration, a female bird clings awkwardly to the stem of a fruiting rice stalk that emerges from the ground, upon which a male bird stands:

> In the beginning of September, while the grain of Rice is yet soft and milky, innumerable flights of these Birds arrive from some remote parts, to the great detriment of the Inhabitants. In 1724, an inhabitant near Ashley river had forty acres of Rice so devoured by them, that he was in doubt, whether what they had left was worth the expence of gathering in.[35]

In Catesby's Carolinas, the annual arrival of the bird could be destructive to the rice crop, but it was also 'esteemed … the greatest delicacy of all other Birds. When they first arrive, they are lean, but in few days become so excessive fat, that they fly sluggishly and with difficulty; and when shot, frequently burst; with the fall: they continue about three weeks, and retire by that time the Rice begins to harden.'[36]

Catesby's observations on the changing sophistication and expansion of rice production in the Carolinas in the 1720s are important eyewitness accounts of a plant that transformed people's lives and landscapes in the Americas:

> Swamps, before they are prepared for Rice, are thick, over-grown with Underwood and lofty Trees of mighty Bulk, which by excluding the Sun's Beams, and preventing the Exhalation of these stagnating Waters, occasions the Land to be always wet, but by cutting down the Wood is partly evaporated, and the Earth better adapted to the Culture of Rice; yet great Rains, which usually fall at the latter Part of the Summer, raises the Water two or three Feet, and frequently cover the Rice wholly, which nevertheless, though it usually remains in that State for some Weeks, receives no Detriment.[37]

In addition to rain-fed rice, where seeds are sown in fields carved out of dry or marshy areas that rely on rain for watering, Catesby mentioned a method of water cultivation that was wholly unfamiliar to European colonists: 'It agrees best with a rich and moist soil, which is usually two feet under water, at least two months in the year. It requires several weedings till it is upwards of two feet high, not only with a hough, but with the assistance of fingers.'[38] Water cultivation involves catching water in reservoirs such as freshwater ponds or swamps, then flooding fields through gravity flow, before flushing the water out into streams. However, such cultivation needs appropriate topography and watersheds, plus vast amounts of labour to flood and weed the fields and to process the harvest. Given the similarity of water cultivation in the Carolinas to some West African practices of crop

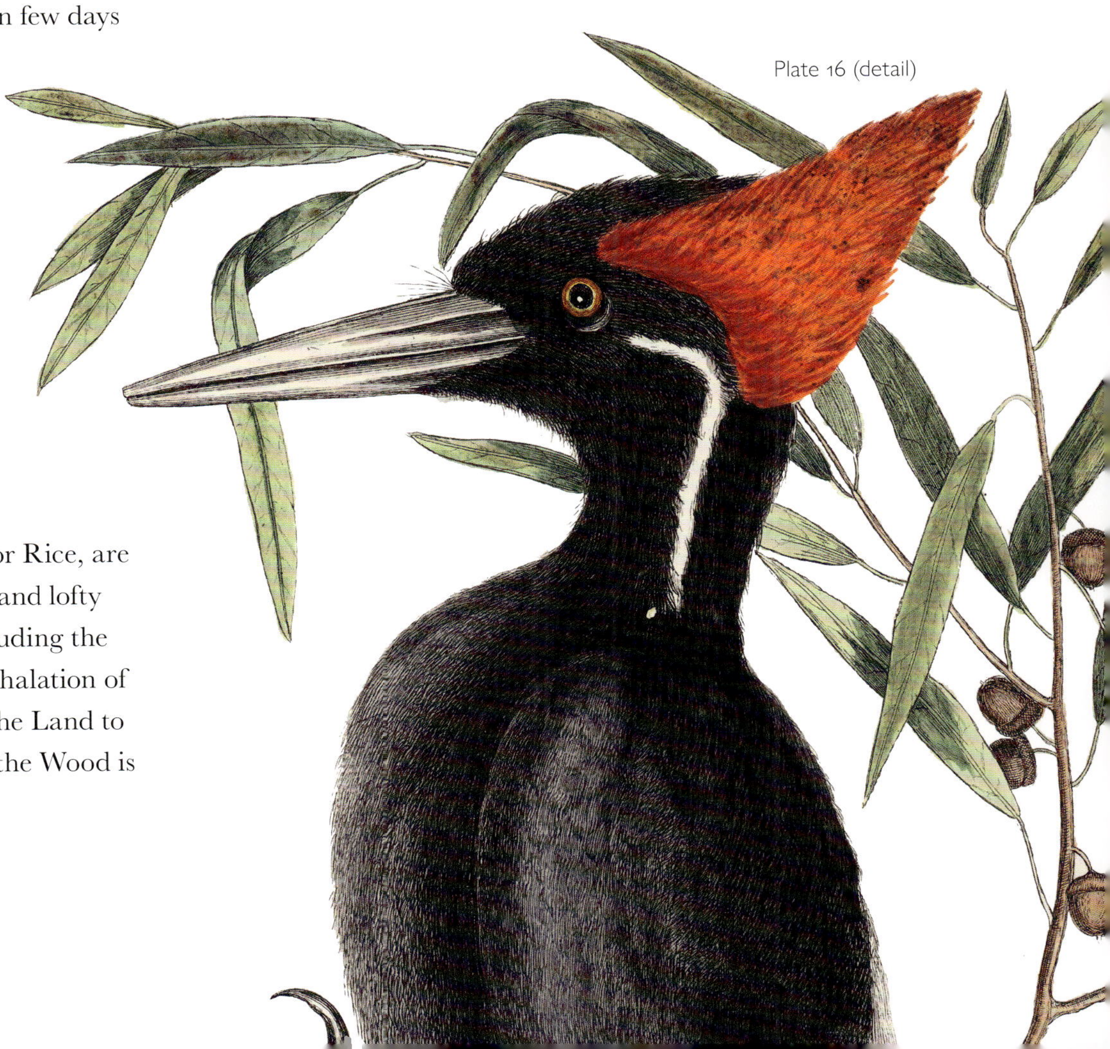

Plate 16 (detail)

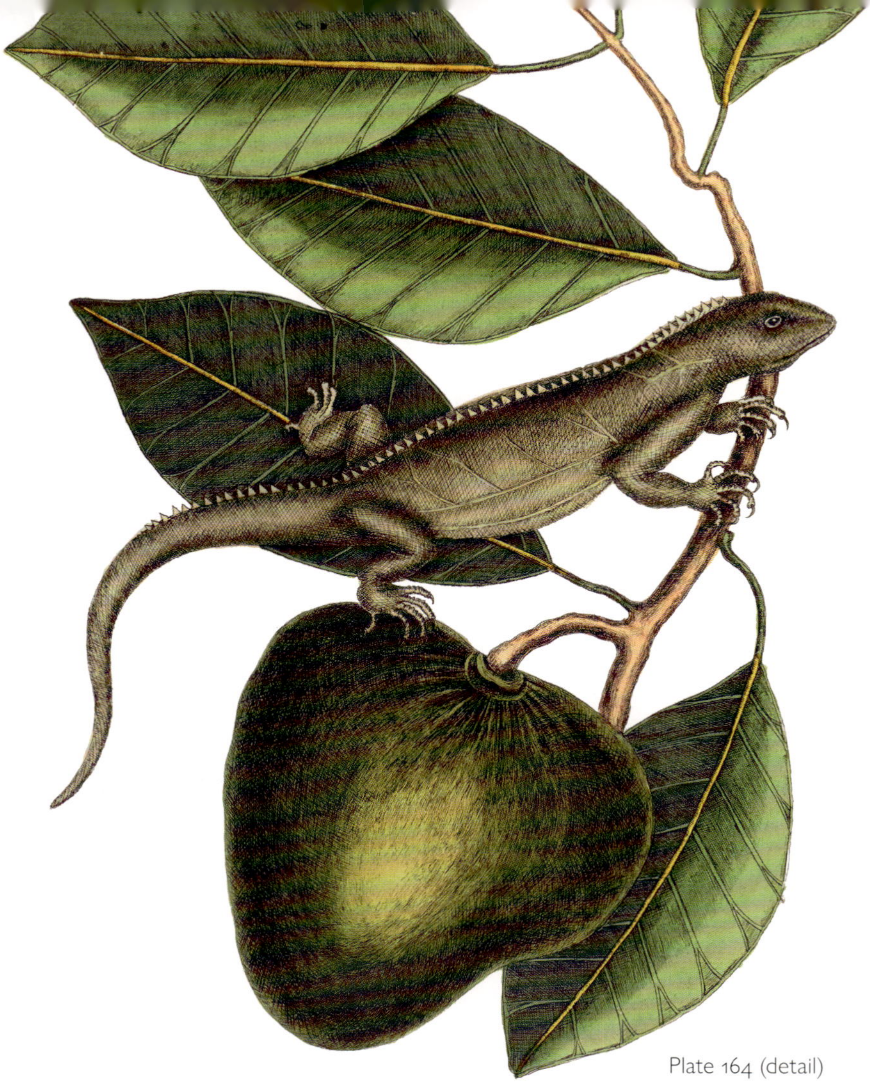

Plate 164 (detail)

cultivation, modern researchers have highlighted the critical role played by enslaved Africans in transferring their knowledge of and expertise in the cultivation and cleaning of rice in creating the rice-based economic powerhouse of the Carolinas.[39]

Catesby also related his understanding of rice introduction into the Carolinas: 'this beneficial grain was first planted in Carolina, about the year 1688, by Sir *Nathaniel Johnson*, then Governor of that Province; ... In the year 1696, a ship touched there from Madagascar by accident, and brought from thence about half a bushel of a much fairer and larger kind.'[40] However, he made no mention of Peter Collinson's report that in 1696 his 'sagacious friend' Charles Dubois 'first put the *Carolinians* on the culture of rice'. Collinson emphasized the serendipity of Dubois's involvement:

> He [Dubois] happened one day, in that year to meet *Thomas Marsh,* a Carolina merchant, at the coffee-house, to whom he said, I have been thinking, from the situation, nature of soil, and climate, that rice may be produced to great advantage in *Carolina:* But, says *Marsh,* how shall we get some to try?. Why, says *Dubois,* I Will enquire for it amongst our *India* Captains – Accordingly, a money bag full of *East India* rice was given to *Marsh,* and he sent it to *South-Carolina;* and in the year 1698, he told his friend *Dubois,* that it had succeeded very well.[41]

Decades of discussion have surrounded the introduction of rice to the Carolinas, but scholars agree that the introduction of rice was not a single event but happened numerous times and that rice was cultivated in the Carolinas before the end of the 1680s.[42]

In addition to mainland North America, Catesby visited the northern Caribbean and islands of the Bahamas, an archipelago of some 3,000 islands and islets. In the 1720s the islands were sparsely inhabited with a population that traded salt, turtles and citrus fruits with the Carolinas and subsisted through 'fishing, striking of Turtle, hunting *Guanas* [iguanas, e.g., *Cyclura cychlura*, **Plate 164**], cutting *Brasiletto* wood [*Caesalpinia bahamensis*, **Plate 151**], *Ilathera* bark [*Croton eluteria*, **Plate 146**], and that of wild cinnamon, or *Winter's* bark [*Canella winterana*, **Plate 150**], for these purposes they are continually roving from one Island to another'.[43] Catesby appreciated the climate of the Bahamas, with their hot, wet summers and warm, dry winters, when he visited in 1725:

> The Bahama Islands are blessed with a most serene air, and are more healthy than most other Countries in the same latitude, they being small, having a dry rocky soil, and pretty high land, are void of noxious exhalations, that lower and more luxuriant soils are liable to. This healthiness of the Air induces many of the sickly inhabitants of Carolina to retire to them for the recovery of their health.[44]

The habitats he found were different to anything he had experienced on the mainland. One of these was maritime swamps dominated by mangroves. Mangroves are trees or shrubs found in tropical and subtropical flooded habitats (sometimes called mangrove swamps) that occur at the interface of the sea and the land.[45] These swamps are periodically covered by tides, so mangroves must flourish in waterlogged soils that vary in their salinity daily. Consequently, different mangrove species occupy different positions relative to the sea.

Catesby illustrated four mangroves, the black (*Avicennia germinans*, **Plate 85**), white (*Laguncularia racemosa*, **Plate 86**[46]), button (*Conocarpus erectus*, **Plate 133**) and red (*Rhizophora mangle*, **Plate 163**). He showed the button mangrove with both flowers and fruits, but referred his readers to Hans Sloane's description in the first volume of his *Voyage to the Islands Madera, Barbados, Nieves, S. Christophers and Jamaica* (1707).[47] He added that the button mangrove, 'being small, I know of no mechanic use they are put to; but for burning it is esteemed the best wood in these latitudes',[48] whilst he emphasized the value of the bark of the black mangrove, which he saw only in fruit, for leather tanning. The red mangrove is the species found closest to the sea. Catesby described (and illustrated) the seed, which germinates before the fruit is released from the tree, as having the 'form of a bobin, with which lace is made'. In addition, he observed the importance of vegetative propagation: 'smaller branches, falling into the mud, strike root, and in a few years become trees'.[49] What he did not notice was the vertical roots (pneumatophores), which mangroves use to get air to their roots.

Young tree stems and pneumatophores, together with aerial roots, contribute to the density of the understorey in mangrove swamps, despite the few tree species they contain. However, mangrove swamps are diverse, species-rich habitats protecting coastal regions from erosion by the sea. They trap soil washed from the land and sediments suspended in the sea, and provide anchor points for invertebrates. Catesby created a vivid portrait of a habitat teeming with life that was unfamiliar to his readers:

> In shallow salt-water, these impenetrable woods of Mangroves are frequented by great numbers of Alligators [*Alligator mississippiensis*, **Plate 163**], which being too big to enter the closest recesses of these thickets, the smaller ones find a secure retreat from the jaws of their voracious parents. These watery woods are also plentifully stored with ravenous Fish, Turtles, and other animals, which prey continually one upon the other, and the Alligators on them all; so that in no place have I ever seen such remarkable scenes of devastation as amongst these Mangroves in *Andros*, one of the *Bahama Islands*, where the fragments of half-devoured carcasses were usually floating on the water.[50]

Plate 40 (detail)

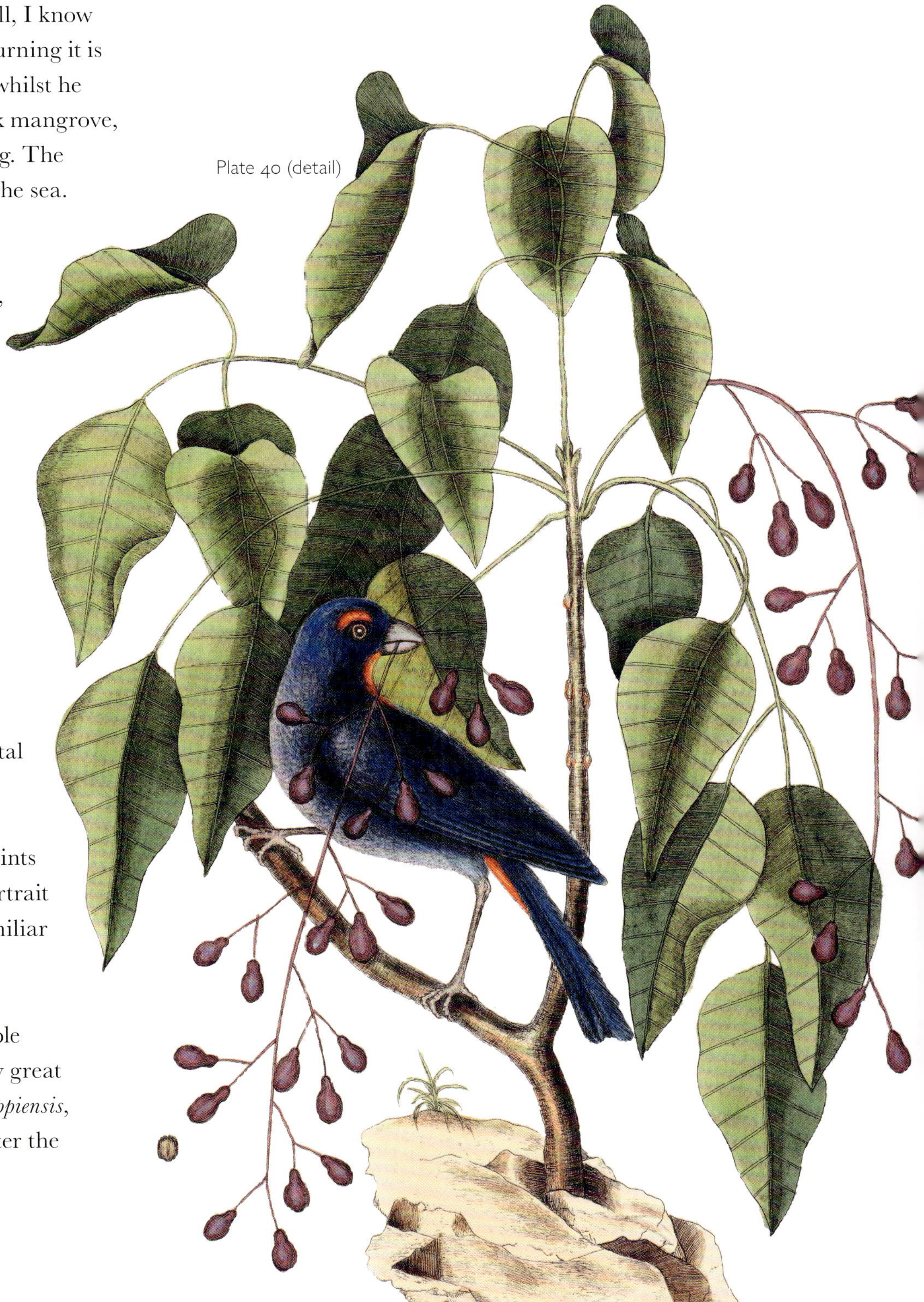

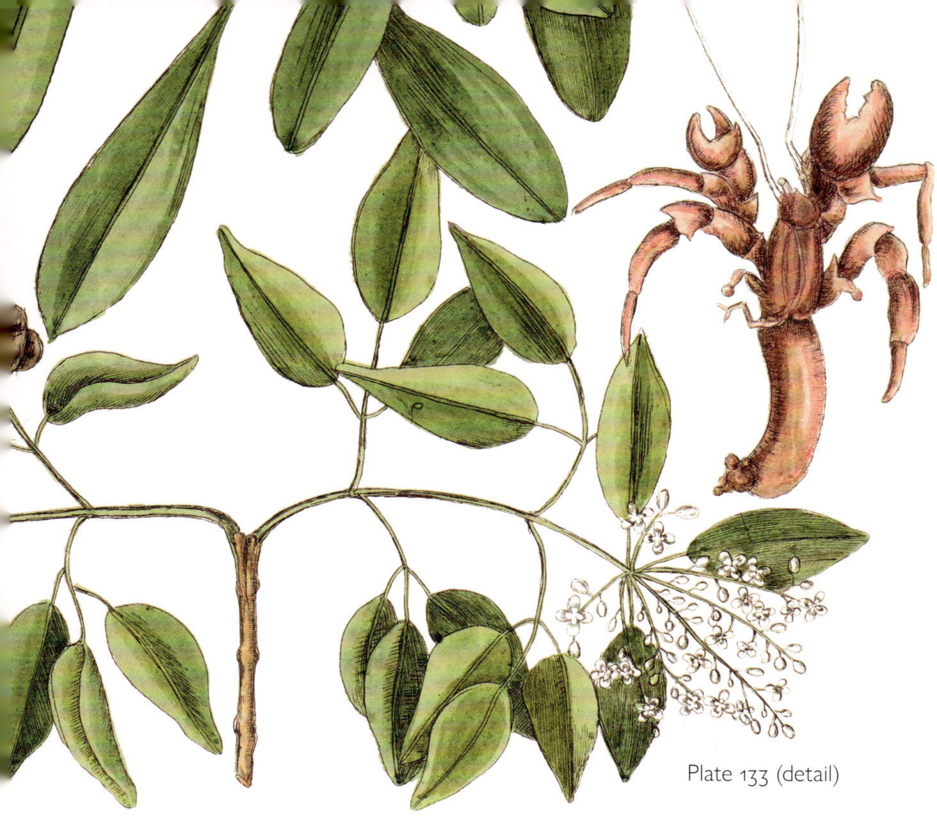

Plate 133 (detail)

One animal inhabiting mangrove swamps, 'where crabs grow',[51] that caught Catesby's attention was the distinctive purple terrestrial Caribbean hermit crab (*Coenobita clypeatus*, **Plate 133**). He describes how the crab occupies and changes shells as it grows, and depicts it both inside a West Indian top shell (*Cittarium pica*) and removed from the shell. However, the animal is shown with a right claw larger than the left, the reverse of what is usually found in nature; this may indicate that Catesby had overlooked that images are reversed when printed from engraved plates.

The other terrestrial crab Catesby illustrated was the nocturnal purple land crab (*Gecarcinus ruricola*, **Plate 132**), a species native to the Caribbean. He reported four colour forms typical of the crab and provided a vivid description of its annual mass migration to spawn in the sea:[52]

> in some places the ground being almost covered with them: so thick they are when out of their holes, that the earth seems to move as they crawl about … They make their holes, like rabbets, in sandy lands, in some of the mountainous islands. They, in breeding time, annually descend the hills, in vast numbers, to lay their eggs near the sea; whatever they meet with in their passage they go over, never going aside, let houses, churches, or what will, stand in their way: they have been known to enter in at a window, and on a bed, where people, who never before had seen any, were not a little surprised.[53]

The illustration is paired with a fruiting blackwood (*Picrodendron baccatum*) branch, which he identified as one of this omnivorous crab's food plants.

Catesby's succinct summary of the Bahamas implies that he visited numerous islands during his stay:

> *Grand Bahama* is the largest of the *Bahama* Islands; it is low, wet, and full of bogs: the Islands of *Andros*, and that of *Abbaco*, being very little better, yet they are all of use, and much frequented for hunting, fishing, and the plenty of excellent timber, and other useful woods they abound in: the Islands of *Exuma*, and *Crooked Island* have many salt ponds, for which they are much frequented: these islands, with *Cat Island*, are said to abound with the most good soil of any other.[54]

The islands however 'may not only be said to be rocky, but are in reality entire rocks, having their surface in some places thinly covered with a light mould; which, in a series of time, has been reduced to that consistence from rotten trees and other vegetables'. Despite the poor prospect of the soil, Catesby observed that the islands were 'always covered with a perpetual verdure, and the trees and shrubs grow as close, and are as thick cloathed with leaves, as in the most luxuriant soil'.[55] The forests he observed were distinctive types of dry forest that include numerous tree species such as pitch apple (*Clusia rosea*, **Plate 199**), mastic ironwood (*Sideroxylon foetidissimum*, **Plate 175**), gumbo limbo (*Bursera simaruba*, **Plate 30**), horseflesh mahogany (*Lysiloma sabicu*, **Plate 142**), bastard torch (*Nectandra coriacea*, **Plate 128**), manchineel (*Hippomane mancinella*, **Plate 195**), pigeon plum (*Coccoloba diversifolia*, **Plate 194**), poisonwood (*Metopium toxiferum*, **Plate 40**), sea grape (*Coccoloba uvifera*, **Plate 196**) and West Indian mahogany (*Swietenia mahogani*, **Plate 181**). Of the numerous palms he found in the Bahamas, he could only lament their exclusion from the *Natural History*: 'I regret my not being able to give their figures, or at least a more accurate description of them.'[56]

Catesby found it 'amazing to see trees of a very large size grow out of rocks, where no soil is visible, and the rock solid and compact before the roots found way to separate them ... All the nourishment that the trees receive, can be only from the ... hollows and chinks of the rocks, where the fibres of the trees insinuate.'[57] One of these trees was the West Indian mahogany (**Plate 181**), the wood of which was in great demand in Britain: the 'Tree increases to a stupendious [*sic*] size in a few years, it being a quick grower ... [with] trunks of them to the thickness of four feet or more in diameter'.[58] West Indian mahogany is one of three true mahogany species in the Americas, 'the excellency of this wood for all domestick uses is now sufficiently known in England: and at the Bahama Islands ... it is in no less esteem for ship-building, having properties for that use excelling oak'. However, eighteenth-century harvesting of West Indian mahogany populations for timber to produce furniture and woodwork in Europe is now thought to be a significant cause of large trees no longer being found in the wild in a type of high grading.[59]

High grading, a sort of genetic erosion, is a form of artificial selection where only the 'best' (i.e., largest, healthiest or most attractive) individuals in populations of commercially important natural resources are harvested. Over generations, populations subject to high grading will gradually lose the 'best' individuals; this changes the population's genetics, which may eventually have long-term adverse effects on species' survival. Catesby recognized at least one potential example of high grading in his description of the Bahama caesalpinia (**Plate 151**):

> the value of this wood has occasioned a scarcity of it on the Bahama Islands, particularly the largest trees being cut down; so that to what dimensions they grow, where largest, I know not; the biggest of those remaining not exceeding two or three inches in thickness, and eight or nine feet in height ... The inhabitants of the Bahama Islands formerly got great part of their subsistance [*sic*] by cutting this wood, but it is now much exhausted.[60]

Bahama caesalpinia, a leguminous tree, was a source of the important eighteenth-century dyewood brasiletto. Other high-value leguminous dyewoods exported to Europe from the Americas from the late 1400s included logwood (*Haematoxylum campechianum*), Nicaragua wood (*Haematoxylum brasiletto*) and brazilwood (*Paubrasilia echinata*), the latter of which gave Brazil its name.[61] Before this period, African and Asian relatives of these species were exotic dyewoods in Europe.[62]

Catesby gives another likely example of high grading in discussing the Virginia snakeroot (*Aristolochia serpentaria*, **Plate 29**), an 'excellent root'.[63] For centuries the peoples of North America and European colonists have valued snakeroot as a means of abortion and a panacea, disinfectant and universal cure for fevers through coughs to snakebites and worms.[64] He observed that the price of snakeroot was 'about six pence a pound when dryed, which is money hardly earned. Yet the Negro slaves, who only dig it, employ much of the little time allowed them by their masters in search of it; which is the cause of there being seldom found any but very small Plants.' However, 'by planting them [snakeroot] in a garden, they increased so in two years time, that one's hand could not grasp the stalks of one plant'.[65]

Catesby's continuous residency in the Carolinas and Bermuda for numerous years and his own curiosity gave him the opportunity to observe seasonal changes in bird populations: 'those which abide in *Carolina* the Winter, necessity drives from the frigid parts of the North, in search of food, with which the more Southern countries, abound; but where Summer birds of passage go at the approach, of Winter, is as little known as to where those of *Europe* go'.[66] Moreover, he inadvertently provided baseline data that facilitated future generations to document bird extinctions.

As he was based in the Carolinas for long periods and moved around, Catesby was able to make seasonal observations of the same species in different places. He summarized some of his conclusions in the *Natural History*, but synthesized them in a paper entitled 'Of Birds of Passage' in the *Philosophical Transactions of the Royal Society of London* (1747). In Europe, the appearance and disappearance of bird species had been commented upon for centuries,[67] but he considered that 'reports of their lying torpid in caverns and hollow trees, and of their resting in the same state at the bottom of deep waters, are notions so ill attested, and absurd in themselves, that they deserve no farther notice'. Yet 'if the immenseness of the Globe be considered, and the vast trails of land remaining unknown but to its barbarous natives, 'tis no wonder we are yet unacquainted with the retreat of these itinerant birds'.[68] His conjecture was that

> the place to which they retire, is probably in the same latitude of the Southern hemisphere, or where they may enjoy the like temperature of air, as in the country from whence they came. By this change they live in perpetual Summer, which seems absolutely necessary for their preservation, because all Summer birds of passage subsist on insects only, and have tender bills adapted to it, and consequently are unable to subsist in a cold country.[69]

One of these birds was the bobolink (**Plate 14**), a remarkable migrant that breeds in northern North America and overwinters in South America, with individuals capable of flying up to 1,800 kilometres per day.[70] Catesby observed that the females, which he ascertained after 'opening some scores prepared for the spit', arrived in the Carolinas in 'infinite swarms' to devour rice. In early spring he found both males and females. Hearing the flights of these birds heading north on three successive nights off Andros Island in September 1725, he conjectured that bobolinks were moving from Cuba to Carolina following the development of the rice crop.[71]

Change is a recurring theme throughout the *Natural History*, especially with almost three centuries of hindsight. Colonists transformed the landscape of North America to accommodate their economies, beliefs and societies. Forests were levelled to make way for fields and pastures, and the trees burnt or converted to timber and naval stores (pitch, tar and resin). Swamps, replete with swamp cypress (**Plate 11**), water tupelo (**Plate 60**) and sweetgum (**Plate 165**), were drained or their hydrology modified:

> *Rice Land* is most valuable, though only productive of that Grain, it being too wet for any thing else. The Situation of this Land is various, but always low, and usually at the Head of Creeks and Rivers, and before they are cleared of Wood are called *Swamps*, which being impregnated by the Washings from the higher Lands, in a Series of Years are become vastly rich, and deep of Soyl, consisting of a sandy Loam of a dark brown Colour. These Swamps, before they are prepared for Rice, are thick, over-grown with Underwood and lofty Trees of mighty Bulk, which by excluding the Sun's Beams, and preventing the Exhalation of these stagnating Waters, occasions the Land to be always wet, but by cutting down the Wood is partly evaporated, and the Earth better adapted to the Culture of Rice.[72]

Within the vastness of the landscape, short-term profligate agricultural practices proliferated, which exhausted the soil of its nutrients and led to new land having to be cleared of forest and swamp:

> The Lands of America from a series of years have accumulated such a coat of prolific soil that tillage is in a manner useless ... [the] fence is removed to another fresh piece of land, some of which yields them plentiful crops twenty years successively without respite ... At a planter's entering on fresh land, he is necessitated first to clear it of a vast burden of large trees and underwood; so much of which as is moveable is piled in heaps, and burned, the trunks being left to rot, which is usually effected in six or eight years.[73]

Yet Native Americans were managing these landscapes long before Europeans arrived; for example, Catesby identified how they burnt grasslands and woodlands to encourage the growth of grass and influence the movement of bison (*Bison bison*, **Plate 220**) herds:

> In *February* and *March* the Inhabitants have a Custom of burning the Woods, which causes such a continual Smoke, that not knowing the Cause, it might be imagined to proceed from Fog, or a natural Thickness in the Air: Likewise the Smoke

of the Tar-Kilns contribute not a little to deceive Strangers, and possesses them with an ill Opinion of the Air of *Carolina*: Add to these, an annual Custom of the *Indians* in their huntings, of setting the Woods on Fire many Miles in Extent.[74]

Two birds illustrated by Catesby highlight issues associated with the consequences of changing distributions, especially overexploitation: the North American passenger pigeon (*Ectopistes migratorius*, **Plate 23**) and the Carolina parakeet (*Conuropsis carolinensis*, **Plate 11**). His illustration of the North American passenger pigeon, nestled in a sprig of south-eastern United States turkey oak, is the first illustration of the pigeon made by a European. He described the bird's migratory habits, observing that in 'some places where they roost … they often break down the limbs of Oaks with their weight, and leave their dung some inches thick under the trees'. Moreover, during his time in Virginia, he had 'seen them fly in such continued trains three days successively, that there was not the least interval in losing slight of them, but that some where or other in the air they were to be seen continuing their flight'.[75] The pigeon – once the most abundant bird in North America – was a game bird for Native Americans: 'The pigeons … afford them some years great plenty of oil, which they preserve for winter use; … they also supply the want of fat in wild turkeys, which in some winters become very lean, by being deprived of their food by the numerous flights of the migratory pigeons devouring the acorns, and other mast.'[76] In the nineteenth century these birds also became cheap food for European settlers.[77] The last wild bird was probably shot in 1901; the last captive individual died in an American zoo in 1914.

The Carolina parakeet, once the most northerly psittacine whose distribution covered nearly half of the present-day United States, is shown perched on a branch of swamp cypress (*Taxodium distichum*). The composition is an ecologically realistic one, emphasizing the bird's use of the tree for feeding and nesting.[78] The colonial American explorer John Lawson described the parakeets as 'mischievous to Orchards', where they 'peck the Apples, to eat the Kernels, so that the Fruit rots and perishes'.[79] Catesby agreed, adding that their 'guts are certain and speedy poison to Cats'.[80] However, the decline of this troublesome bird coincided with rapid agricultural and industrial expansion in the nineteenth century; it was extinct by the mid-1940s.[81]

Plate 60 (detail)

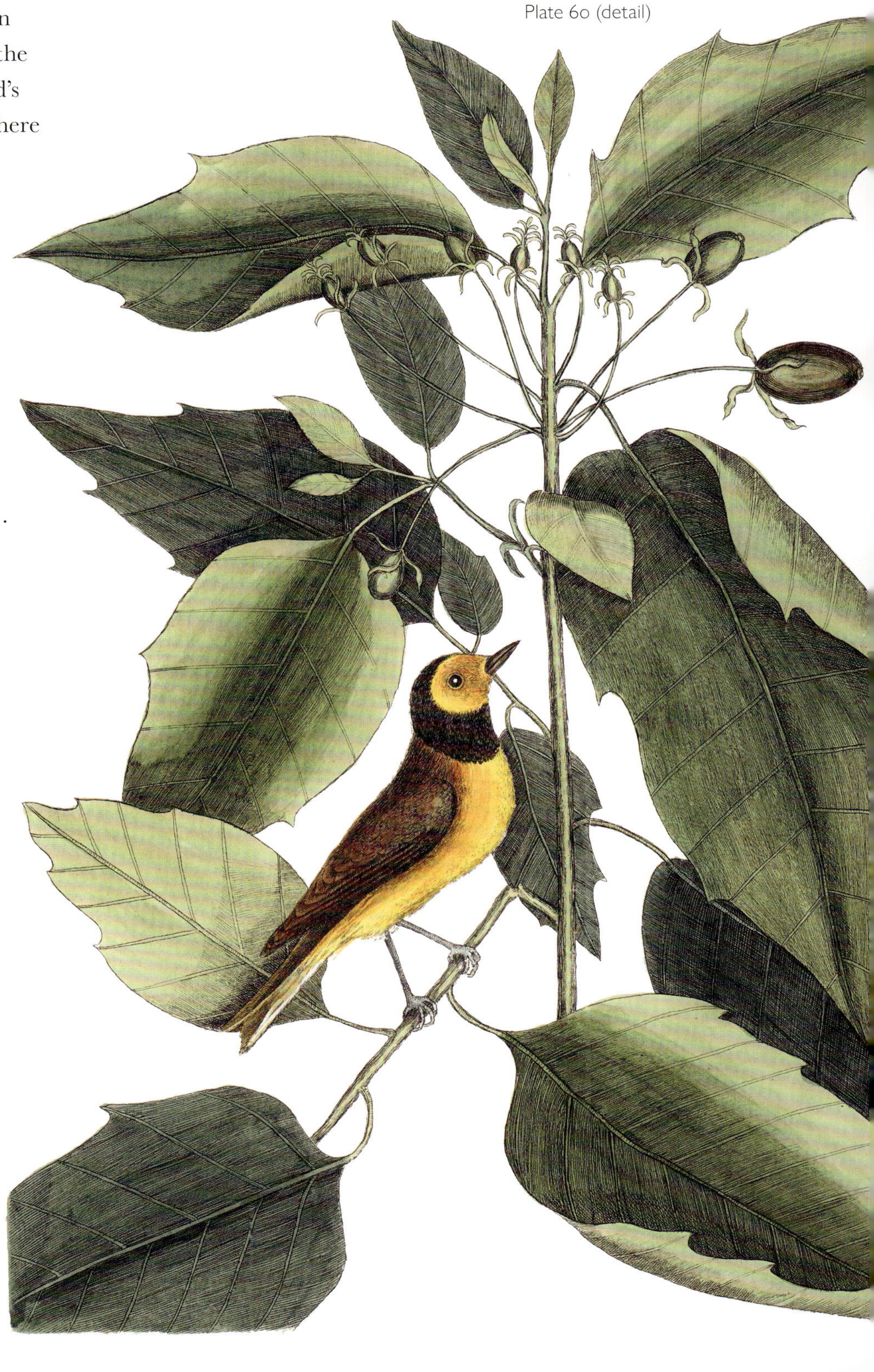

67

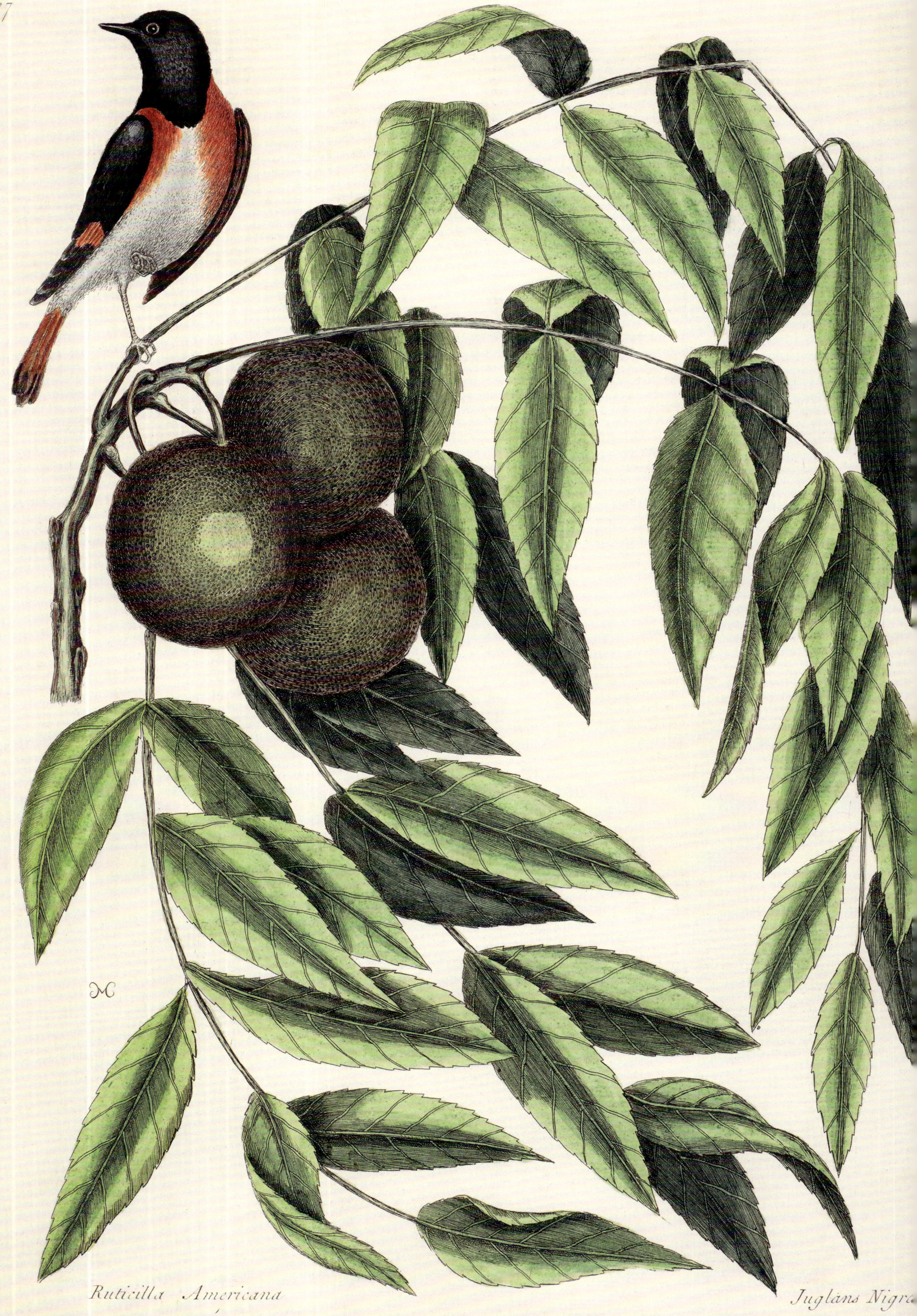

Ruticilla Americana

Juglans Nigra

On Indigenous and Enslaved Peoples 4

To the Hospitality and Assistance of these Friendly Indians, I am much indebted, for I not only subsisted on what they shot, but their First Care was to erect a bark hut, at the approach of rain, to keep me and my Cargo from wet.

MARK CATESBY[1]

The economic, social and political positions of Catesby's sponsors in England overshadow the often unacknowledged contributions made by others to his fieldwork and ultimately the success of the *Natural History*, for example, the colonists, Native Americans and black African slaves who acted as supports and guides, and provided him with information about the plants and animals he saw.

Catesby was fascinated by Native Americans, as were the intended audience for his seven-page account of their lives contained in the *Natural History*. In addition to his own observations, he drew liberally on those of Lawson. Catesby recognized that Indigenous peoples were essential to the success of his travels through the Carolinas. During his excursions into montane regions of the Carolinas, he 'employed an *Indian* to carry my Box, in which, besides paper and materials for Painting, I put dry'd Specimens of Plants, Seeds, &c. as I gather'd them'.[2] He also obtained information from them about plants and animals, but was conscious that this information might have been partial, as Native Americans were 'reserv'd and averse to reveal their secret mysteries to *Europeans*, that the relations of the most inquisitive can be but little depended on'. This was why he tried to confine himself 'to what I have learn'd by a personal knowledge of them [Native Americans]; and as Natural History is the subject of this book, I conceive it impertinent to relate tedious, narratives of religious ceremonies, burials, marriages, &c. which are too often the product of invention, or credulity in the relater'.[3]

Plate 67

One of the Indigenous peoples highly regarded by Catesby were the 'Chigasaws' (Chickasaw):

> With five of these *Indians* [Chickasaw] and three white men we set out to hunt; after some days continuance with good success, at our returning back, our Indians being loaded with skins, and barbacued buffalo [*Bison bison*], we espied at a distance a strange *Indian*, and at length more of them appeared following one another, in the same tract as their manner is: our five in perceiving these to be *Cherokee Indians*, and their enemies, being alarm'd, squatted, and hid themselves in the bushes, while the rest of us rode up to the *Cherokees*, who were then increased to above twenty: after some parley, we took our leave of each other, they marching on towards their country, and we homewards; in a short time we overtook our *Chigasaws*, who had hid their loads, and were painting their faces, and tripping up every little eminence, and preparing themselves against an assault. Though the *Cherokees* were also our friends, we were not altogether unapprehensive of danger, so we separated from our *Indian* companions, they shortening their way by crossing swamps and rivers, while we with our horses were necessitated to go further about, with much difficulty, and a long march, for want of our *Indian* guides.[4]

Catesby considered Native Americans to be remarkably healthy compared to Europeans: 'no people have stronger eyes, or see better in the night or day', which 'to this happy constitution of body is owing their little use of physic'.[5] On the negative side, 'few amongst these Americans [are] so robust, and of so athletic a form as

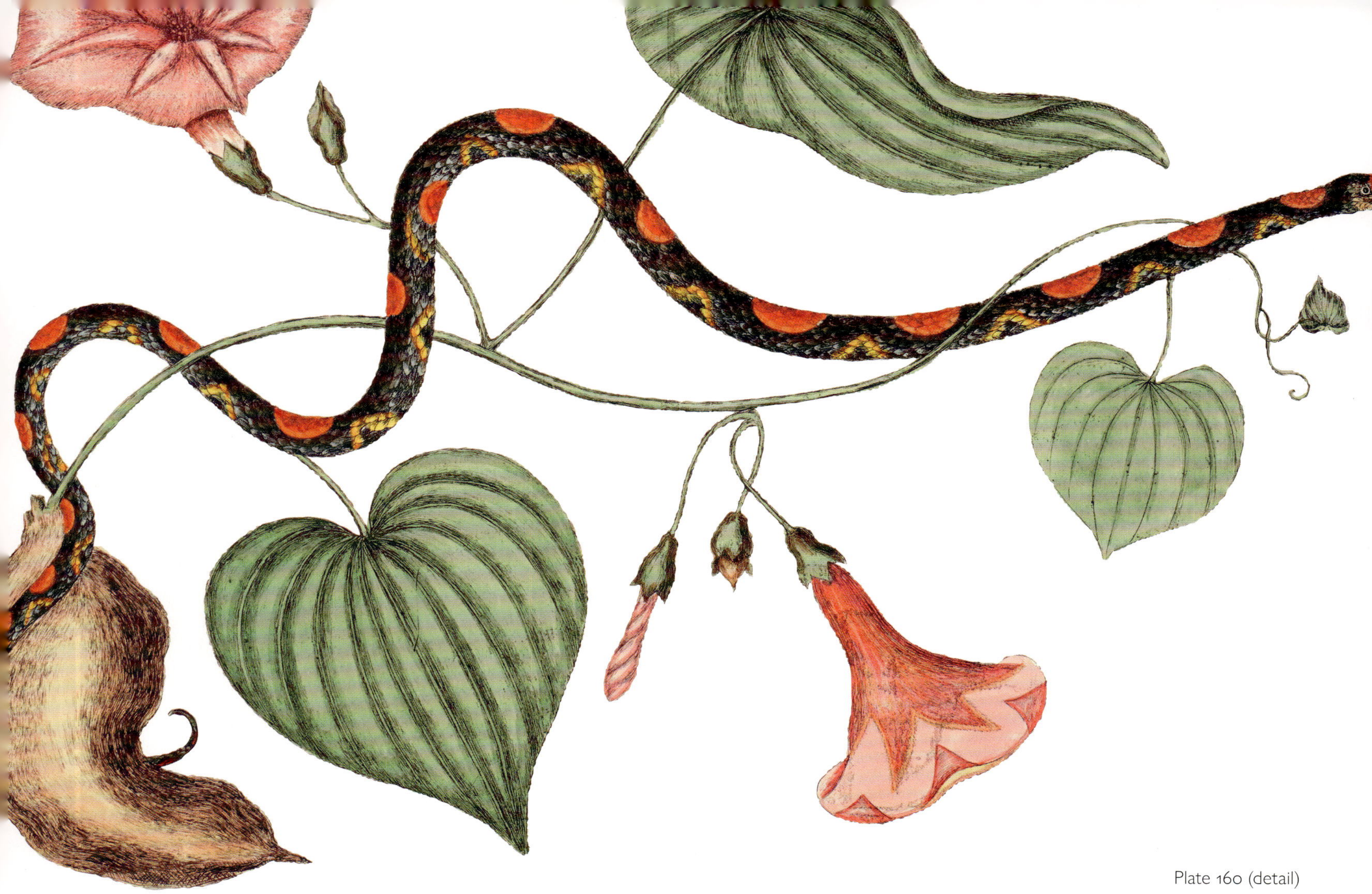

Plate 160 (detail)

amongst Europeans, nor are they so capable of lifting great burthens, and enduring so hard labour'.[6] These were characteristics he associated with an apparent reluctance to till the land:

> They have a vast country to range in, and the choice of the most delightful and fertile parts of it to inhabit, by which with little labour they might indulge the greatest luxury. Yet so little are they inclined that way, or even make so little use of these blessings, that, by depending wholly on Providence, they are sometimes drove to necessity.[7]

Maize (*Zea mays*) was the only grain Catesby saw Native Americans cultivating, although they consumed many of the animals and plants around them, including deer, pumas, buffalo (**Plate 220**), black walnuts (*Juglans nigra*, **Plate 67**), hiccory nuts (**Plate 38**), chinquapin (*Castanea pumila*, **Plate 9**) and live oak acorns (**Plate 17**). Despite Native Americans' apparent 'aversion to pepper and mustard',[8] Catesby observed fruits of green arrow arum (*Peltandra virginica*, **Plate 83**), which he thought 'excessive hot and astringent in my mouth',[9] used to flavour boiled venison. Another food plant they used, the Mexican vanilla orchid (*Vanilla mexicana*, **Plate 207**), whose native range extended from southern Florida into tropical America, did not impress him. He reported that 'the Spaniards perfume their chocolate, and employ Indians to cure the pods, which they do, by laying them in the sun to dry, then dipping them in an oil drawn from the kernel of the Acajou [*Anacardium occidentale*] nut' (**Plate 209**). However, 'this perfume is so little agreeable to an English palate, that it is rarely made use of any more in our American Plantations than at home, and therefore not cultivated by us'.[10] Nearly three decades after he wrote these words, according to the author of *The Gardeners Dictionary* (1768), the situation had not changed, 'though it [vanilla orchid] might be easily carried thither [to the American colonies] and propagated'.[11]

Catesby enjoyed hunting; for example, in the Bahamas he hunted and ate American flamingo (*Phoenicopterus ruber*, **Plates 73–74**): 'the flesh is delicate, and nearest resembles that of a Patridge in taste … A Man, by concealing himself from their sight, may kill great numbers of them, for they will not rise at the report of a gun; … they stand gazing, and as it were astonish'd, till they are most or all of them kill'd.'[12] He was in awe of Native Americans as hunters: 'they are indefatigable, and will travel further, and endure more fatigue than a European is capable of.'[13] To procure some specimens, he relied explicitly on the skills of Native Americans. For example, the yellow-breasted chat (*Icterus virens*, **Plate 50**) 'are very shy Birds, and hide themselves so obscurely, that after many hours attempt to shoot one, I was at last necessitated to employ an *Indian*, who did it not without the utmost of his skill',[14] whilst the sora (*Porzana carolina*, **Plate 70**), a small water rail, 'become so very fat in Autumn, by feeding on Wild Oats, that they can't escape the *Indians*, who catch abundance by running them down'.[15] In the case of the whooping crane (*Grus americana*, **Plate 75**), his entire description was taken from a skin presented to him 'by an *Indian*, who made use of it for his tobacco pouch'.[16]

Native Americans' sophisticated and subtle knowledge of medicine and medicinal herbs was founded upon knowledge systems alien to eighteenth-century Europeans.[17] Catesby identified a type of holly called *yaupon* (*Ilex vomitoria*, **Plate 157**), which was traded over great distances across North America, as important for the well-being of Native Americans through its ceremonial uses: 'it restores lost appetite, strengthens the stomach, giving them agility and courage in war.'[18] Although 'by the sour faces the *Indians* make in drinking this salubrious liquor', he observed, 'it seems as little agreeable to an *Indian* as to an *European* palate, and consequently that the pains and expences they are at in procuring it from remote distances, does not proceed from luxury (as tea with us from *China*) but from its virtue, and the benefit they receive by it'.[19] He compared the properties of *yaupon* with those of the South American holly *yerba maté* (*Ilex paraguayensis*), the dried leaves of which are used as a stimulant, and which he presumably knew through imports into London by Spanish traders or had seen being consumed in Florida. Another tree with medicinal properties was sweetgum (**Plate 165**) which, when the bark is damaged, produces a fragrant resin that 'congeals into transparent resinous drops, which the *Indians* chew, esteeming it a preservative of their teeth'.[20]

An an observer Catesby had to remain sceptical. As he reported information from others, he was constantly in the position of filtering and assessing new data for their quality. He therefore preferred to use his own observations where possible. Where this was not possible, he made value judgements on the veracity of his sources. For example, he was uncertain about the information received from a 'Colonel Moore, a Gentleman of good reputation in *Carolina*' that he had

> seen an *Indian* daub himself with the juice of this Plant [saltmarsh morning glory, *Ipomoea sagittata*, **Plate 35**]; immediately after which, he handled a Rattle-Snake [**Plate 141**] with his naked hands, without receiving any harm from it … others affirm, that they have seen the *Indians* use a plant to guard themselves against the venom of this sort of Snake; but they were not observers nice enough to inform me of what kind it was.[21]

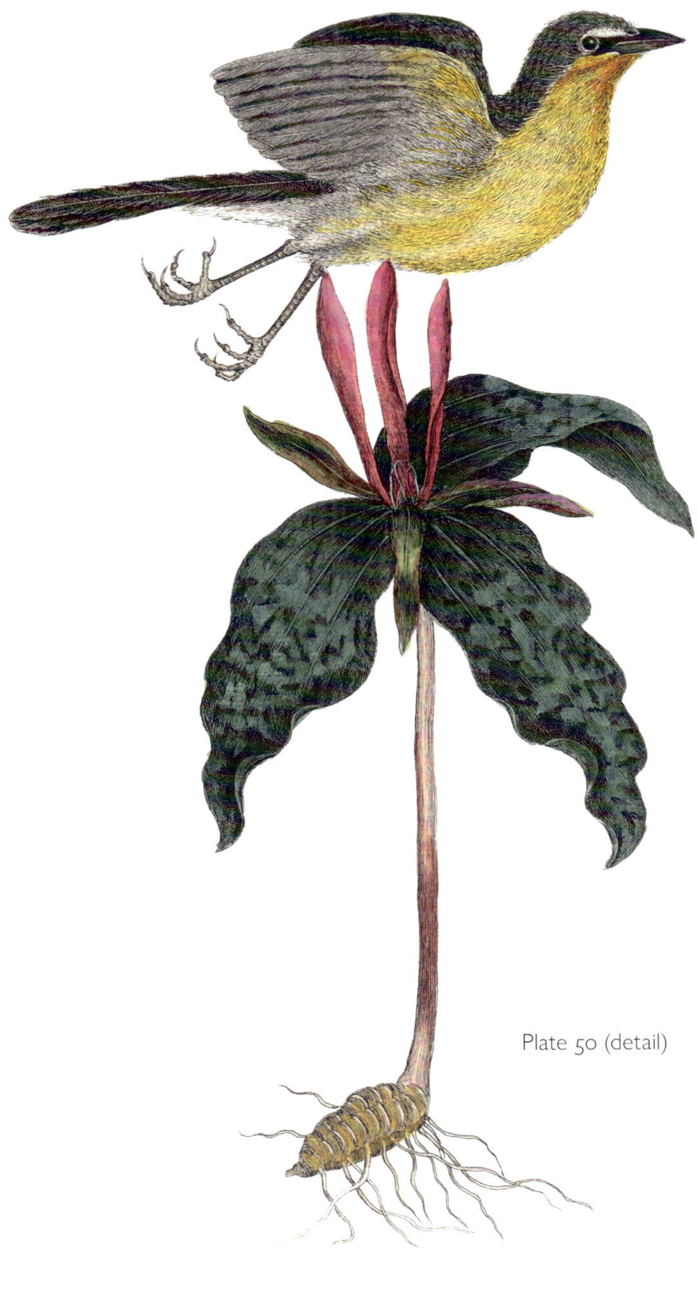

Plate 50 (detail)

He was dismissive of the 'silly belief' that the eastern coachwhip (*Masticophis flagellum*, **Plate 154**), a slender, harmless snake up to one and a half metres in length and native to the southern United States, would 'by a jirk of their tail, separate a man in two parts'.[22]

Slaves transported across the Middle Passage from West Africa were essential for the commercial success of the Thirteen Colonies.[23] Enslaved black Africans would have been familiar to Catesby in Virginia. He was unsurprised to find similar practices in the Carolinas, where the slaves cultivated rice rather than tobacco as their masters' cash crops. When he arrived in the Carolinas in the early 1720s, there were an estimated 12,000 slaves.[24] In 1725 in the Bahamas, he estimated that there were fewer than 1,000 inhabitants, 'besides about [blank space] Negro slaves';[25] it is perhaps telling that the number of slaves was accidentally or deliberately omitted. Shockingly, he wrote to William Sherard about six months after his arrival in the Carolinas requesting permission to 'buy a Negro Boy which I cannot be without'[26] to help him meet his obligations as a collector. We do not know whether Sherard's permission was forthcoming or whether Catesby went through with the purchase. We do not know what Sherard's attitude to slavery was, but he would have been familiar with the practice from his time as a diplomat at Smyrna (İzmir) in Ottoman Turkey.[27] In the *Natural History*, Catesby emphasizes the unfamiliar or observations likely to appeal to his intended audience, which may explain why the few mentions of enslaved people are focused on food culture.

In the *Natural History*, Catesby appears curious about the food that colonists had learnt to appreciate in their daily diets, together with the unusual foods of Native Americans and of African slaves. The most important cereal consumed in the colonies was maize,[28] which

> is eat[en] not only by the Negro slaves, but by the generality of white people. Its easy culture, great increase, and above all its strong nourishment, adapts it to the use of these countries as the properest food for Negro slaves; some of which, at a time when by the scarcity of this grain they were obliged to eat wheat, found themselves so weak that they begged of their master to allow them Indian corn [*Zea mays*] again, or they could not work.[29]

Two cereal grains mentioned by Catesby, 'bunched Guinea corn' (*Sorghum*) and 'spiked Indian corn' (millet, *Panicum*), were mostly used for 'feeding fowls' but were also important foods for the 'Negroes, who brought [them] from Guinea, their native country, and are therefore fond of having it'. Another important food he thought 'of great use for feeding Negroes, being a strong hearty food', were 'eight or ten different sorts'[30] of beans he considered native to the Americas. He recognized that they had been cultivated by Native Americans long before the arrival of the English and were 'better' than peas and beans introduced from England. He argued that the English plants 'degenerated' after a few years unless regularly refreshed with new seed from England.

Besides maize as a food plant, Catesby reserved his greatest praise for the Virginian, or sweet, potato (*Ipomoea batatas*, **Plate 160**), which despite his suggestions, is native to the Americas:[31]

> This excellent root seems to merit the preference of all others; not only in regard to the wholesomeness and delicacy of its food, but for its more general use to mankind than any other root; it being one great part, if not the principal subsistence [*sic*] of the greater part of *Africa*; and is likewise in great use, both in *America* and in the southern parts of *Asia*. They being of so easy culture, so quick of growth, and of so vast an increase, that the propagating it seems more agreeable to the indolence of the *Barbarians*, than cultivating grains, which require a longer time, with more labour and uncertainty. In all our Colonies of *America*, as well Islands as Continent, these roots are in great esteem and use; the common white people, as well as the negro slaves, subsisting much upon them: nor are they thought unworthy a place at principal tables.[32]

He identified another plant as a useful carbohydrate source, which he suggested was also of African origin, the tropical and subtropical American herb tannia (*Xanthosoma sagittifolium*, **Plate 145**), which,

> not caring to increase much in *Carolina*, and will grow no where north of that colony; yet the negroes there (who are very fond of them) by annually taking up the roots to prevent rotting, get a small increase. They are of so acrimonious a quality, that there is a necessity of boiling them eight or ten hours before they are eatable. A little before I left *Carolina*, there was introduced a new kind, wholly without that bad quality, and requiring no more than common time to boil them; and may be eat raw, without offending the throat or palate. This was a welcome improvement among the negroes, and was esteemed a blessing; they being delighted with all their African food, particularly this, which a great part of Africa subsists much on.[33]

Among the other foods used by enslaved Africans that attracted Catesby's attention were pawpaw fruits (*Asimina triloba*, **Plate 185**) and purple land crabs (**Plate 132**). The pawpaw is a tree of low-lying, fertile swamps, which was widely consumed by Native Americans.[34] The fruit 'contains a yellow pulp of a sweet luscious taste; in the middle of which lie, in two rows, twelve feeds, divided by so many thin membranes. All parts of the Tree have a rank, if not a fetid smell: nor is the fruit relished but by very few, except negroes.'[35] In the Bahamas, migratory purple land crabs, which populated mangrove swamps, were of 'no small help to the negro slaves, who on many of the islands would fare very hard without them'.[36]

Two other animals, the chigoe flea (*Tunga penetrans*, **Plate 210**) and the velvet ant (*Dasymutilla occidentalis*, **Plate 215**), which caused problems 'especially to

Plate 145 (detail)

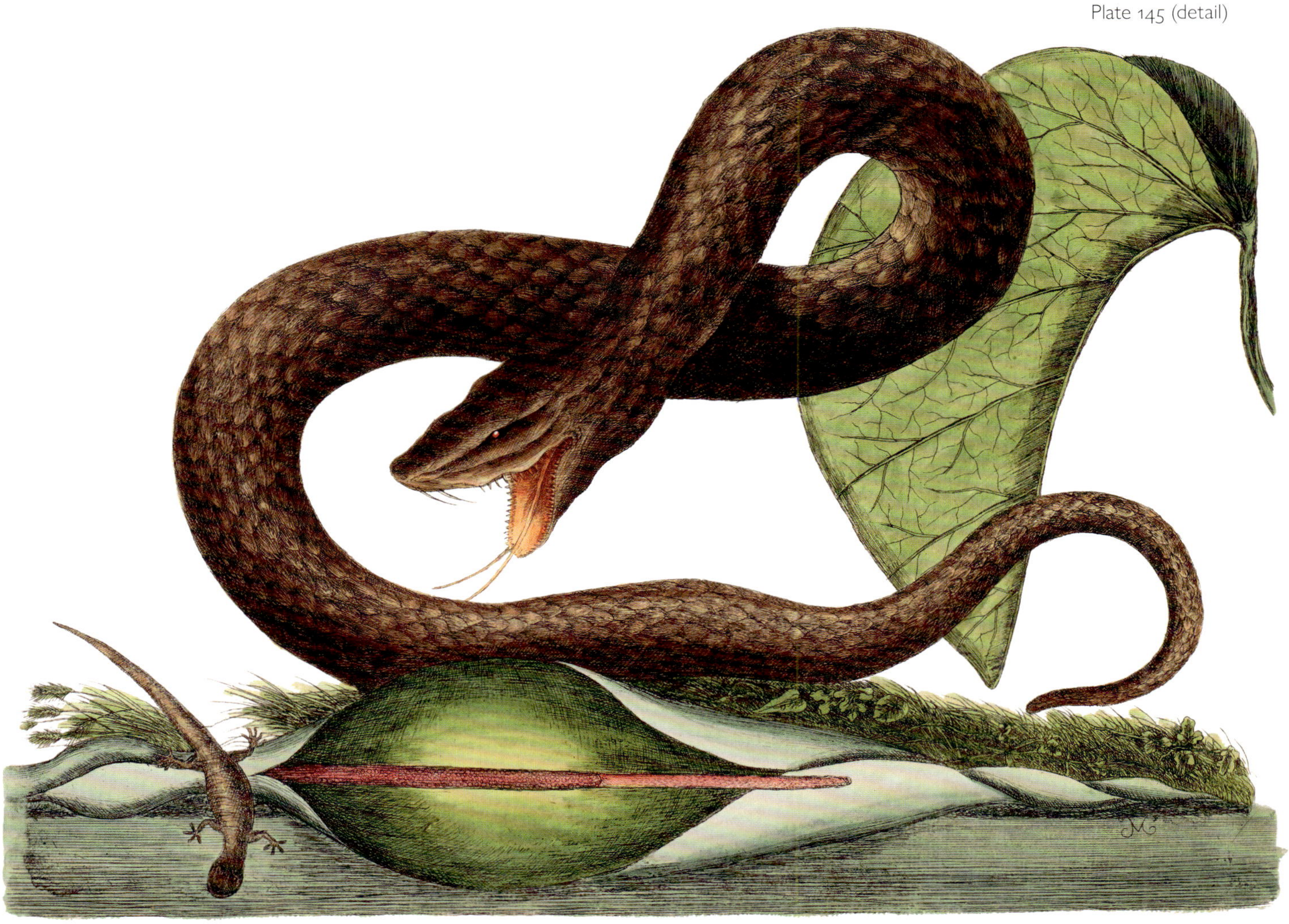

Negroes and others that go bare-foot, and are slovenly',[37] merited Catesby's attention and illustration. The chigoe flea, which he depicts as it would appear under a microscope, 'penetrate[s] the skin, under which they lay a bunch or bag of eggs, which swell to the bigness of a small pea or tare, and give great pain till it is taken out; to perform which, great care is required for fear of breaking the bag, which endangers a mortification, and the loss of a leg, and sometimes life itself'.[38] The velvet ant, despite its name, is a parasitoid wasp that lays its eggs inside the bodies of other arthropods. In his illustration of a female velvet ant, Catesby identifies three of its defence behaviours: the exoskeleton which is 'is of so strong and hard a contexture, that being trod upon by men or cattle, they receive no harm'; their ability to run fast and evasively; and their venom released from 'a long sting in their tails, which causes inflammation and great pain, for half an hour to those who are stung by them'.[39] He admits to being perplexed by this insect: 'what they feed on, and in what manner they breed, and where they secure themselves in winter, is to me unknown.'[40]

Rarely, Catesby deferred to the greater knowledge of the enslaved population. For example, on the River Stono, southwest of Charleston in the Carolinas, which is famous for its Slave Rebellion in September 1739, 'three or four teeth of a large animal' were found, 'which, by the concurring opinion of all the *Negroes*, native *Africans*, that saw them, were the grinders of an Elephant'. Invoking popular eighteenth-century ideas for the origin of fossils, Catesby was fascinated by deposits discovered in North America: 'there is no part of the Globe where the signs of a deluge more evidently appears.'[41] He was working with the belief that the earth was 'universally replenished with all animals from *Noah*'s ark after the general deluge, and that those *European* birds which are in *America*, found their way thither at first from the Old World'.[42]

below Fig. 4 Letter (dated 11 August 1745) from Peter Collinson in Chelsea asking the Sherardian professor of botany in Oxford, Johann Dillenius, for advice on the classification of a garden shrub introduced to Britain from the Carolinas by Mark Catesby – the Carolina allspice (*Calycanthus floridus*, Plate 46).

right Plate 185 (detail)

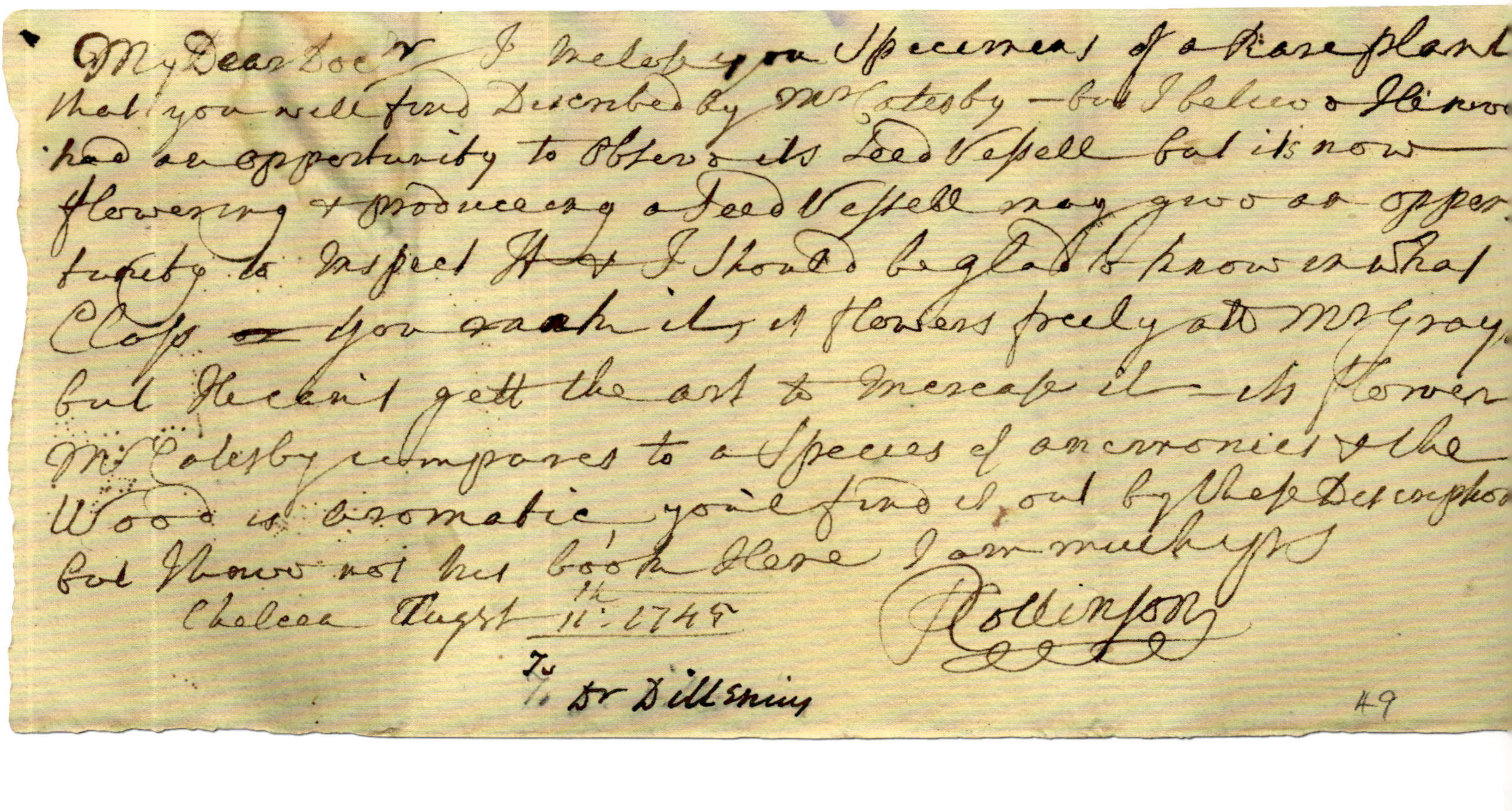

My Dear Docr I inclose you Specimens of a Rare plant that you will find Described by Mr Catesby — but I believe he never had an opportunity to Observe its Seed Vessell but its now flowering & produceing a Seed Vessell may give an oppertunity to Inspect It & I should be glad to know in what Class you rank it, it flowers freely att Mr Gray's but He can't gett the art to Increase it — its flower Mr Catesby compares to a Species of Anemonies & the Wood is Aromatic you'l find it out by these Descriptions but I know not his book Here I am [illegible] yrs

Chelsea Augst 11th 1745

P Collinson

To Dr Dillenius

49

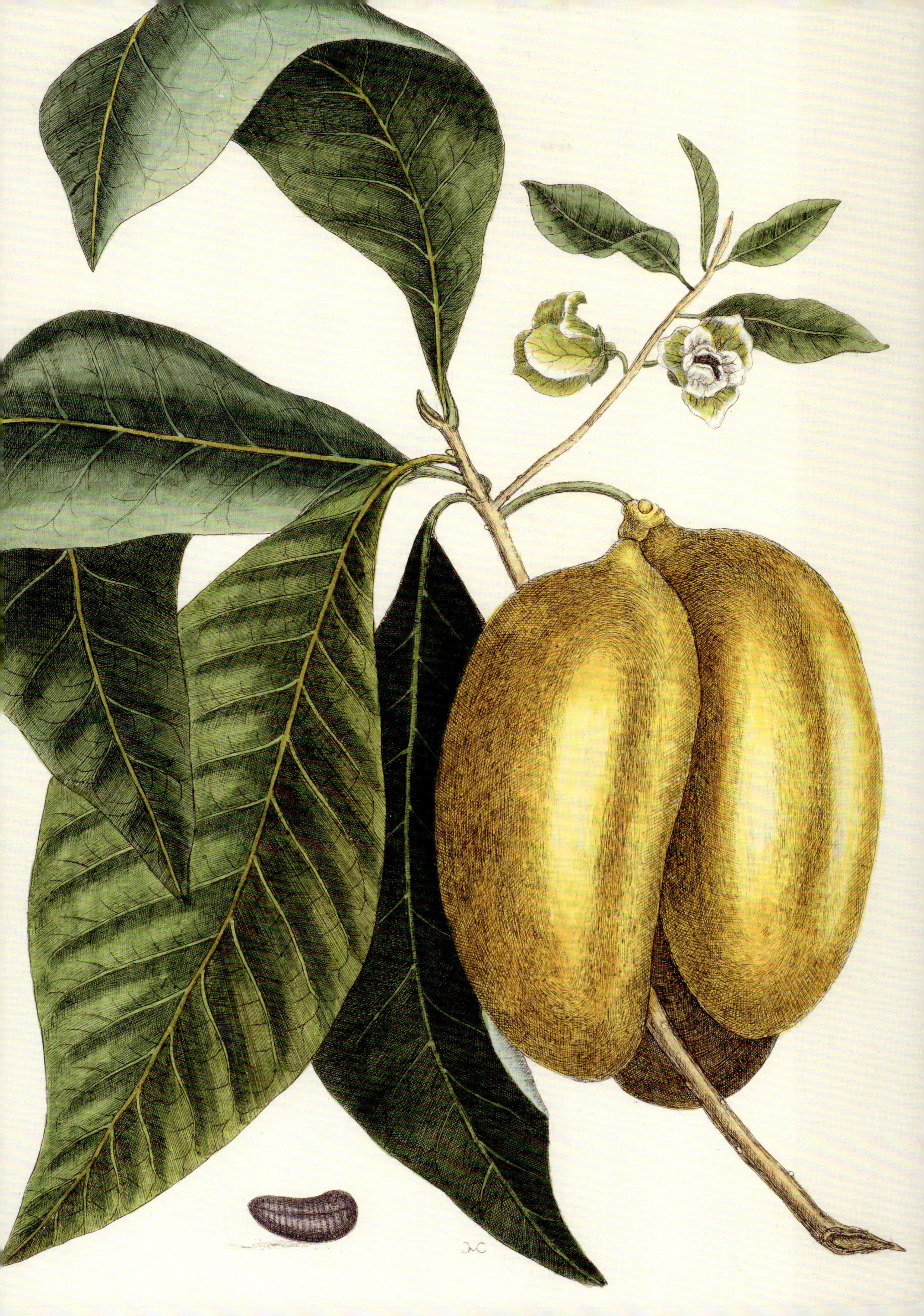

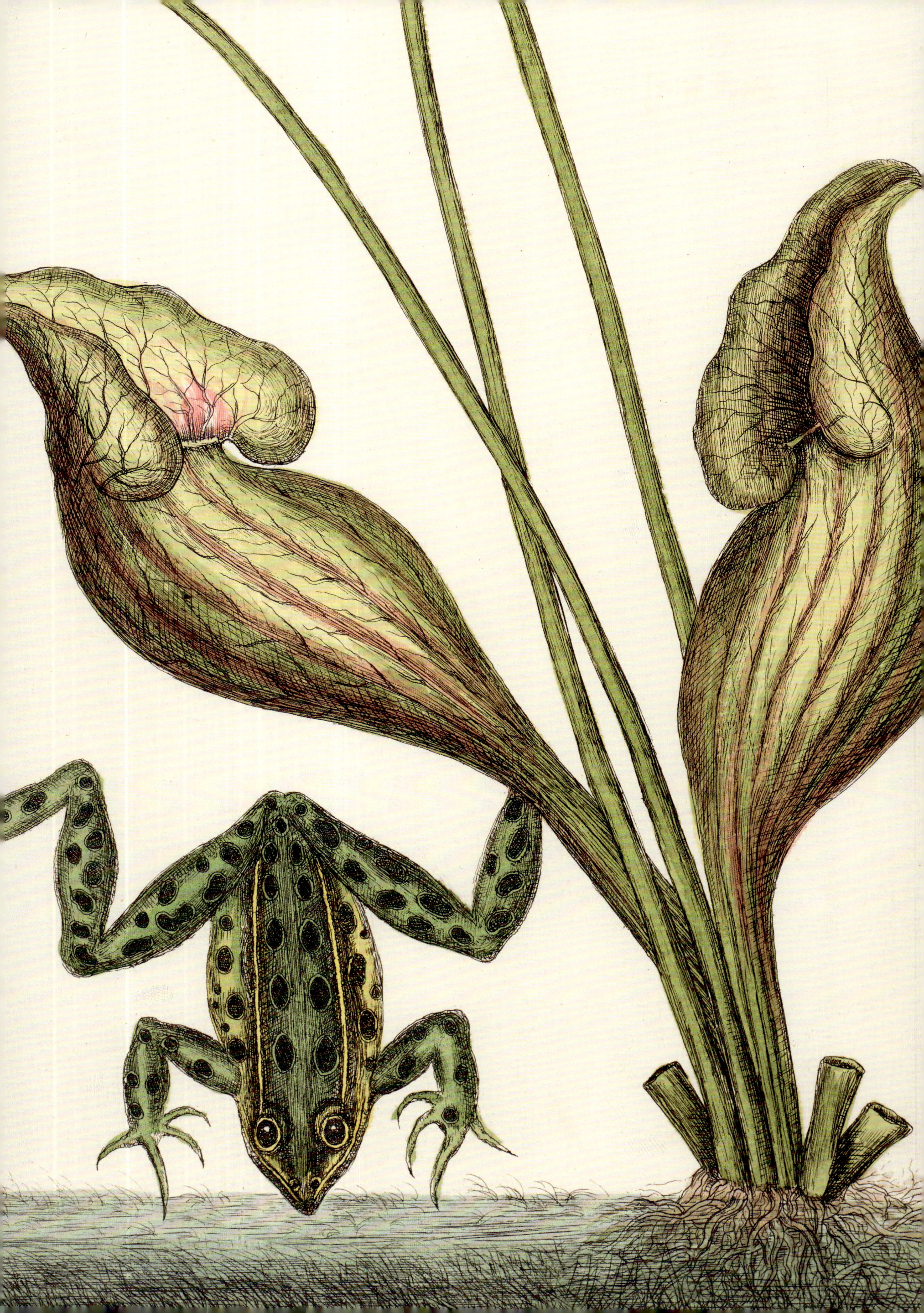

On Collection and Cultivation 5

It will easily be imagined that a forest of a thousand miles in length, extending twenty degrees in latitude from north to south (for such is the territory of the crown of Great Britain on the continent of America) must afford a plentiful variety of trees and shrubs, that may be usefully employed to inrich and adorn our woods by their valuable timber and delightful shade; or to embellish and perfume our gardens with the elegance of their appearance and the fragrancy of their odours; in both which respects they greatly excel our home productions of the like kind.

MARK CATESBY[1]

Catesby contributed to the eighteenth-century interest in North American plants by collecting specimens, introducing species for cultivation and collaborating with commercial nurserymen to foster their propogation. His work promoted the grandeur of landscapes in south-eastern North America as he sought to sell the continent's plants to a British audience keen to fill their gardens with novelties. Moreover, he wanted to change landscapes by encouraging the planting of American trees to enrich the diversity of British landscapes, bringing the Thirteen Colonies to British shores.[2]

Natural history collecting, the process of getting a specimen from the field into a scientific collection, is a complex, intricate process that requires collectors to be efficient and organized. It is hard, physically challenging and dangerous work, far removed from the familiar British stereotype of a refined pursuit for the clergy, women and children.

As Catesby travelled, he observed and collected different types of animals and plants from areas that were poorly known to Europeans, for sponsors with multifarious desires and little understanding of, and perhaps interest in, the conditions under which he worked. If he was to succeed, he had to manage his sponsors' expectations, preserve animals and plants, protect his collections from pests, fungi and damp, and record what he observed. The longer he spent in North America, the more knowledgeable he became at refining his art to satisfy the desires of his sponsors, the realities of his working conditions and his own interests.

Plate 170 (detail)

A preserved specimen is evidence linking an organism's identity to the time and place of its collection. Unlike records of names or some sorts of images, which rely on the authority of the list maker or interpretations made by an artist, specimens may be critically re-evaluated by future generations.[3] Physical specimens, and the associated field observations, that were once preserved in private collections today fill the world's natural history museums, where they may be used for research, teaching and public engagement.[4] In addition to preserved specimens, Catesby also made illustrations and collected seeds that he hoped might one day be grown in British gardens.[5]

Careful attention to detail is essential if specimens are to have permanent scientific value. In the case of plants, fresh plants, with fruits and/or flowers, are spread out between absorbent papers and dried quickly under pressure to keep the samples flat. Since Catesby left no details of his field equipment and practices, inferences must be drawn from his specimens themselves and the practices of contemporary natural historians. Cursory broadsheets for natural history collecting had been published by James Petiver in about 1700. Petiver's instructions for the preservation of all animals, vegetables and minerals came to fewer than one thousand words. For example,

> in Collecting PLANTS, Pray observe to get that part of either Tree, or Herb, as hath its Flower, Seed or Fruit on it; but if neither, then gather it as it is, and if the Leaves which grow near the Root

> of any Herb, differ from those above, be pleased to get both to Compleat the Specimen; these must be put into a Book, or Quire of Brown Paper stich'd (which you must take with you) as soon as gathered; You must now and then shift these into fresh Books, to prevent either rotting themselves or Paper.[6]

Petiver emphasized that all available species should be collected – 'As amongst Forreign Plants, the most common Grass, Rush, Moss, Fern, Thistle, Thorn, or vilest Weed you can find will meet with Acceptance, as well as a scarcer Plant: So in all other things, gather whatever you meet with' – but there were limits: 'if very common or well known, the fewer of that Sort, will be acceptible'. Despite the simplicity of the instructions, the technique is time consuming and considerable practice is essential to prepare scientifically useful specimens that stand the test of time.

Animal remains, especially fish, reptiles, birds and mammals, are likely to have been more problematic to preserve than plants. Catesby does not report how he preserved the animals he collected, but standard protocols of the period recommended flaying vertebrates, drying their skins and boiling their carcasses to retrieve bones. It is hardly conceivable that he spent hours surrounded by stinking pots of macerating wildlife. It is more likely that everything he collected was prepared as quickly and easily as possible and stored dried, using the poorly developed methods of the eighteenth century.[7] It is therefore not surprising that most of the natural history specimens he returned to London were flattened, dried plants, which when glued to a sheet of paper together with their associated field data became herbarium specimens. The quality demanded of elite herbarium specimens, especially their labelling, has changed dramatically since the sixteenth century, although the technology to produce them has changed little. Kept free of pests and damp, herbarium specimens will survive hundreds of years, and in the twenty-first century have been exploited by researchers in many different areas of the sciences, arts and humanities, often in manners that the original collectors could never have envisaged.[8]

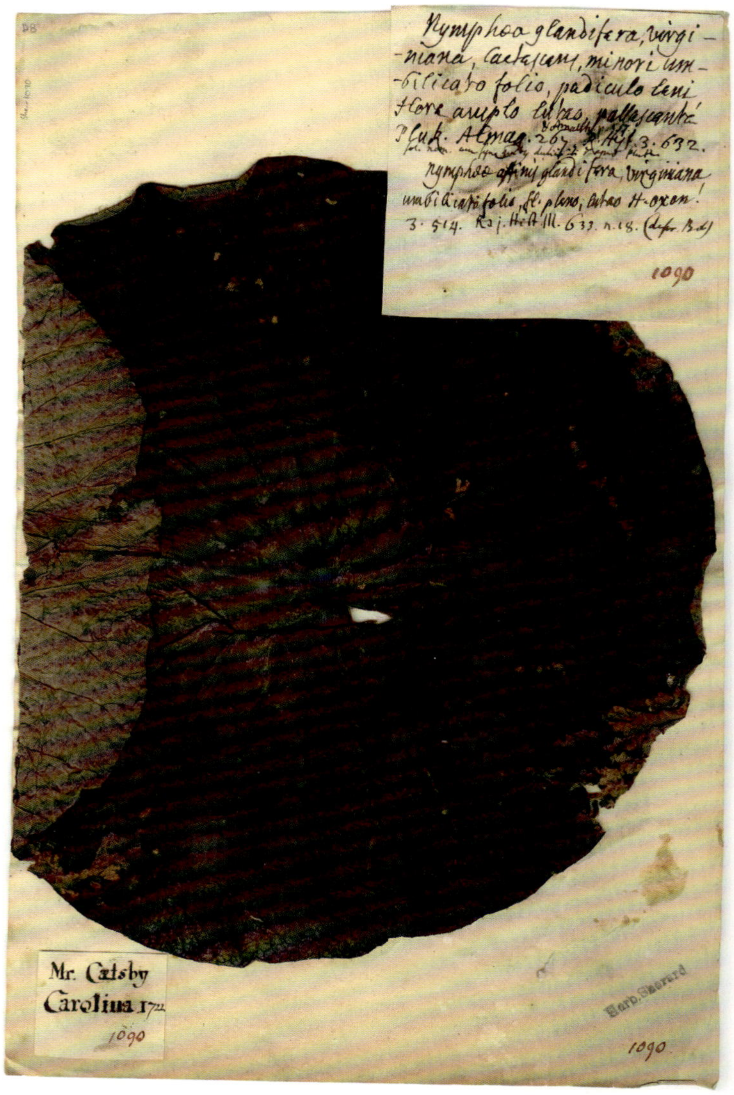

Fig. 5a An American lotus (*Nelumbo lutea*) leaf collected by Mark Catesby in Carolina in 1722. Preserving the flower of this water plant proved too much even for Catesby's skills as a field naturalist.

Catesby had to treat his specimens carefully since they were indirectly funding his travels. Poor specimens would give poor returns to patrons who wanted new species and good-looking specimens to add to their collections. Moreover, since his patrons were powerful men, their good favour could be useful to him after he eventually returned to London. He was often on the move, collecting as he went, and would therefore have had material at various stages of preparation. Consequently, there would have been a constant, no doubt eventually habitual, routine of changing absorbent papers between the drying specimens and of drying damp papers over a fire or in the sun. Such labour was particularly arduous on rainy or cloudy days.

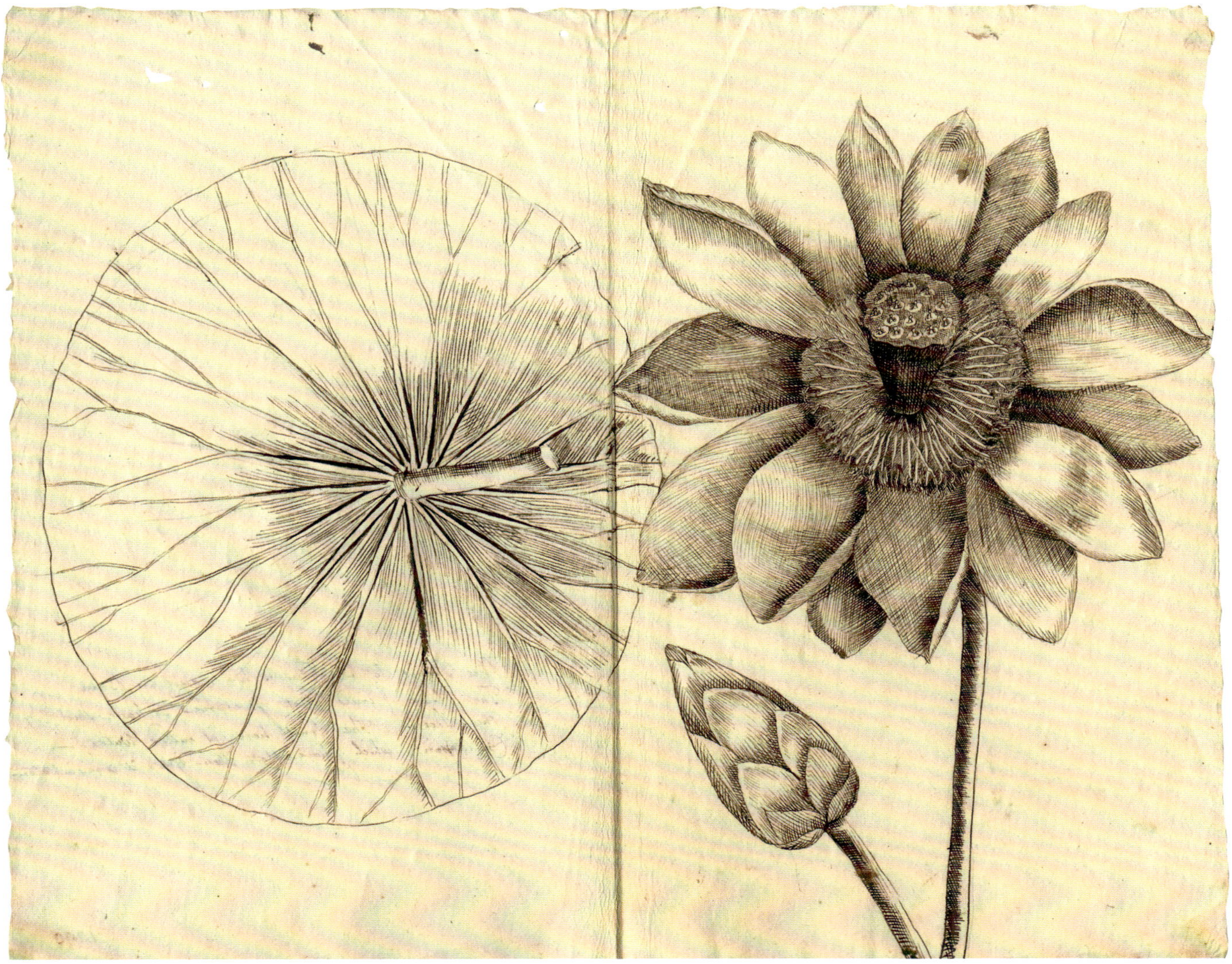

Fig. 5b Pen and ink sketch of an American lotus (*Nelumbo lutea*) (a) flower and (b) leaf made by Mark Catesby in the field in Carolina in 1722. He sent the illustration to William Sherard with the comment 'The flowr I could not preserve so have sent this scetch'.

In the field, Catesby made choices as to what he would collect and observe, and particularly of the species he chose to present in his *Natural History*. In the case of the plants, he focused on 'Forest-Trees and Shrubs, shewing their several mechanical and other Uses … [and] those Plants, that will bear our English Climate'.[9] Among the snakes and the birds ('except some Water Fowl'), he considered that few had escaped his notice. In the Carolinas, he found few fish; most of those illustrated in the *Natural History* came from the Bahamas because he 'had been often told they were very remarkable, yet … surprised to find how lavishly Nature had adorned them with *Marks* and *Colours* most admirable'.[10] He admitted that he was overwhelmed by the diversity of insects across the Carolinas and the Bahamas.[11]

Catesby's approach to collection was simple: 'never to be twice at ~~the same~~ one place in the same season'.[12] Most of his plant specimens have either flowers or fruits, although occasionally he resorted to sterile specimens. For example, a slippery elm (*Ulmus rubra*) specimen bears the rather apologetic label: 'I never could discover either Seed or blossom this tree has.'[13] In the field, he collected specimens into 'botanical books', which were returned to his sponsors.[14] His extant specimens are well flattened, indicating that he may have completed the drying process in a plant press rather than relying solely on a botanical book.

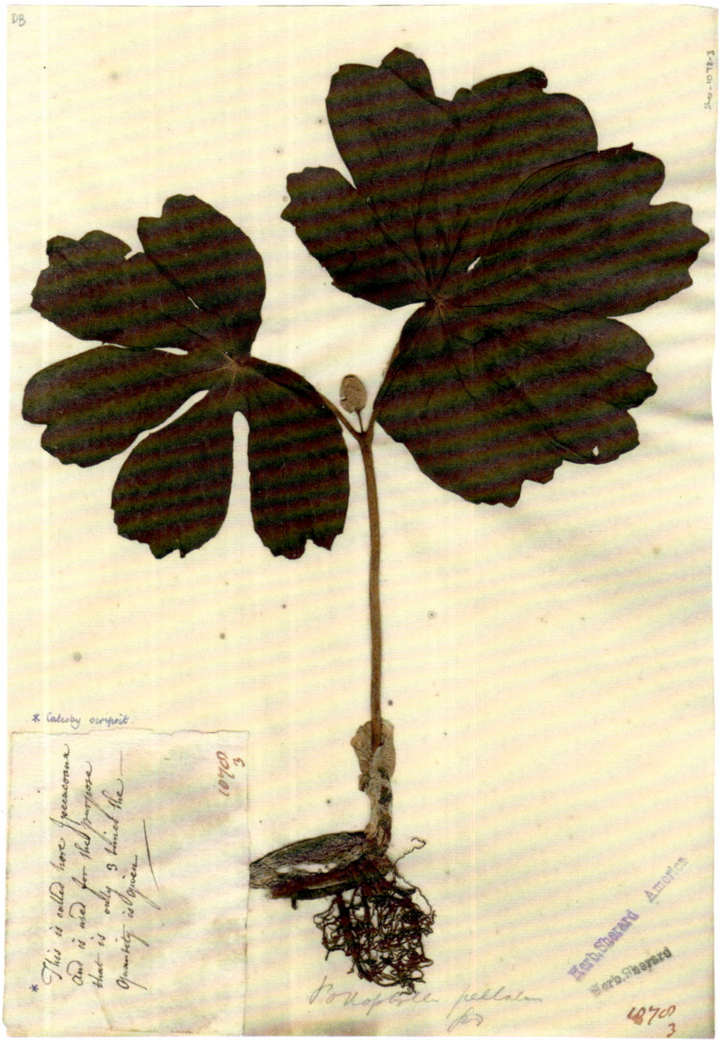

Fig. 6 Specimen of the North American medicinal plant mayapple (*Podophyllum peltatum*) collected and labelled by Mark Catesby. This was used as an emetic, hence Catesby's reference to 'Ipecacoana' (ipecacuana, *Psychotria*).

However, some specimens challenged even a collector as determined as Catesby. He successfully preserved waterlily flowers[15] but met his match with an American lotus (*Nelumbo lutea*), collected in the Carolinas in 1722. On the reverse of a pen and ink drawing he sent to William Sherard (fig. 6), he stated: 'This seems to be Clusius his Egyptian Bean The flowr I could not preserve so have sent this scetch. The fruit here is called Water Chinkpens [*Nelumbo lutea*] which I have not seen yet. it grows in Water.'[16] Included on the drawing of the flower was also a sketch of the preserved leaf he sent to Sherard at the same time. The sketch was an elegant solution to the problem of providing a permanent record of a flower that was difficult to preserve.

With many sponsors to satisfy, Catesby had to collect duplicates of specimens: 'out of the last I sent I reserved some which I now send so that those with the triplicates I presume may be spared, not that I know who they will be acceptable to except Mr Dale, for I have sent Books to Sr Hans, so pleas to doe with them what you think fitt.'[17] To keep track of the data he collected about each specimen, he used a numbering system. A specimen without data such as its appearance in life, its habitat and where and when it was collected has little scientific value: 'If to any ANIMAL, PLANT MINERAL & c. you learn its Nature, Nature, Vertue or Use, it will be still the more Acceptible.'[18] A specimen of sand heath (*Ceratiola ericoides*) is labelled as 'This Shrub grows on barren Hills usually or foot high[.] Those with bare leaves was – gathered in Aprile the other in September',[19] whilst a specimen of clammy locust (*Robinia viscosa*) is labelled 'This kind of Acacia I never saw but on one Spot of highland near Savanna river. This was a Shrub – occasioned by the fire. The flower is a little tinctur'd with purple. The Stalks not hairy as the precedent but very glutinous and clammy like Catch fly Lichnis.'[20] Catesby also took Petiver's advice by collecting information about common names and uses. Thus, mayapple (*Podophyllum peltatum*, **Plate 24**) 'is called hore Ipecacoana and is used for the purpose that is only times the quantity is given', and basswood (*Tilia americana*) 'is called Wahoo or the rope-Tree from the toughness of its Bark being much Made use of instead of Rope – A kind of Tilia I concieve'.[21] In the case of crabgrass (*Digitaria sanguinalis*), a species that arrived in the Americas with European colonists, his label is simple and direct: 'The plague of the planters, as it is in Virginia'.[22] Keen to satisfy his sponsors, Catesby even asked Sherard's advice as to 'whether you require to know the soyl [in which] every plant and tree grows'.[23] However, most of his specimens lack associated data, suggesting it may have been mislaid once the specimens were returned to Britain.

Catesby was not above mislabelling specimens. Apparently responding to a request for information from Sherard, he added the note 'This is what produces ye Square Nut you desire to know concerning, it is a Shrub' to a specimen of the American snowbell (*Styrax americanus*) from South Carolina. Johann Dillenius,

Sherard's secretary, annotated Sherard's specimen, 'this must be a mistake for ye flower does not agree. there is a Specimen of this very like if not ye same amongst ye Styrax', to which an unknown hand later added 'its [*sic*] a mistake'.[24]

Specimens had to be kept dry and free from all manner of insect, fungal and vertebrate pests; specimens eaten by 'Vermin which often breed in them'[25] were useless. Catesby found that some specimens were more easily preserved than others:

> in looking over the Books I now send I find or plants mouldy, tho' they were perfectly cured and well dryed, when I put them up and are kept in a dry, upper, South Room. In turning over those for Sr Hans I find the same plants mouldy and only those, so that I find some plants tho' not succulent are perticularly subject to it. The weather is indeed, and has for some time been, extream Moist Cloudy and Sultry.[26]

Once Catesby had collected sufficient specimens, he sent them on their risky journey to London and the collections of Sloane, Sherard and Dubois. Catesby's specimens were eventually integrated into large institutional collections in London and Oxford, but drifted into obscurity because they were rarely studied.[27] Some specimens he collected were clearly associated with seeds he was sending back to Britain. For example, a label on a specimen of whip-poor-will flower (*Trillium cuneatum*) states: 'this kind of Saterion (as I think it is) is of a pale red has ~~at~~ between a Bulbose and Tuberus root[.] The seed is No 8.'[28]

Catesby had, 'by a long acquaintance with the trees and shrubs of America, and a constant attention since for several years to their cultivation here', become an authority on the plants of North America. However, he was aware that the safe arrival of seed or a living plant on European shores from North America was not enough for it to flourish. Seeds had to germinate, seedlings to thrive and survive, and adults to be propagated if new plants were to become widely planted. Yet, with a little care and attention, he argued, American plants could 'brave our winters as stoutly as if they were our own productions'.[29] In eighteenth-century Britain, coaxing exotic plants to grow outside of their native ranges, for economic or aesthetic reasons, was a luxury for few because it demanded time and money to experiment and get results.

Consider the case of passion flowers, members of the largely American genus *Passiflora*, which first came to European attention in the mid-sixteenth century, attention that read into them a Christian Crucifixion symbology that was eventually fossilized in the plant's common and scientific names.[30] The symbolism, together with their distinctive, unusual flowers and curious, often edible, fruits, made passion flowers desirable to wealthy European gardeners. During the seventeenth century, four species were cultivated in British gardens and a century later there were at least nineteen cultivated species.[31] One of these species was a yellow passion flower (*Passiflora lutea*) Catesby

Plate 24 (detail)

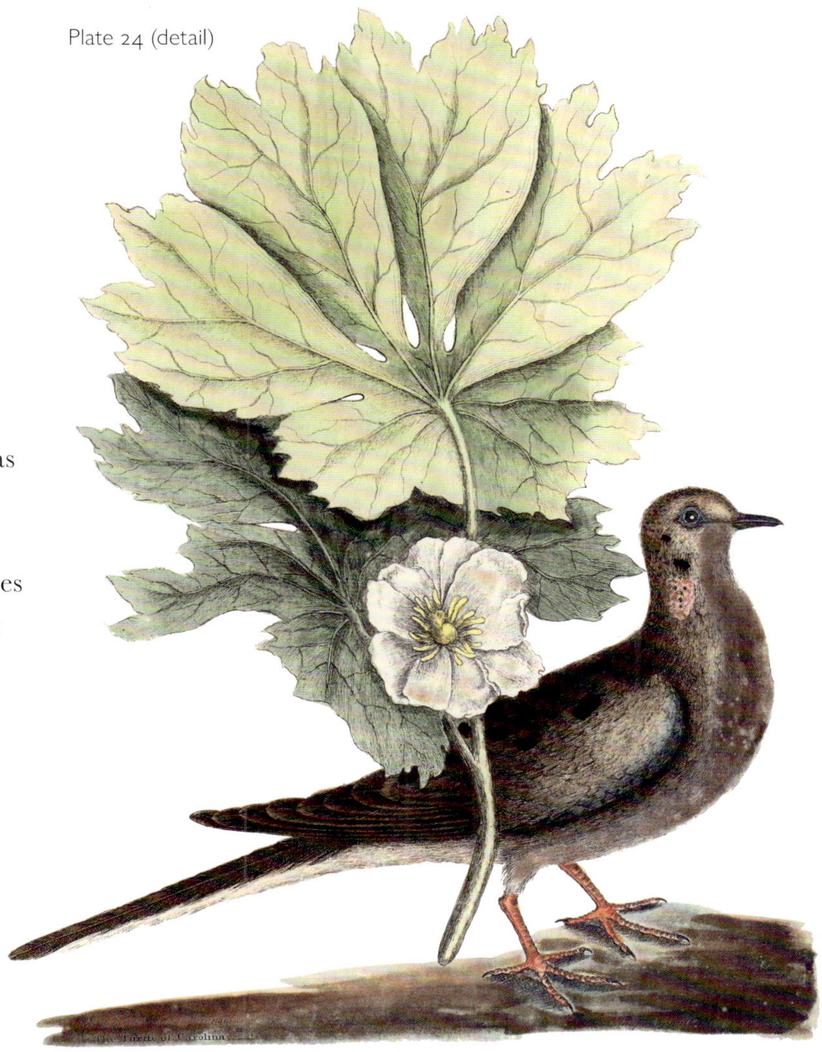

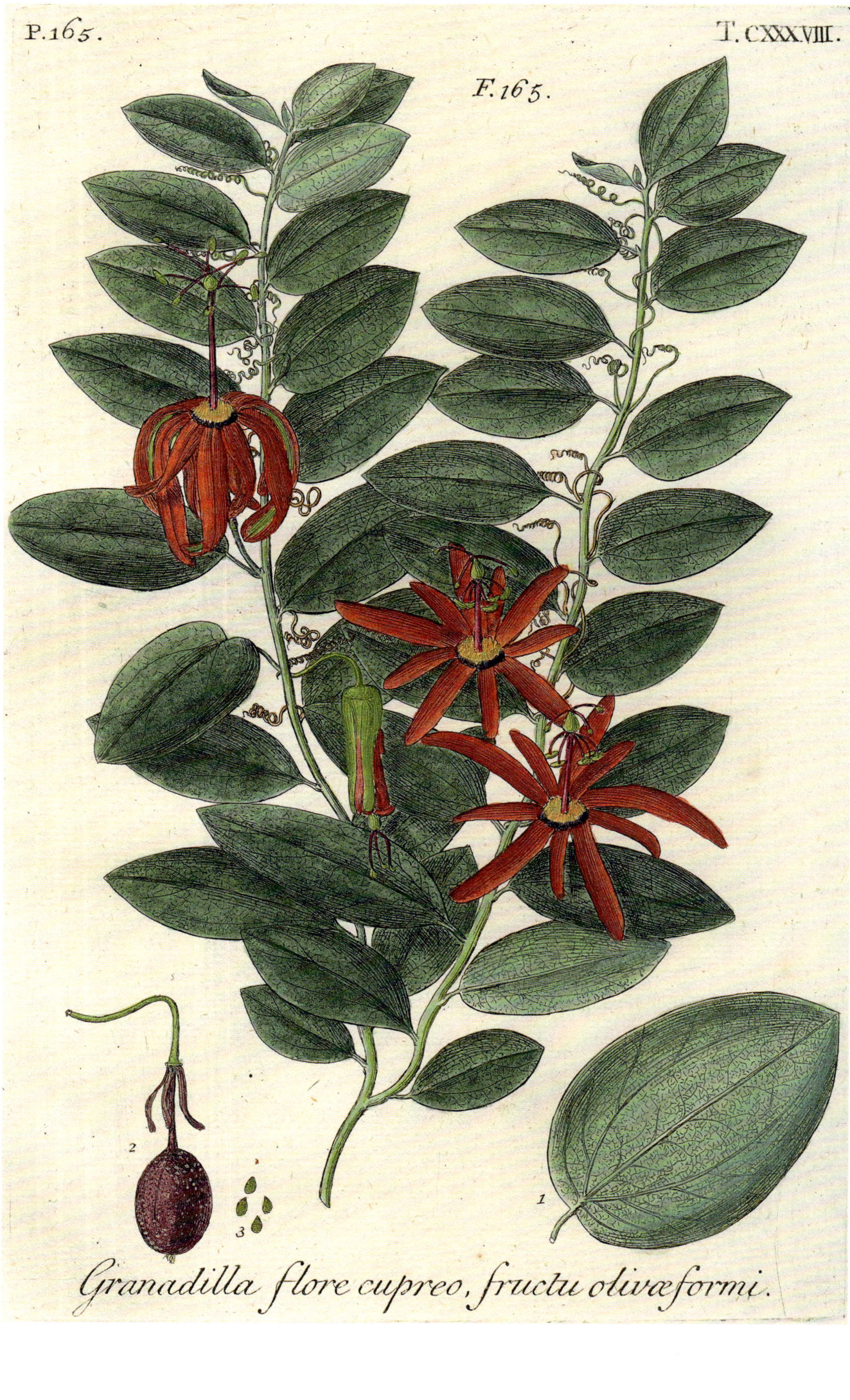

Fig. 7 The devil's pumpkin (*Passiflora cupraea*) as drawn, engraved and hand coloured by Johann Dillenius for his *Hortus Elthamensis* (1732). This plant was grown in James Sherard's garden at Eltham, from seeds collected by Mark Catesby on the island of Providence in 1724.

sent to Britain from Virginia in about 1715.[32] In the *Natural History*, he illustrated two passion flowers. In 1737 the corky-stemmed passion flower (*Passiflora suberosa*) is shown as a minor part of one composition (**Plate 151**), where it twines around the stem of the Bahama caesalpinia (*Caesalpinia bahamensis*), along with a harmless common garter snake (*Thamnophis sirtalis*). The other illustration, published in 1743, is of the devil's pumpkin (*Passiflora cupraea*, **Plate 193**), which, despite its diabolical common name, is also a passion flower. In this case, the devil's pumpkin twines around the Caribbean frangipani tree (*Plumeria obtusa*), a species now cultivated worldwide in glasshouses or in tropical and subtropical urban landscapes. Catesby's engraving of the devil's pumpkin (fig. 8) was based on collections he made in Providence in the Bahamas in 1724.

At least fifteen of the more than 400 species grown in James Sherard's Eltham garden were derived from seed collected by Catesby.[33] However, Catesby's vision of American plants in Britain was not one of tender plants conserved inside glasshouses 'owing to the Difficulties which Gentlemen have met with in preserving them in the Winter'[34] or of species that need constant attention.[35] He promoted an idea of carefully selected species for planting by the many, that is, the plants had to be hardy:

> Many kinds of American plants … have been procured and raised from thence; which, though hitherto principally in the possession of the opulent and curious, they, it is to be hoped, will for the benefit of their country be excited to encourage their propagation and increase … as well for the benefit of our woods, as for ornaments to our gardens.[36]

Publishing the *Natural History* meant that Catesby had to make choices about what to include and exclude. One of his foci was forest trees and shrubs that were useful to people, 'shewing their several mechanical and other Uses, as in Building, Joynery, Agriculture, Food, and Medicine'.[37] He also took 'notice of those Plants, that will bear our English Climate, which I have experienced from what I have growing at Mr. *Bacon*'s,

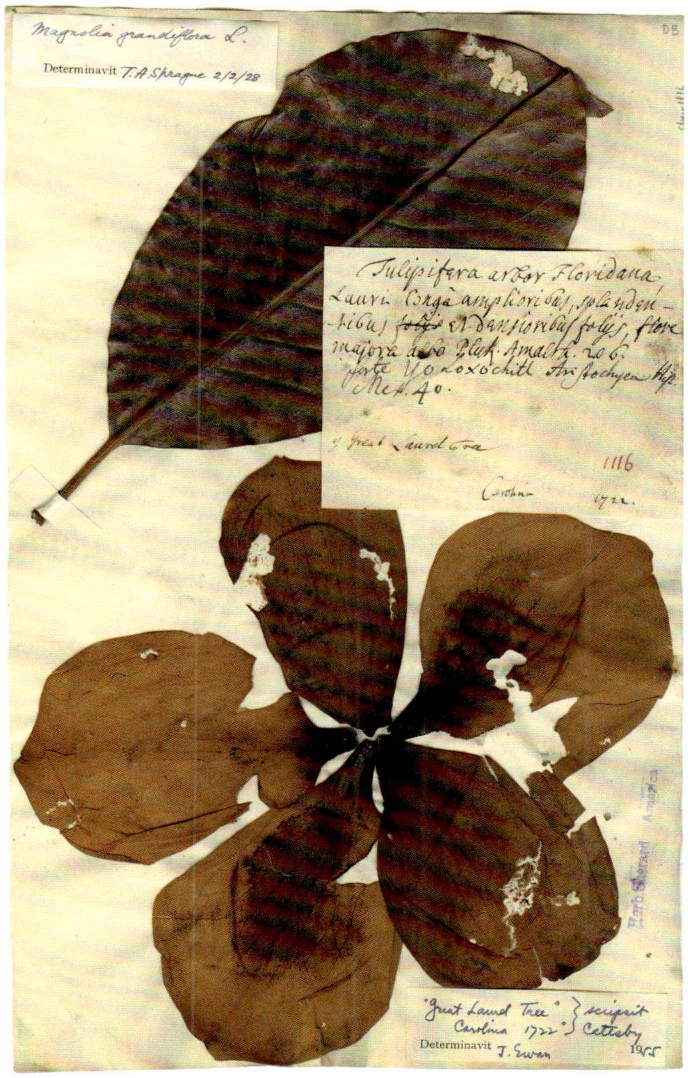

Fig. 8 One leaf and five flower petals of a specimen of southern magnolia (*Magnolia grandiflora*) collected by Mark Catesby in 1722, which he called the 'Great Laurel Tree'.

Successor of the late Mr. *Fairchild* at *Hoxton*, where many have withstood the Rigour of several Winters, without Protection, while other Plants, tho' from the same Country, have perished for Want of it'.[38] His inclusion of horticultural and arboricultural plants is unsurpising given the importance of gardening to his intended audience and a ready market for the right plants from the overseas possessions of European empires.[39]

Consider the southern magnolia (**Plate 161**), the 'laurel-tree of Carolina', which is an evergreen forest tree of the south-eastern United States. The plant fascinated Catesby and he promoted its cultivation in Britain: 'of all the tree able to endure our climate, that have yet been introduced to England, there is none

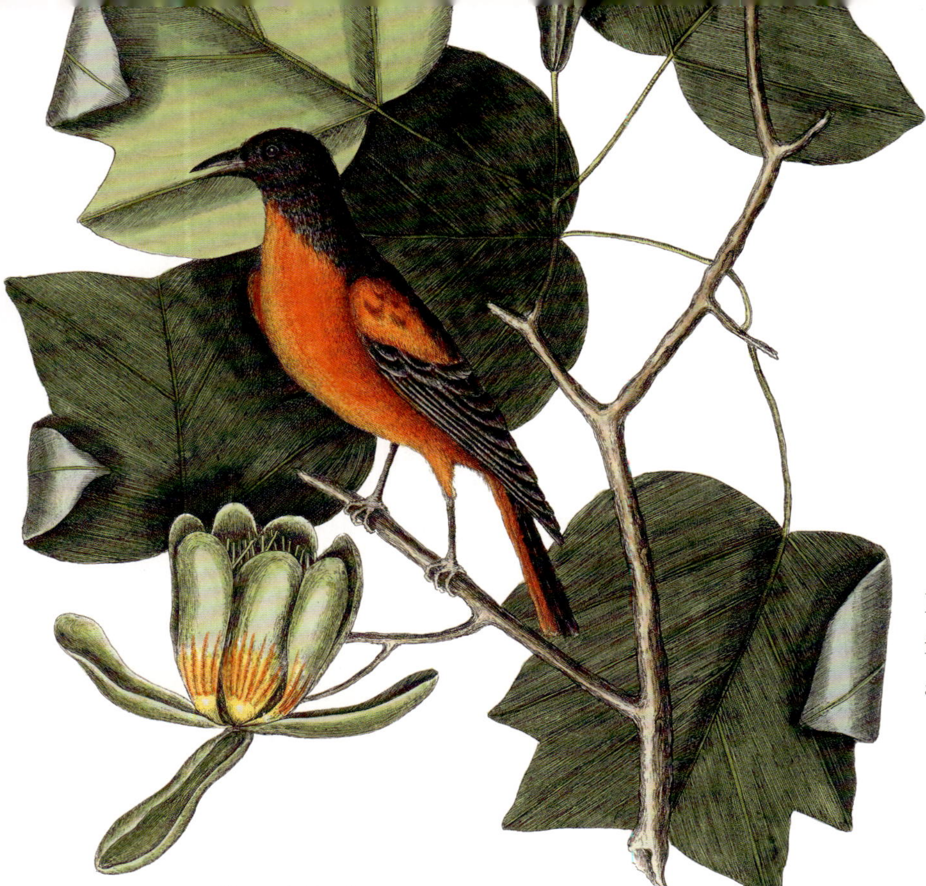

Plate 48 (detail)

that can equal this magnificent ever-green', although he warned that 'they are now become almost extinct in many parts of the country, where they abounded before the introduction of cattle'.[40] This magnolia surpassed other magnolias grown in early eighteenth-century Britain through its height, its large, shiny evergreen leaves with russet-coloured lower surfaces, its 'fragrant and ample blossoms' and its 'glittering fruits',[41] from which red-jacketed ripe seeds hung. Catesby's etching in the *Natural History*, based on an original illustration by the German botanical illustrator Georg Dionysius Ehret (1708–1779), gives a sense of the splendour of this tree. Today, Ehret is regarded as one of the foremost botanical illustrators in mid-eighteenth-century Europe through his collaborations with Carolus Linnaeus (1707–1778), Hans Sloane and Margaret Cavendish Bentinck (1715–1785), Duchess of Portland.[42]

Introduced to early eighteenth-century Europe, the southern magnolia became a prized horticultural trophy.[43] However, before the development of the Wardian case – the mobile greenhouse rather like a modern terrarium – in the nineteenth century, the challenge of transporting such prizes was immense.[44] As Catesby highlighted, southern magnolia seed collection was fickle:

> success depends in a great measure on their being kept in such a degree of heat and moisture as is requisite to preserve them in their long passage; for if they are put up too dry, their juices will remain inactive and make no effort towards vegetation; if they are kept too warm and moist, they will sprout in the box and perish; and too much moisture and cold rots them.[45]

But even if seed were harvested at the optimal time, its survival during passage across the Atlantic meant adherence to specific instructions:

> Prepare a square box of the size of a bushel or less, at the bottom of which put a layer of light earth two inches deep, spread thereon a single layer of seeds, then again a layer of earth, and so dispose your seeds and earth alternately, stratum super stratum, until the box be full; then nail down the lid and let it be placed between decks. As soon as you get the box from the ship after it comes to England, separate the seeds from the earth through a wire sieve, spread them out and let them remain so a day or two till they are dry; then put them in a bason of lukewarm water; by which means the sound seeds will be proved by their sinking to the bottom, and the bad ones by their floating.[46]

The tulip tree (*Liriodendron tulipifera*, **Plate 48**), with its magnificent tulip-shaped blooms, 'was first brought over by John Tradescant under the name of the Tulip-tree … I wish we had more of them'.[47] Nearly a century after John Evelyn's appeal, Catesby observed that

> though few trees in America are more prolific of seeds than these, yet the seeds of many of them are so apt to prove abortive, that a large quantity collected from different trees will admit of a better chance for raising a number of them: put them up in a box of sand, and stow them in a dry place; for moisture in their passage is apt to rot them.[48]

Another American plant cultivated in seventeenth-century Britain was the trumpetcreeper (*Campsis radicans*, **Plate 65**), which 'climb trees, on which they run a great height, and are frequently seen to cover even the dead trunks of very tall trees: … they produce bunches of red flowers, somewhat like the fox-glove'.[49] The apothecary John Parkinson grew it in his London garden before 1640, but it 'never bore flower with mee, nor any other that hath it in our country',[50] despite having flowered in Paris at least as early as 1635.[51] Catesby recommended that the 'seeds should be sent to England in their cods [fruits], and will grow very readily when sown in pots, with the moderate heat of a hot bed; … they will abide the rigour of our sharpest winters … they will produce plenty of their beautiful blossoms'.[52] His trumpetcreeper illustration creates an ecological association by incorporating an unrealistically posed ruby-throated hummingbird (*Archilochus colubris*) visiting a flower, together with a curious comment that 'by thrusting themselves too far into the flower, are sometimes caught'.[53] The hummingbird is the trumpetcreeper's primary pollinator, but bees may also play a significant role.[54]

In other cases, transportation and propagation were more straightforward. Catesby heaped great praise upon the sweetgum (**Plate 165**): 'the regular form and beauty of this tree deserves the regard of the Curious, as none of the American trees affect more our soil and climate … In a warm aspect it will stand our severest frosts in the open ground: the hardiness of this tree, with its beauty and regular form, recommends it as extremely fit for avenues.' Moreover, it could be readily propagated from seed: 'the vessels [fruits] containing the seeds should be gathered a little before they open, and sent with the seeds in them; or sow them in a box of moist mould, in which let them be brought, and they will come up well in a virgin soil without any assistance.'[55] On the title page of his posthumously published *Hortus Europae Americanus* (1767), Catesby identified his concern with trees and shrubs 'adapted to the climates and soils'[56] of Britain and Europe. However, he also realized that the limitation of growing American trees in Britain was access to living plants:

> As those who are desirous and have it in their power to procure large quantities of seeds and plants from America, may be at a loss what instructions to send their correspondents abroad, I have been particular in giving an account where the several kinds of plants are to be found that are uncommon, and in directing how they are to be collected, packed up, and secured, so as to preserve them in good condition during their passage; which are matters of the utmost consequence, though less known even than their culture.[57]

For those interested in North American plants but with lesser means, Catesby recommended the nursery at Fulham owned by Christopher Gray, who 'for many years made it his business to raise and cultivate the plants of America (from whence he has annually fresh supplies) in order to furnish the Curious with what they want; and that through his industry and skill a greater variety of American forest-trees and shrubs may be seen in his gardens, than in any other place in England'.[58]

Other nurserymen associated with the cultivation of Catesby's introductions were Thomas Fairchild in Hoxton and Robert Furber in Kensington. The sales catalogue of 'English and foreign trees' available from Furber's nursery in 1727[59] lists many plants from Virginia and Carolina that are likely to have been introduced by Catesby, two of which were specifically attributed to him: 'Acacia of Carolina, by Mr. Catesbe [*sic*]' and 'Mr Catesby's new Climber'; the latter is probably the American wisteria (*Wisteria fructescens*).[60]

One plant available from both Furber's and Gray's nurseries was the southern catalpa (**Plate 49**). This tree, native to eastern North American forests and now widely planted across Europe, has broad pyramidal floral clusters composed of many large, white, foxglove-like flowers.[61] Each flower 'is powdered [on the inside] as it were with purple specks, through which run two parallel chains of larger spots of a yellow colour'.[62] The flowers eventually produce long, bean-like fruits containing flat, winged seeds. This spectacular horticultural discovery was claimed by Catesby to have been 'unknown to the inhabited parts of *Carolina*, till I brought the seeds from the remoter parts of the country'. So enthralled by the tree's 'uncommon beauty' were the people of the Carolinas that the tree

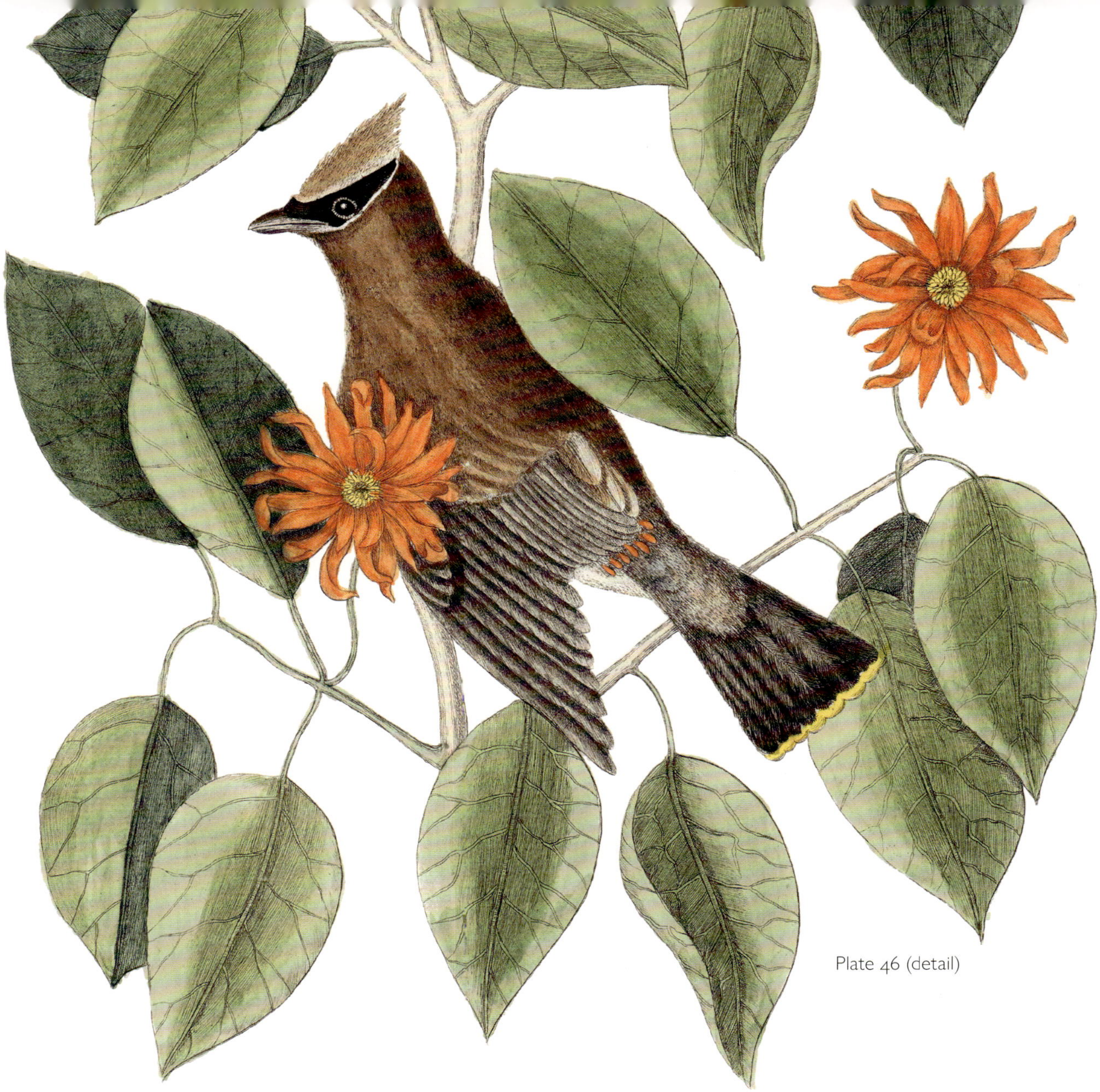

Plate 46 (detail)

became 'an ornament to many of their gardens, and probably will be the same to ours in England'.[63] By 1768, Philip Miller was able to state that 'it is now very plenty in the English gardens, especially near London, where there are some of them near twenty feet high, with large stems, and have the appearance of trees'.[64] However, fruits had yet to be produced so, to meet the demand, seeds had to be imported annually from South Carolina.

From 'some hundred miles on the back of Charles Town',[65] Carolina allspice (*Calycanthus floridus*, **Plate 46**, was introduced to cultivation in Britain by Catesby. This deciduous, aromatic, red-flowered understorey shrub, native to the forests of south-eastern North America, is still cultivated in Britain, although not as commonly as the southern catalpa. Catesby published his illustration in 1730. Fifteen years later, Miller published a coloured engraving of Carolina allspice,[66] commenting that, 'after I had given a figure of this plant … I received Monsieur Du Hamel's [Henri-Louis Duhamel du Monceau] book of the trees and shrubs,[67] which will grow in the open air in Paris, in which he had given a bad figure of this plant, under the title of Butneria'. Miller insisted on using his own name 'Basteria' for the plant 'to avoid confusion, which must attend the frequent alteration of the names of plants, which is too much in fashion at present'.[68] Linnaeus used the illustrations of Catesby, Miller and Duhamel when he formally gave Carolina allspice its scientific name in 1759.[69] In the mid-1700s Miller thought the Carolina allspice very rare in British gardens, despite Catesby's assertion that it can be readily increased by layering.[70] Peter Collinson 'believe[d] He [Catesby] never had an opportunity to Observe its Seed Vessel [fruit]',[71] suggesting that Catesby imported rooted plants rather than seeds. However, under the paws of an eastern chipmunk (*Tamias striatus*, **Plate 175**) illustrated by Catesby there is a mature Carolina allspice fruit.

For an introduction to succeed, multiple introductions may be necessary. The hoptree (*Ptelea trifoliata*, **Plate 183**) has a native range that extends through eastern North America into Mexico but was known to Catesby only from the upper parts of the Savannah River in the Carolinas.[72] This small tree, with tripartite leaves and aromatic, elm-like fruits, was introduced to England from Virginia in the early eighteenth century by John Banister, but several severe winters resulted in the experiment failing. In contrast, the seeds Catesby sent from the Carolinas in 1724 'succeeded so well as to furnish many gardens with the plants'.[73] Sometimes introductions fail because seed from the wrong part of a species' range was used; for example, Catesby recommended that tulip trees from the 'most northward [in the native distribution] are best for our climate'.[74] Some plants are general horticultural failures because they need constant attention and never find their way out of the collections of connoisseurs. The lilac-blue-flowered boxwood (*Jacaranda caerulea*, **Plate 42**), a sort of jacaranda, which Catesby found on Providence in the Bahamas and introduced to England in 1724, was growing in gardens near London in 1768.[75] This horticultural experiment was ultimately a failure, but contributed to ensuring that gardens, and the plants they contain, are in a constant state of change.

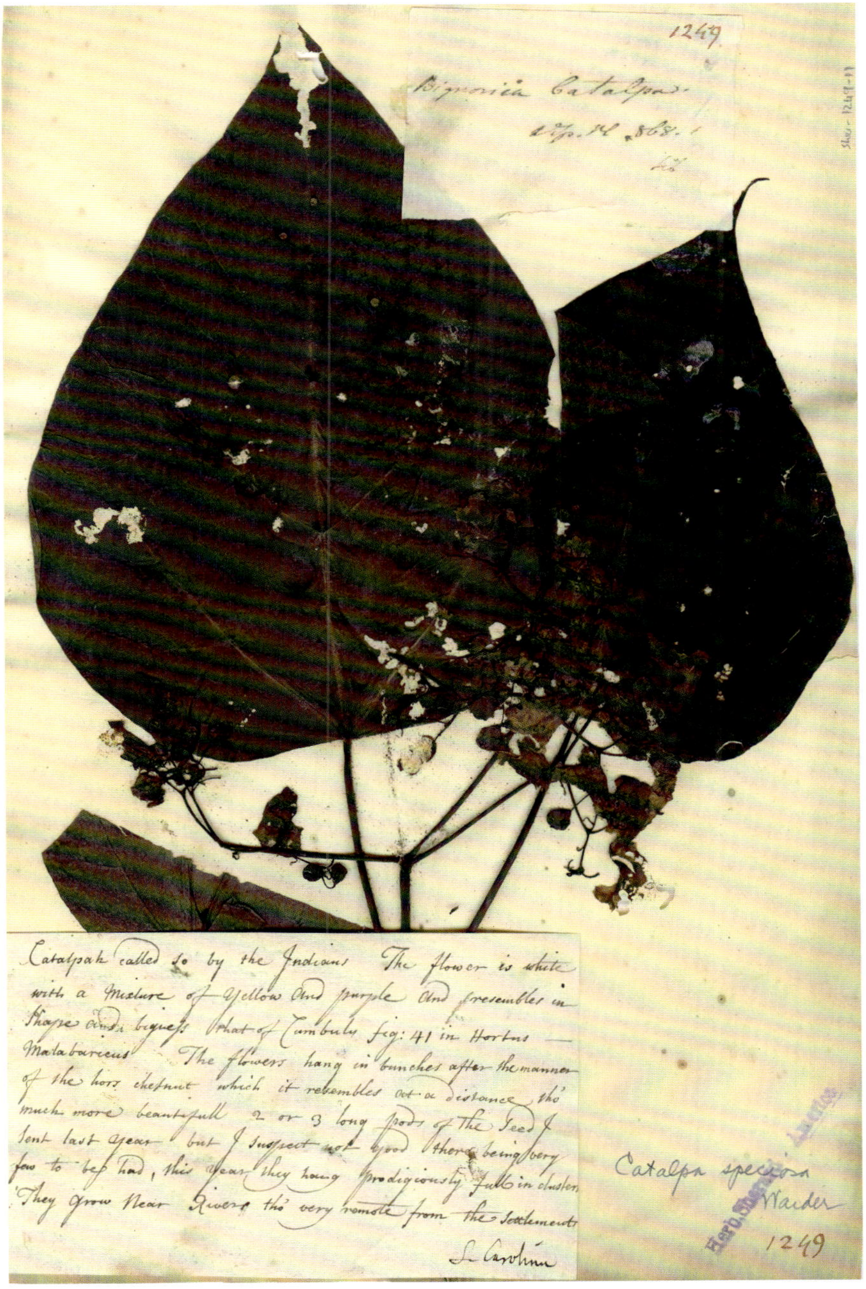

Fig. 9 A flowering specimen of southern catalpa (*Catalpa bignonioides*) collected by Mark Catesby in the Carolinas, probably in 1723, and sent to William Sherard for his personal herbarium. Catesby's autograph label states: 'Catalpah called so by the Indians[.] The flower is white with a Mixture of Yellow and purple … The flowers hang in bunches after the manner of the hors[e] chestnut which it resembles at a distance tho' much more beautiful[.] 2 or 3 long pods of the Seed I Sent last year but I suspect not good there being very few to be had, this year they hung prodigiously full in clusters[.] They grow Near Rivers tho' very remote from the Settlements[.]'

Changes in garden landscapes also happen through disease, loss of interest or changes in fashion. Catesby was enthusiastic about the ornamental potential of the American sycamore (*Platanus occidentalis*, **Plate 56**), 'a denison of England, and in some soils propagates its offspring by seeds and suckers, as our ashes and elms do: they grow to be lofty and stately trees … the bark smooth, glossy, and so variegated with white, green, &c, that they produce a fine effect amongst the other trees'.[76] However, this splendid native of eastern and southern North American forest has been judged by modern gardeners to have no value in British gardens because it is very prone to spring frost and is adversely affected by a fungal leaf blight.[77] Instead, the hybrid between the American sycamore and the oriental plane (*Platanus orientalis*), the so-called London plane (*Platanus* × *acerifolia*), has become a popular street tree.

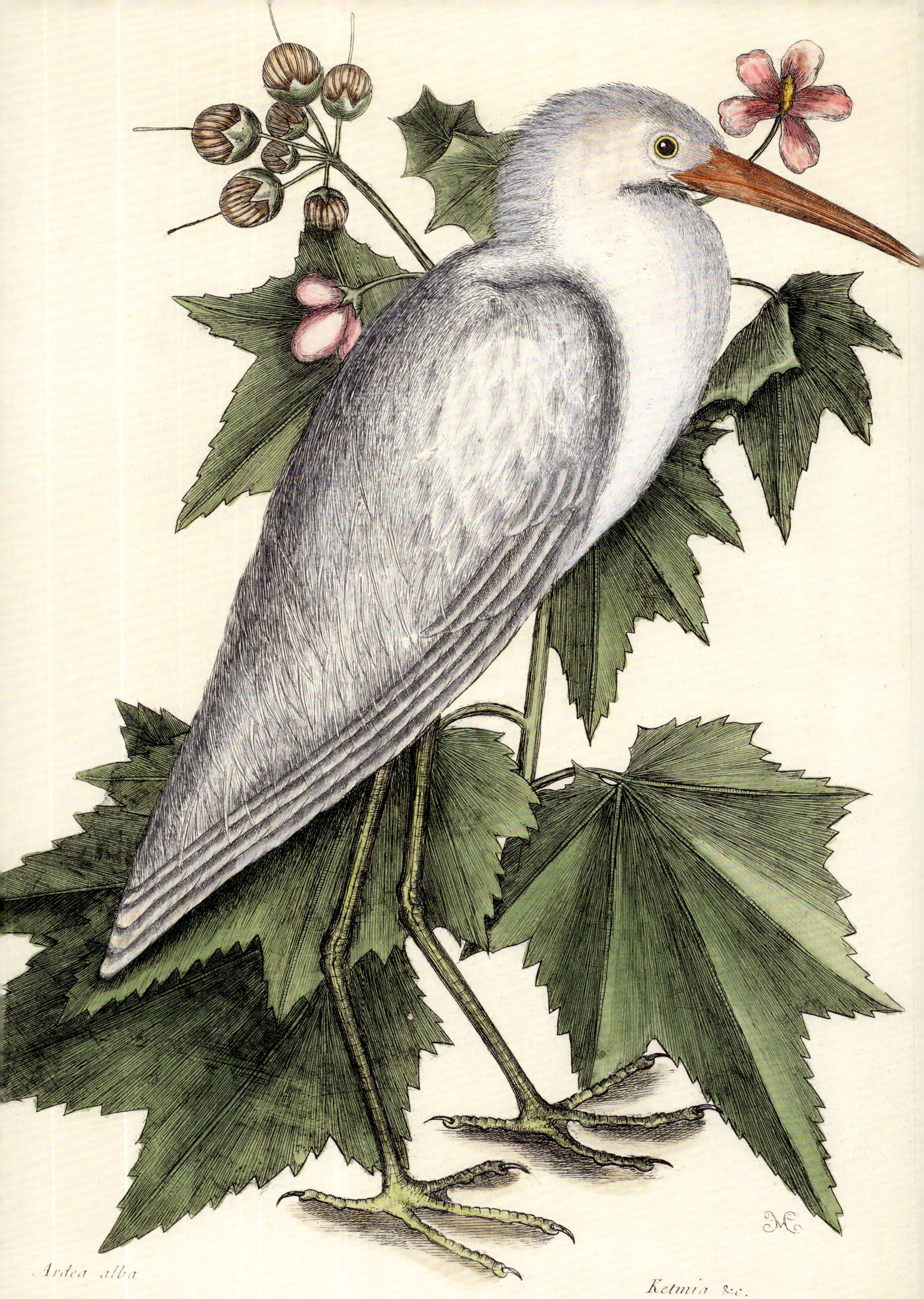

Ardea alba *Ketmia &c.*

On Illustrations 6

For as all different Birds have gestures peculiar to them, it is requisite they should be drawn from the living Birds, otherwise it is impracticable to give them their natural air; which method, except in a few Birds, has been practised through the whole Collection.

MARK CATESBY[1]

A specimen – the what, when and where of natural history – facilitates re-evaluation and reinterpretation of data in the light of new evidence. However, the processes used to preserve plants and animals often distort them such that they would be barely recognizable alive. In contrast, natural history illustrations enable features to be preserved, such as colour, habit and posture, that are readily lost and difficult to capture effectively in field notes. As with physical specimens, natural history illustrators are concerned with scientific data preservation, information presentation and knowledge proliferation. Critical factors in effective scientific natural illustration are therefore naturalism and accuracy based on detailed, first-hand observation.[2]

Accuracy in a natural history illustration is often not enough. For an illustration to become widely circulated, used and remembered, it must also be aesthetically pleasing. The nineteenth-century naturalist-artist William Burchell considered other aspects of illustration, emphasizing that those who

> propose to visit countries little known, or seldom frequented … [should] regard the art of drawing as of the highest importance; not merely as the means of giving their friends an idea of those scenes and objects which they have beheld, but for their own gratification, and for the pleasure of a renewal of past impressions far more lively than any pen can render a written journal.[3]

Whoever illustrators are or wherever they work, they make choices about how to record their subjects, choices that are influenced by their mastery of techniques, the working conditions in which they find themselves, the type of subject they are depicting, their objectives and perhaps how the image will be published. In 1869 the botanical illustrator Walter Fitch, 'more accustomed to the pencil than the pen',[4] was forthright in his view that draughtsmanship was essential in illustration: 'facility in colouring is easily acquired, but a correct eye for drawing is only to be rendered by constant observation'. Moreover,

> correct drawing and colouring should be found in the same work, for the absence of the former cannot be compensated by any excellence in the latter. Most beginners in flower drawing are desirous of rushing onto colour before they can sketch – unaware that the most gorgeous daub, however laboured, if incorrectly drawn is only a crude effort at 'paper staining'.

However, 'easily acquired' facility still requires attention to detail, as Johann Dillenius noted: 'Mr Catesby tells me he coloured his plant [Bahamian phymosia, *Phymosia abutiloides*, **Plate 77**] … by memory, which is ye reason of his mistake in making it glaucous'.[5]

Importantly, the editing of original models happens at each step of the process as features are removed, rotated, infilled, obscured, emphasized and resized according to the purpose of the illustration, the intention of the author, and available technologies. Moreover, each step of this editorial process erodes or accentuates detail. Catesby was keenly aware of

Plate 77

these issues as he explicitly set out his methodology for making illustrations:

> In designing the Plants, I always did them while fresh and just gather'd: And the Animals, particularly the Birds, I painted them while alive (except a very few) and gave them their Gestures peculiar to every kind of Bird, and where it would admit of, I have adapted the Birds to those Plants on which they fed, or have any Relation to. Fish which do not retain their Colours when out of their Element, I painted at different times, having a succession of them procur'd while the former lost their Colours: I dont pretend to have had this advantage in all, for some kinds I saw not plenty of, and of others I never saw above one or two: Reptiles will live many Months without Sustenance, so that I had no difficulty in Painting them while living.[6]

The scientific 'interpretations' of natural history illustrations, and hence the limitations placed on conclusions drawn from them, are determined by the purpose of an illustrator and their skills. Consequently, one immediate criterion to assess the quality of a natural history illustration is whether the organism depicted can be readily identified based on the illustration alone. As Carl Linnaeus stated: 'a draughtsman, an engraver, and a botanist are equally necessary to produce a praiseworthy figure; if any of these is at fault, the figure turns out flawed.'[7] However, despite collecting specimens, Catesby does not mention them in the *Natural History*, and therefore the explicit link between a physical object and his interpretation of that object is lost, potentially making unambiguous identification of the *Natural History* illustrations a challenge for modern users.[8]

Catesby as an illustrator and artist has been comprehensively investigated through the research of the British fine art scholar Henrietta McBurney.[9] Catesby's method was to make pencil and ink sketches on paper in the field, which he later augmented by observing objects and illustrations in collections such as those of Hans Sloane. His intention appears, initially,

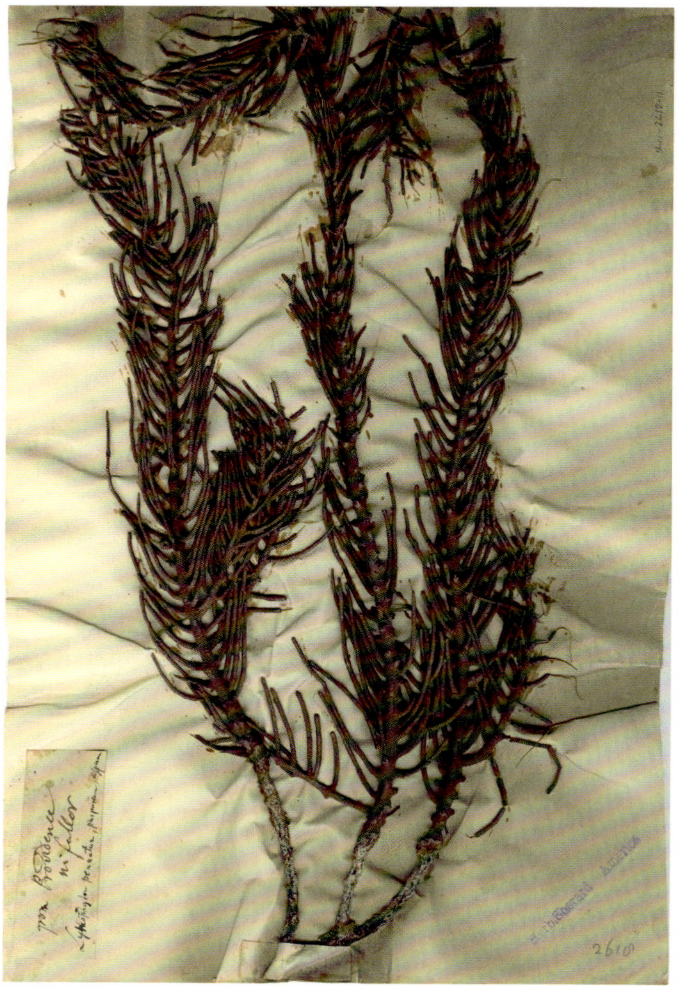

Fig. 10 Specimen of the purple sea plume (*Pseudopterogorgia acerosa*), a soft coral, probably collected by Mark Catesby on the island of Providence, and sent to William Sherard. Catesby illustrated this species (Plate 135) in the *Natural History* in 1736.

to have been to record what he saw for himself and his sponsors; only when he returned to England did his attention turn to the quite different task of creating illustrations that were suitable for publication.

Catesby represented at least 408 species of animals, plants and fungi in compositions arranged across the 220 plates of the *Natural History*. These organisms consist of invertebrates (44 plates); fishes (48 plates); amphibians and reptiles (35 plates); birds (118 plates); mammals (10 plates); fungus (1 plate); gymnosperm (1 plate); and flowering plants (178 plates). Butterflies and moths (**Plates 35, 183–84, 186, 188, 190–91, 194–97, 200**) are obvious and distinctive American invertebrates upon which Catesby could focus, but he also drew attention to more cryptic groups of organisms such as zoophytes (e.g., corals), which some have argued combine the characteristics of plants and animals.[10] In 1761 Linnaeus had no doubt that 'zoophyta … [live] by a mere vegetable life, and are increased every year under their bark, like trees, as appears from the annual rings in a section of the trunk of a *Gorgonia*. They are therefore vegetables, with flowers like small animals.'[11] Seven of Catesby's illustrations (**Plates 73–74, 113, 121, 134–135, 137**) include soft corals, members of the family Gorgoniidae. Catesby found these corals in shallow waters, 'defended from the impetuosity of the waves by rocks', on the coast of Virginia, the Carolinas and especially the Bahamas, in clear waters up to 27 metres deep. It is tempting to think that some of these corals were collected by Catesby as he fished for spotted morays (*Gymnothorax moringa*, **Plate 121**), 'pulled out of the water with a hook and line, … [which] by their violent twistings, break off, and often bring up with them branches of coral or other sea vegetables'.[12]

The focus on flowering plants, and their presence throughout most of the plates in the *Natural History*, is hardly surprising given the interests of his patrons, the economic potential of plants and, through our modern understanding, their key ecological roles. Catesby's justification for including birds as the largest animal group was similarly pragmatic:

> As there is a greater Variety of the feather'd Kind than of any other Animals, (at least to be come at) and as they excel in the Beauty of their Colours, and have a nearer relation to the Plants which they feed on and frequent; I was induced chiefly (so far as I could) to compleat an Account of them, rather than to describe promiscuousty Insects and other Animals.[13]

Regarding the reptiles, Catesby was confident of the comprehensiveness of his work and the accuracy of his illustrations: 'upon shewing my Designs of them [snakes] to several of the most intelligent persons, many of them confessed that they had not seen them all, and none of them pretended to have seen any other kinds.'[14] Most of his illustrations of fish were from the Bahamas, as he found few fish in the Carolinas.

Appearances of organisms differ between and within species. Consequently, since the earliest days, illustrators have been faced with choices and compromises. Choices mean deciding which individual or stage of a plant's life to represent. Compromises come with deciding whether to create an epitome from individuals to highlight key features, or a portrait of a specific individual that includes the blights and ravages of growth, age and handling. Moreover, natural history illustrators rarely ignore published images, and Catesby was no exception.[15]

He prided himself as a scrupulous observer keen to understand the natural world through direct, objective observation and to test reports based on unsubstantiated rumour and hearsay. He was not just recording what he observed or what he was told, but tried to synthesize and filter his observations. Some of this filtering happened in his illustrations as he tried to correct or challenge misunderstandings.

The cashew tree (*Anacardium occidentale*, **Plate 209**) is native to dry regions in central and north-eastern Brazil where, thousands of years before Europeans arrived in the Americas and transported it across the

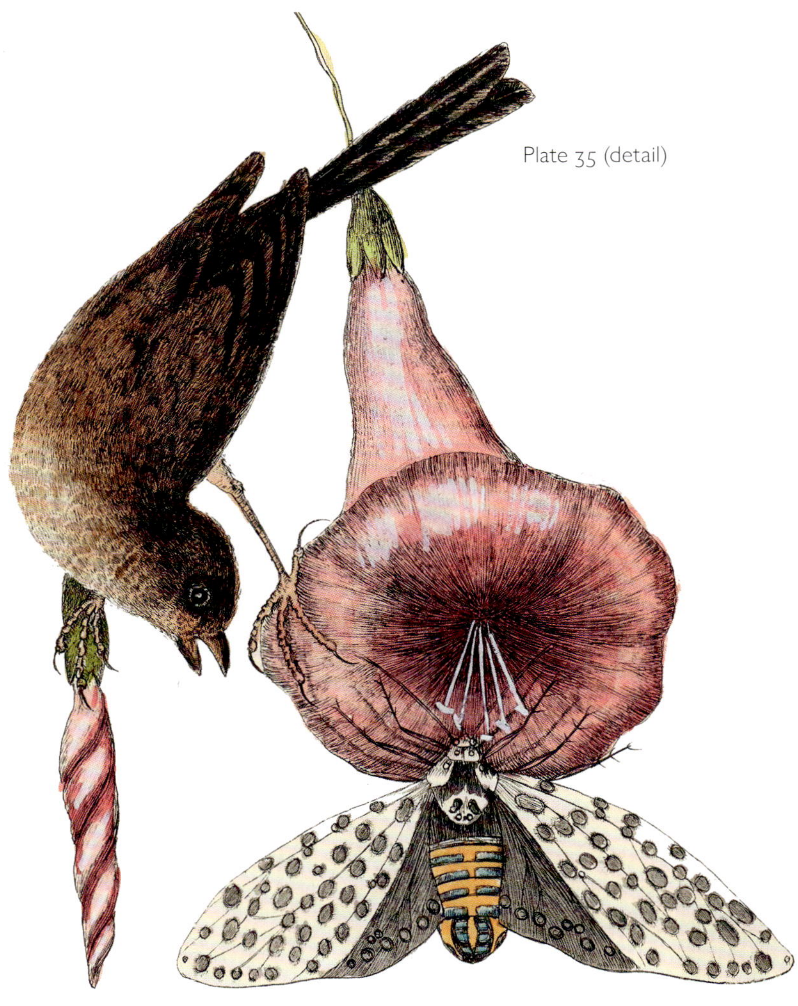

Plate 35 (detail)

globe, it was used by Indigenous peoples for food, medicine and fuel.[16] The name 'cashew' appears to derive from the Portuguese *caju* or the Tupí *acaju*, and was apparently first used by the English privateer and navigator William Dampier (1659–1728).[17] Catesby probably found the tree growing in coastal areas of islands in the Bahamas, although he states that the tree grows in 'Jamaica, Hispaniola, and many other places of America, within the Tropicks'. As he points out, one of the tree's usual features is that it has a hard fruit with a fleshy stalk. The fruit, 'not unlike, in form and size, to a Hare's kidney', contains an edible seed surrounded by a hard shell containing 'acrimonious caustick, inflammable oil'. Below the fruit is a fleshy, brightly coloured, pear-shaped stalk, the 'cashew apple'.[18]

Superficially, Catesby's illustration gives a good sense of the appearance of a branch of the cashew tree, especially regarding the colours of the flowers and fruits. However, details of the plate show liberties being taken with the plant's biological features. For example, in life the flowers are usually tightly clustered, but Catesby shows them laxly arranged. Moreover, the flower clusters are a mixture of many male flowers with a few hermaphrodite flowers; this means that the clusters of fruits he depicted are unlikely to be found. On the right-hand side of the plate, he has depicted a stalk where one fruit sits above another, near a flower cluster unconnected to the rest of the branch, indicating some inattention to detail.

Given such subtle points, it is ironic that Catesby criticized the German artist-naturalist Maria Sibylla Merian (1647–1717) for 'unluckily' reversing the obvious positions of the fruit and 'apple' in her depiction of the cashew in her *Metamorphosis insectorum Surinamensium* (1705).[19] One explanation may be that she knew only mature fruits and their attached 'apples' from marketplaces rather than attached to trees. This may also explain Merian's odd placement of mature fruits arising directly from a stem at the base of a cluster of leaves rather than from the position of the flowers.

Our familiarity with the habits and behaviours of common mammals and birds often makes identifying poor illustrations of them straightforward. The proportions may be wrong or the stances unnatural, giving the illustration the appearance of an animal having suffered at the hands of an indifferent taxidermist or having been cobbled together from second-hand reports. However, perhaps because we may not observe them as closely as we do animals, we often overlook the individual differences of plants. As Linnaeus emphasized about botanical illustrations, a few years after Catesby's death, 'all parts should be recorded in their natural position and size'.[20] Catesby was keen to ensure that his audience was aware that, as far as possible, his subjects were drawn from living (or recently dead) individuals rather than from museum

Plate 209 (detail)

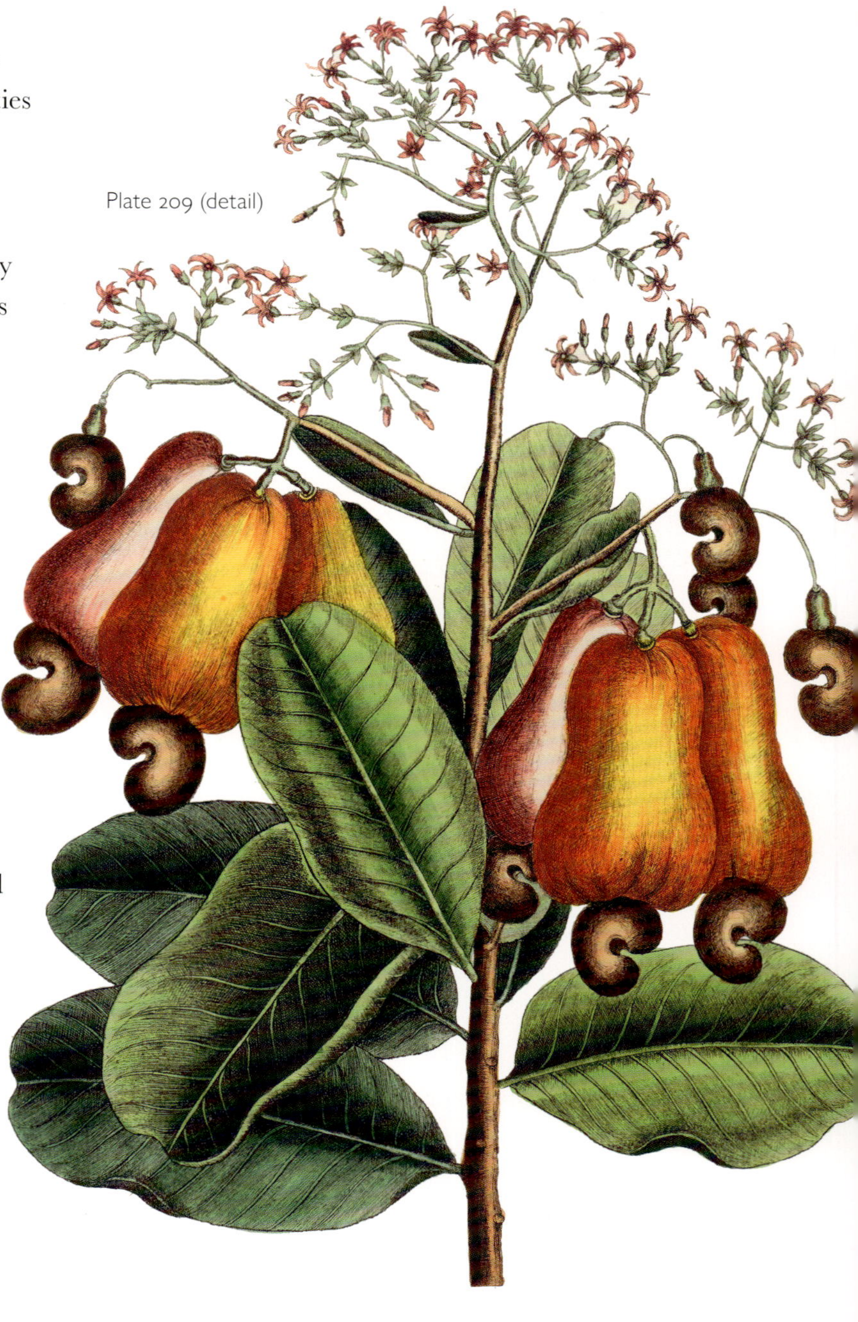

objects. However, his subject choices were sometimes eclectic, driven as much by market novelty as by a desire to represent plant and animal diversity in the Carolinas and the Bahamas. One example is the Cuban parrot (*Amazona leucocephala*, **Plate 10**), a species found in Cuba, the Bahamas and the Cayman Islands. There is no evidence that Catesby saw this species in the Bahamas; rather, the specimen he illustrated was apparently stuffed, having originated in Cuba: 'It was shot by an Indian, on the Island Cuba; and being only disabled from flying, he carried it to the Governour of the Havanna, who presented it to a Gentlewoman of *Carolina*, with whom it liv'd some years, much admir'd for its uncommonness and beauty.' He was concerned that the bird, having not been painted from life, would have an unnatural stance: 'the figure of this Bird has the disadvantage of all the rest, it being painted only from the case.'[21] Yet he was not above placing birds in distinctly unnatural poses; for example, an American robin (*Turdus migratorius*, **Plate 29**) is shown, on a tree stump, lying on its back with its feet in the air, whilst a yellow-rumped warbler (*Setophaga coronata*, **Plate 58**) is suspended by its left leg on a thread hanging from the flower of the spreading pogonia orchid (*Cleistesiopsis divaricata*).

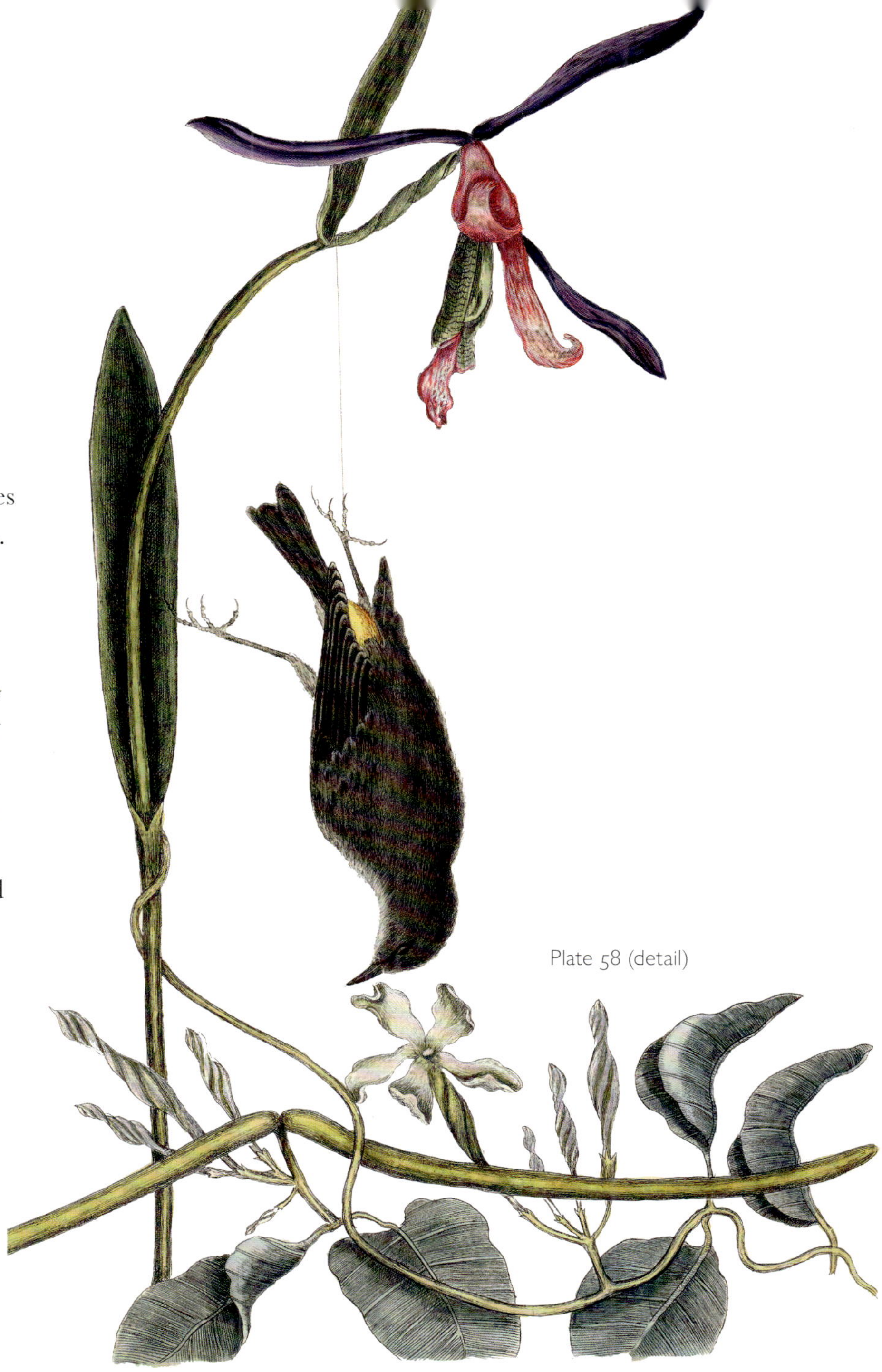

Plate 58 (detail)

Some of Catesby's published illustrations appear as they might in an artist's notebook, as apparently disconnected studies, as, for example, the leaves and flowering and fruit structures of the goldenclub (*Orontium aquaticum*, **Plate 82**) and the green arrow arum (**Plate 83**). However, in his illustration of the eastern hog-nosed snake (*Heterodon platirhinos*), the mole salamander (*Ambystoma talpoideum*) and the tannia plant (**Plate 145**) the confusion was deliberate:

> The subject of this Plate is, as it appeared to me, at a great inundation, where, by the violence of the current, fish, reptiles, with other animals and insects, were dislodged from their holes, &c. floating upon heaps of vegetable refuse; where the voracious and larger Serpents were continually preying upon the smaller, as well as those of their own kind, as others, which, in that confusion, were more easily surprised.[22]

Indeed, many of Catesby's compositions encourage the viewer to interpret them as biological scenes, showing how organisms live or interact. For example, the Carolina moonseed (*Cocculus carolinus*, **Plate 51**) is shown twining around a post, where it 'runs up the palisades of Fort More [near Savannah, Georgia] in Carolina'.[23] This approach can have unintended consequences, one of which is that plants may appear more robust than they are because of the birds perching upon them; for example, a chipping sparrow straddling a flower and flower bud of the delicately twining saltmarsh morning glory (**Plate 35**), a 'Nine penny-weight' (*c*.14 grammes) painted bunting perched on a petal of a loblolly bay flower (**Plate 44**) and a smooth-billed ani perched on the stem of pink lady's slipper (*Cypripedium acaule*, **Plate 203**). Another

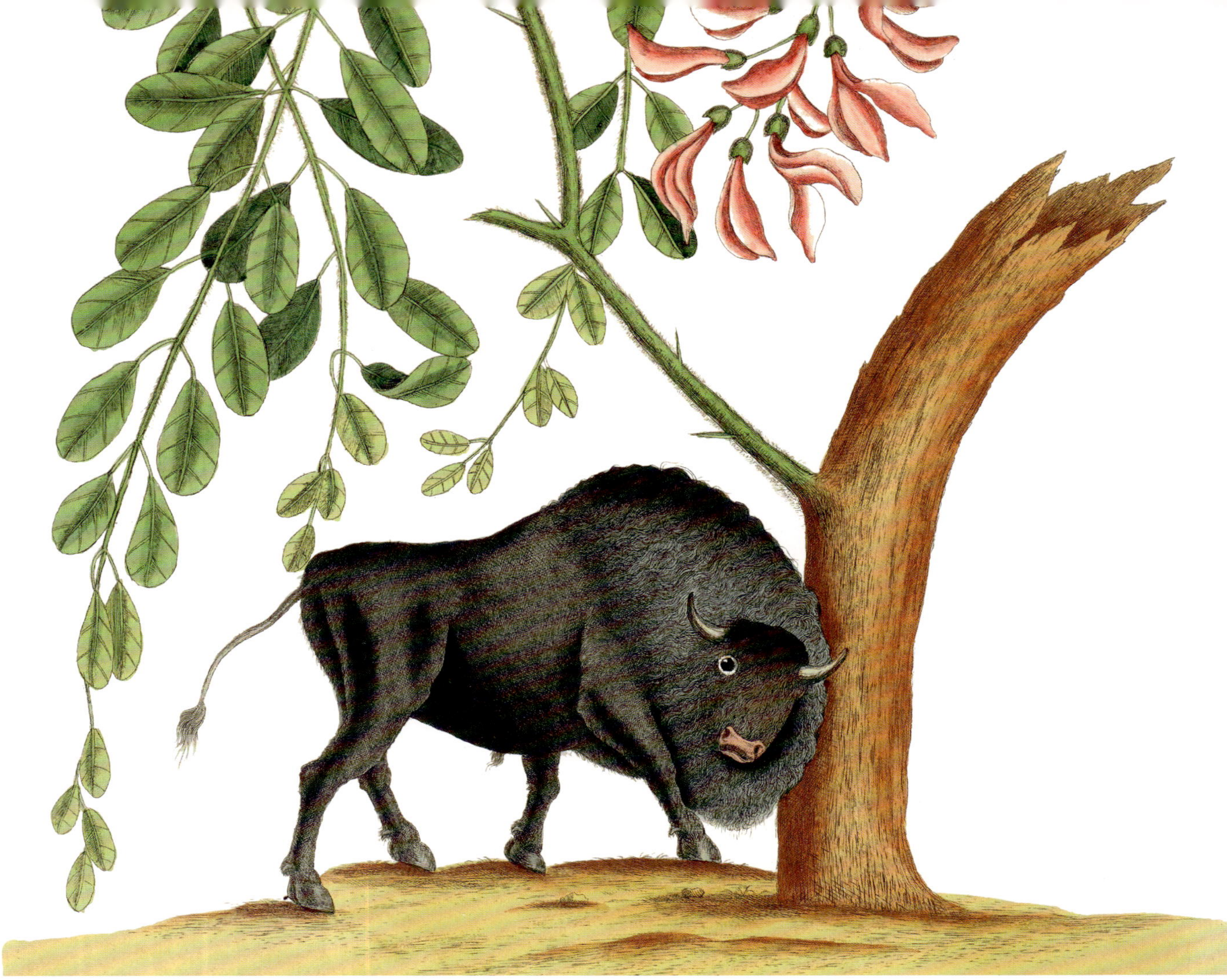

Plate 220 (detail)

consequence is to do with relative scale. Faced with apparently integrated images, and with no other information, viewers are forced to create internal scales that may lead to visual dissonance.

Catesby's illustration of the American bison (**Plate 220**), 'a perfect likeness of this awful creature', shows this iconic mammal rubbing itself against the broken trunk of a rose locust tree (*Robinia hispida*). Catesby 'never saw any of these trees but at one place near the Apalatchian mountains where Buffelos had left their dung'.[24] His integration of the animal and the plant into this scene, together with their size disparity, creates dissonance. According to the internal scale of the bison the flowers appear huge, but when the flowers are used as the scale the bison becomes miniscule. Either way, the trunk seems incapable of sustaining the attentions of the bison or of supporting the flowering shoot. Moreover, the flowering shoot incorrectly emerges from a leaf, despite his claim that his model was a living plant rather than a dried specimen.[25] Another example of such visual dissonance is the alligator (**Plate 163**). We assume that the illustration is of a mature individual, in which case the associated plant would be huge. However, Catesby states that the animal depicted has just emerged from an egg, despite having features associated with an adult animal.[26] Another case where his illusion of a complete scene creates confusion is in his depiction of the American flamingo (**Plates 73–74**), a species of southern Florida and the Caribbean. This bird, which 'when it stands erect, is five feet [*c*.1.5 metres] high', is shown the same height as the double-forked plexaurella (*Plexaurella dichotoma*) gorgoniid, which reaches about one metre in height. However, Catesby appears to have seen only gorgoniid specimens a fraction of this size 'growing to the white rocks in above ten fathom [*c*.18 metres] water',[27] despite his illustration implying that it has dryland associations. One often forgets how rare independently produced illustrations once were in the modern world, which today is awash with natural history images. If natural history illustrations are to have scientific value, associations, scale and proportion must be accurate, especially if few other comparative illustrations are available.

When Catesby published his engraving of the devil's pumpkin (**Plate 193**), two other illustrations had been published based on his seed collection from the Bahamas. Catesby's seed samples, dutifully sent to Britain, germinated successfully in the gardens of his sponsors – the first records of the species' cultivation in Britain. An illustration of one of these successes, from the Chelsea Physic Garden, was published in John Martyn's *Historia plantarum rariorum* (1728).[28] Four years later an illustration of the plant raised in James Sherard's garden at Eltham, Greater London (formerly Kent), was published in Johann Dillenius's *Hortus Elthamensis* (1732),[29] where differences between the cultivated and the wild plants were emphasized: 'from the said island [Providence] we have a dry specimen of this plant, the leaves of which are larger and rounder than those of ours.'[30] Catesby made no reference to these earlier works when he published his own illustration. Instead, he carefully emphasized that his illustration was made in the species' natural habitat: 'plants grow plentifully on many of the Bahama Islands, where I painted them in the natural appearance as is here represented.'[31] Martyn, Dillenius and Catesby used different polynomial names for the devil's pumpkin, which illustrates a familiar natural history problem: the confusion created by nomenclature. In the early eighteenth century, William Sherard tried to resolve the confusion, but he (and his immediate successors) failed.[32] It was Carl Linnaeus, who gave the devil's pumpkin its formal scientific name in 1753 based on all three illustrations and descriptions, who began to resolve scientific naming confusion.[33]

Arbiters of taste may relegate natural history illustration to mere technical drawing constrained by strict rules, in which artistic expression has no business. However, an illustrator's mastery of their medium and their artistic choices adds dramatically to the overall appeal, visual impact and scientific purpose of an illustration, which in turn contribute to the cultural nature of science. In skilled hands, even the most arcane natural history objects are rendered informative, intriguing, thought provoking and aesthetically pleasing.

In 1990 the American botanist Robert Wilbur gave a forthright assessment of Catesby's illustrations and text: '[his] abilities verbally to describe the plants were if anything even less developed than his talents as a biological draftsman.'[34] As Wilbur acknowledged, his assessment was harsh, especially given the extent to which Linnaeus relied upon them as he classified and named the natural world in the mid-eighteenth century.[35] Yet, despite changes between the eighteenth century and today in the standards demanded for the recording of scientific data, Catesby's illustrations have scientific value as data, and perhaps more importantly as objects for portraying species, disseminating information and engaging people with the natural world in ways that cannot be achieved simply by detailed measurements and images.

Plate 163 (detail)

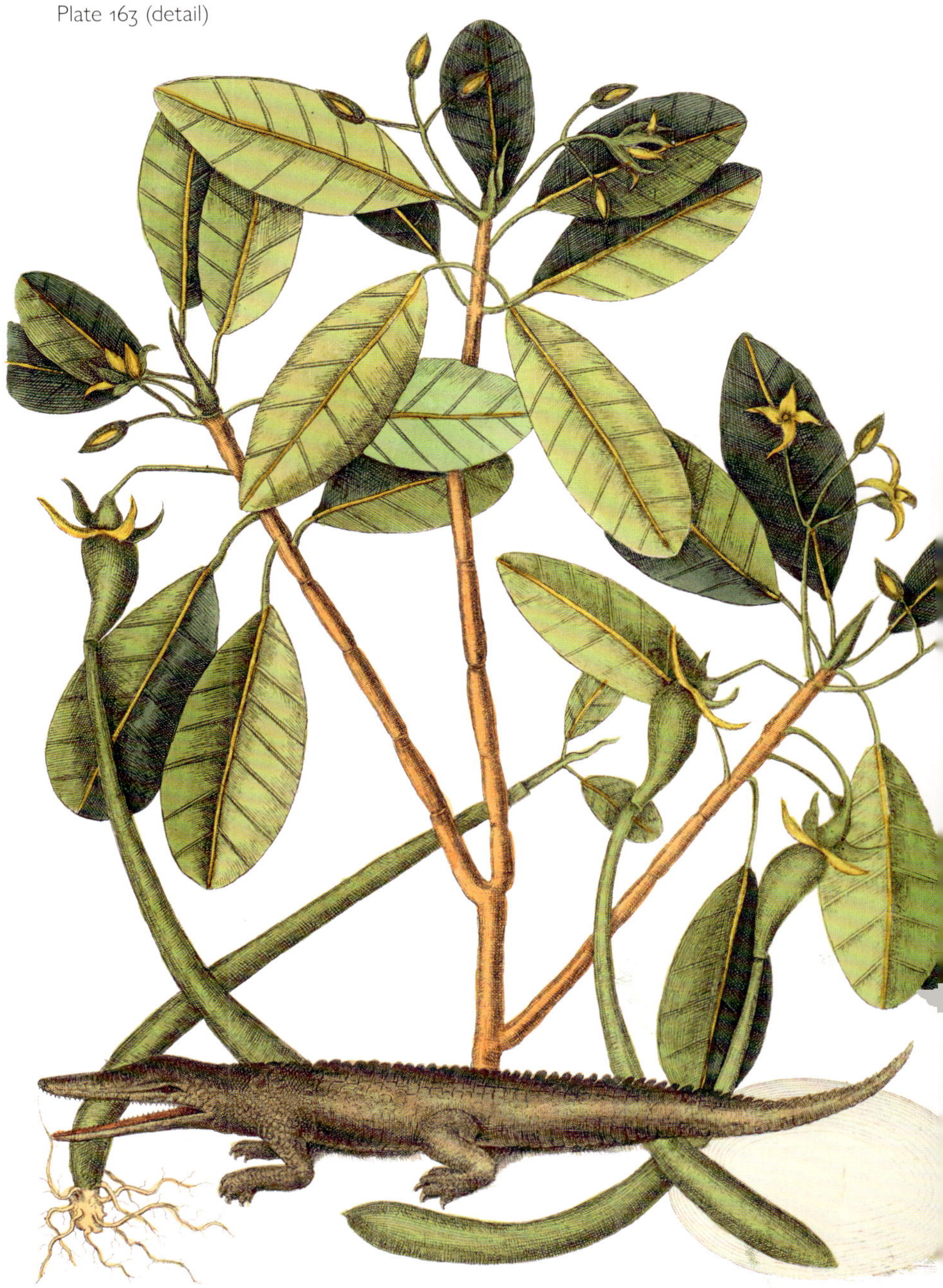

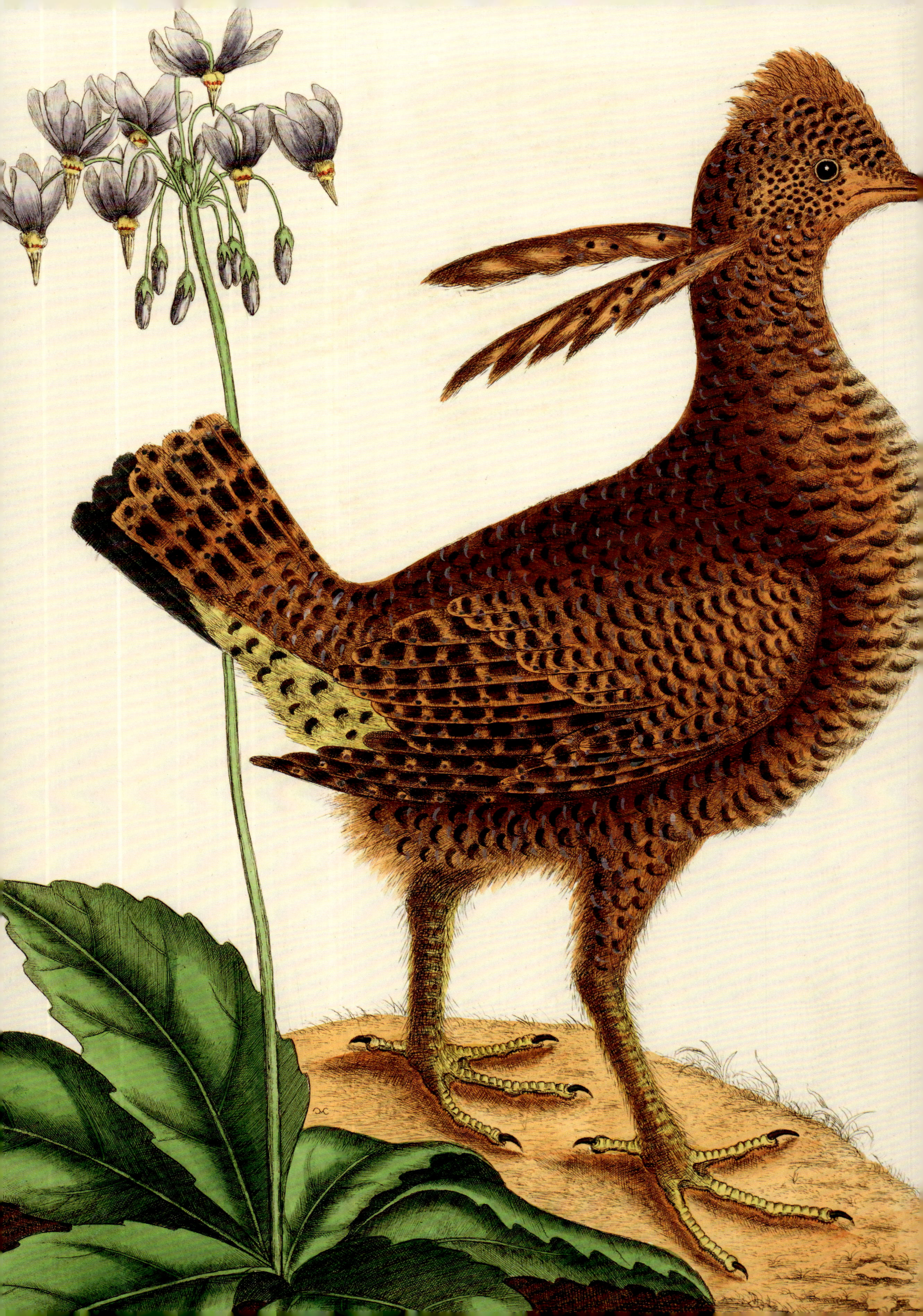

On Publishing the *Natural History* 7

As Figures convey the strongest Ideas, and determine the Subjects treated of in Natural History, the Want of which hath caused so great Uncertainty in the Knowledge of what the Antients have described barely as words; in order to avoid such Confusions, we shall take care to exhibit every thing drawn by the Life, as well as described in the most particular manner'.

MARK CATESBY[1]

Publication is fundamental to the circulation and legacy of natural history observations, and to the process of naturalists becoming recognized as authorities outside their immediate circle of contacts. Without circulation, observations cannot be compared and evaluated against existing knowledge or incorporated into general understandings of the natural world. However, the publication of results, though no less arduous and time consuming than fieldwork, requires a different set of skills from those of the field naturalist.

In eighteenth-century Britain, relief printing of text and intaglio printing of illustrations facilitated communication through print.[2] In relief printing, ink is carried on the surface of a raised pattern, whilst in intaglio printing, ink is carried in furrows and dimples marked into a surface. Grooves in sheets of soft metal, most commonly copper, are made using two primary methods: engraving and etching. Metal engraving involves removing those pieces of the plate that form the inked line using a graver (burin). The engraver pushes the sharp-pointed, V-shaped graver into the surface of the plate, then forces it through the metal to scoop out a sliver of copper. Curved grooves are created by rotating the plate against the graver, whilst varying pressure on the graver creates different groove thicknesses and depths. The thicker or deeper the groove, the more ink is held and the darker the line will appear on the printed page. Etching, in contrast, is a chemical rather than a mechanical process, allowing the etcher to work more naturally than the engraver. The etcher coats a metal plate in acid-resistant wax, then scratches lines into the surface to expose the metal beneath. When the wax-coated plate is immersed in an acid bath, only exposed metal is dissolved. The longer the acid 'bites', the deeper and wider the line and hence the darker the print. Intaglio printing is associated with the so-called eighteenth-century 'golden age' of botanical illustration, of which the etching in Catesby's *Natural History* is an example. Such works, often fantastically expensive 'trophy' volumes, were more likely to adorn the libraries of institutions and wealthy connoisseurs rather than be in the hands of practical naturalists wrestling with recording and understanding the diversity of life on earth.

Catesby's transformation of his field collections, observations and drawings into the published work was a protracted, laborious process. He returned to London in 1726 with plans to publish his observations and drawings of the plants and animals of the Thirteen Colonies as folio-sized (approximately 37 × 53.8 centimetres), hand-coloured engraved portraits.[3] The vision was grand and consequently would be expensive to realize. Catesby had three funding routes[4]: find patrons; sell his work to a printer who would take both the risks and the rewards; or pay to have the work done himself. Although the sponsors of his fieldwork were keen for him to publish, they were unable to finance the enterprise.[5] Eventually, he got a loan from Peter Collinson, the Quaker cloth merchant at the centre of a network of naturalists, travellers and wealthy collectors interested in the natural history of the Americas,[6] and the English artist and art master to the aristocracy, Joseph Goupy (*c.*1689–1769) taught him to etch copper plates and colour prints. However, still more funds were necessary to secure a market for Catesby's

Plate 201 (detail)

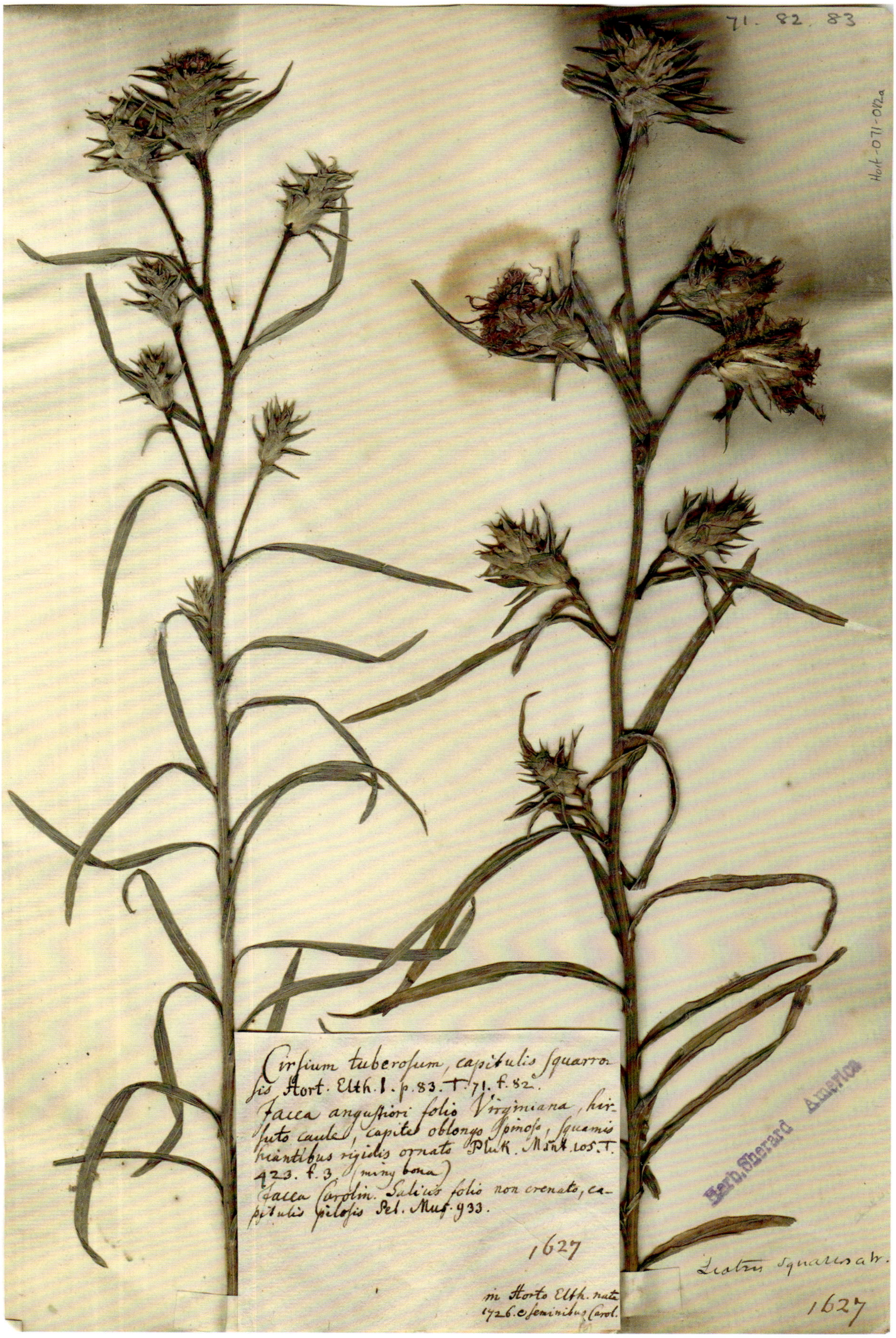

Fig. 11a (above) Specimen of the scaly blazing star (*Liatris squarrosa*) raised in James Sherard's garden at Eltham, Kent, in 1726 from seed sent by Mark Catesby from the Carolinas. Fig. 11b (right) This species was illustrated in Johann Dillenius's account of the garden's contents, *Hortus Elthamensis* (1732).

P. 83. F. 82. T. LXXI.

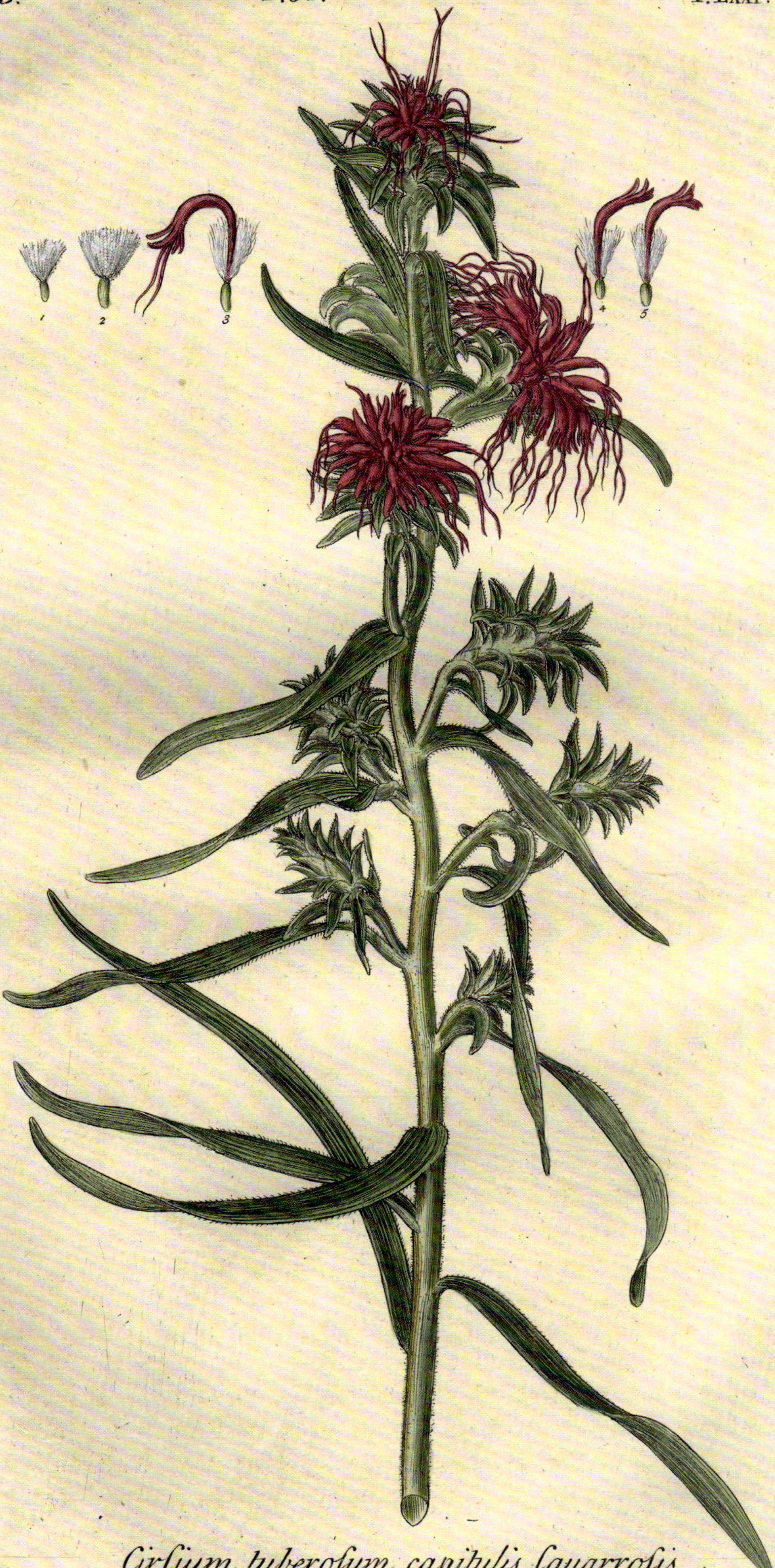

Cirsium tuberosum, capitulis squarrosis.

risky venture, so he had to attract 'Encouragers' (subscribers). During the eighteenth century, the subscription model emerged as the preferred way for authors to produce high-quality, well-illustrated limited editions of natural history works, although it frequently led to multivolume publications coming to a premature end when the subscriptions dried up.[7]

Catesby advertised for subscribers by publishing a handbill, *Proposals, for Printing an Essay Towards a Natural History of Florida, Carolina and the Bahama Islands*, in the late 1720s. The bill emphasized his credentials as an experienced traveller in Carolina, Florida and the Bahamas, as well as the seriousness of his plan: 'The Encouragers of this Work are only desired to give their Names and Places of Abode to the Author and his Friends … no Money being desired to be paid 'till each Sett is deliver'd; that so there may be no Ground to suspect any Fraud, as happens too often in the common way of Subscription.'[8]

In addition, the *Proposals* describe the contents of the work and its price: 'it is intended to publish every Four Months TWENTY PLATES, with their Descriptions, and printed on the same paper as these PROPOSALS. The Price of which will be one Guinea.'[9] The length and content of each bilingual (English and French) description appears to have been determined by Catesby's knowledge of, or interest in, the organisms in each illustration and their likely curiosity value to his intended audience.

For Catesby, images had precedence over the written word:

> The Illuminating Natural History is so particularly Essential to the perfect understanding of it, that I may aver a clearer Idea may be conceiv'd from the Figures of Animals and Plants in their proper Colours, than from the most exact Description without them: Wherefore I have been less prolix in the Description, judging it unnecessary to tire the Reader with describing every Feather, yet I hope sufficient to distinguish them without Confusion.[10]

Consequently, a high degree of accuracy in terms of both draftsmanship and colouring was essential if the publication was to succeed: 'of the Paints, particularly Greens, used in the Illumination of Figures, I had principally a regard to those most resembling Nature, that were durable and would retain their Lustre, rejecting others very specious and shining, but of an unnatural Colour and fading Quality.'[11] Catesby, aware of his limitations as an artist, tried to pre-empt criticism:

> As I was not bred a Painter I hope faults in Perspective, and other Niceties, may be more readily excused, for I humbly conceive Plants, and other Things done in a Flat, tho' exact manner, may serve the Purpose of Natural History, better in some Measure than in a more bold and Painter like Way.[12]

The first twenty hand-coloured plates, published as a group called a fascicle, were issued in 1729 to very positive reviews, and set the standard for the remaining 200 plates. However, even with the best intentions, Catesby quickly discovered that his time frame was too optimistic: he simply lacked the financial resources to publish quickly. The first fascicle of twenty plates was presented to the Royal Society on 22 May 1729, whilst the fifth fascicle (the end of the first volume of the *Natural History*) was presented three and a half years later. The five fascicles of the second volume of the *Natural History* took over nine years to complete (1734–43), with the last part – a single-fascicle appendix – being published in 1747.[13] However, financial resources do not guarantee rapid publication of field observations. For example, there was an eighteen-year gap between the publication of the two volumes of Hans Sloane's *A Voyage to the Islands Madera, Barbados, Nieves, S. Christophers and Jamaica* (1707, 1725), which were based on Sloane's field observations made thirty-eight years earlier. By combining the roles of author, engraver, colourist and ultimately financier, Catesby faced fewer delays with his own publication.

His bid to win subscribers was successful. His 'List of the Encouragers'[14] shows that approximately

180 copies of the *Natural History* were sold.[15] Moreover, his subscribers were from broad sweeps of British, colonial American and European society, from royalty, through social and scientific aristocracy, to gardeners, naturalists and booksellers.[16] Supporters who had helped him earlier in his career subscribed, for example Samuel Dale, Charles Dubois, Hans Sloane and Richard Mead. He honoured the latter by calling the shooting star 'Meadia' (*Primula meadia*, **Plate 201**), a name later adopted by Carl Linnaeus, after him. However, there is no evidence that Catesby saw this plant in the wild. The plant he depicted was from a specimen flowering in Collinson's garden at Peckham in September 1744, which was grown from seed collected by John Bartram (1699–1777), the Philadelphia-based botanist and horticulturalist who also subscribed to the *Natural History*.[17]

Other colonial subscribers included the planter, lawyer and man of letters William Byrd II (1674–1774), who lived outside Williamsburg and whom Catesby had known in Virginia; two governors of South Carolina, Francis Nicholson and Robert Johnson (1682–1735); Charles Calvert (1637–1715), who inherited the colony of Maryland from his father; and the English sailor, privateer and first Royal Governor of the Bahamas, Woodes Rogers. In addition to Caroline of Brandenburg-Ansbach (1683–1737), Queen of Great Britain and Ireland, and her daughter-in-law Augusta of Saxe-Gotha-Altenburg (1719–1772), there were five female subscribers. In Britain, these were Mrs Bridgeman, Mrs Canon, Mrs Gray (wife of the nurseryman Christopher Gray) and Mary Howard (*c.*1712–1773), Duchess of Norfolk. The only woman to subscribe from colonial America was 'Mrs Holloway of Virginia'.

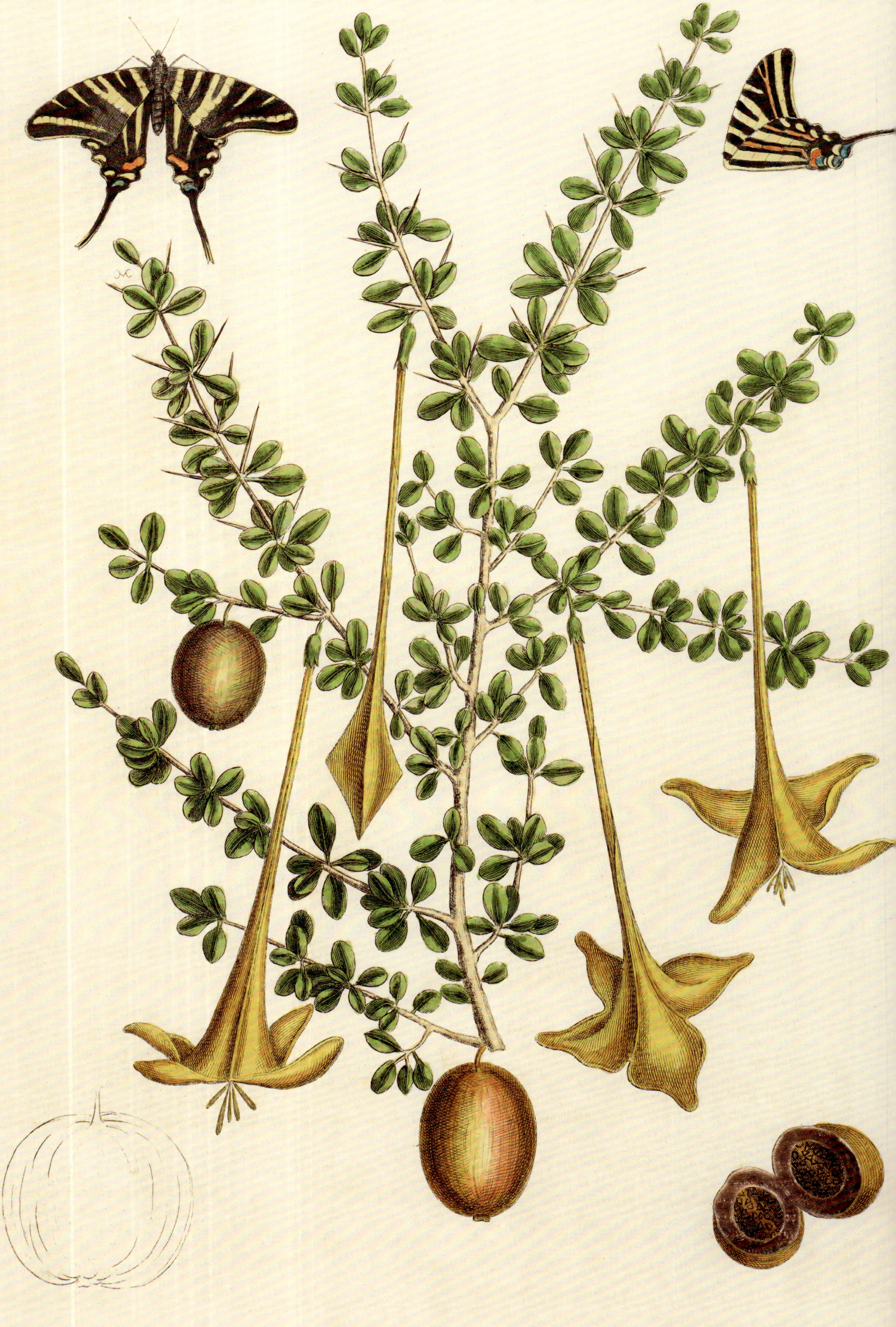

Legacy 8

... by the concurrent endeavours of the philosopher and artisan, I question not but many of them will be found useful to purposes, of which at present we have not the least conception.

MARK CATESBY[1]

An essential part of Catesby's work was his attempt to separate fact from fiction whilst working within a framework that emphasized the importance of objective, verifiable evidence. However, when judging and interpreting Catesby's illustrations, we must realize that many of his images were the de facto authorities of the period. Most of his readers would never have seen the organisms he depicted 'in the flesh', and would have seen few, if any, images of them other than those presented in the *Natural History*. It is only with the benefit of nearly three centuries of accumulated natural history research in the field, in laboratories and in libraries, with access to hundreds of samples and images, that effective judgements can be made about Catesby's work.

With an apparent lack of false modesty, Catesby was aware of the potential future value of his observations and specimens. He hoped to stimulate others' curiosity in the natural history of North America through his illustrations and the plants he returned to European gardens. The rediscovery of Catesby's work in the mid-twentieth century resonated with issues surrounding landscape and habitat change, changes in species' distribution and species extinction – issues that are prominent in current political and scientific discussions.[2]

Catesby's time in the Americas showed that he was comfortable in the company of colonial governors as well as of Indigenous peoples and colonists. Catesby's character, as much as his authority and the timeliness and circulation of his observations, helped his *Natural History*, despite its idiosyncrasies and inaccuracies, to stand the uncompromising test of time. Even if his ideas might not always have been right, they were a catalyst for others to investigate and refute his interpretations. Especially important for the scientific legacy of the *Natural History* were that the Swedish naturalist Carl Linnaeus used it when he described new species, and the serendipity of Linnaeus's work being adopted as the baseline for modern scientific naming of the natural world.[3] For example, when Linnaeus formally described the purple land crab (**Plate 132**) in 1758, he cited Catesby's illustration, summarizing his observations of the crab's behaviour: 'leaving the forests in great numbers every year to seek the shores of the sea'.[4] Moreover, the lily thorn (*Catesbaea spinosa*, **Plate 200**), a native shrub of northern Caribbean dry, evergreen woodlands, was introduced to English glasshouses in 1726 from Nassau (Bahamas), where Catesby found only two plants. The plant's flowers, shaped like a 'Roman trumpet', produce fruits about the size of 'Pullet's egg[s]' with an 'agreeable tartness and good flavour', which Catesby considered 'capable of being improved by cultivation'.[5] The eighteenth-century Dutch botanist Jan Frederik Gronovius (1690–1762) coined the plant's scientific name, *Catesbaea*, to commemorate Catesby. Indeed, Catesby's own illustration, published in 1743, is the type of the species.

Catesby was not a collector who briefly visited an area and then left. He wrote and illustrated from personal knowledge of the plants and animals of the Thirteen Colonies and the Bahamas, where he lived for more than a decade. He was a critical observer who saw how seasons changed the natural history of the areas in which he lived. This is especially obvious in his extensive scrutiny of birds and their annual movements, and the conclusions he drew from these

Plate 200 (detail)

observations.[6] Indeed, Catesby's most singular and long-lasting biological contribution has been his objective observation of bird migration. His *Natural History* escaped the ignominy of becoming a trophy volume. Once the entire work was complete in 1747, it soon went into a second (1754) and then a third (1771) edition, and it was translated into Dutch, German and Latin.[7]

His first-hand knowledge also made him an authority for other colonial naturalists. For example, the Virginian clerk and botanist John Clayton (*c.*1694–1773) got to know Catesby during his stay in the colony in the 1710s.[8] They maintained a friendship and correspondence when Catesby returned to England. In the 1730s Clayton sent Catesby dried specimens and a manuscript describing the plants of Virginia. Catesby was unable to use the material, so passed it on to Johan Gronovius, who published it in his *Flora Virginica* (1739–43).

The son of an East Anglian yeoman, Catesby had a unique set of skills that enabled him to be accepted at the heart of the early eighteenth-century British scientific establishment. The observations he recorded in the *Natural History* remain vivid snapshots of the flora and fauna of colonial North America and of the curiosity of an extraordinary naturalist-artist.

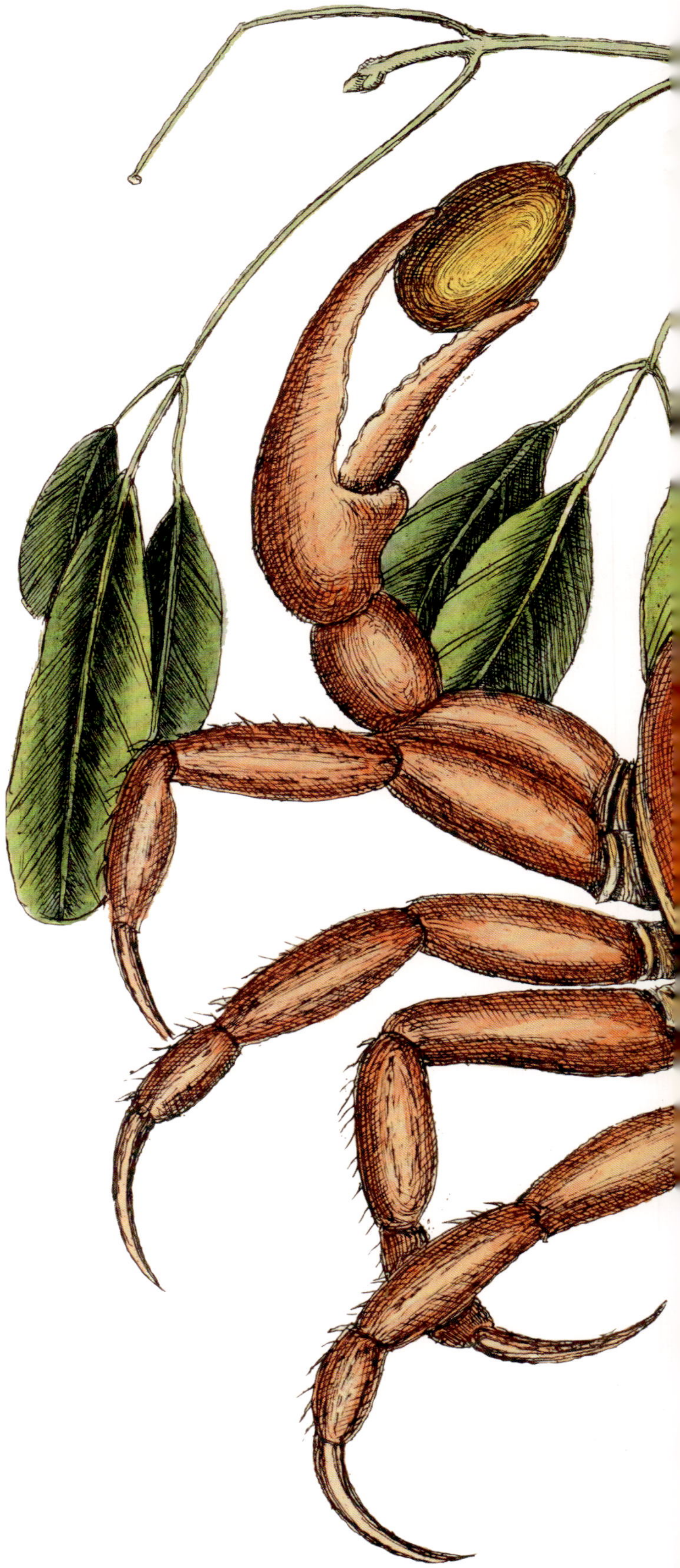

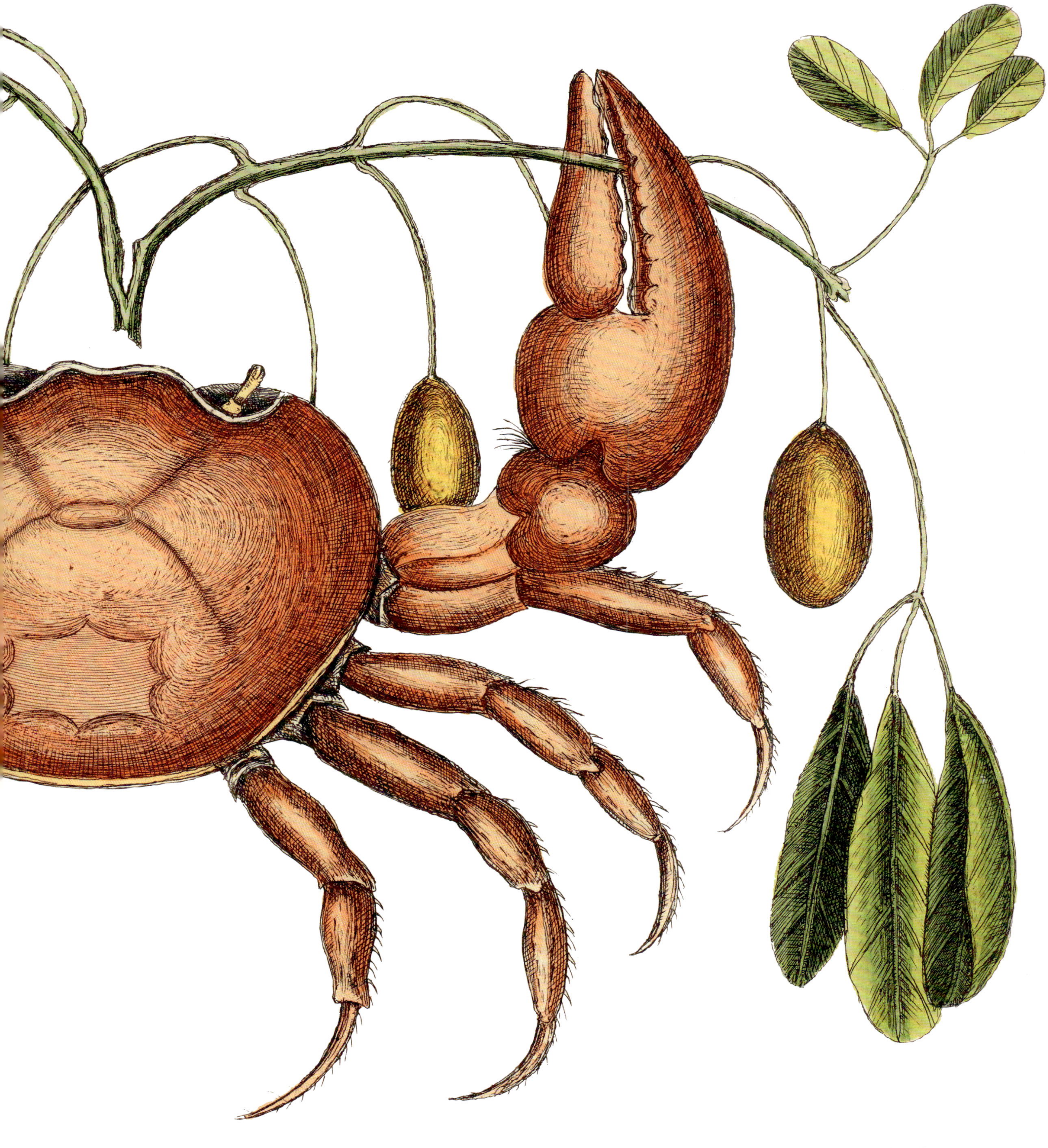

Plate 132 (detail)

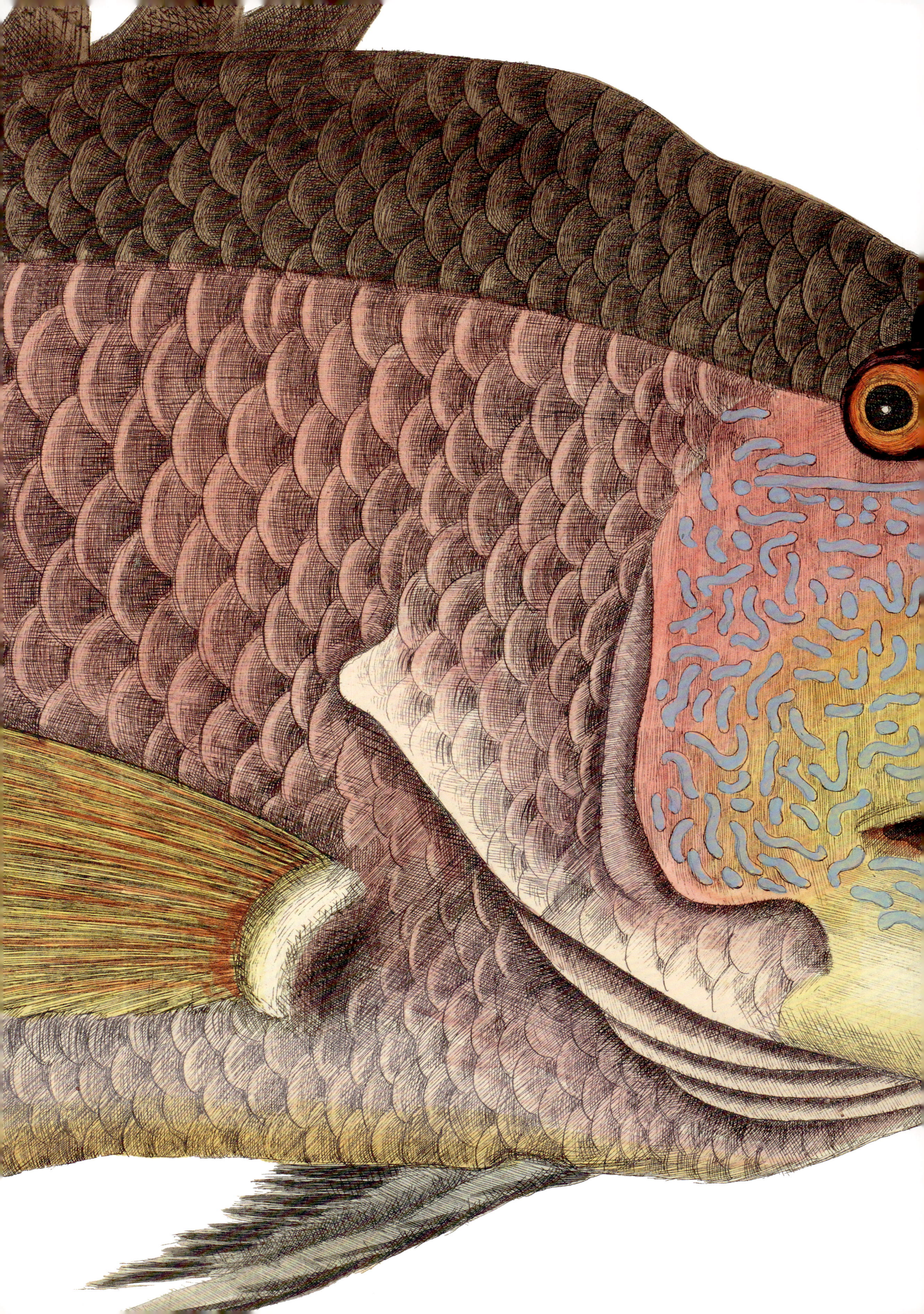

Plates

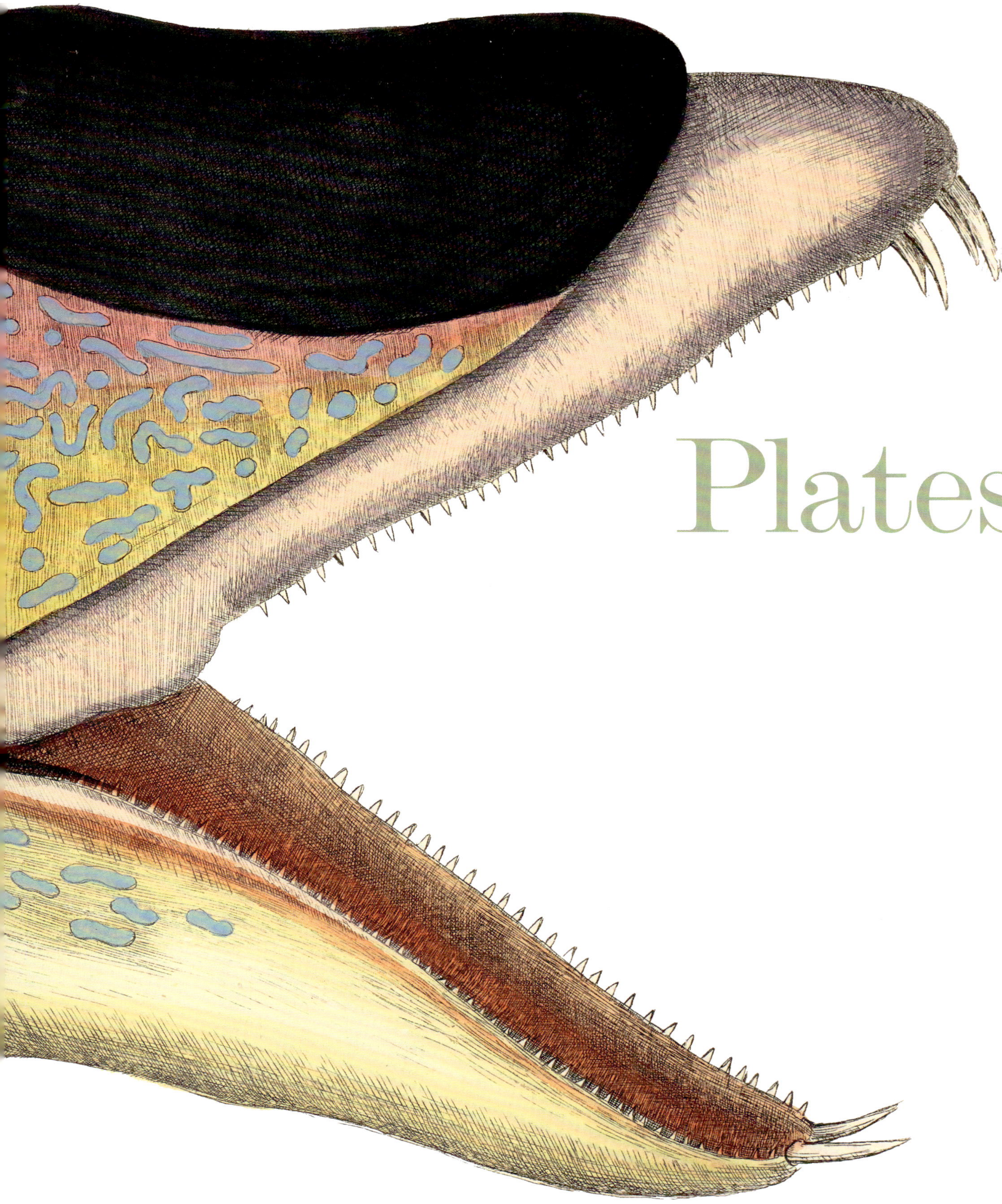

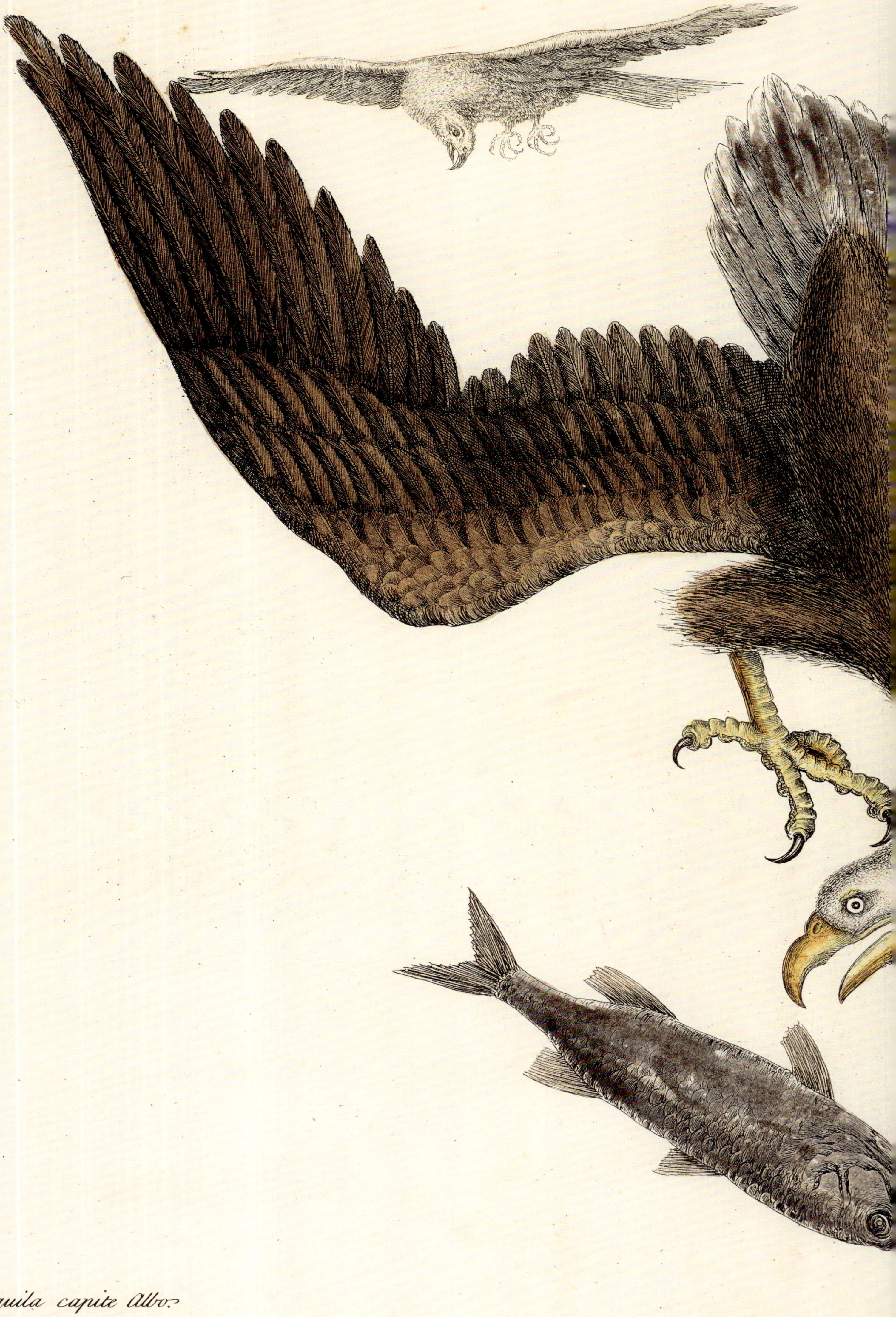

Aquila capite Albo.

The White headed Eagle.

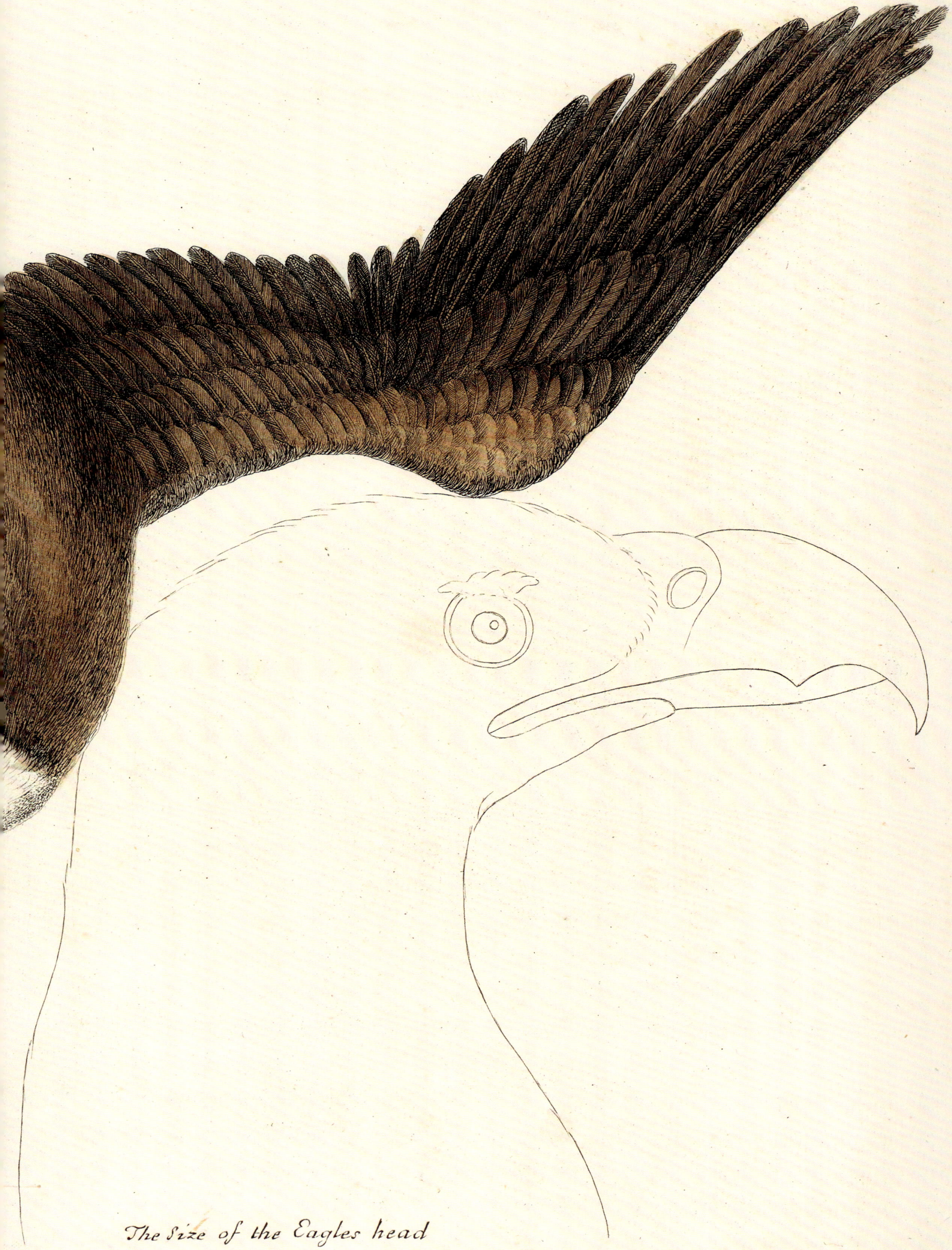

The Size of the Eagles head

Accipiter Piscatorius.

The fishing Hawk.

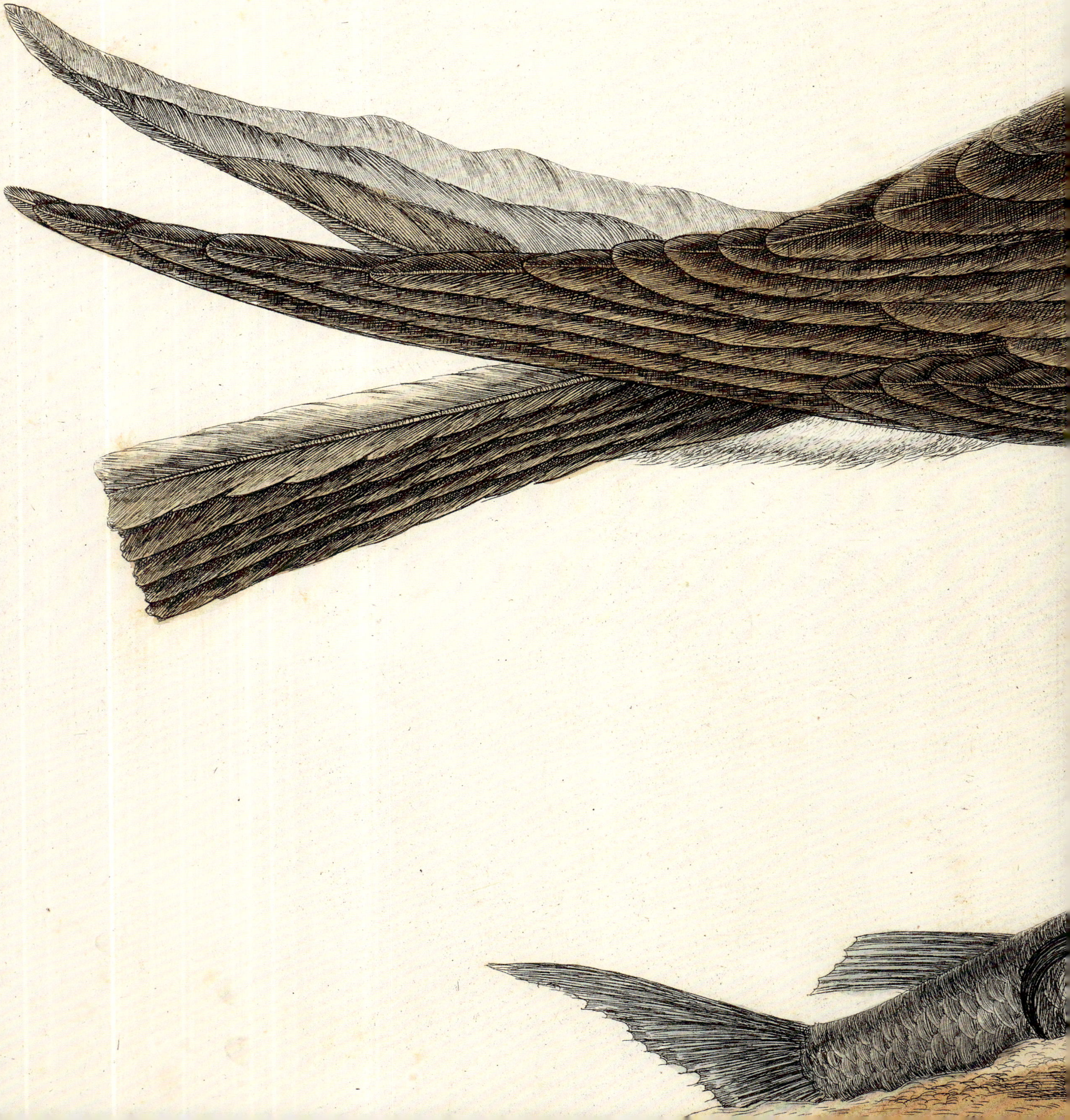

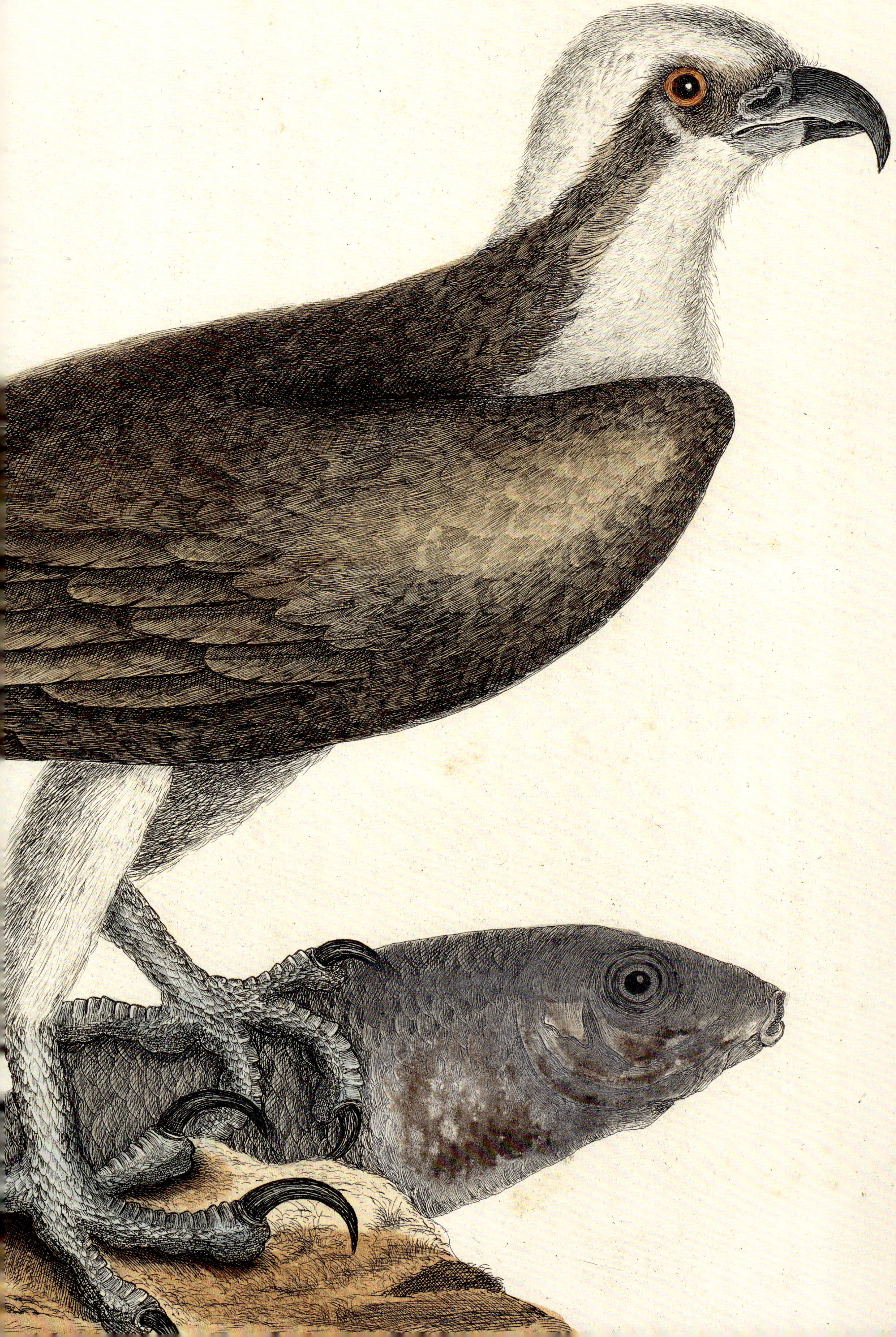

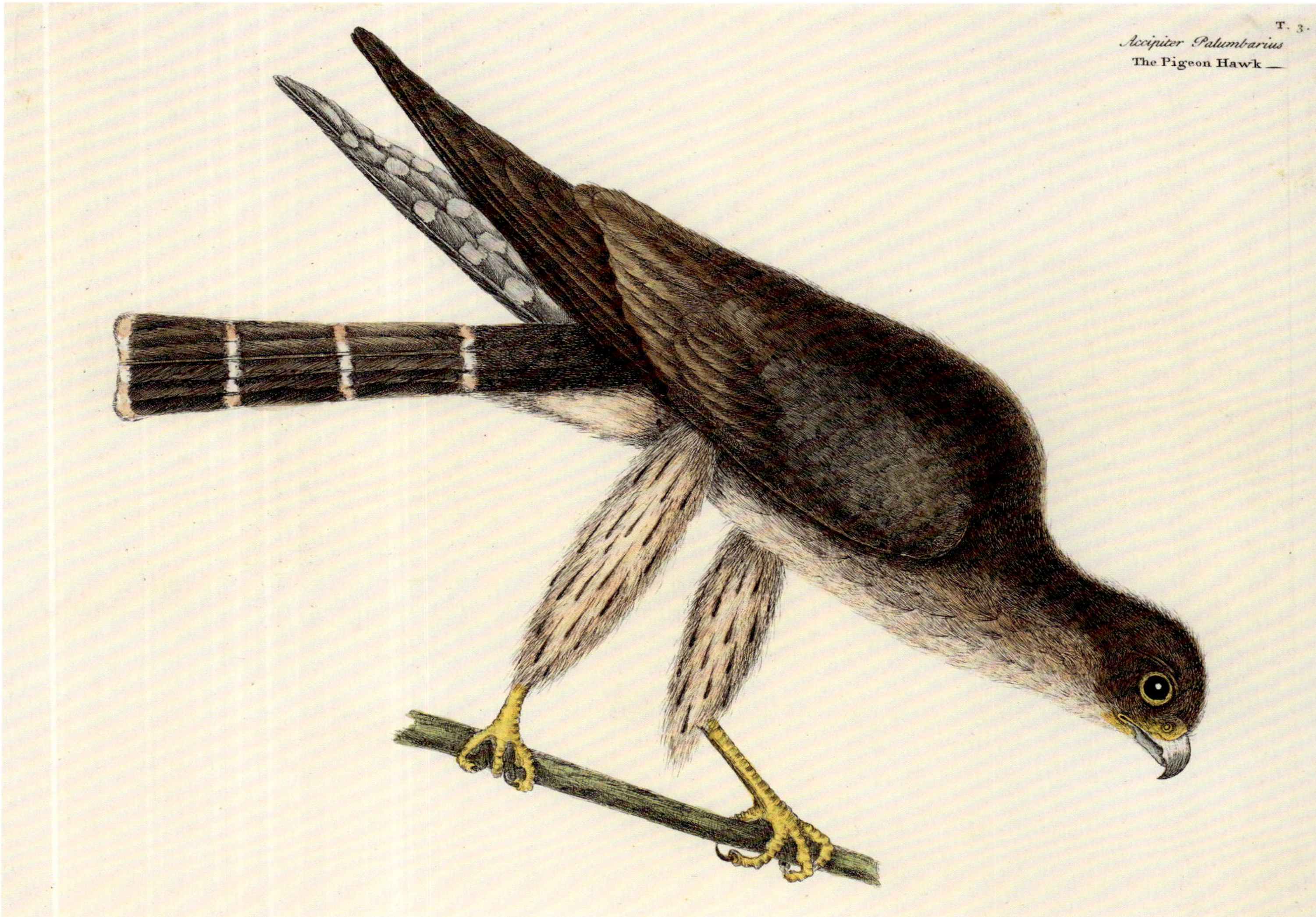

previous pages

1

Bald eagle (*Haliaeetus leucocephalus*); flathead grey mullet (*Mugil cephalus*); osprey (*Pandion haliaetus*; in background) (vol. 1, t. 1)

2

Osprey (*Pandion haliaetus*); flathead grey mullet (*Mugil cephalus*) (vol. 1, t. 2)

these pages

3 *above*

Merlin (*Falco columbarius*) (vol. 1, t. 3)

4 *right*

Swallow-tailed kite (*Elanoides forficatus*) (vol. 1, t. 4)

T. 4

Accepiter cauda forcipater furcata.

The Swallow tail'd Hawk.

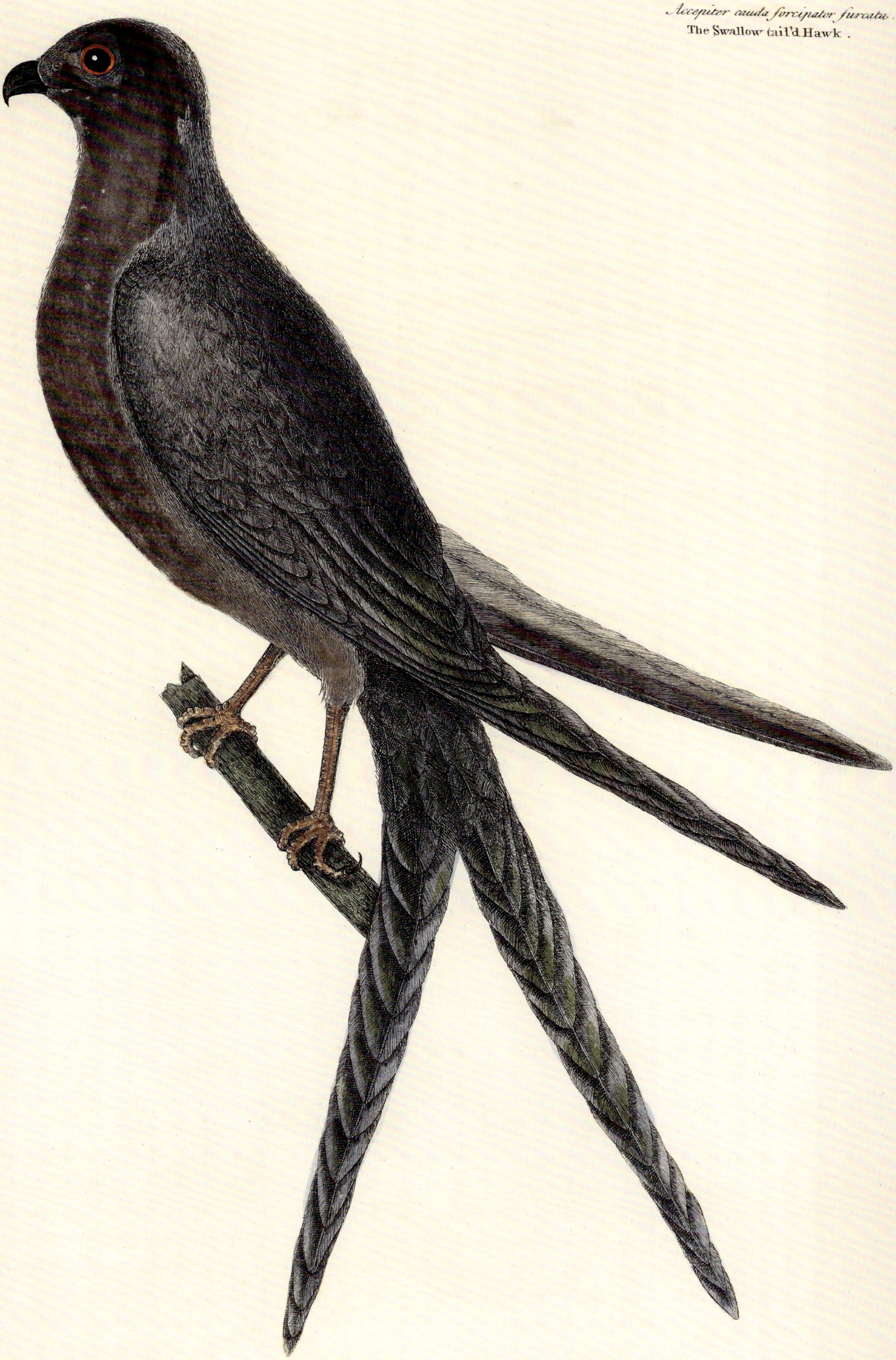

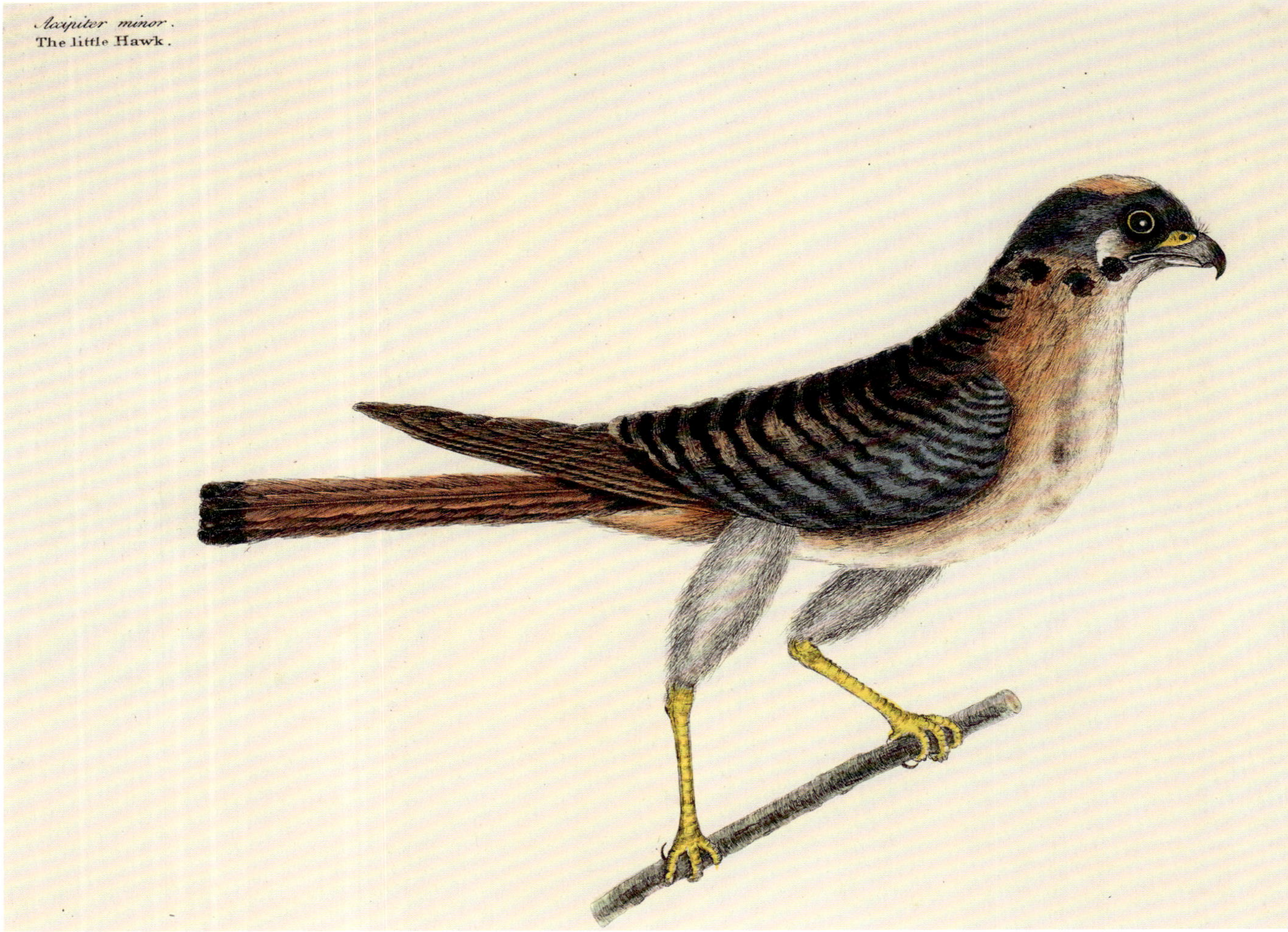
Accipiter minor.
The little Hawk.

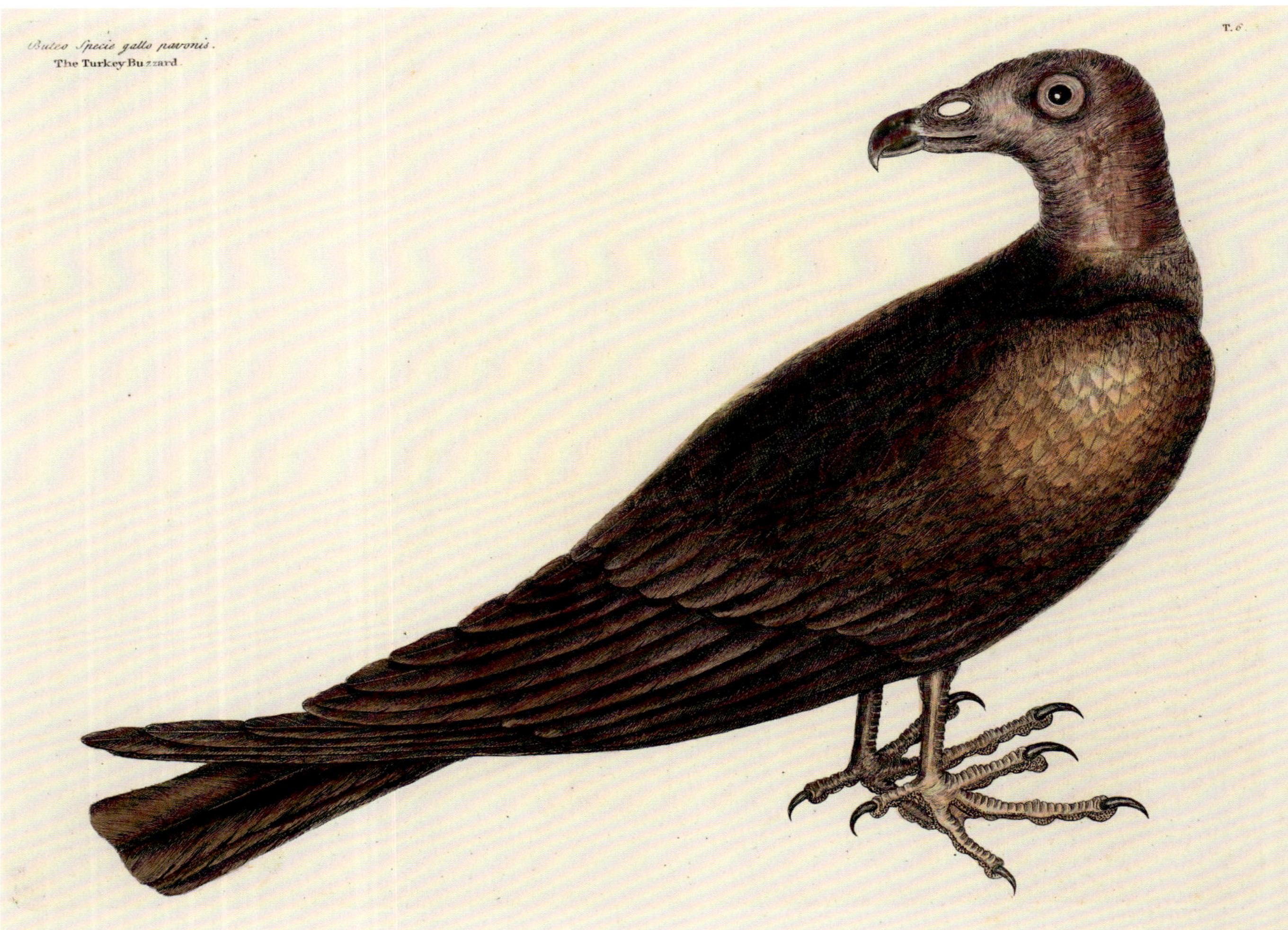
T. 6.
Buteo Specie gallo pavonis.
The Turkey Buzzard.

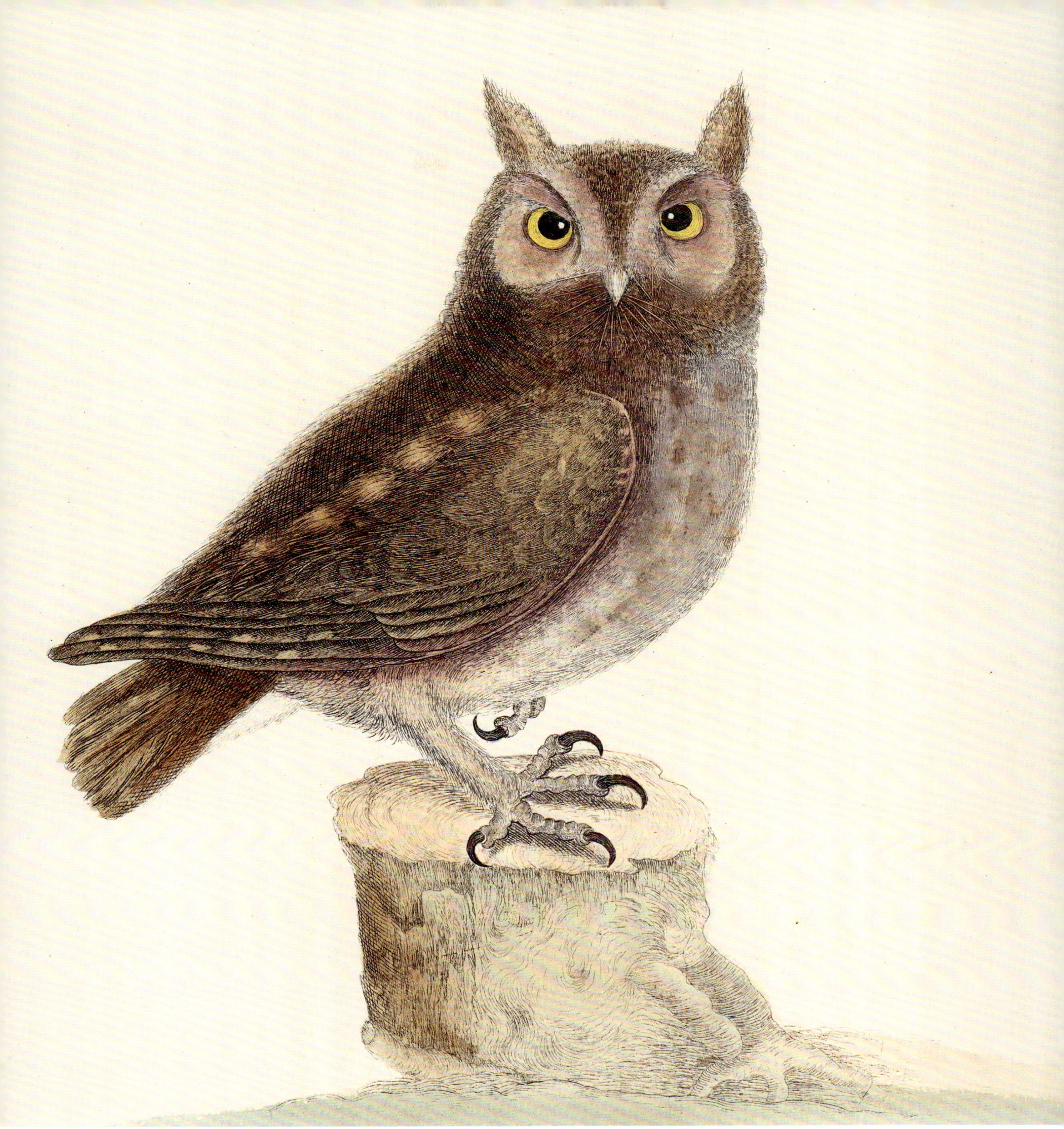

5 *top left*
American kestrel (*Falco sparverius*) (vol. 1, t. 5)

6 *bottom left*
Turkey vulture (*Cathartes aura*)

7 *above*
Eastern screech owl (*Megascops asio*) (vol. 1, t. 7)

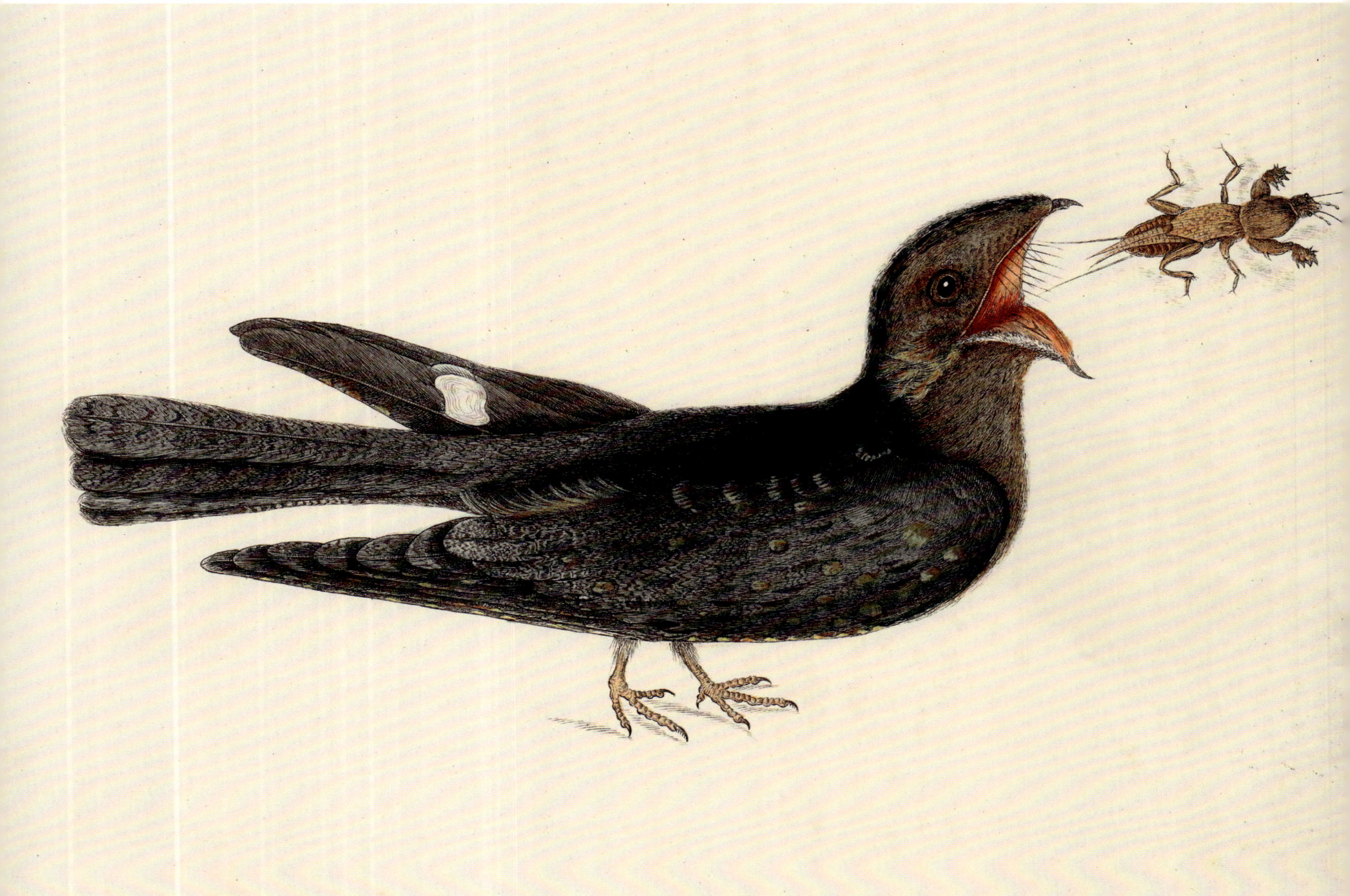

8

Appears to be a chimaera of three species, comprising common nighthawk (*Chordeiles minor*), chuck-will's-widow (*Caprimulgus carolinensis*) and eastern whip-poor-will (*Caprimulgus vociferus*), plus northern mole cricket (*Neocurtilla hexadactyla*) (vol. 1, t. 8)

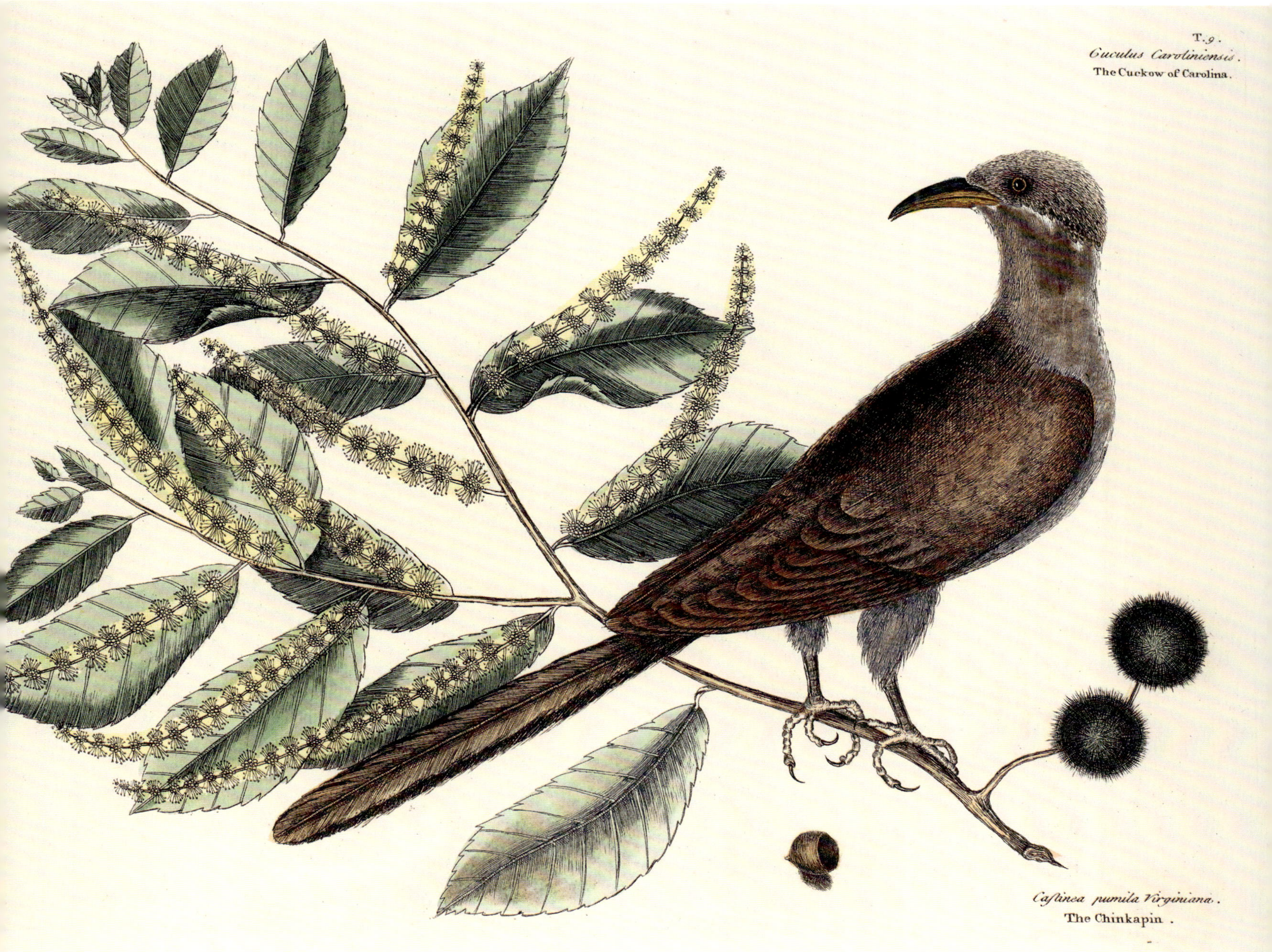

9

Yellow-billed cuckoo (*Coccyzus americanus*); chinquapin (*Castanea pumila*) (vol. 1, t. 9)

Frutex Lauri folio pendulo, fructu tricocco Semine nigro Splendente.

Red Wood.

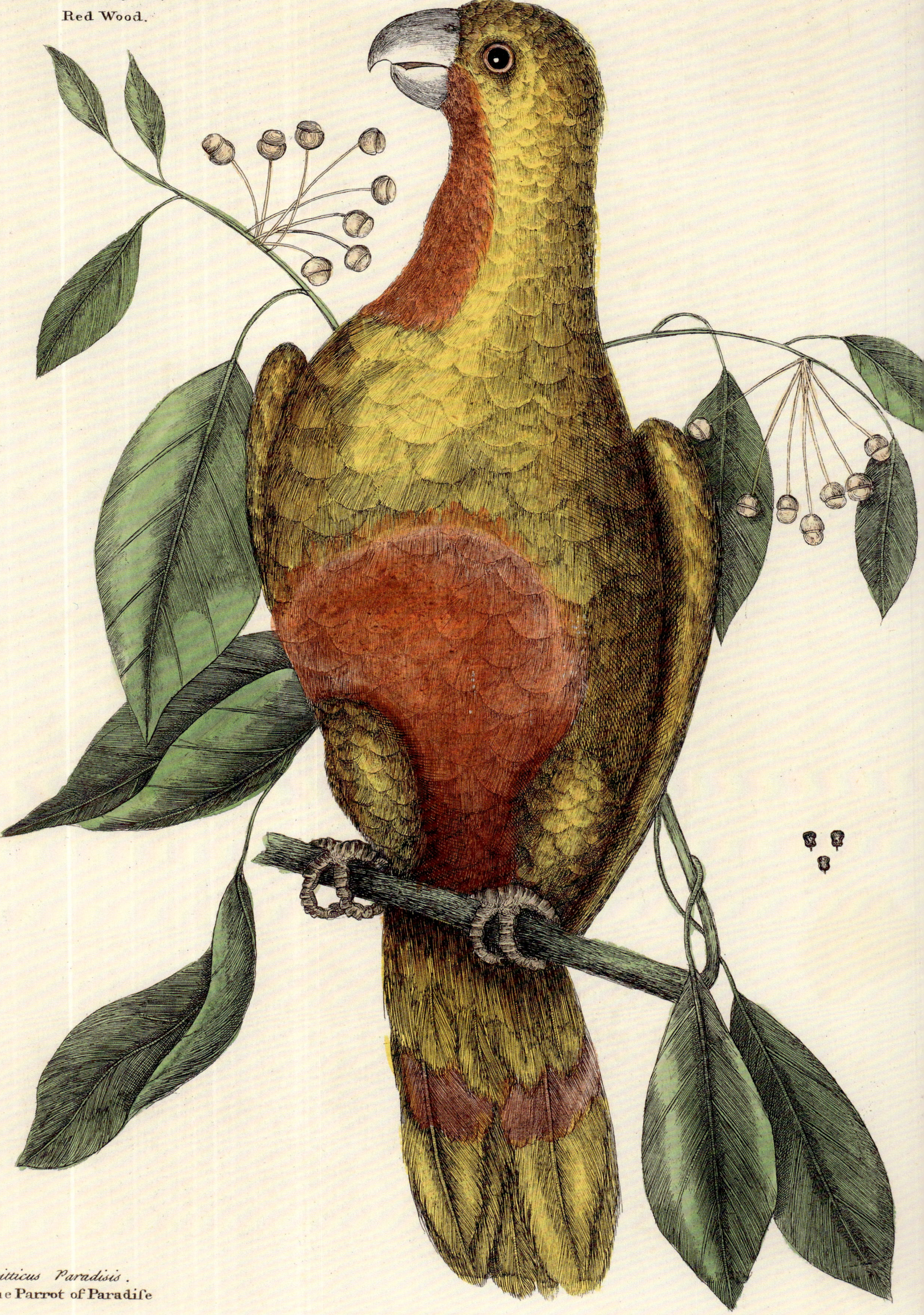

Psitticus Paradisis.

The Parrot of Paradise

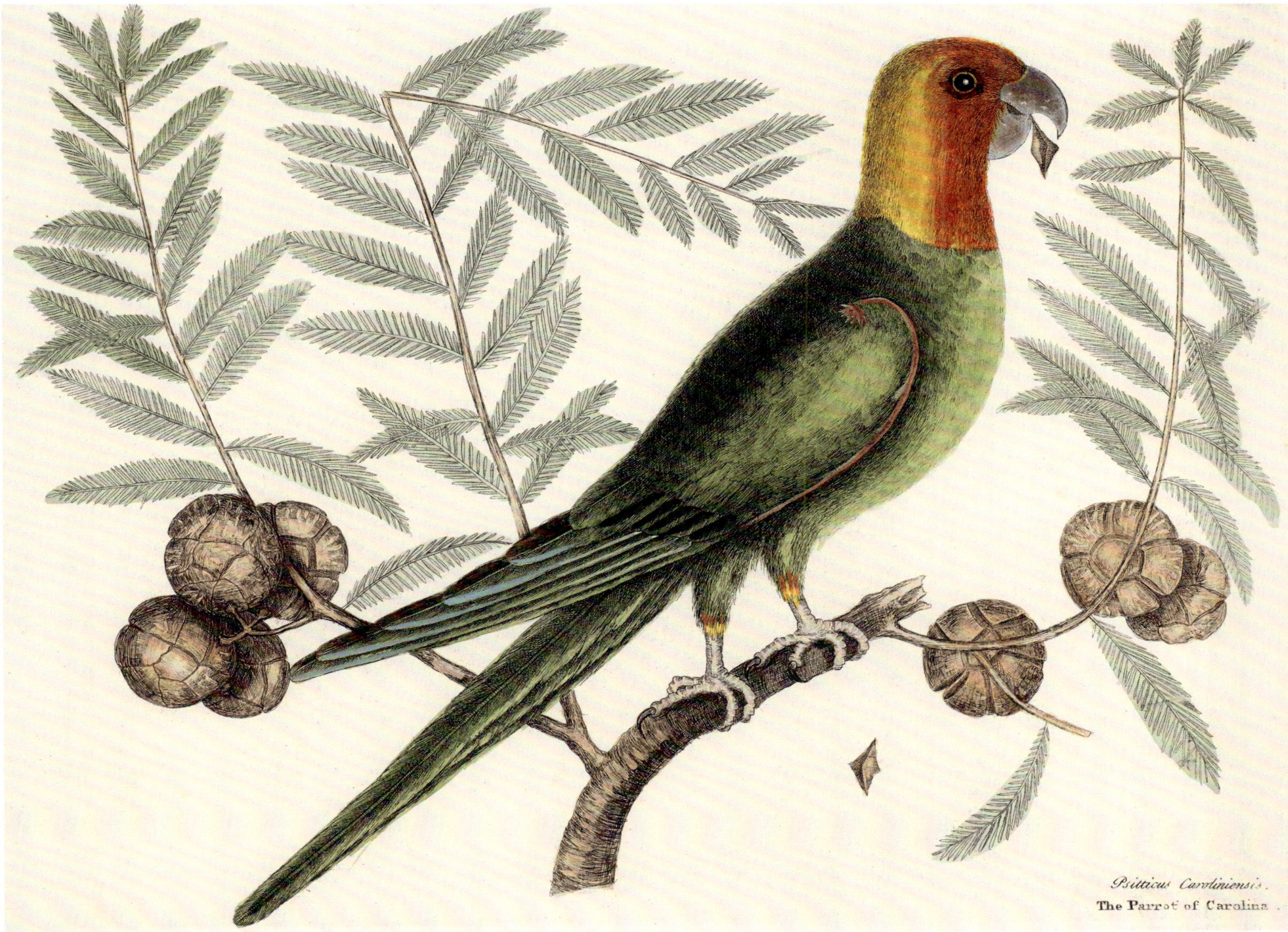

10 *left*
Cuban parrot (*Amazona leucocephala*);
soldierwood (*Colubrina elliptica*)
(vol. 1, t. 10)

11 *above*
Carolina parakeet (*Conuropsis carolinensis*);
swamp cypress (*Taxodium distichum*)
(vol. 1, t. 11)

Monedula purpuria.
The purple Jackdaw.
T. 12.

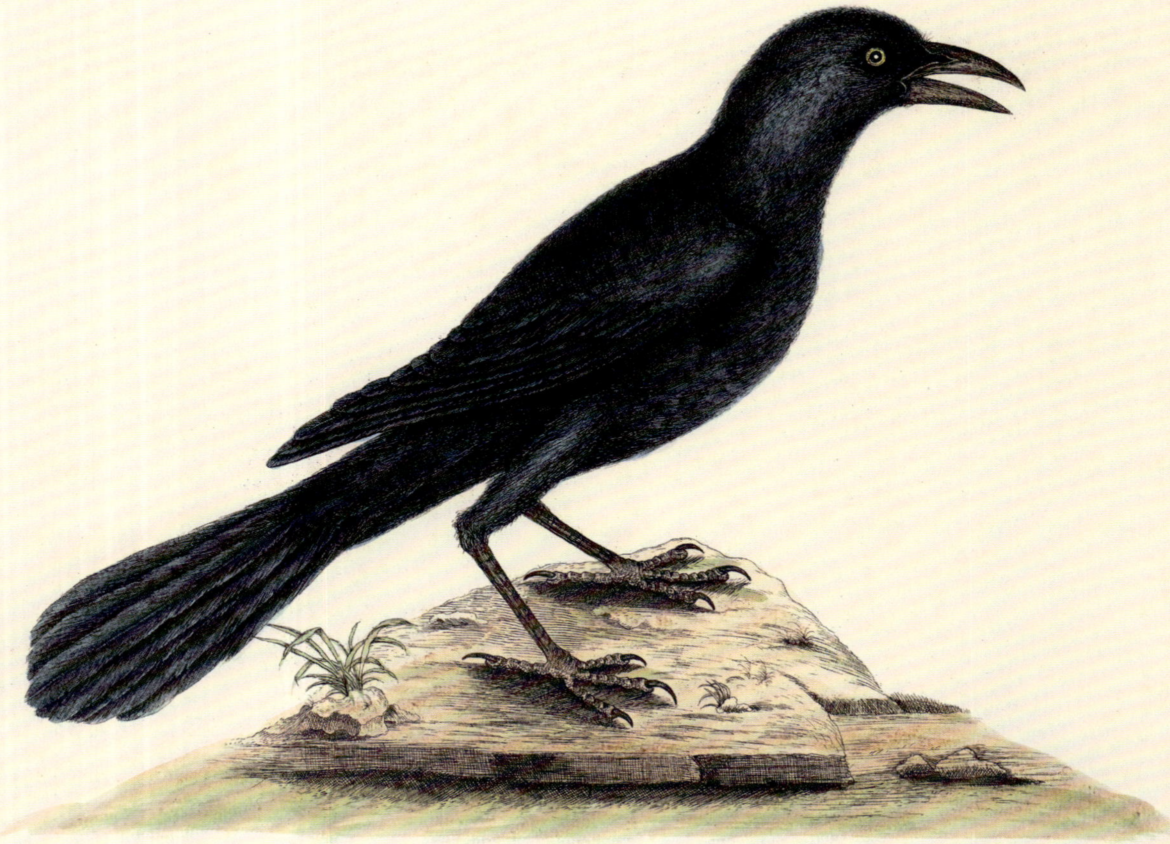

T. 12.
Sturnus niger alis supernis rubro colore.
The red Wing'd Starling.

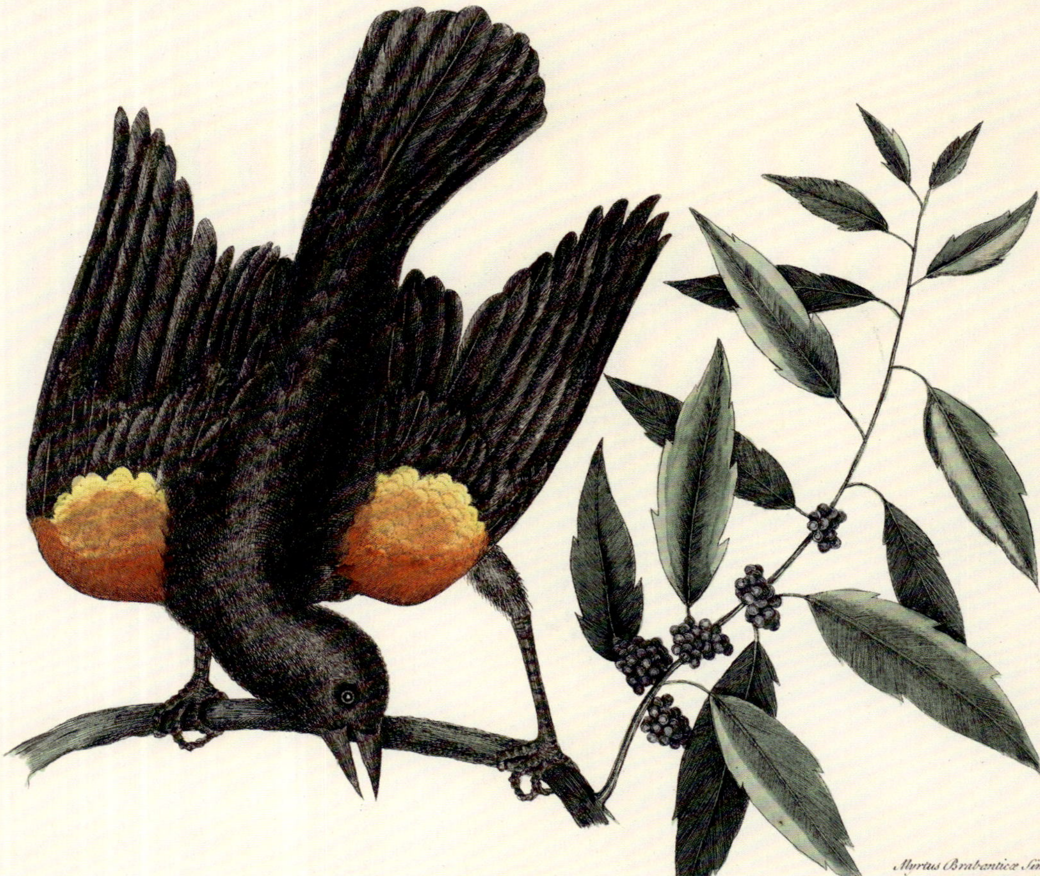

Myrtus Brabanticæ Similis Caroliniensis humilior; foliis latioribus, et magis Serratis.
The broad leaved candle-berry Myrtle.

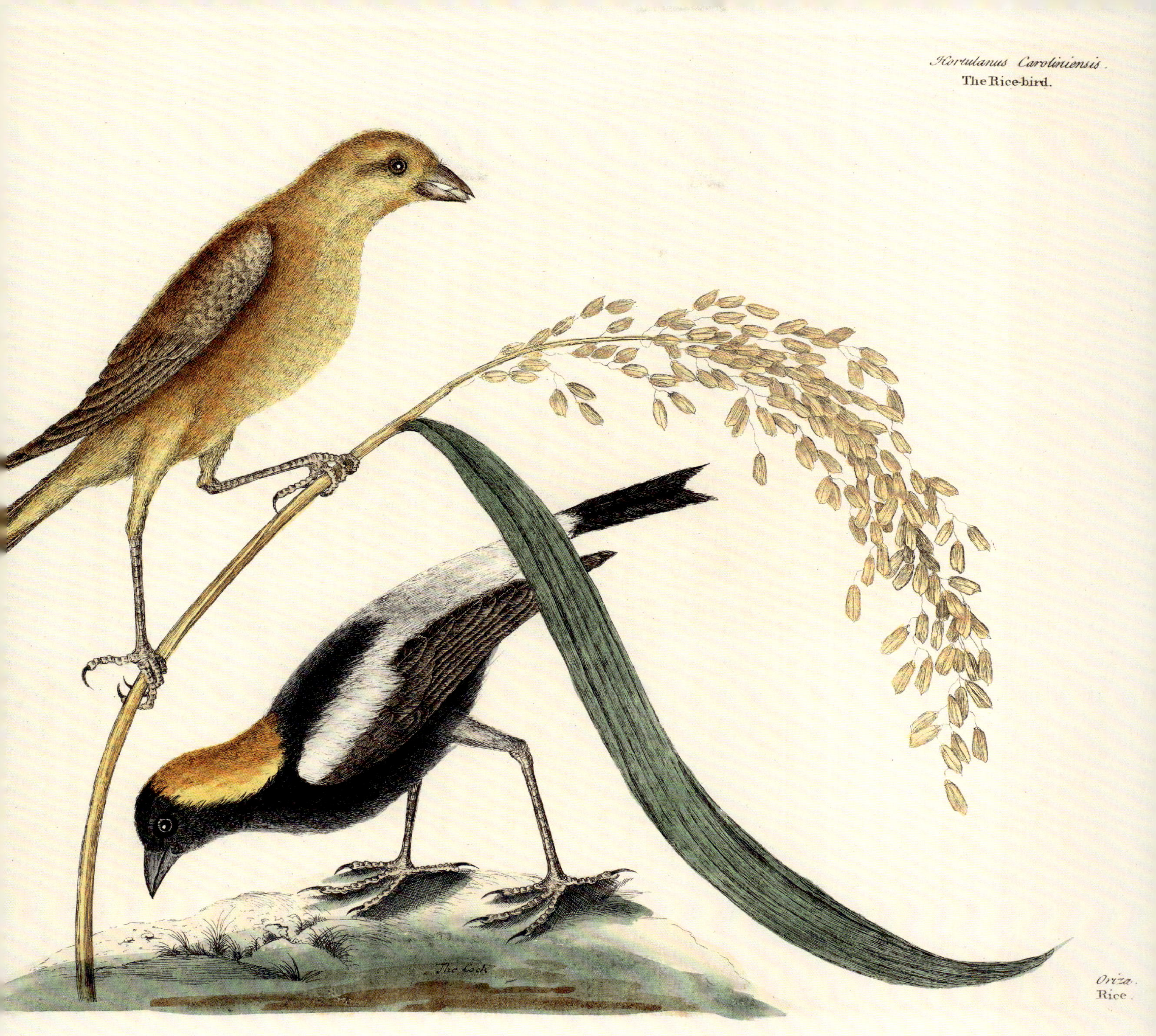

12 *top left*
Common grackle (*Quiscalus quiscula*) (vol. 1, t. 12)

13 *bottom left*
Red-winged blackbird (*Agelaius phoeniceus*); southern bayberry (*Myrica heterophylla*) (vol. 1, t. 13)

14 *above*
Bobolink (*Dolichonyx oryzivorus*; female perched, male standing); rice (*Oryza sativa*) (vol. 1, t. 14)

15 *following pages*
Blue jay (*Cyanocitta cristata*); laurel greenbrier (*Smilax laurifolia*) (vol. 1, t. 15)

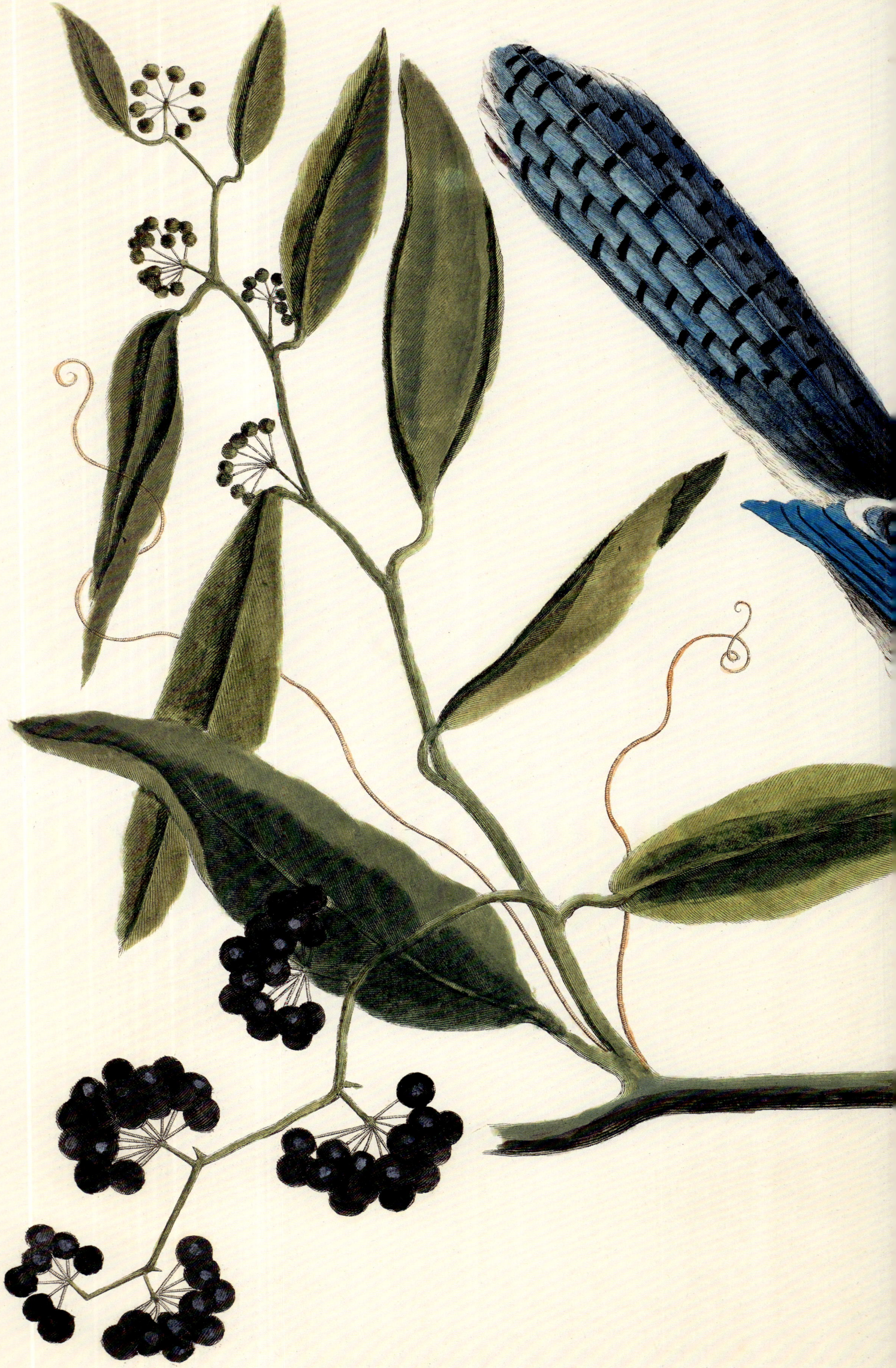

Smilax lævis Lauri folio non Serrato, baccis nigris.

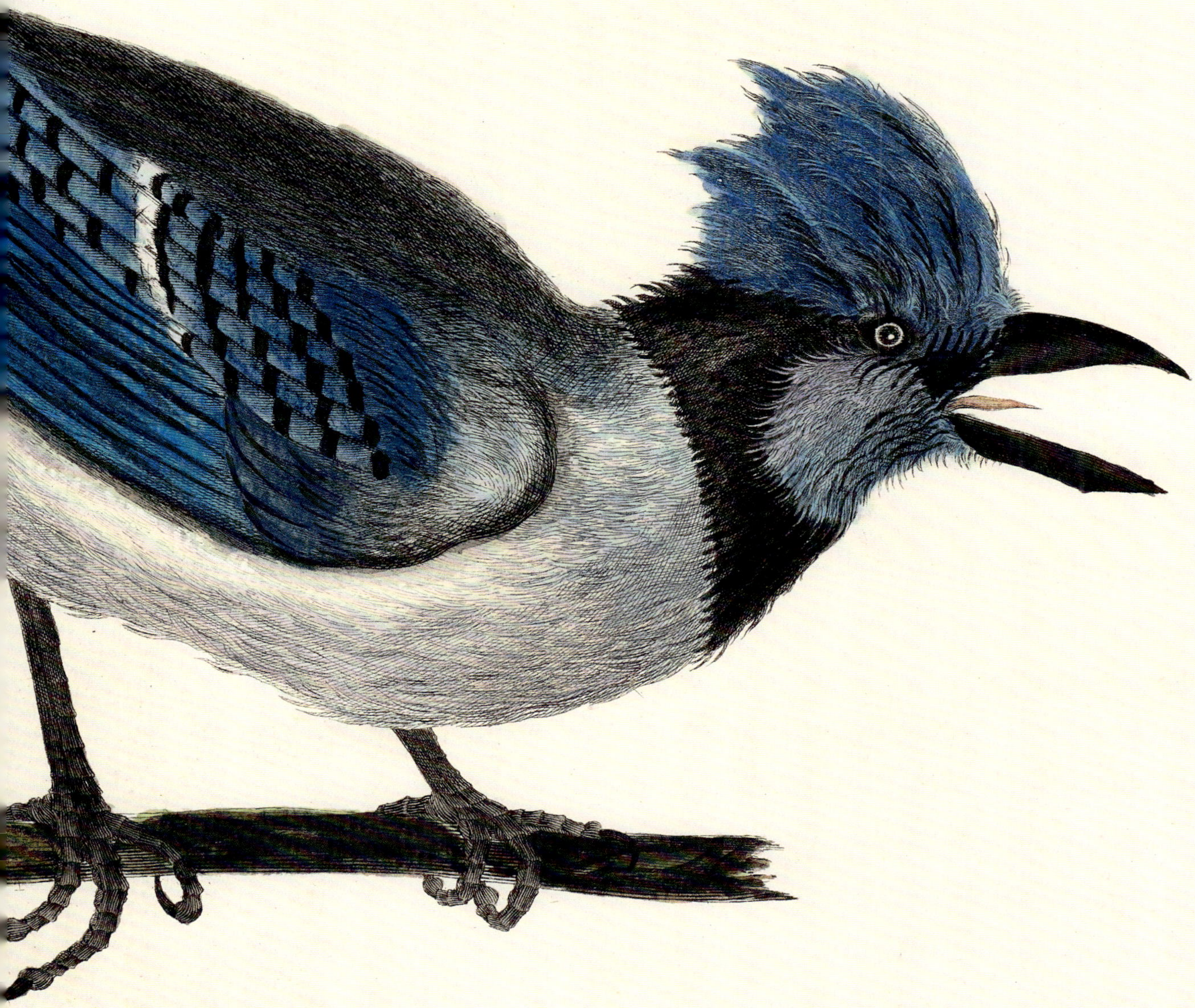

T. 15.

Pica cristata cærulea.
The crested Jay.

Quercus, anpotius; Ilex Marilandica Folio longo angusto Salicis: Ray: Hist:

Willow Oak.

Picus maximus rostro albo

Largest White Bill'd Woodpe

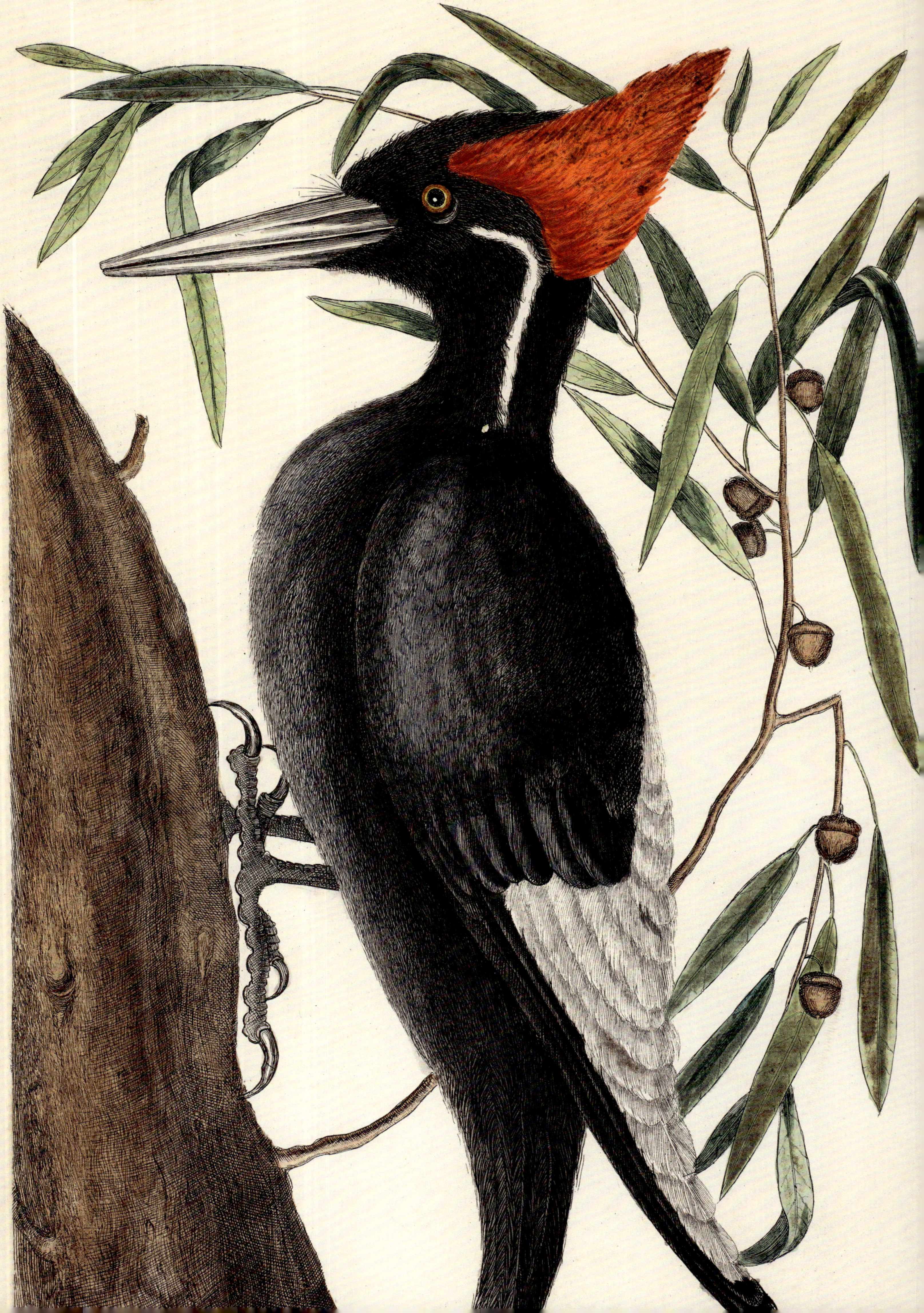

Sempervirens folijs oblongis non sinuatis.

ister Live Oak.

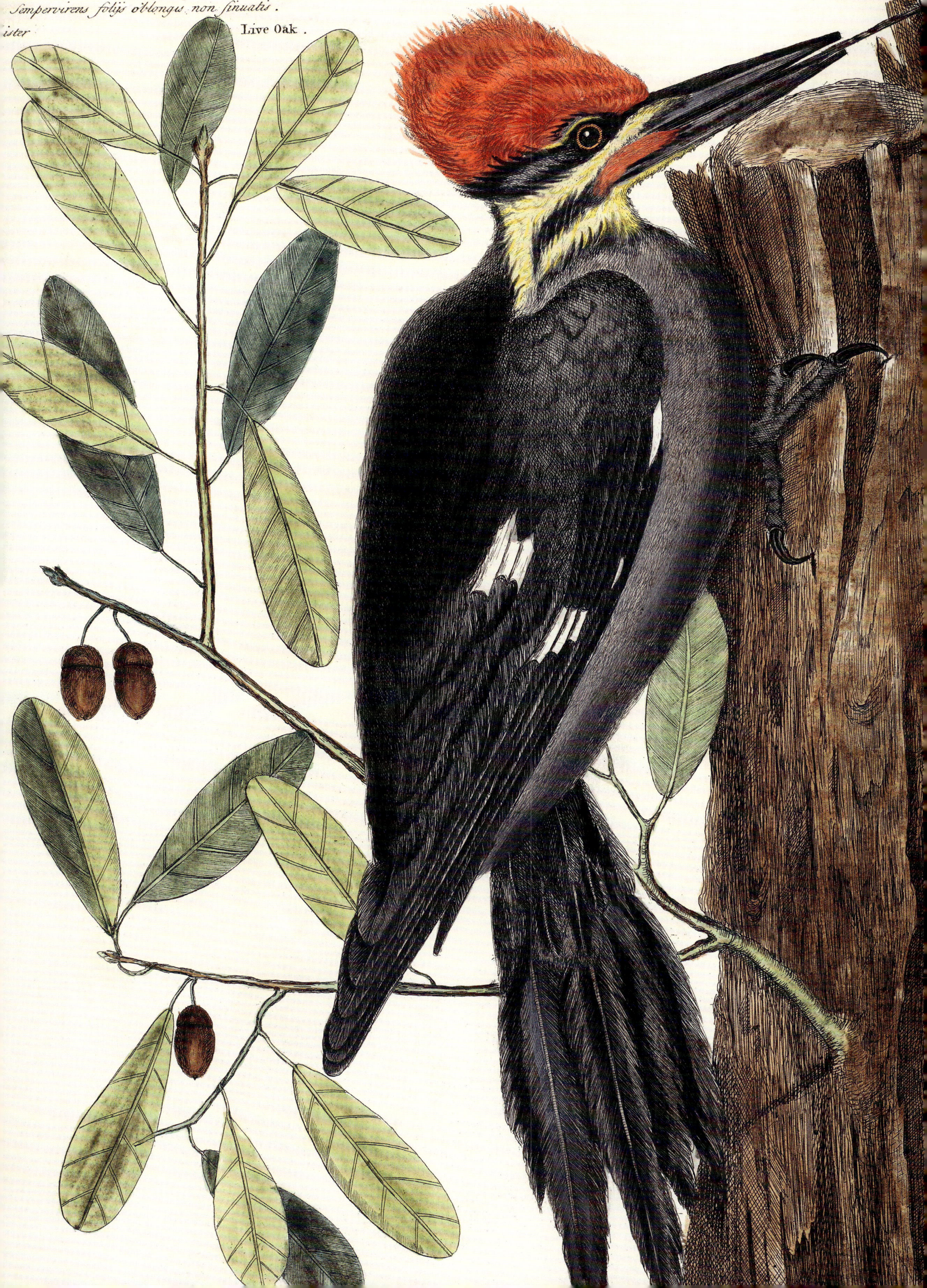

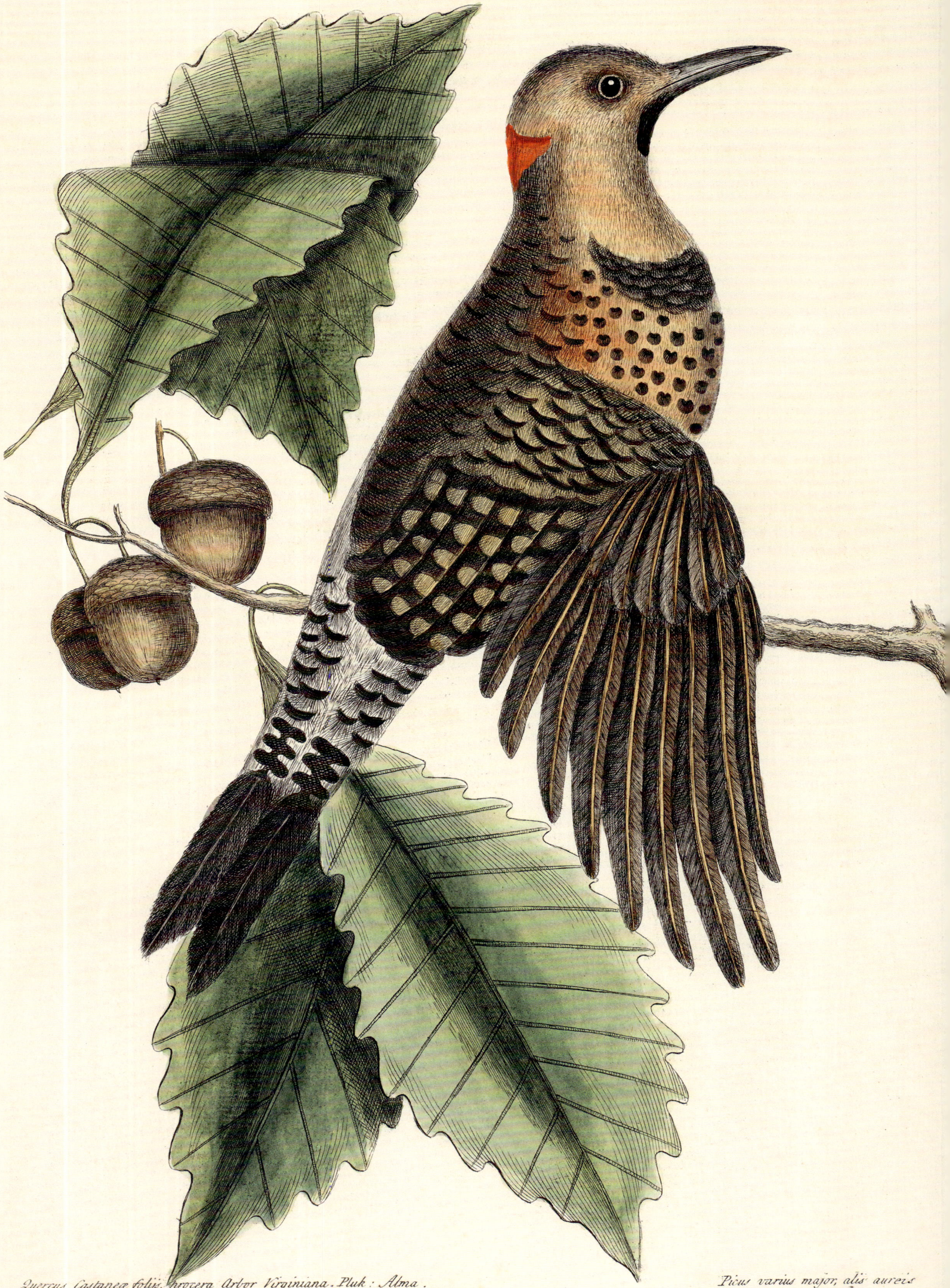

Quercus Castaneæ foliis, procera Arbor Virginiana. Pluk: Alma.
Chesnut Oak.

Picus varius major, alis aureis
The golden Wing'd Woodpecker.

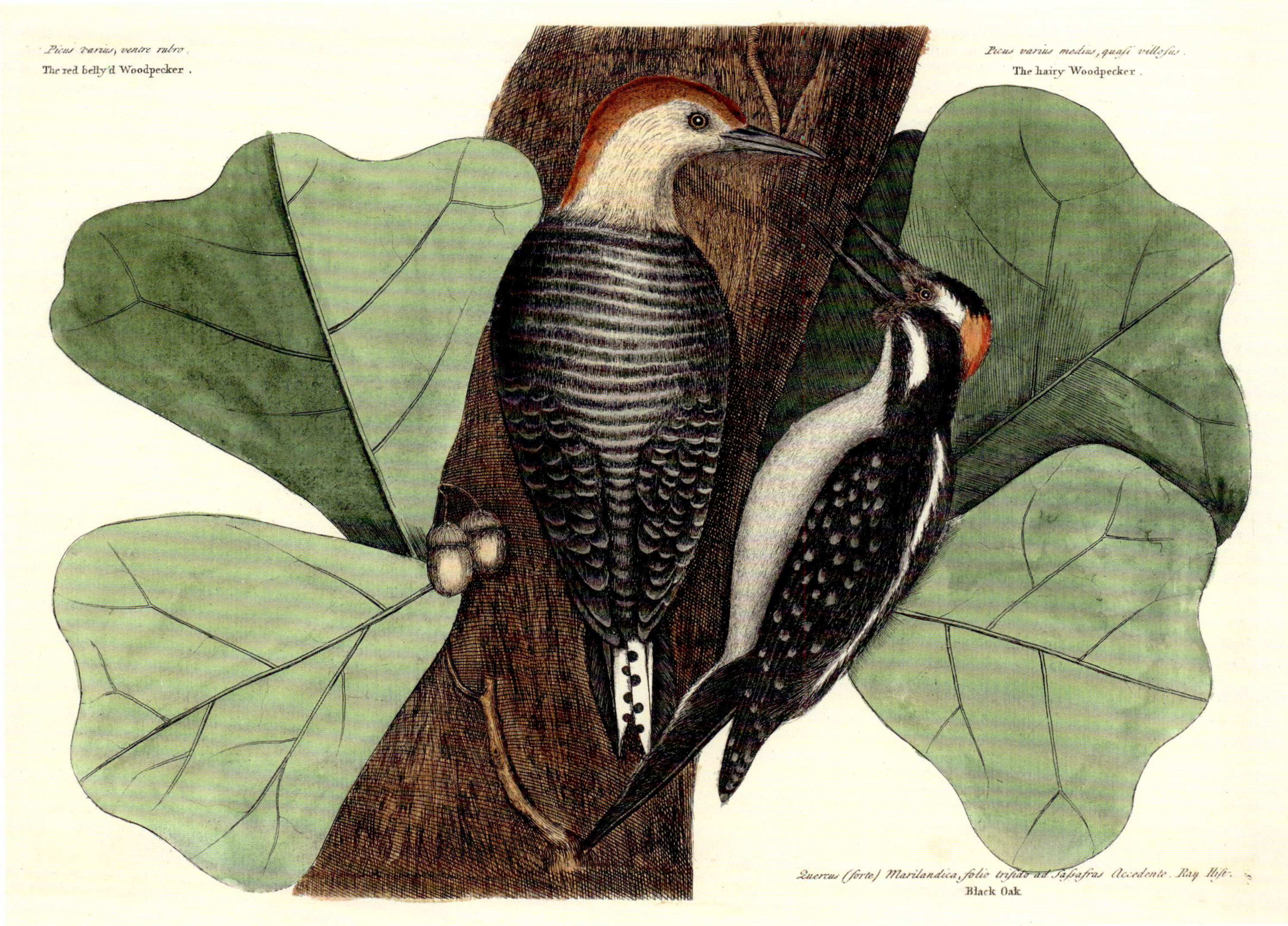

previous pages

16 *left*
Ivory-billed woodpecker (*Campephilus principalis*); willow oak (*Quercus phellos*) (vol. 1, t. 16)

17 *right*
Pileated woodpecker (*Drycopus pileatus*); live oak (*Quercus virginiana*) (vol. 1, t. 17)

18 *left*
Northern flicker (*Colaptes auratus*); swamp chestnut oak (*Quercus michauxii*) (vol. 1, t. 18)

19 *above*
Red-bellied woodpecker (*Melanerpes carolinus*; left); hairy woodpecker (*Picoides villosus*; right); blackjack oak (*Quercus marilandica*) (vol. 1, t. 19)

Quercus folio non serrato, in Summitate quasi triangu

Water Oak.

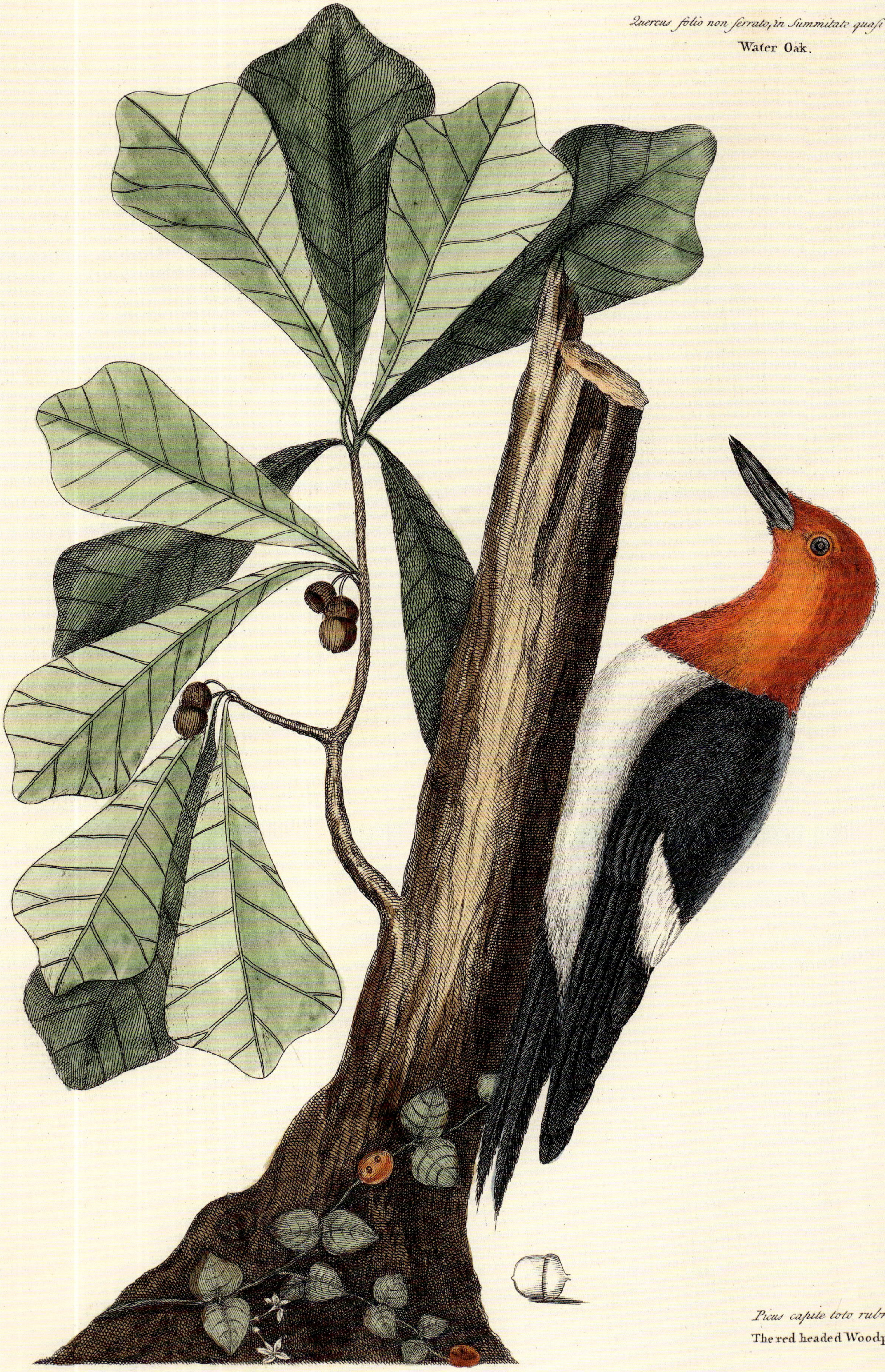

Picus capite toto rubro

The red headed Woodpecker

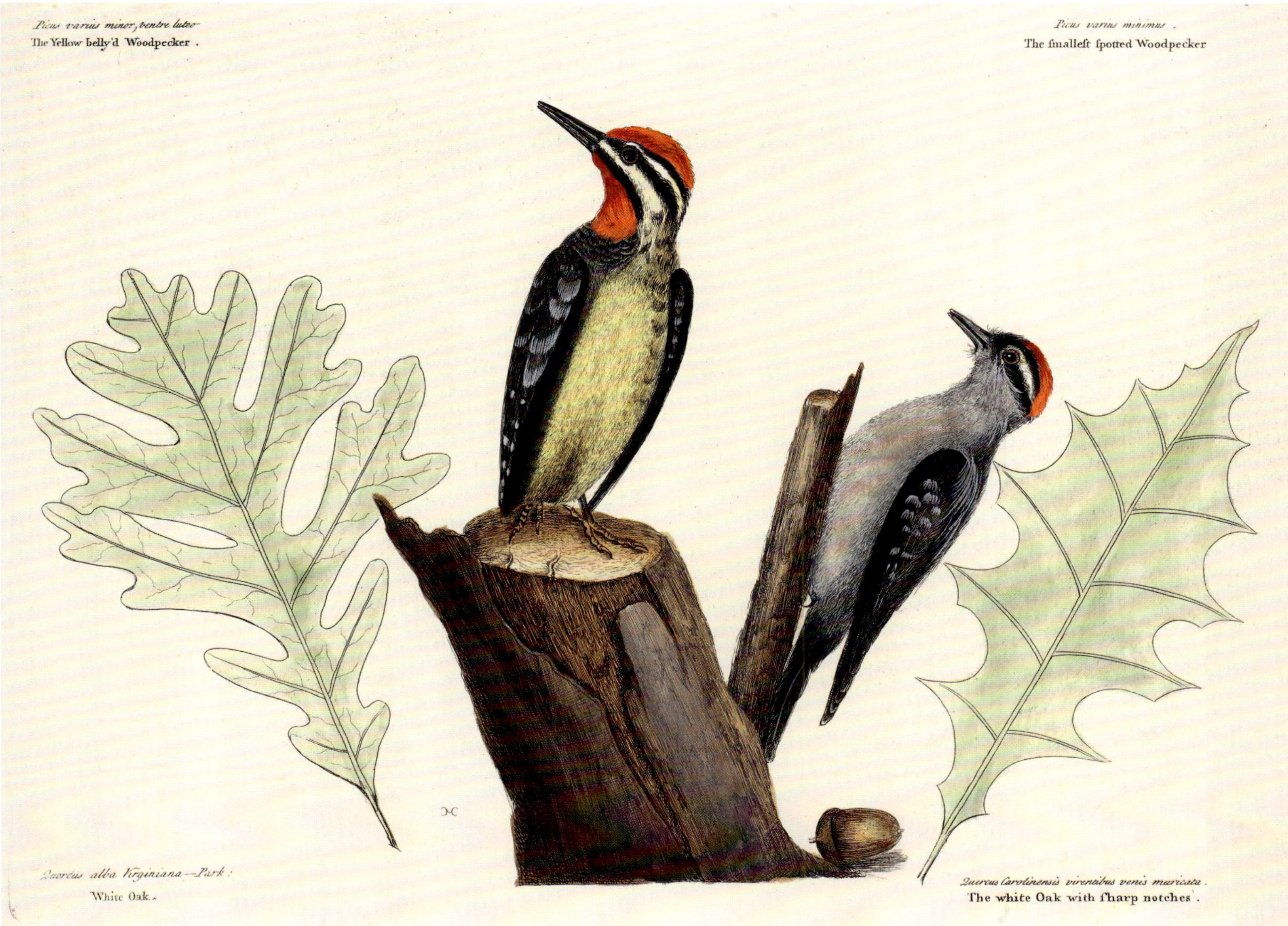

20 *left*
Red-headed woodpecker (*Melanerpes erythrocephalus*); water oak (*Quercus nigra*); partridge berry (*Mitchella repens*; around trunk base) (vol. 1, t. 20)

21 *above*
Yellow-bellied sapsucker (*Sphyrapicus varius*; left); downy woodpecker (*Picoides pubescens*; right); white oak (*Quercus alba*; left leaf plus acorn); southern red oak (*Quercus falcata*, right leaf) (vol. 1, t. 21)

T. 22.

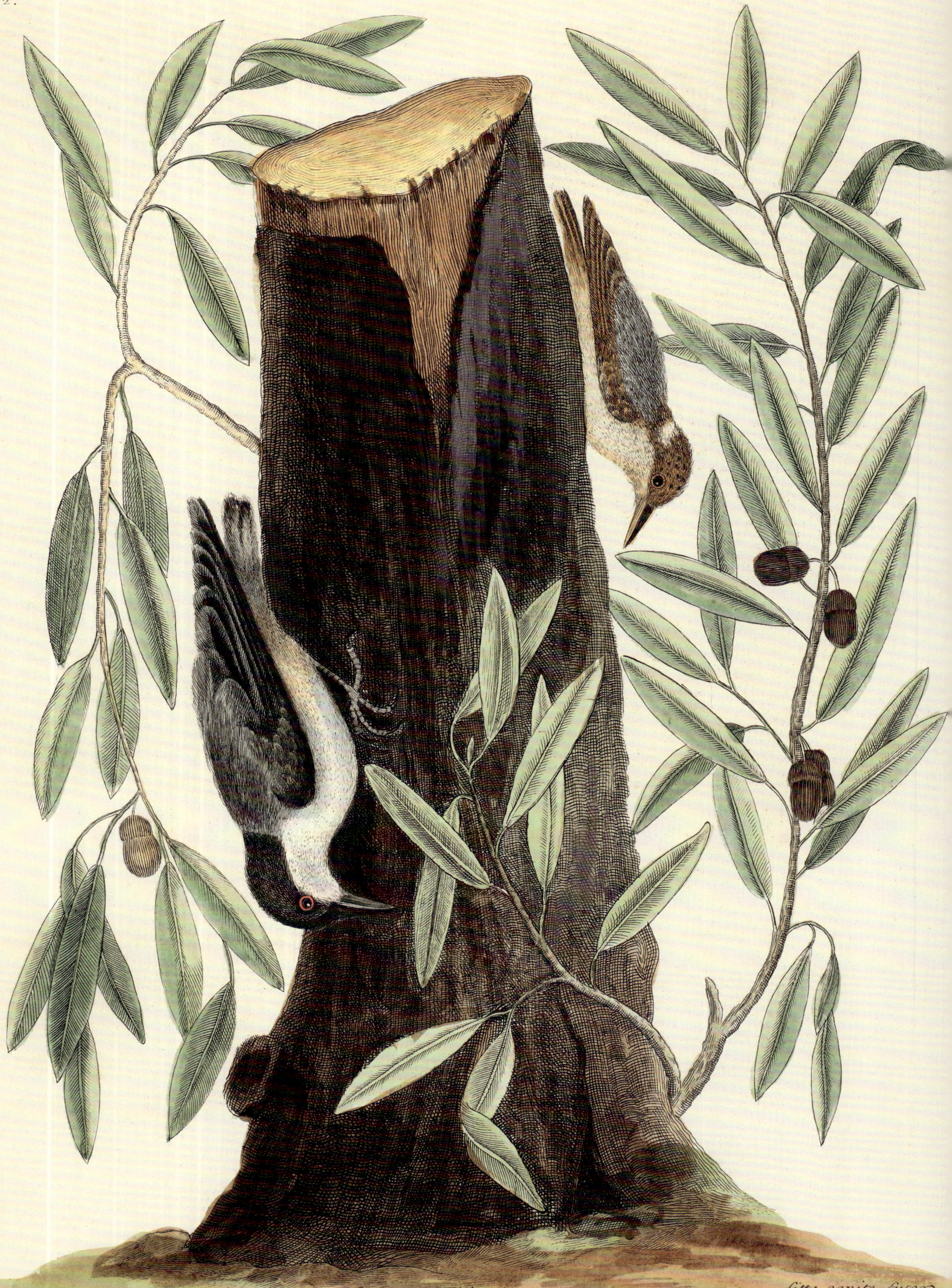

Sitta capite nigro. The Nuthatch.

Quercus humilior &c. The highland Willow Oak.

Sitta capite fusco. The Small Nuthatch.

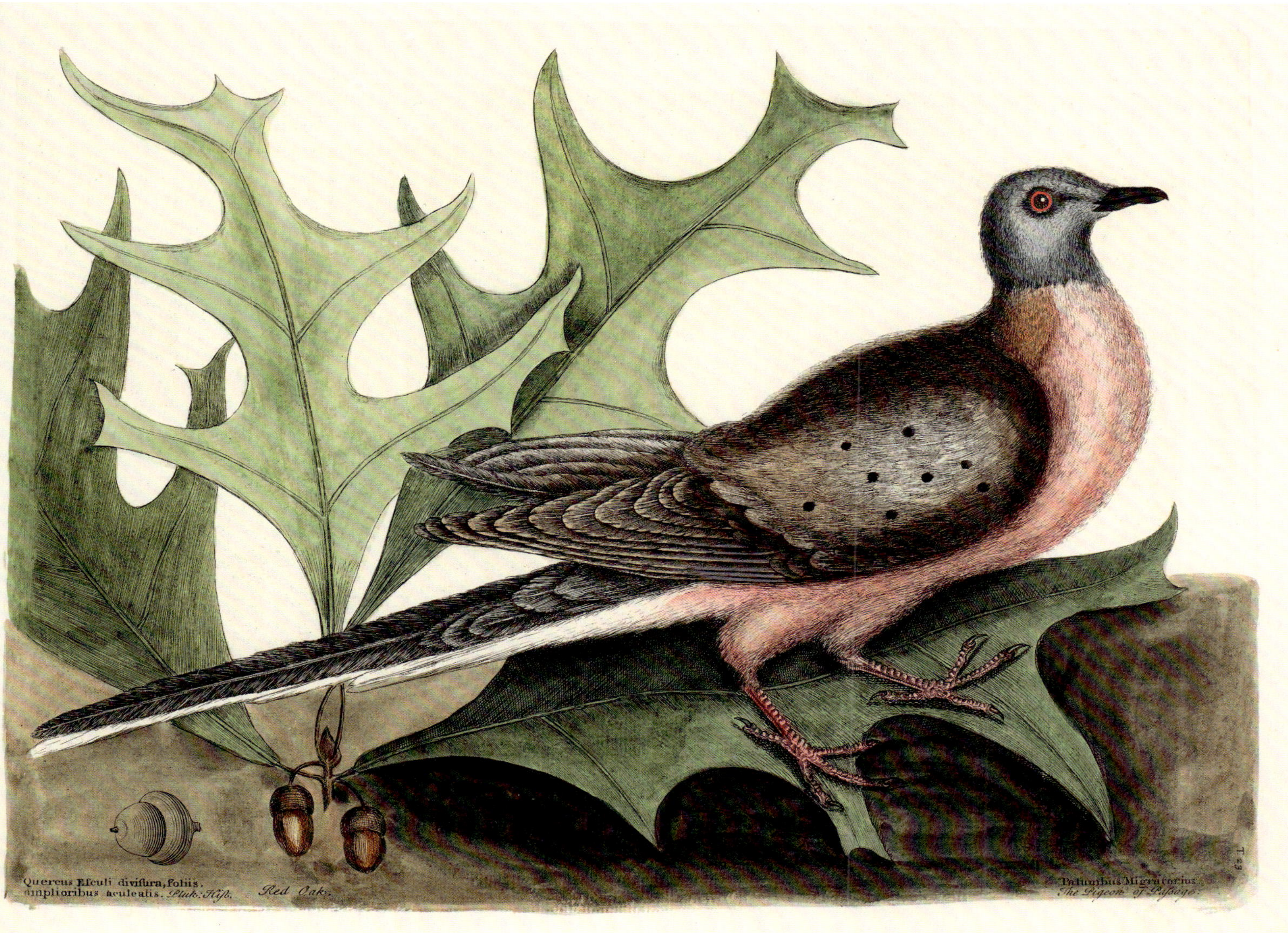

22 *left*
Bluejack oak (*Quercus incana*); white-breasted nuthatch (*Sitta carolinensis*; lower); brown-headed nuthatch (*Sitta pusilla*; upper) (vol. 1, t. 22)

23 *above*
Passenger pigeon (*Ectopistes migratorius*); American turkey oak (*Quercus laevis*) (vol. 1, t. 23)

T. 24.

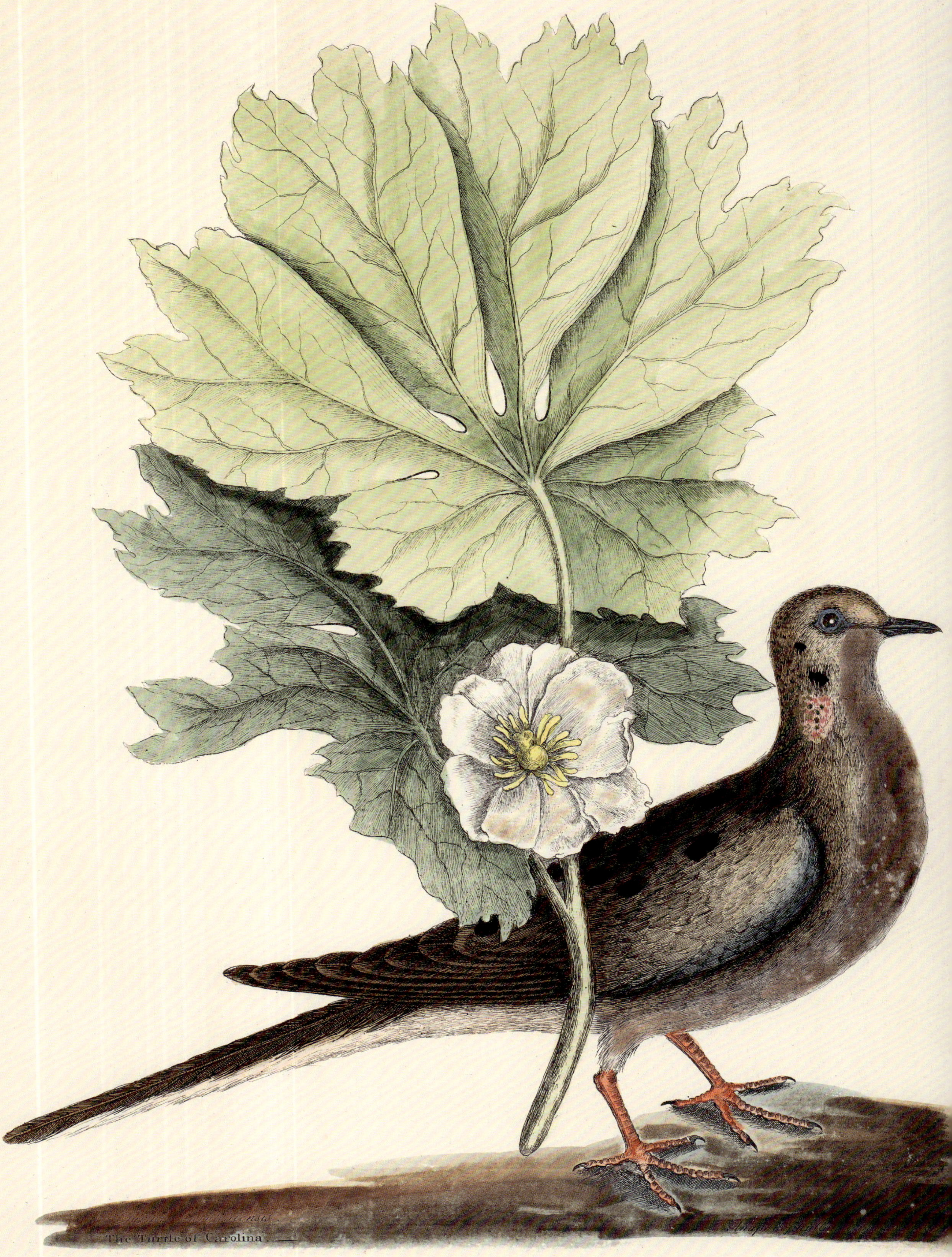

The Turtle of Carolina.

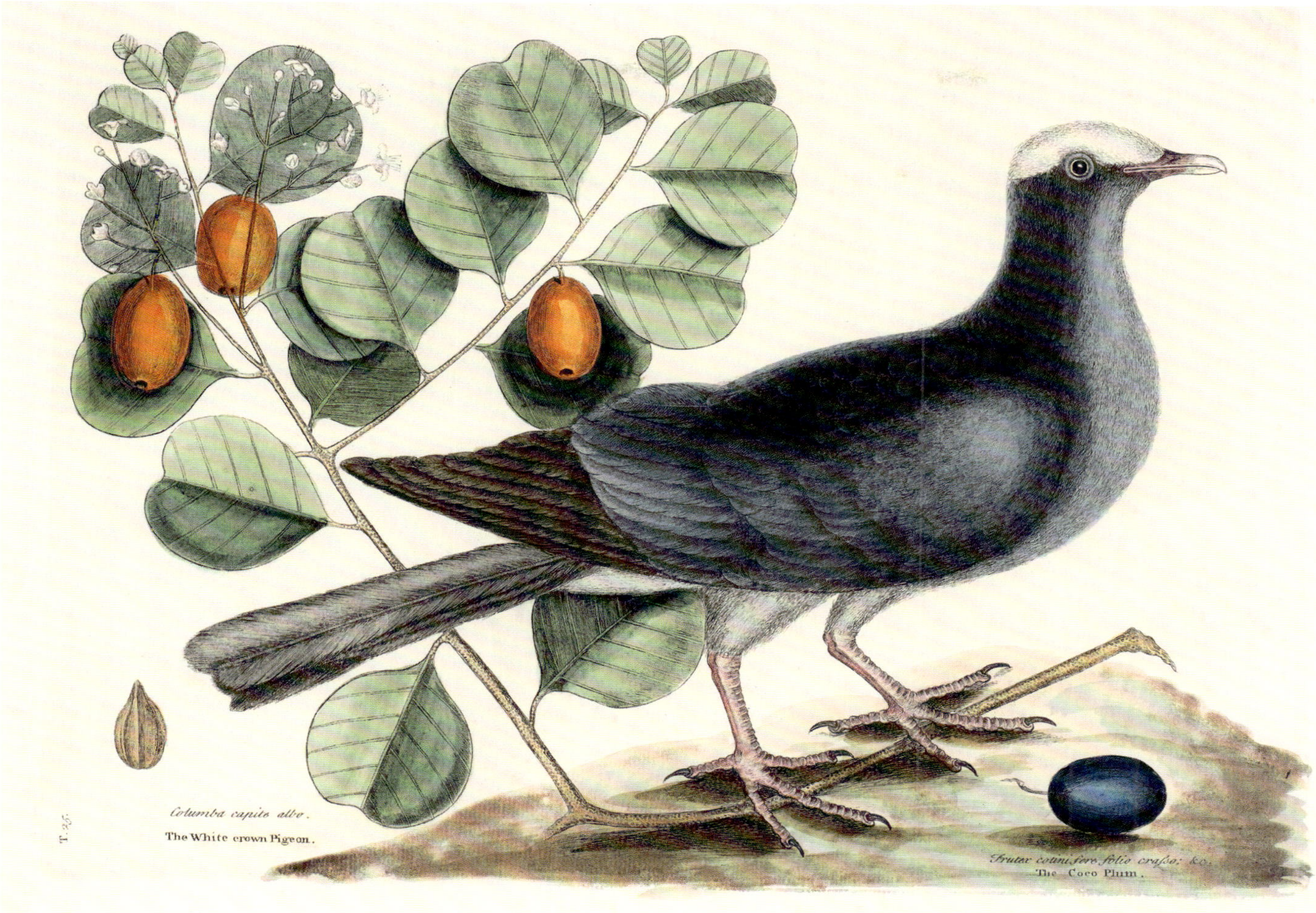

24 *left*
Mourning dove (*Zenaida macroura*); mayapple (*Podophyllum peltatum*) (vol. 1, t. 24)

25 *above*
White-crowned pigeon (*Patagioenas leucocephala*); coco plum (*Chrysobalanus icaco*) (vol. 1, t. 25)

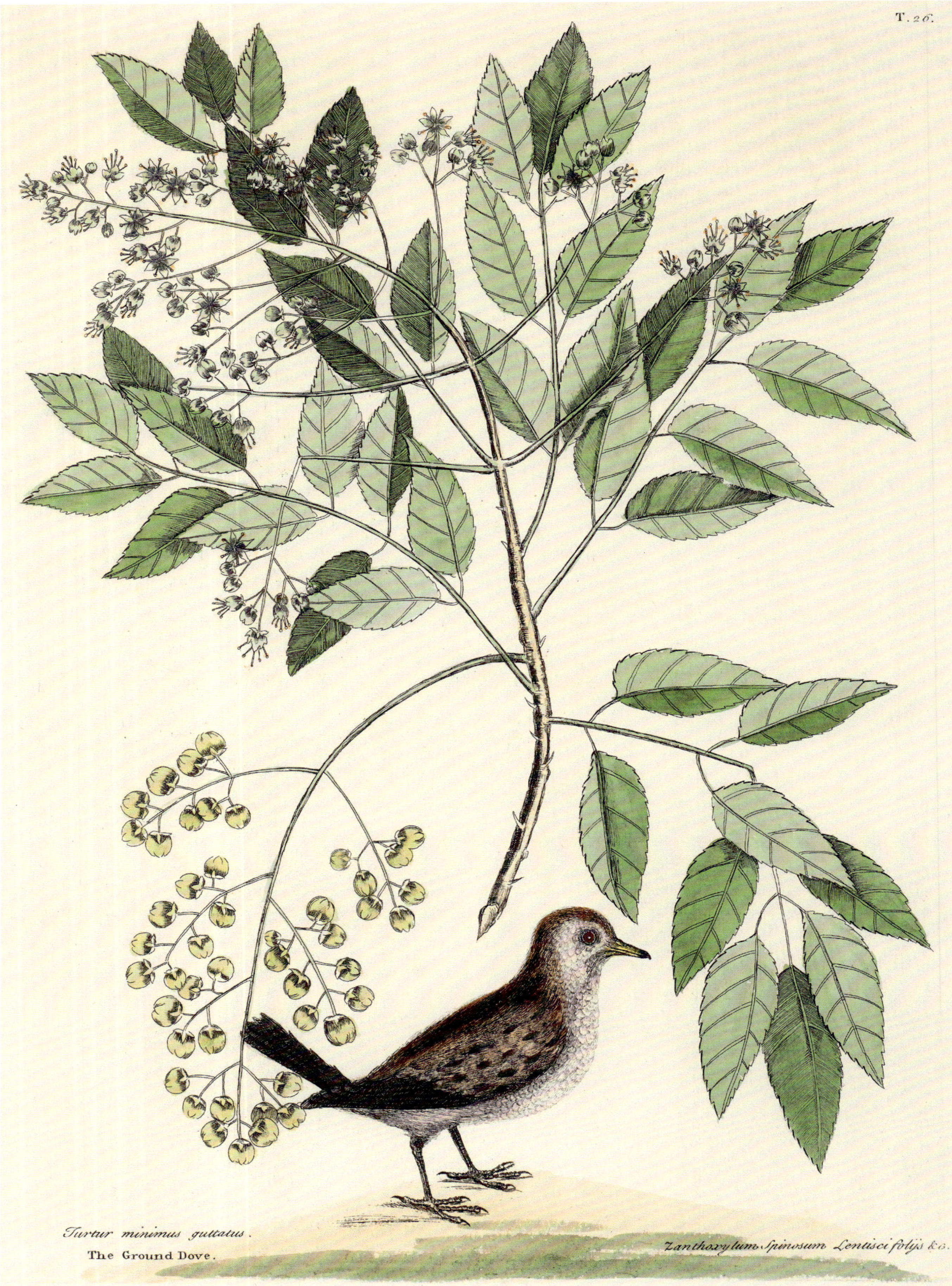
T. 26.
Turtur minimus guttatus.
The Ground Dove.
Zanthoxylum Spinosum Lentisci foliis &c.

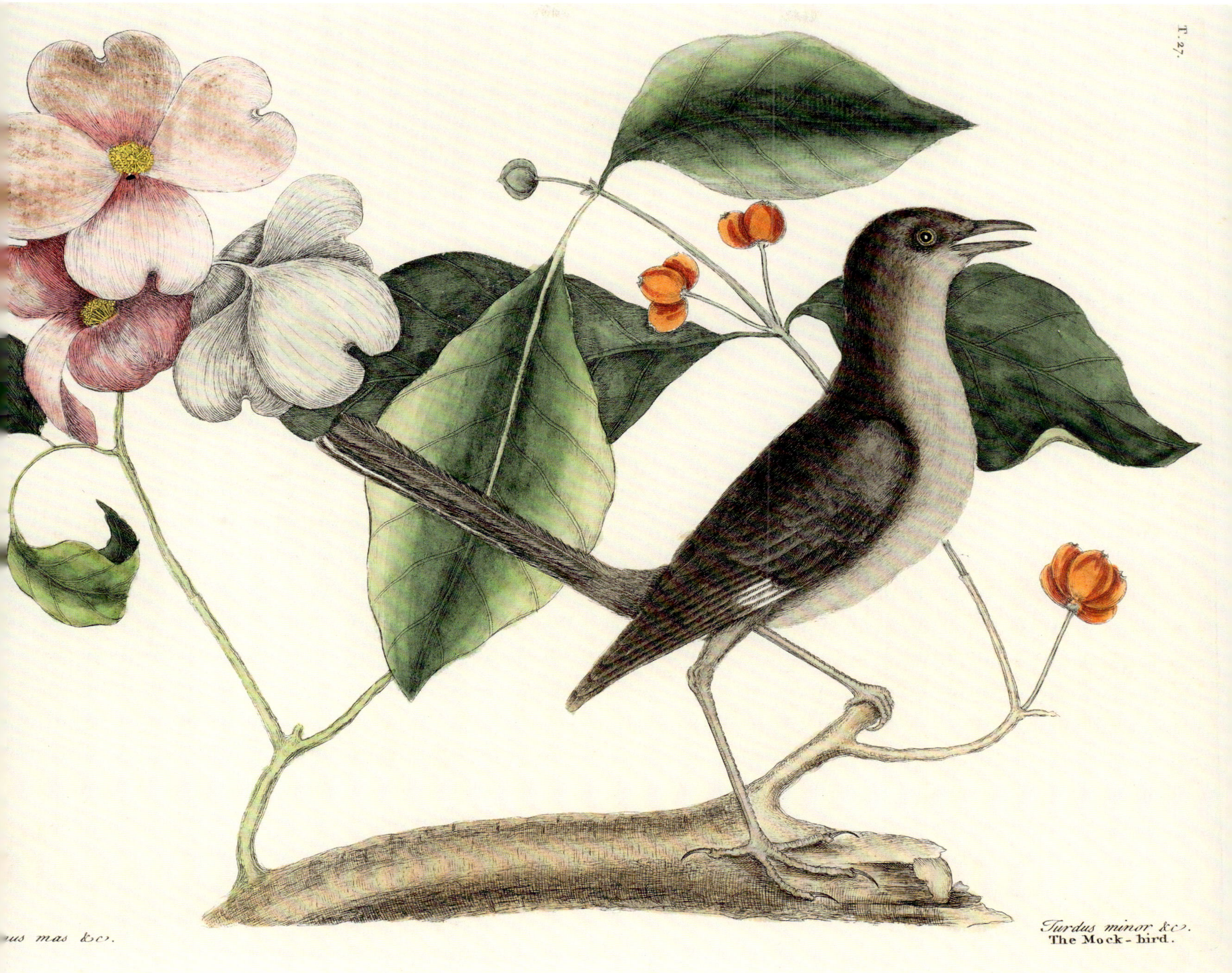

26 *left*
Common ground dove (*Columbina passerina*); Hercules' club (*Zanthoxylum clava-herculis*) (vol. 1, t. 26)

27 *above*
Northern mockingbird (*Mimus polyglottos*); flowering dogwood (*Cornus florida*) (vol. 1, t. 27)

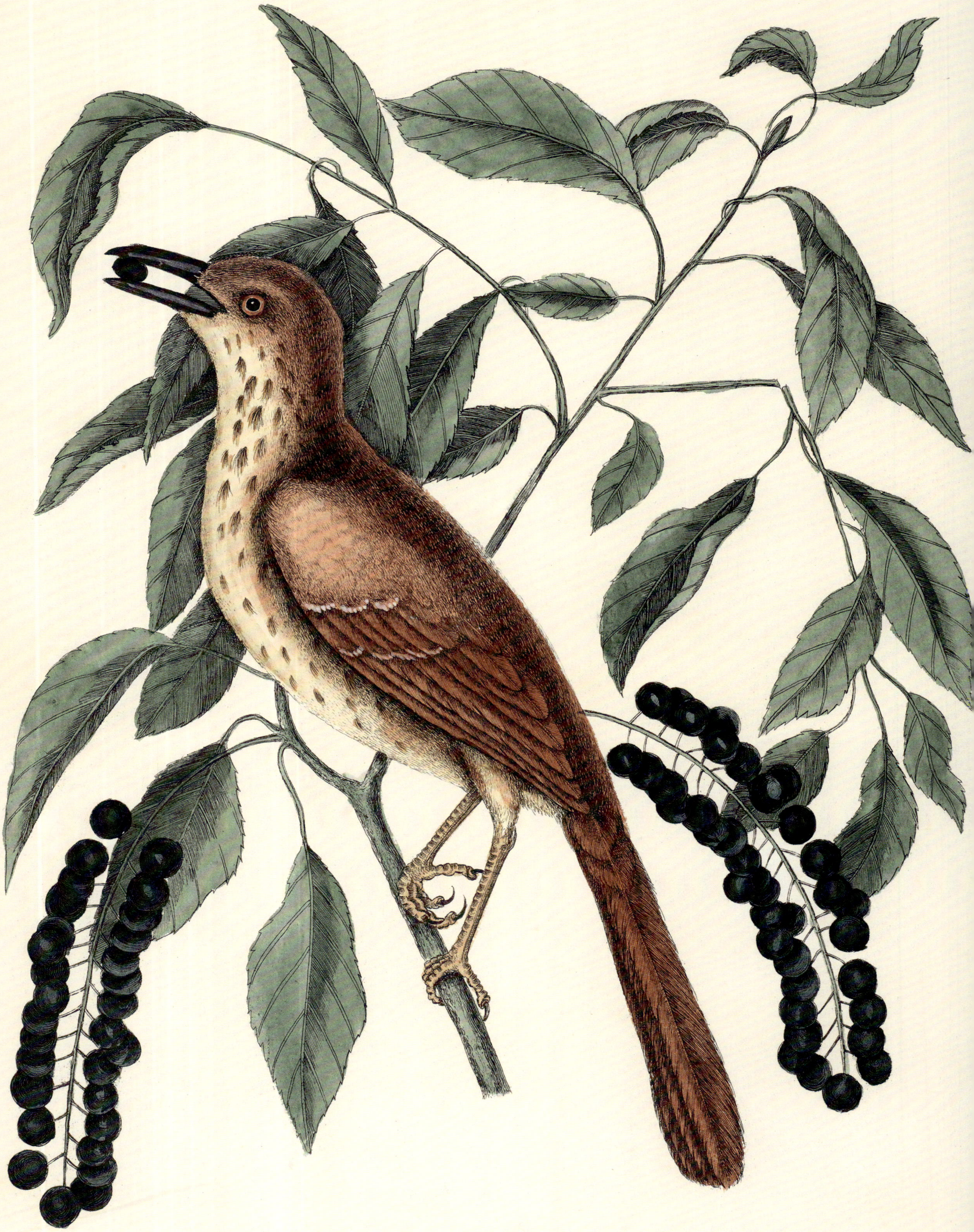

Turdus ruffus.

The fox coloured Thrush.

Cerafo Similis arbuscula &c

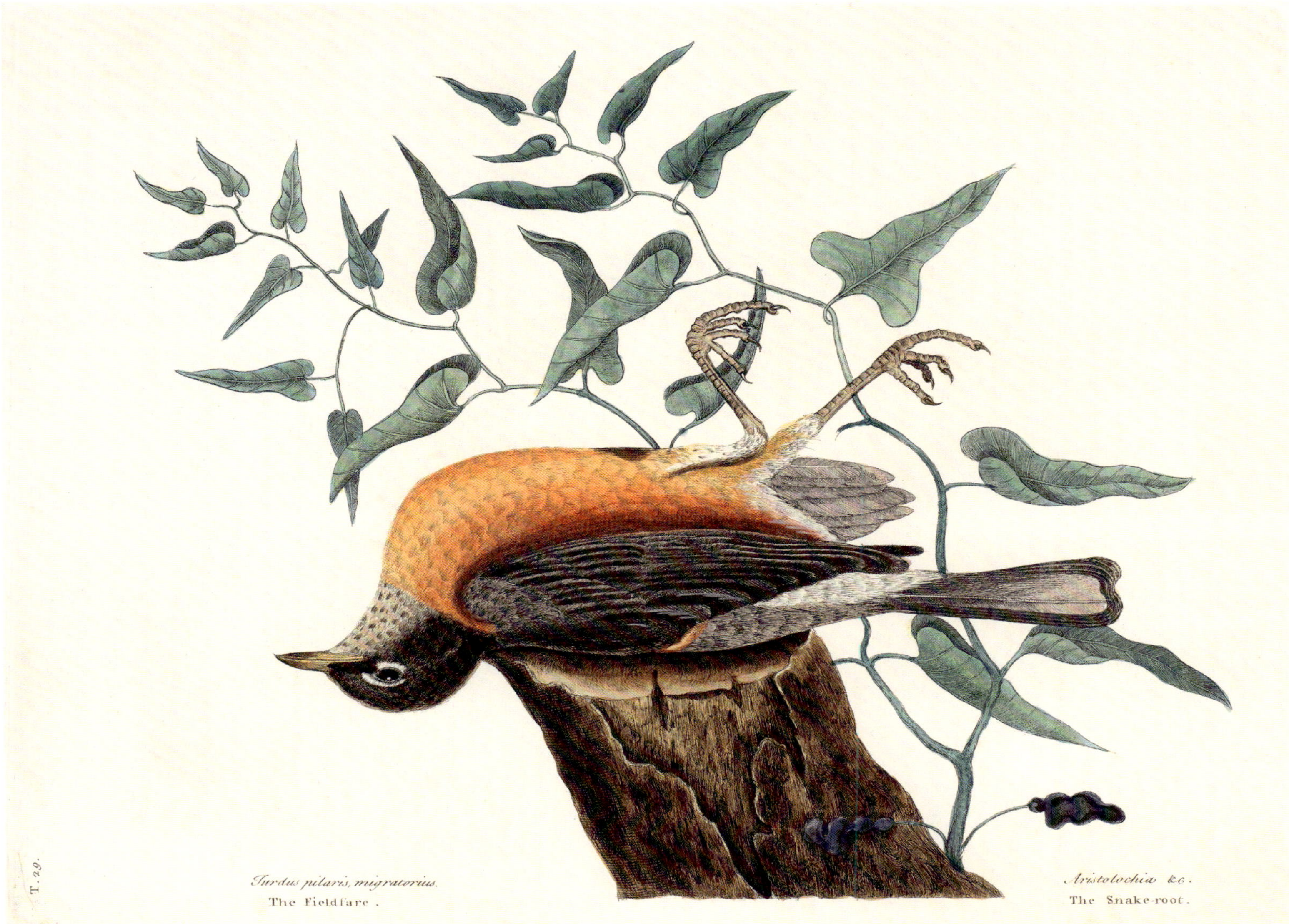

28 *left*
Brown trasher (*Toxostoma rufum*); probably black cherry (*Prunus serotina*) based on geographic distribution rather than features of illustration (vol. 1, t. 28)

29 *above*
American robin (*Turdus migratorius*); Virginia snakeroot (*Aristolochia serpentaria*) (vol. 1, t. 29)

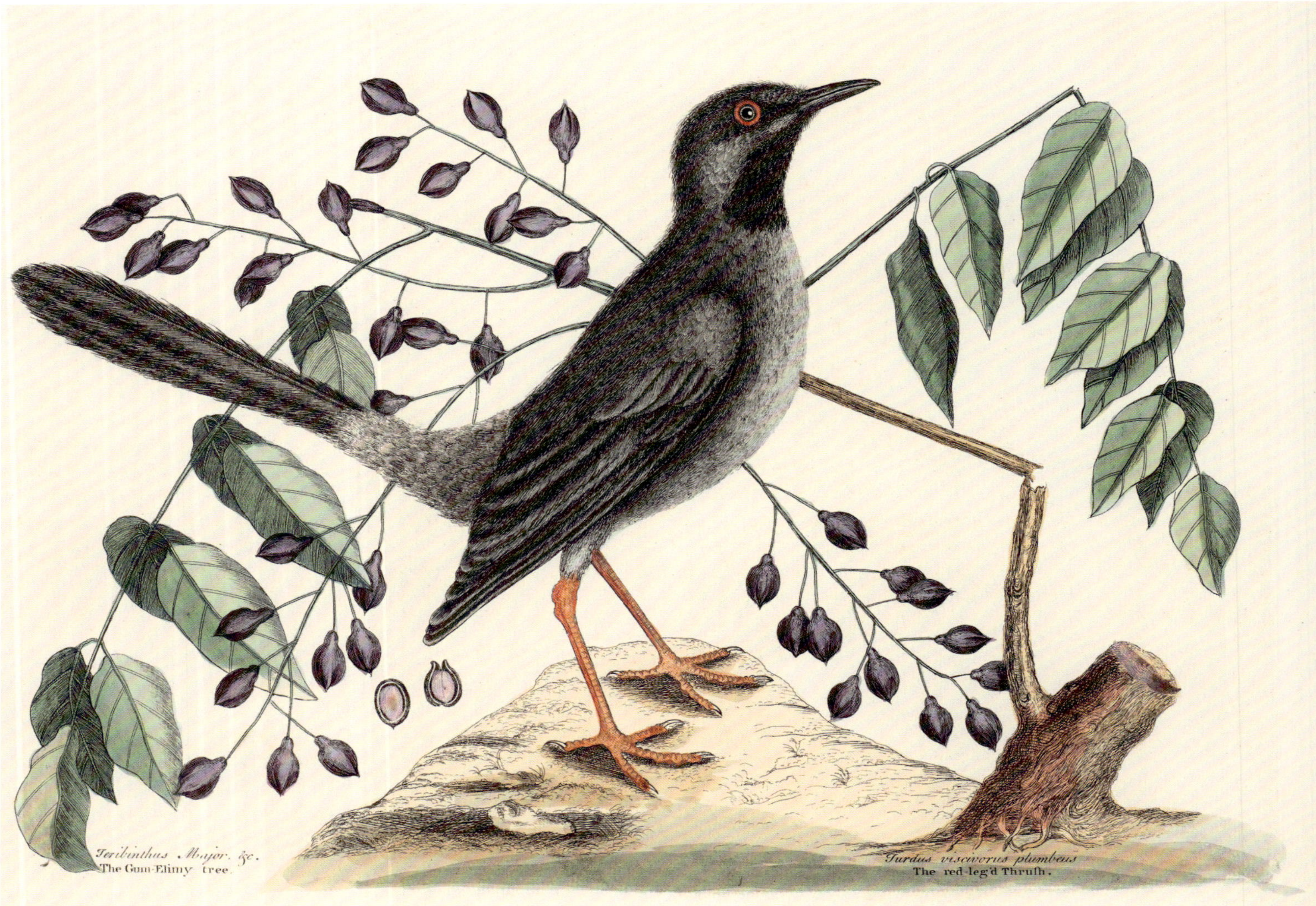

30 *above*

Red-legged thrush (*Turdus plumbeus*); gumbo limbo (*Bursera simaruba*) (vol. 1, t. 30)

31 *right*

Dahoon (*Ilex cassine*); probably hermit thrush (*Catharus guttatus*) (vol. 1, t. 31)

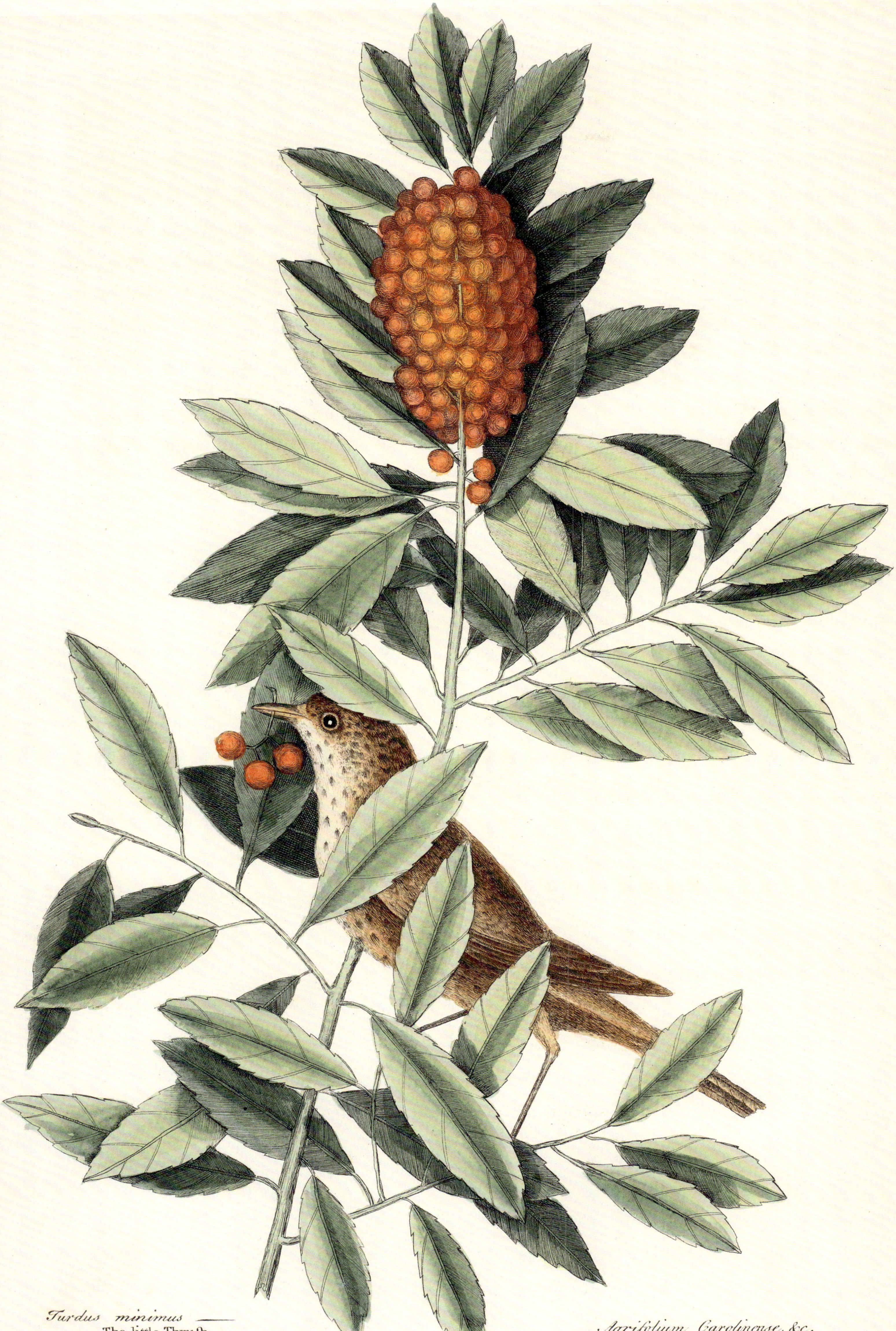

Turdus minimus —
The little Thrush.

Agrifolium Carolinense, &c.
The Dahoon Holly.

T.

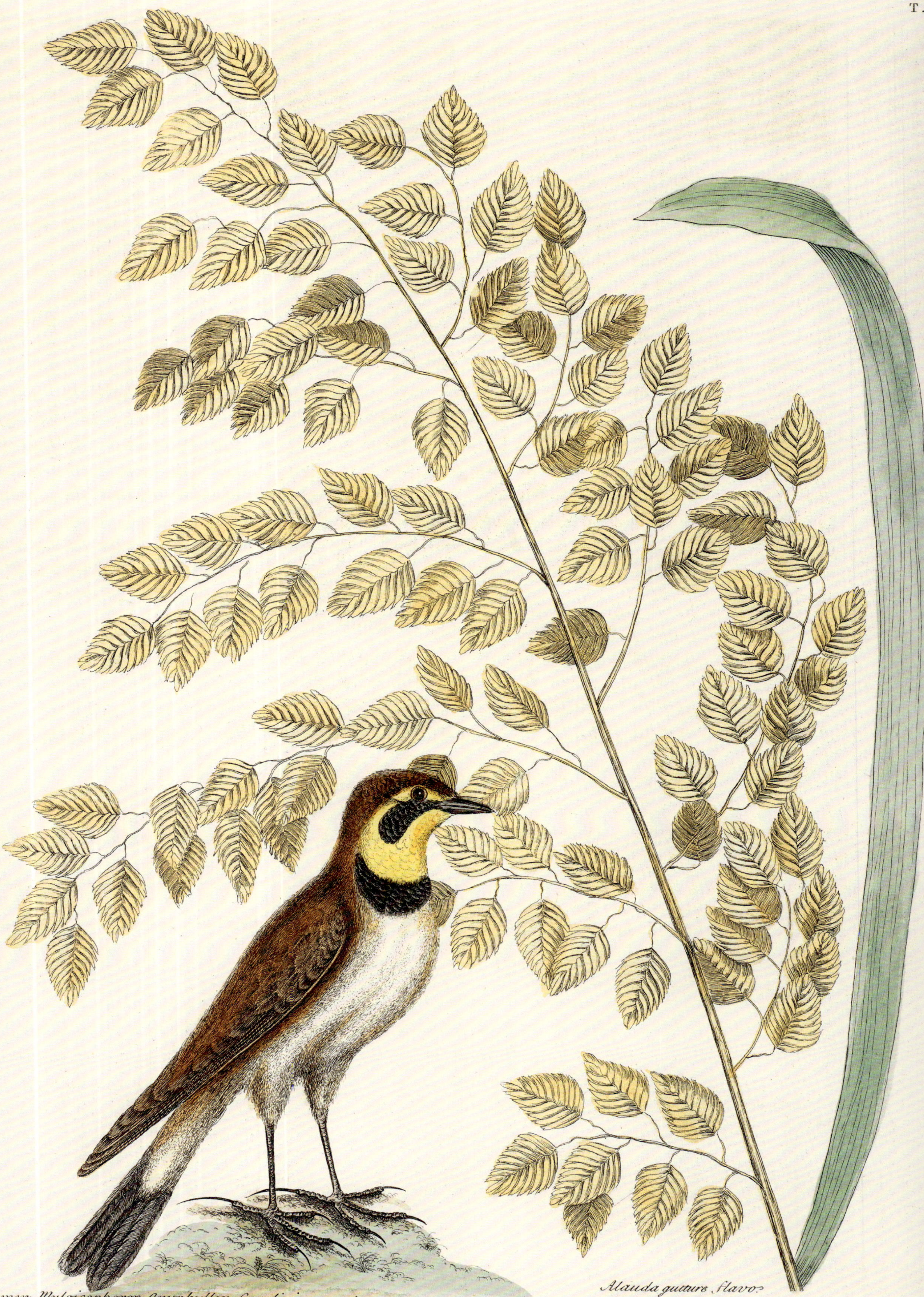

Gramen Myloicophoron Oxyphyllon Carolinianum &c.
Pluk: Almag: *p. 173.*

Alauda gutture flavo.
The Lark.

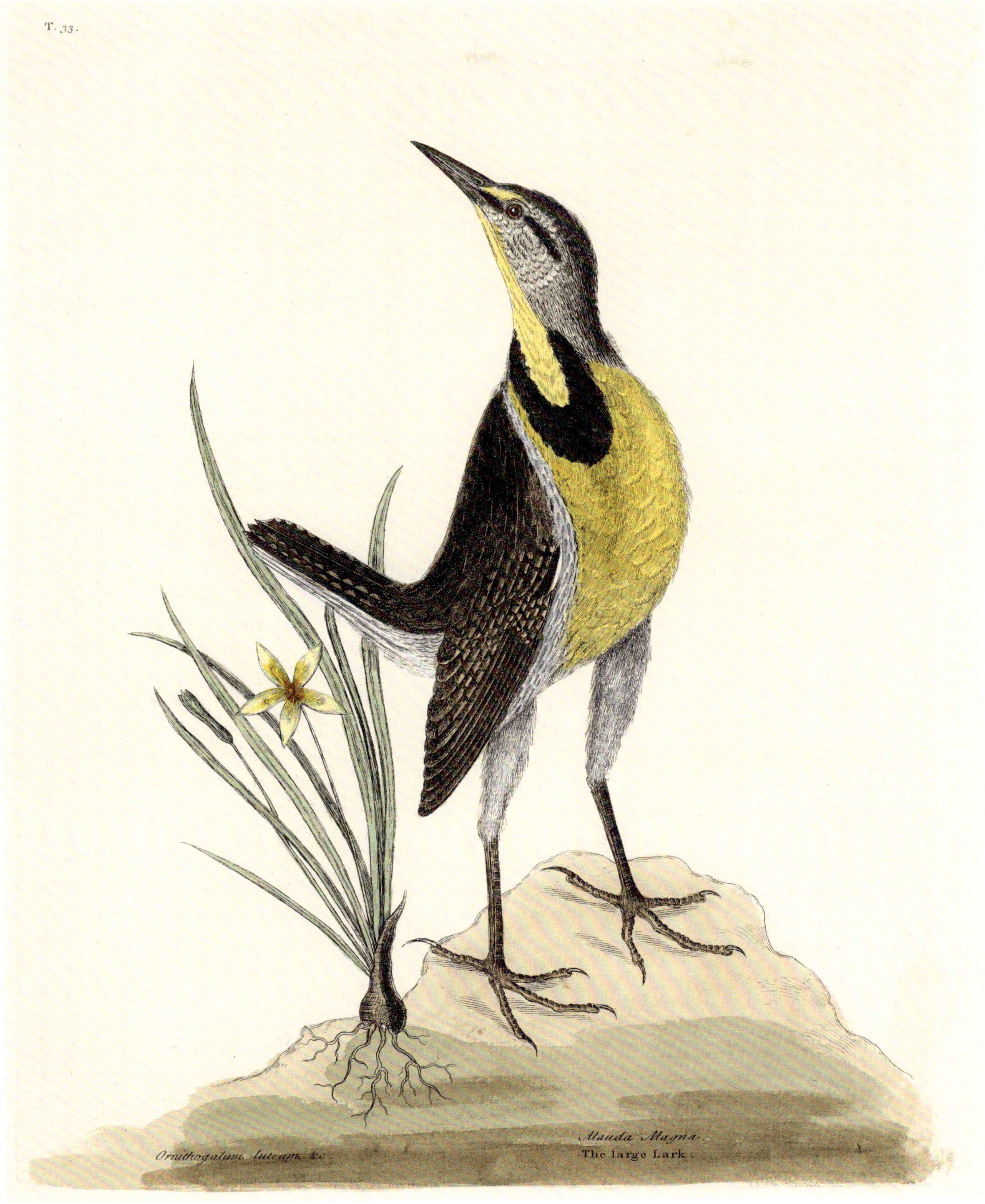

32 *left*

Horned lark (*Eremophila alpestris*); sea oats (*Uniola paniculata*) (vol. 1, t. 32)

33 *above*

Eastern meadow lark (*Sturnella magna*); common goldenstar (*Hypoxis hirsuta*) (vol. 1, t. 33)

T. 34.

Populus nigra folio maximo &c.

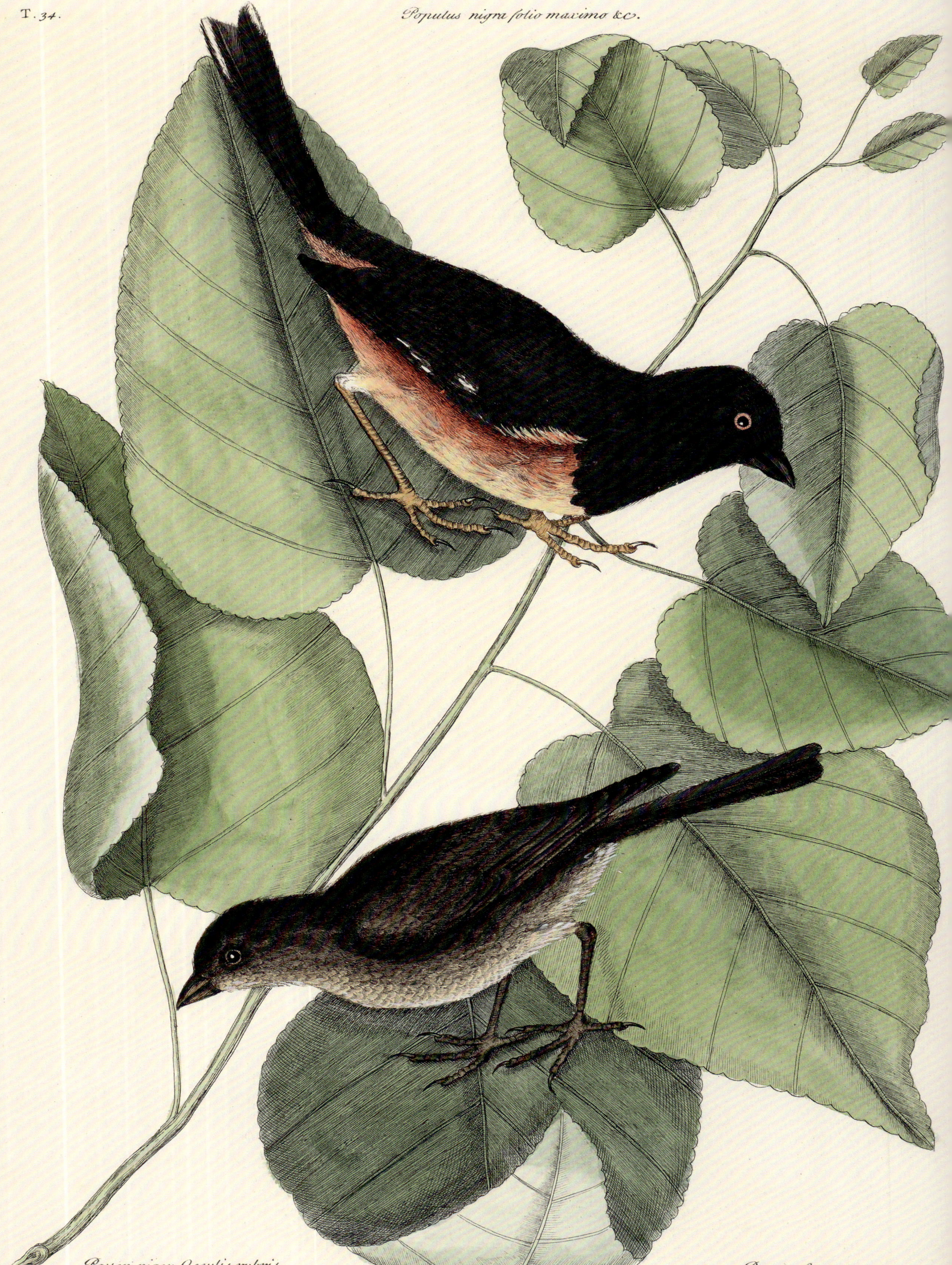

Passer niger, Occulis rubris.
The To-whe Bird.

Passer fusca.
The cow-pen Bird.

T. 35.

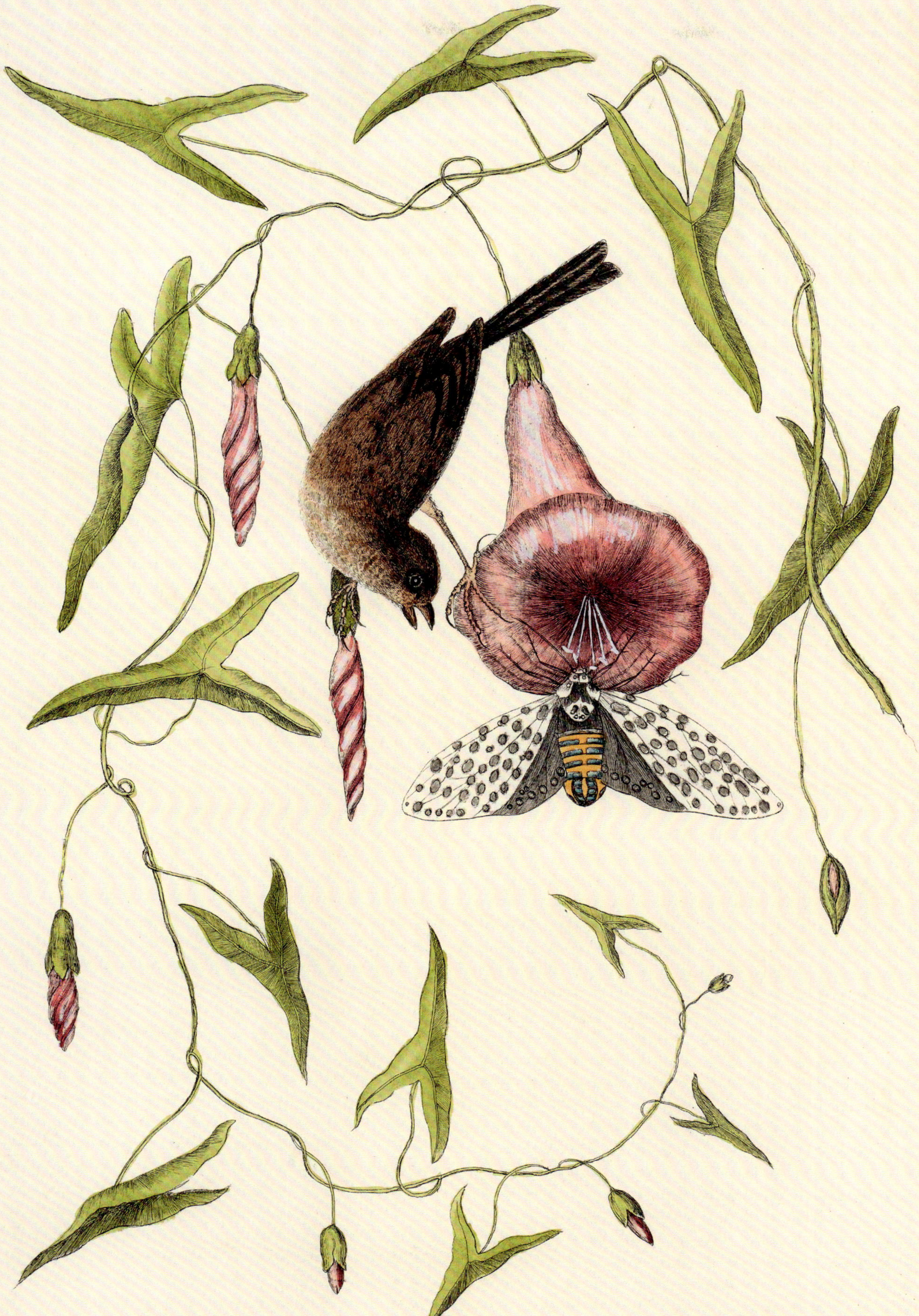

Passerculus.
The little Sparrow.

Convolvulus Carolinensis &c.

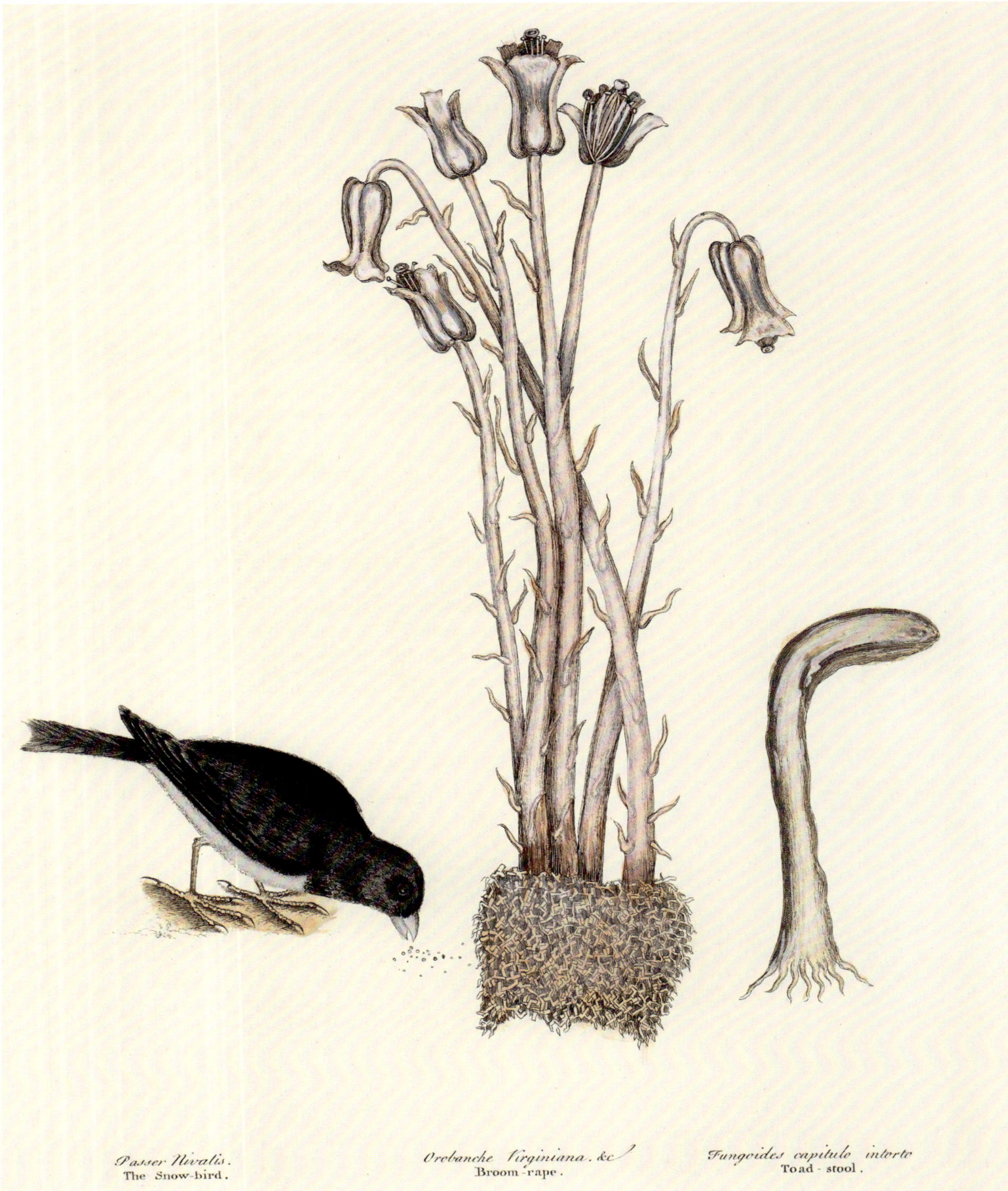

previous pages

34 *left*

Brown-headed cowbird (*Molothrus ater*; lower); eastern towhee (*Pipilo erythrophthalmus*; upper); swamp cottonwood (*Populus heterophylla*) (vol. 1, t. 34 January 1730)

35 *right*

Chipping sparrow (*Spizella passerina*); saltmarsh morning glory (*Ipomoea sagittata*); giant leopard moth (*Hypercompe scribonia*) (vol. 1, t. 35)

36 *above*

Dark-eyed junco (*Junco hyemalis*; left); one-flower Indian pipe (*Monotropa uniflora*; middle); black adder-tongue fungus (*Geoglossum glabrum*; right) (vol. 1, t. 36)

37 *right*

Black-faced grassquit (*Tiaris bicolor*); beef bush (*Tabebuia bahamensis*) (vol. 1, t. 37)

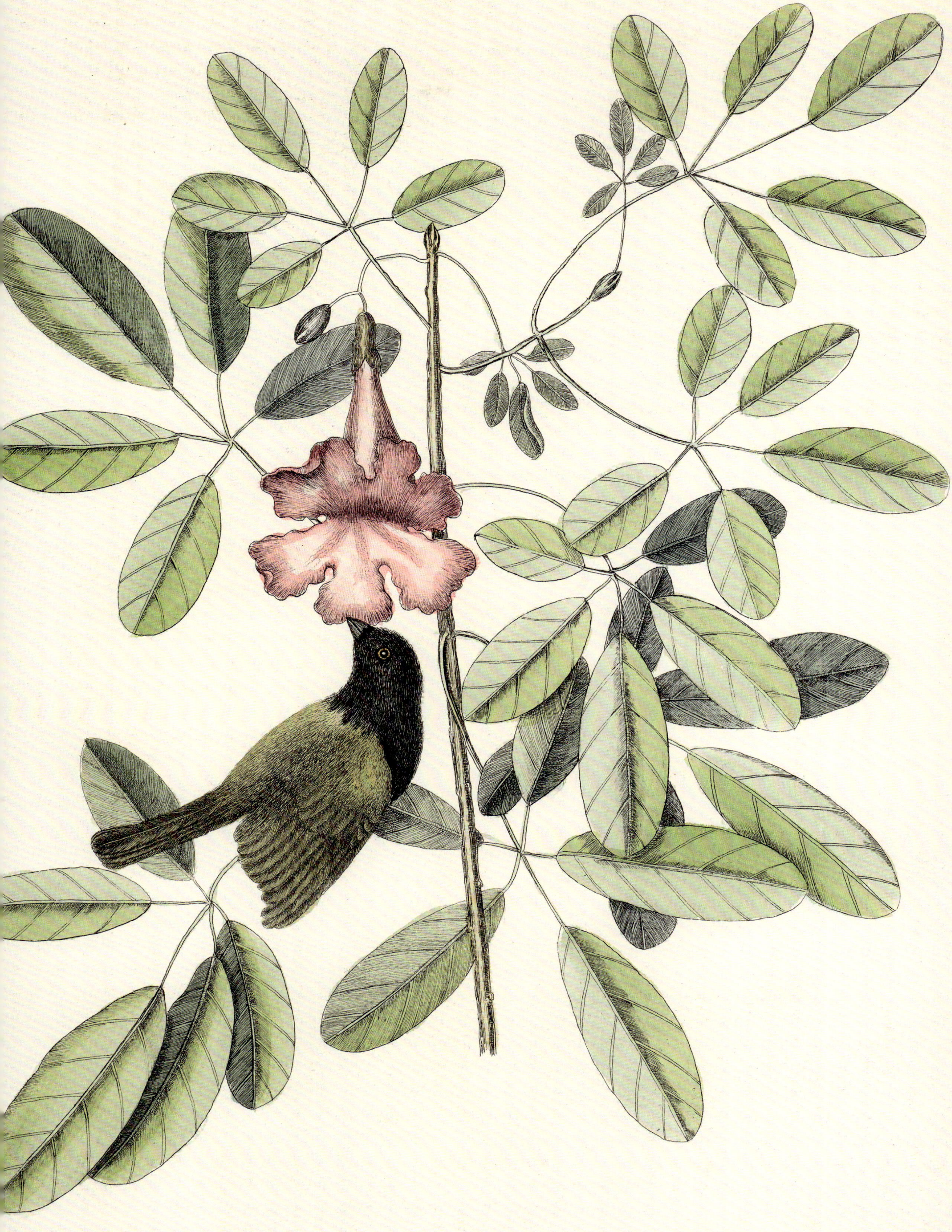

Passerculus bicolor Bahamiensis.

The Bahama Sparrow

Bignonia Arbor Pentaphylla. &c.

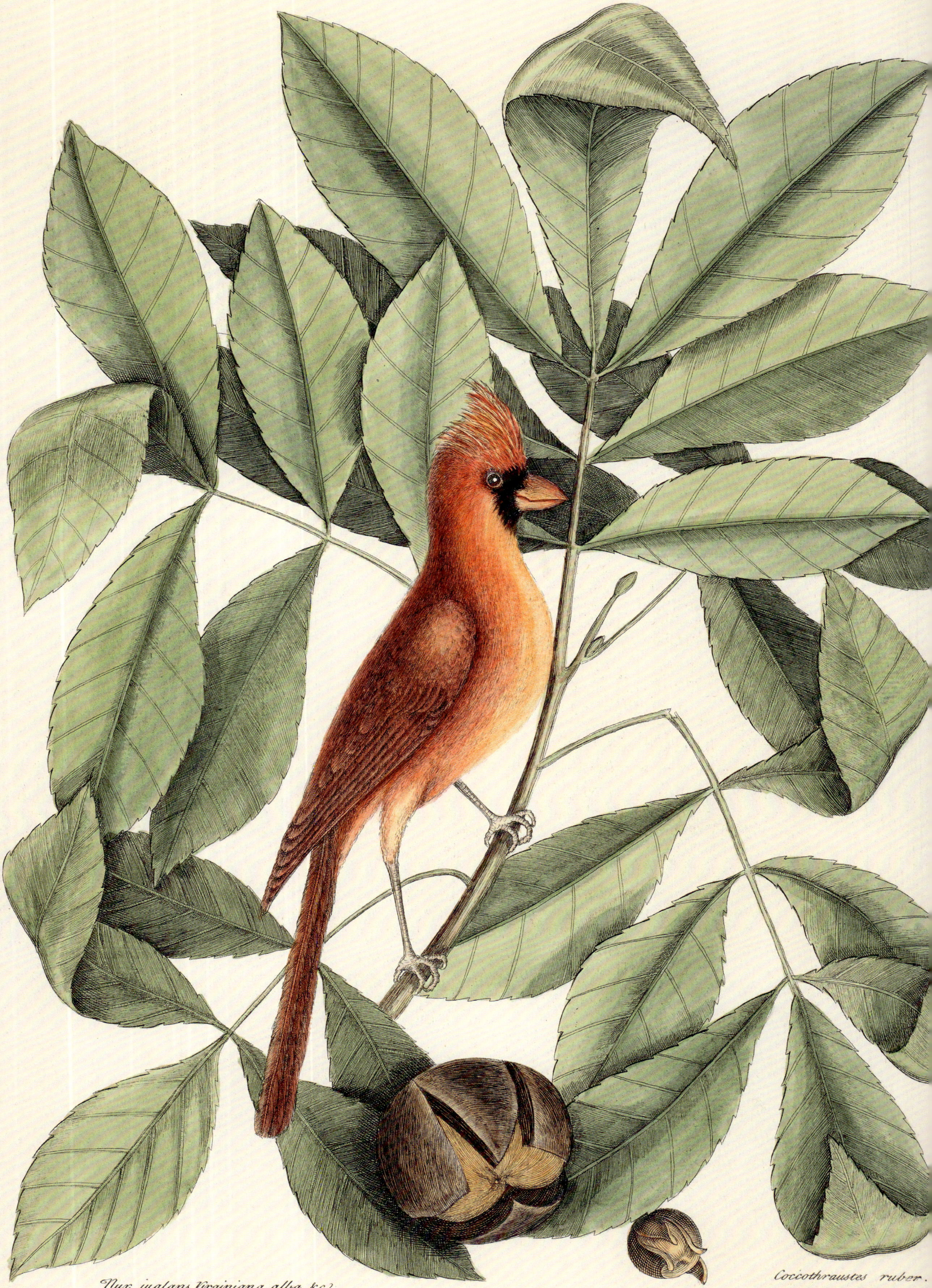

Nux juglans Virginiana alba &c.
The Hiccory Tree.

The Pig-nut.

Coccothraustes ruber.
The red Bird.

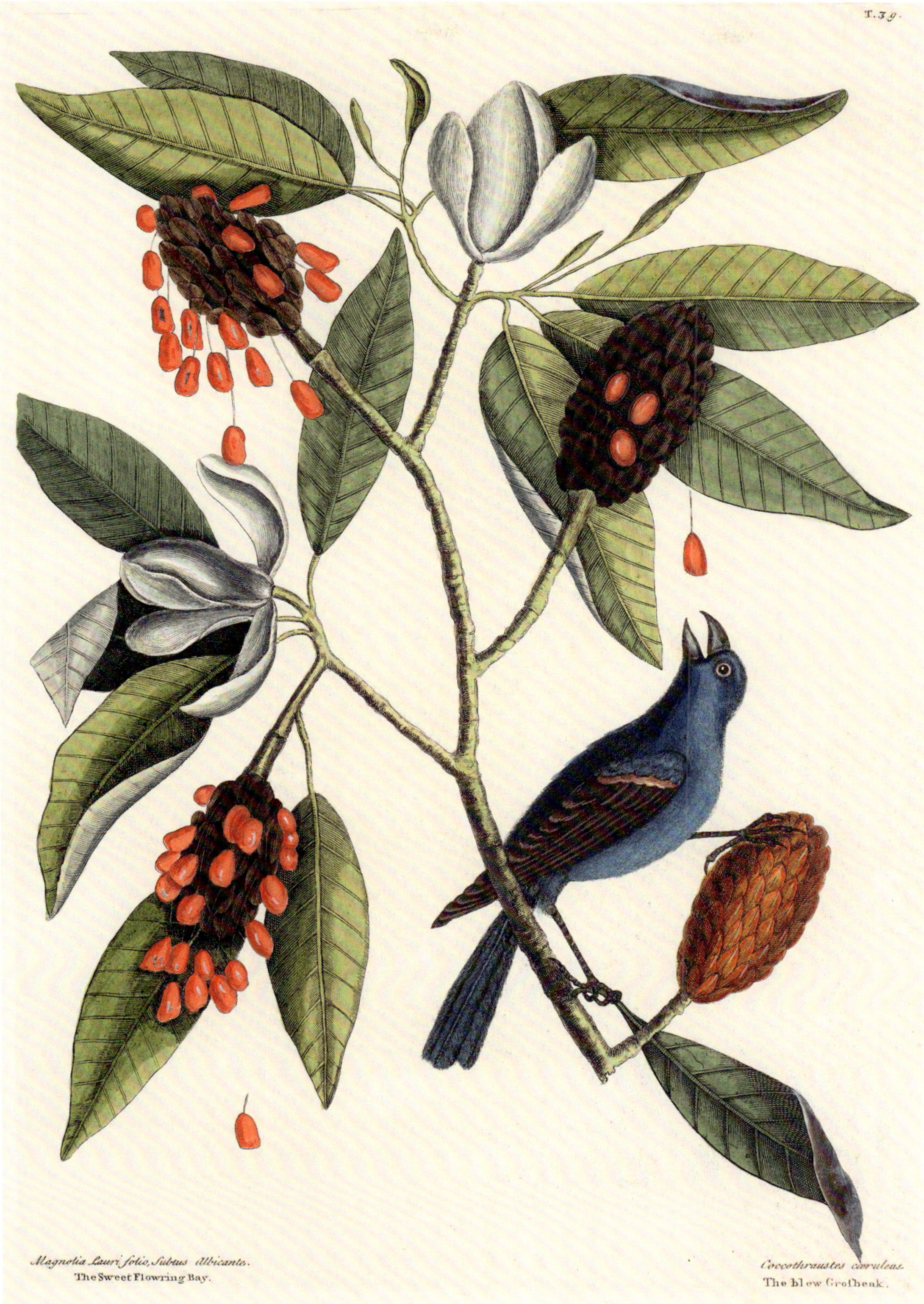

38 *left*
Northern cardinal (*Cardinalis cardinalis*); probably mockernut hickory (*Carya tomentosa*; shoot and large nut); probably pignut hickory (*Carya glabra*; small nut) (vol. 1, t. 38)

39 *above*
Blue grosbeak (*Passerina caerulea*); sweetbay magnolia (*Magnolia virginiana*) (vol. 1, t. 39)

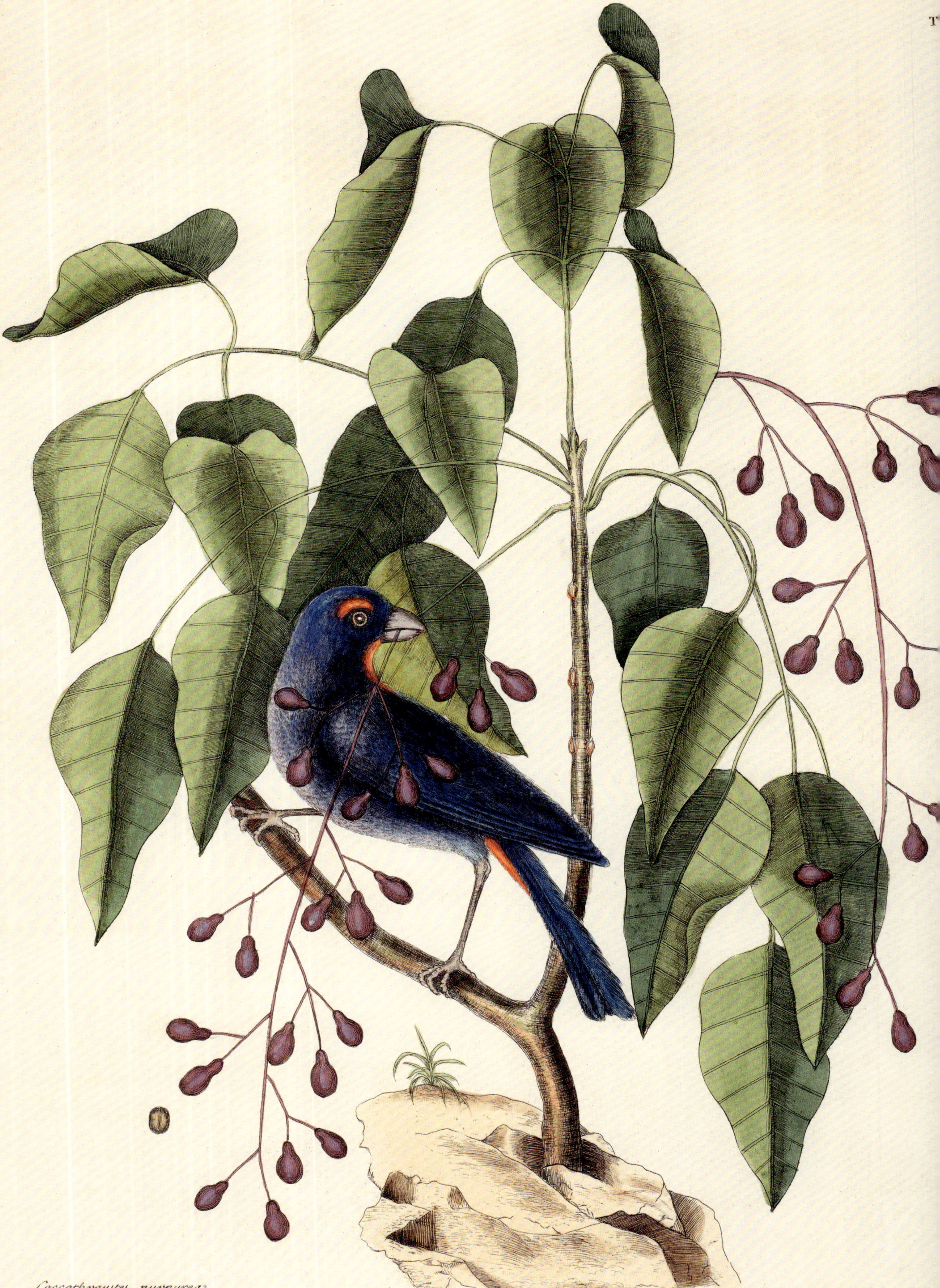

Coccothraustes purpurea.
The purple Grofs-beak.

Toxicodendron folijs alatis

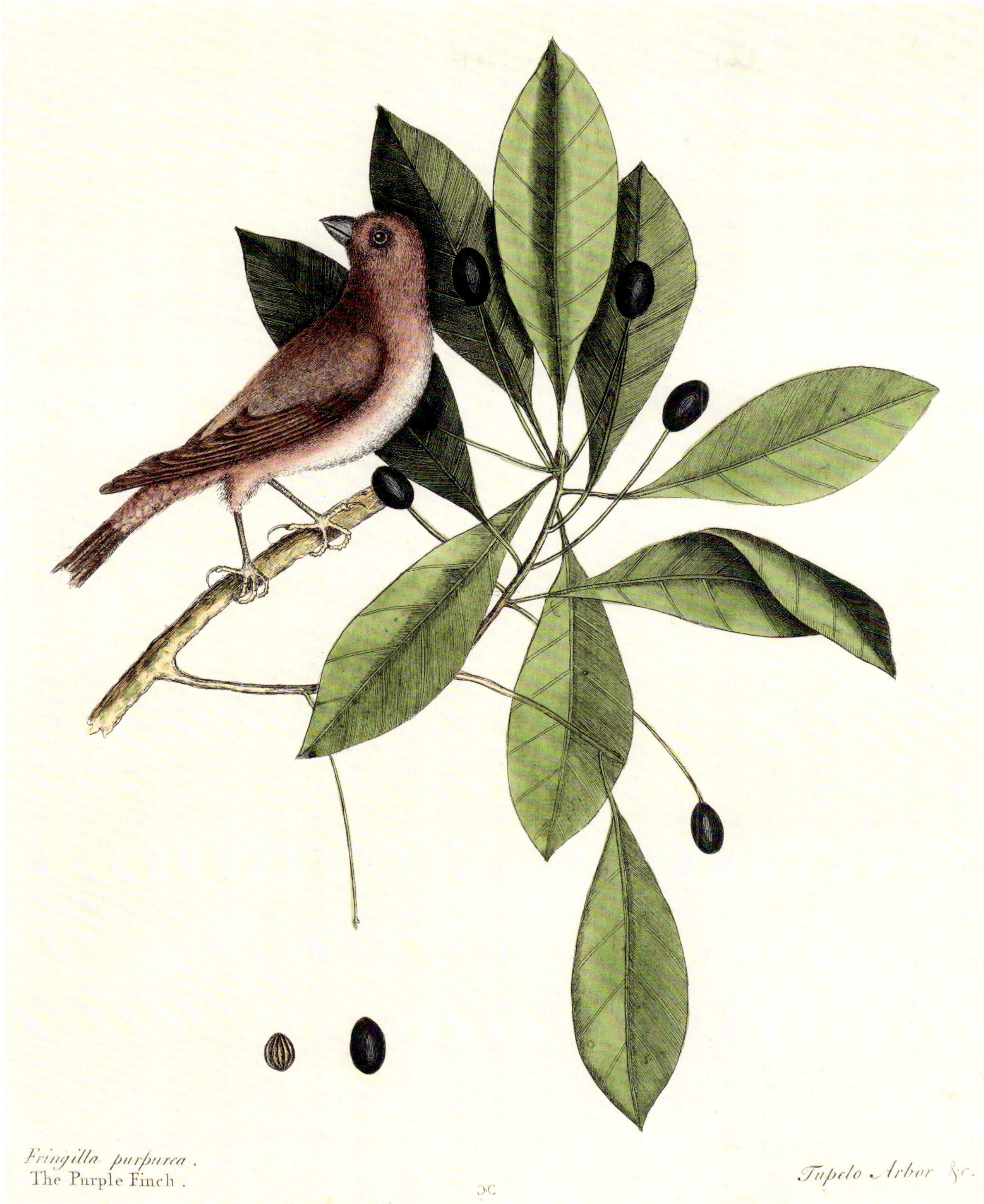

40 *left*
Greater Antillean bullfinch (*Loxigilla violacea*); poisonwood (*Metopium toxiferum*) (vol. 1, t. 40)

41 *above*
Purple finch (*Carpodacus purpureus*); black tupelo (*Nyssa sylvatica*) (vol. 1, t. 41)

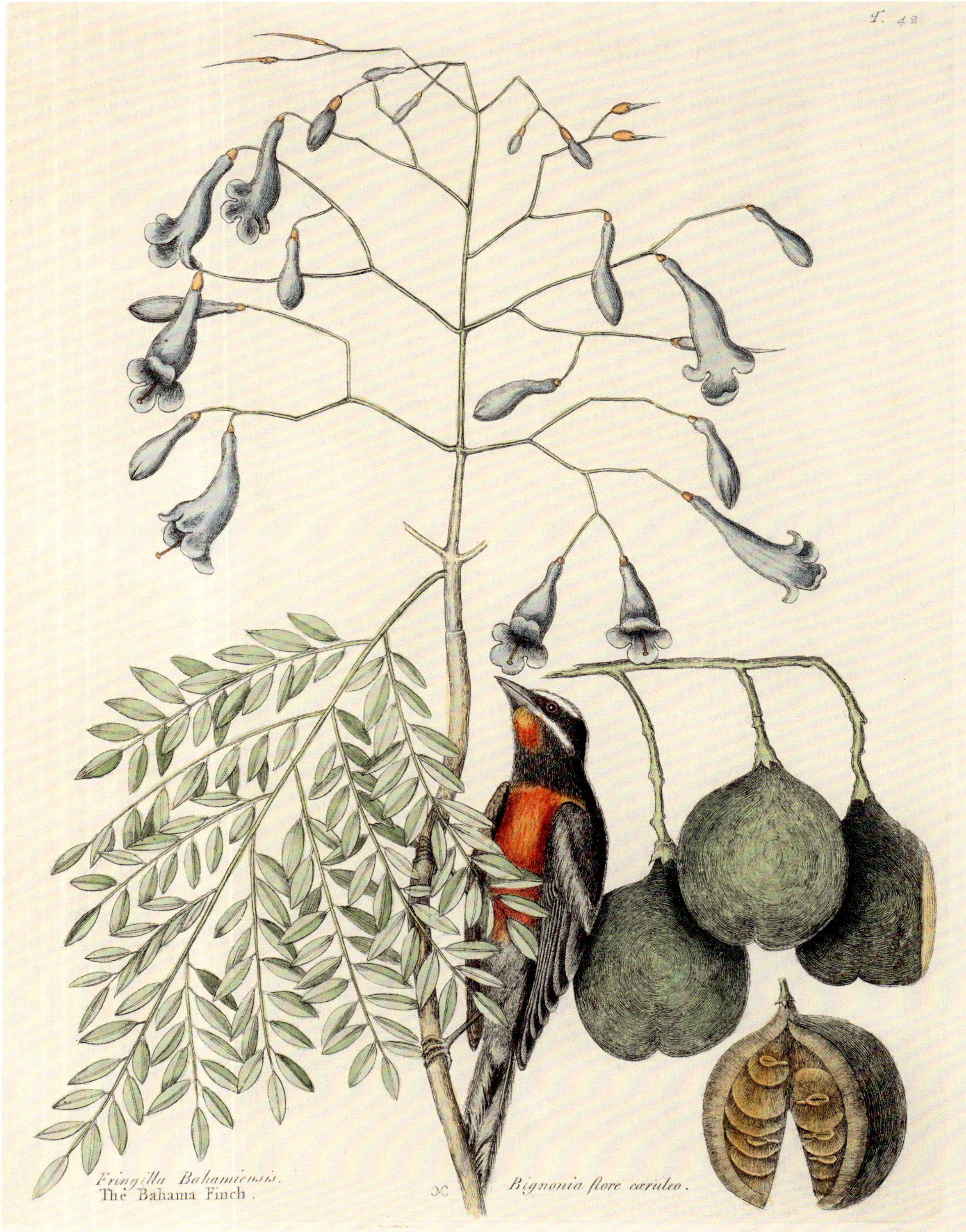

42 *above*
Stripe-headed tanager (*Spindalis zena*); boxwood (*Jacaranda caerulea*) (vol. 1, t. 42)

43 *right*
American goldfinch (*Carduelis tristis*); water locust (*Gleditsia aquatica*) (vol. 1, t. 43)

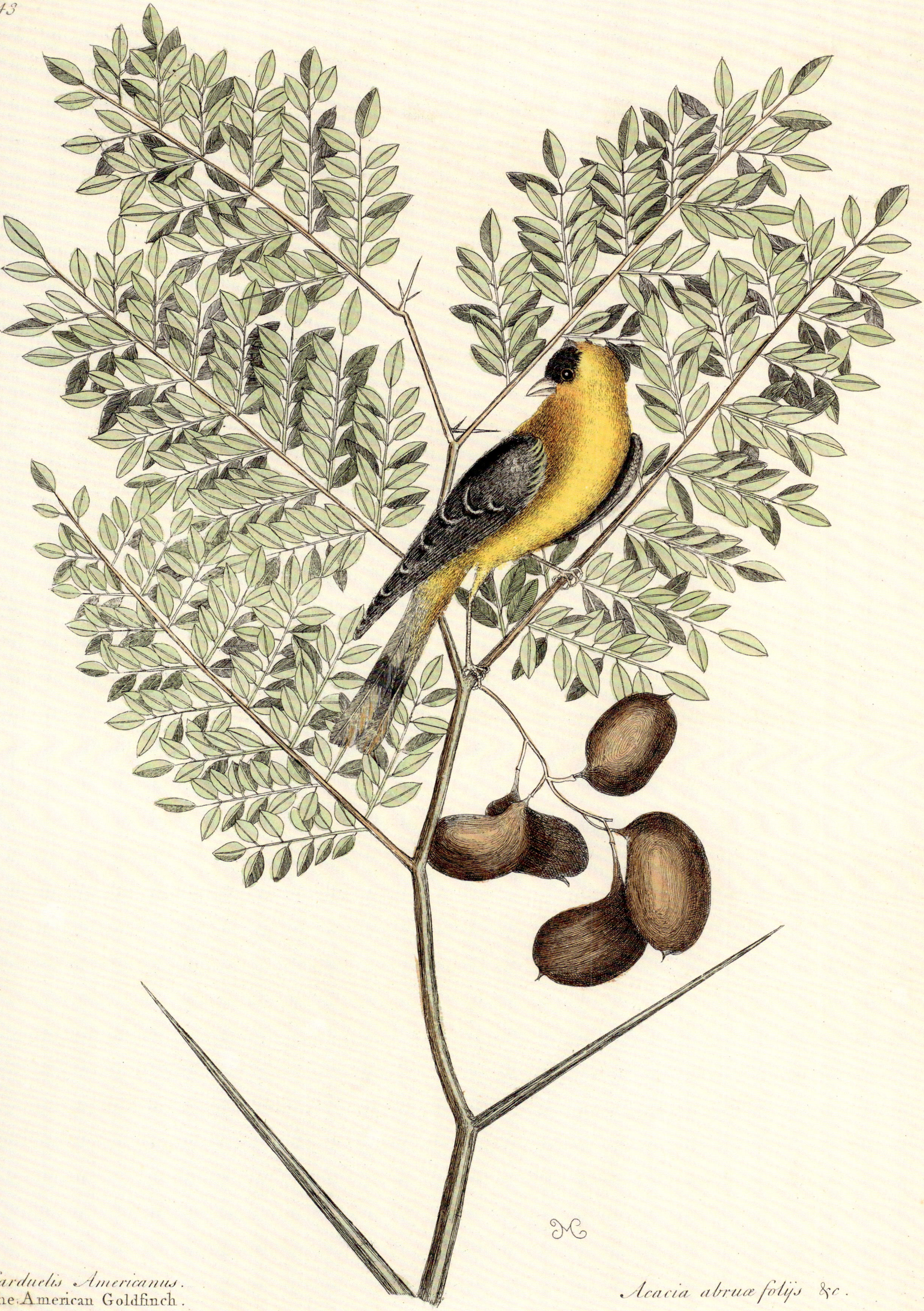

Carduelis Americanus.
The American Goldfinch.

Acacia abruæ folijs &c.

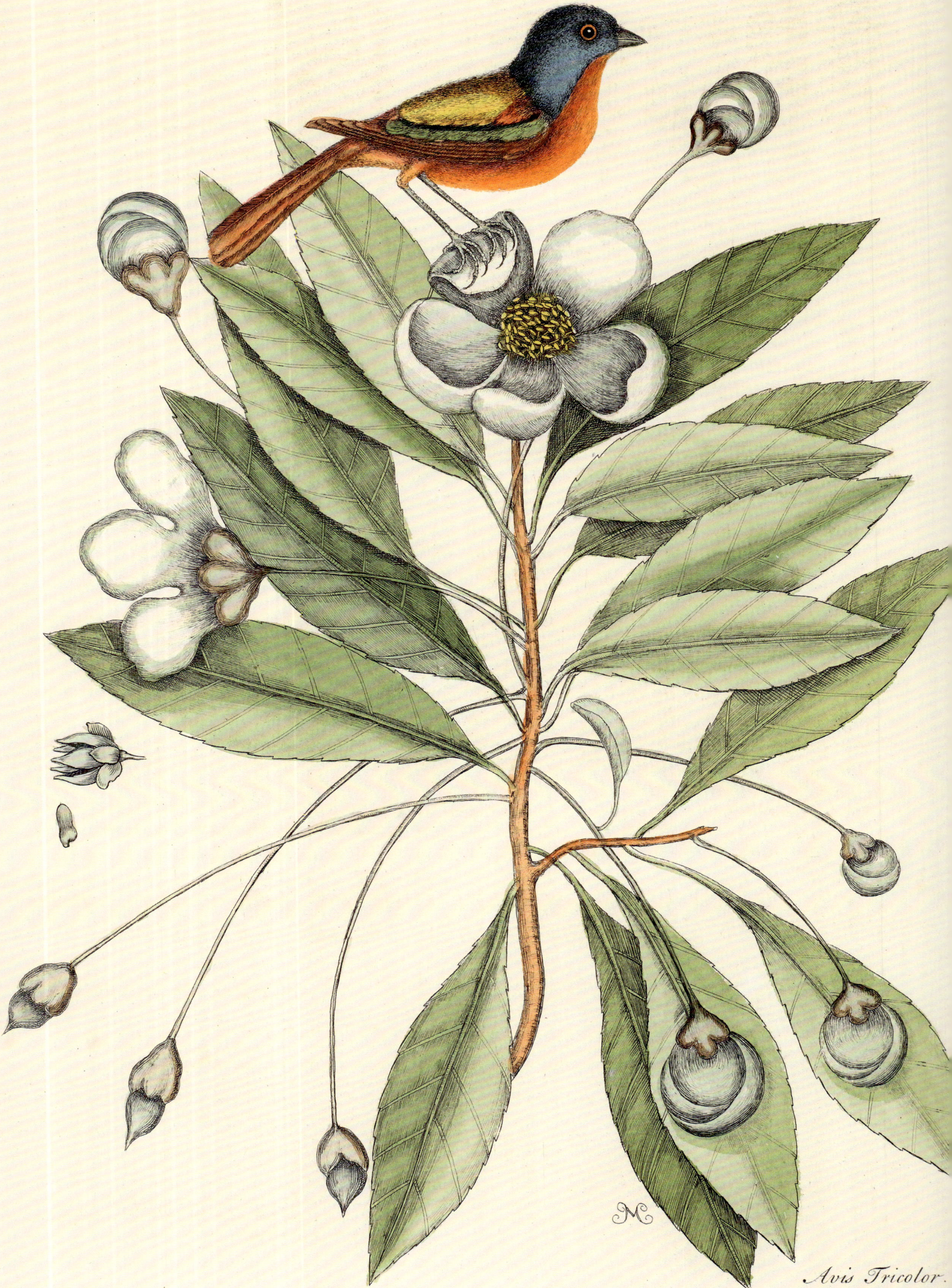

Althea Floridana.

Avis Tricolor.
The Painted Finch

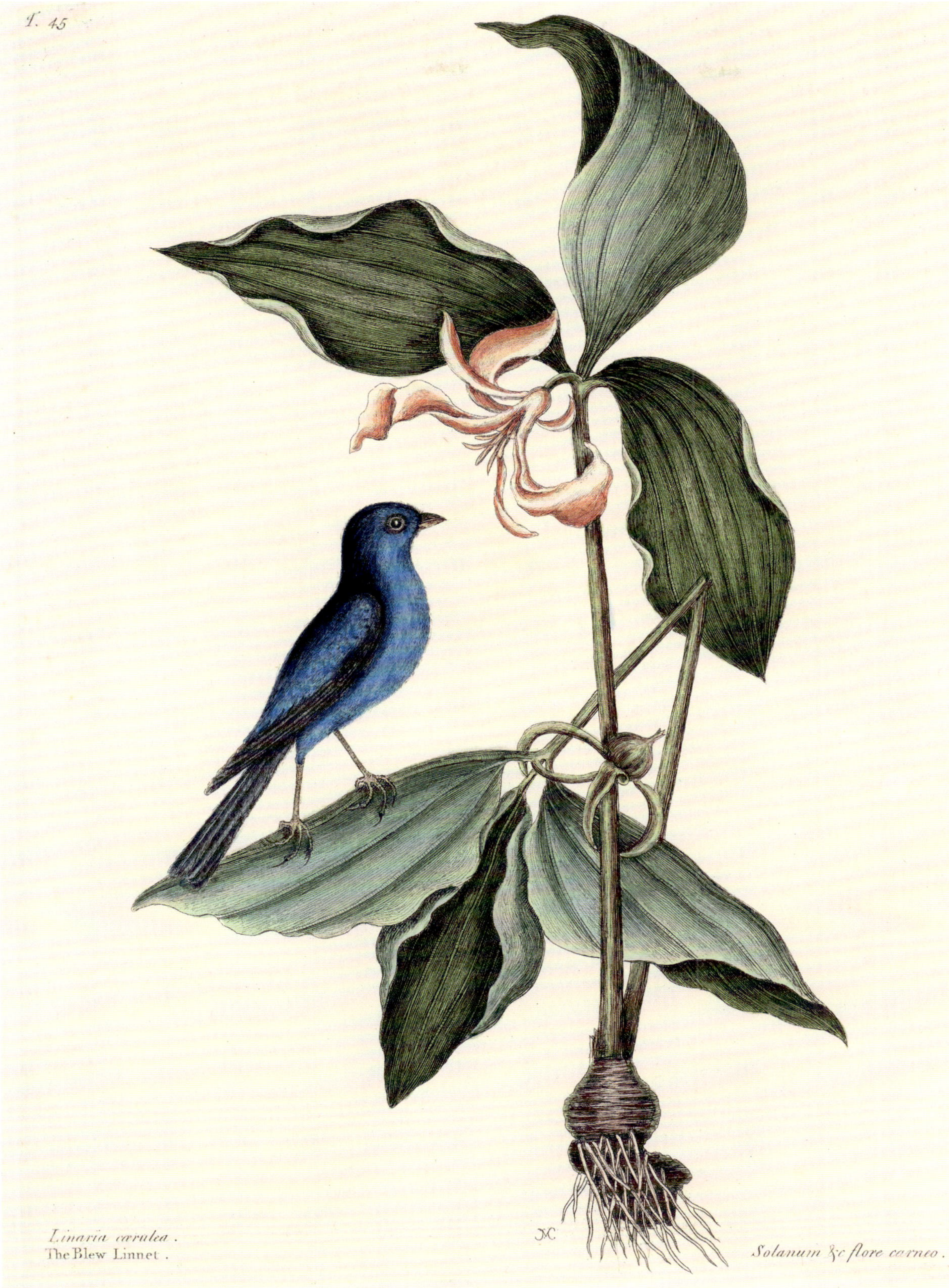

44 *left*
Painted bunting (*Passerina ciris*); loblolly bay (*Gordonia lasianthus*) (vol. 1, t. 44)

45 *above*
Indigo bunting (*Passerina cyanea*); bashful wakerobin (*Trillium catesbaei*) (vol. 1, t. 45)

I.

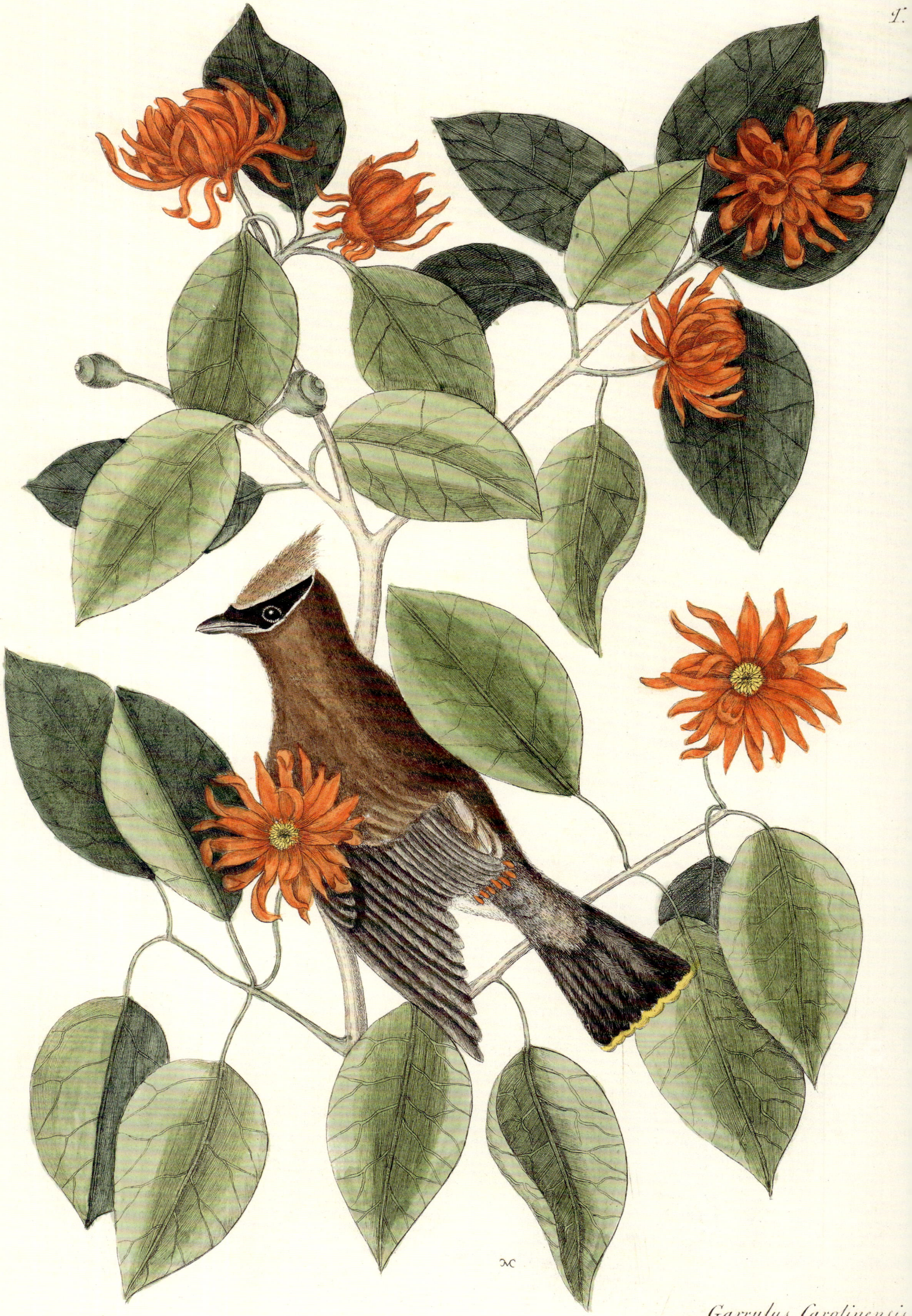

Frutex Corni foliis &c.

Garrulus Carolinensis
The Chatterer.

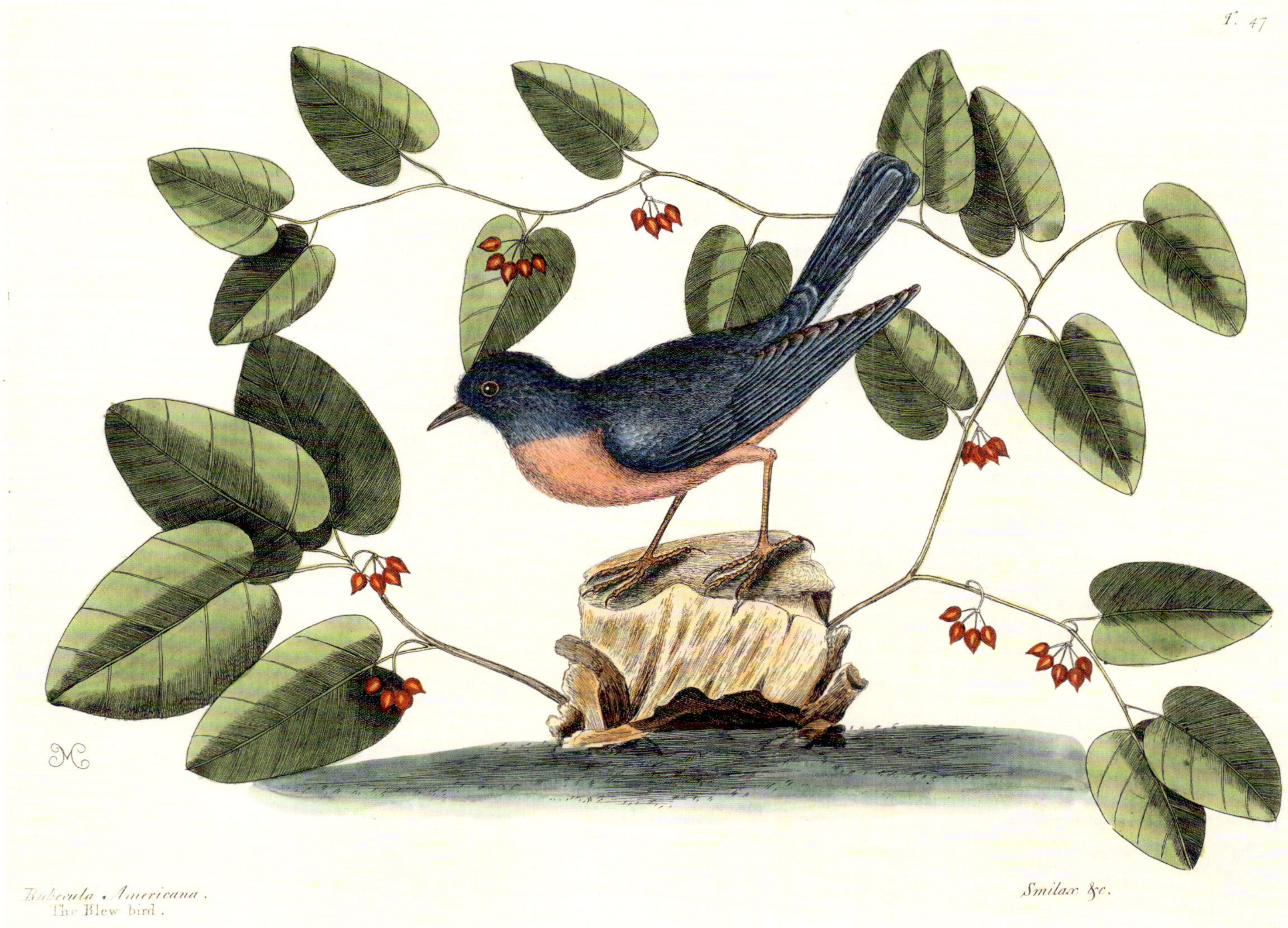

46 *left*
Cedar waxwing (*Bombycilla cedrorum*); Carolina allspice (*Calycanthus floridus*) (vol. 1, t. 46)

47 *above*
Eastern bluebird (*Sialia sialis*); sarsaparilla vine (*Smilax pumila*) (vol. 1, t. 47)

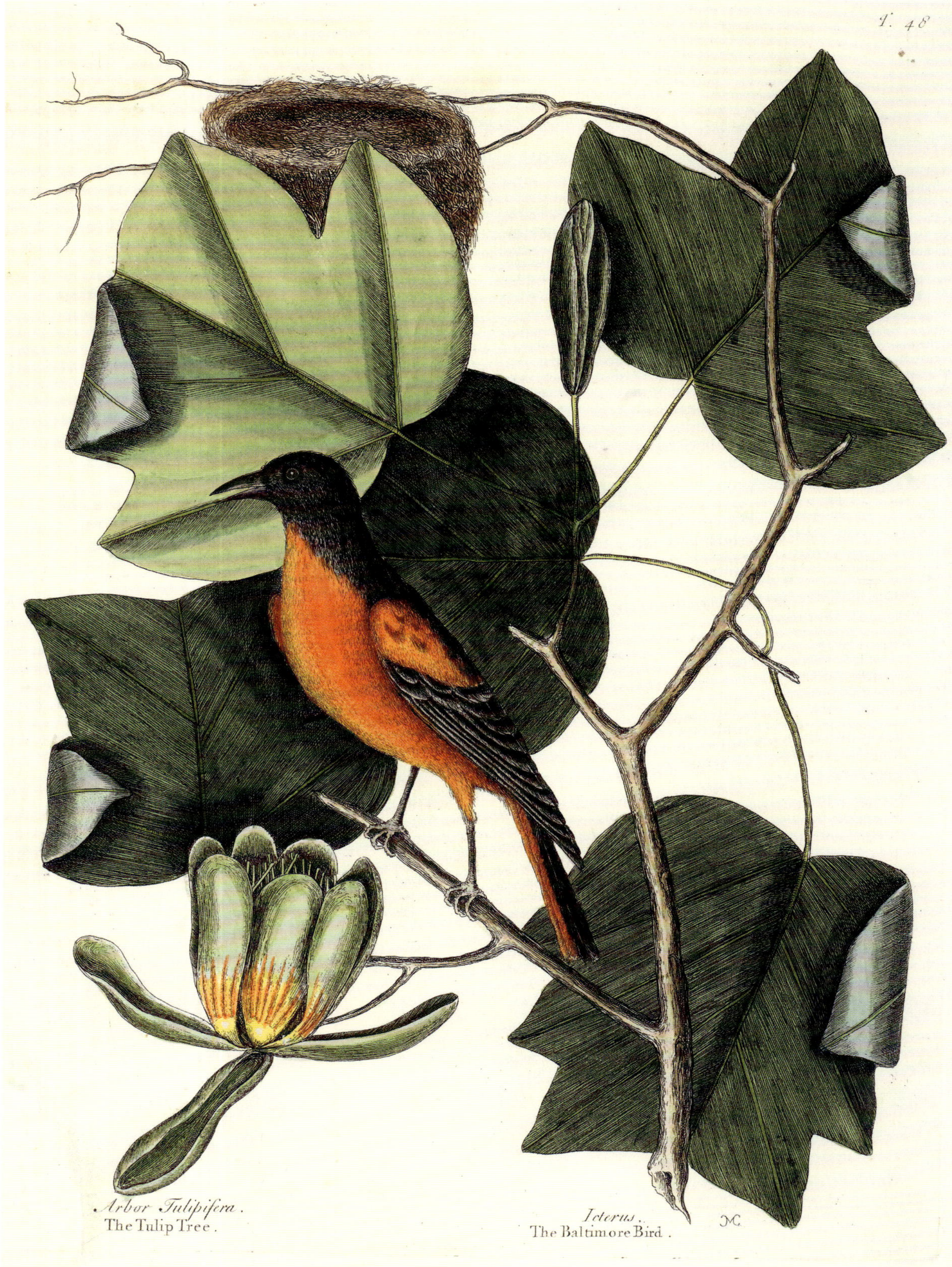

48 *above*

Baltimore oriole (*Icterus galbula*); tulip tree (*Liriodendron tulipifera*) (vol. 1, t. 48)

49 *right*

Orchard oriole (*Icterus spurius*); southern catalpa (*Catalpa bignonioides*) (vol. 1, t. 49)

49

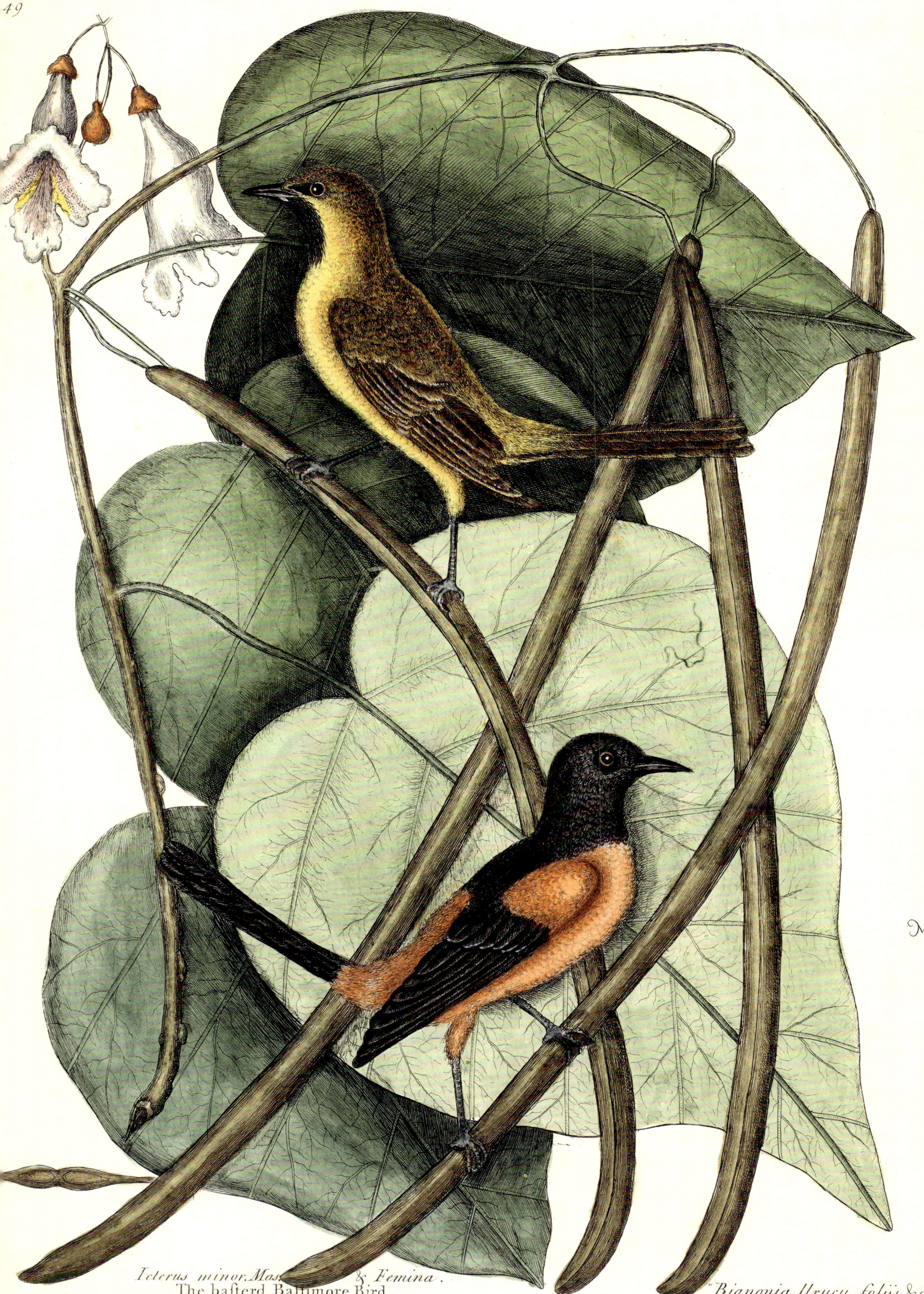

Icterus minor. Mas & Femina.
The basterd Baltimore Bird.

Bignonia Urucu folijs &c.
The Catalpah Tree.

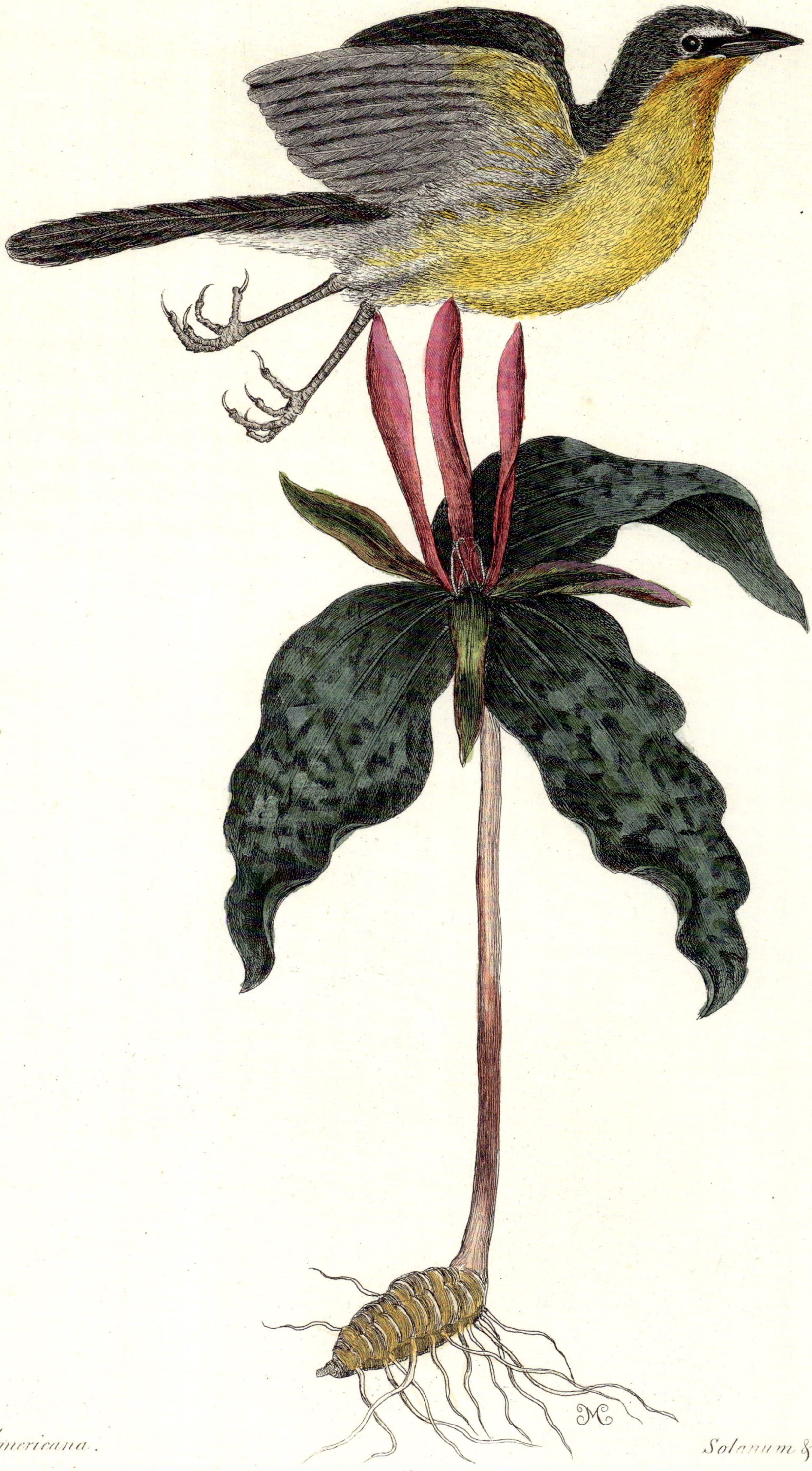

Œnanthe Americana.

Solanum &c fl: purpureo

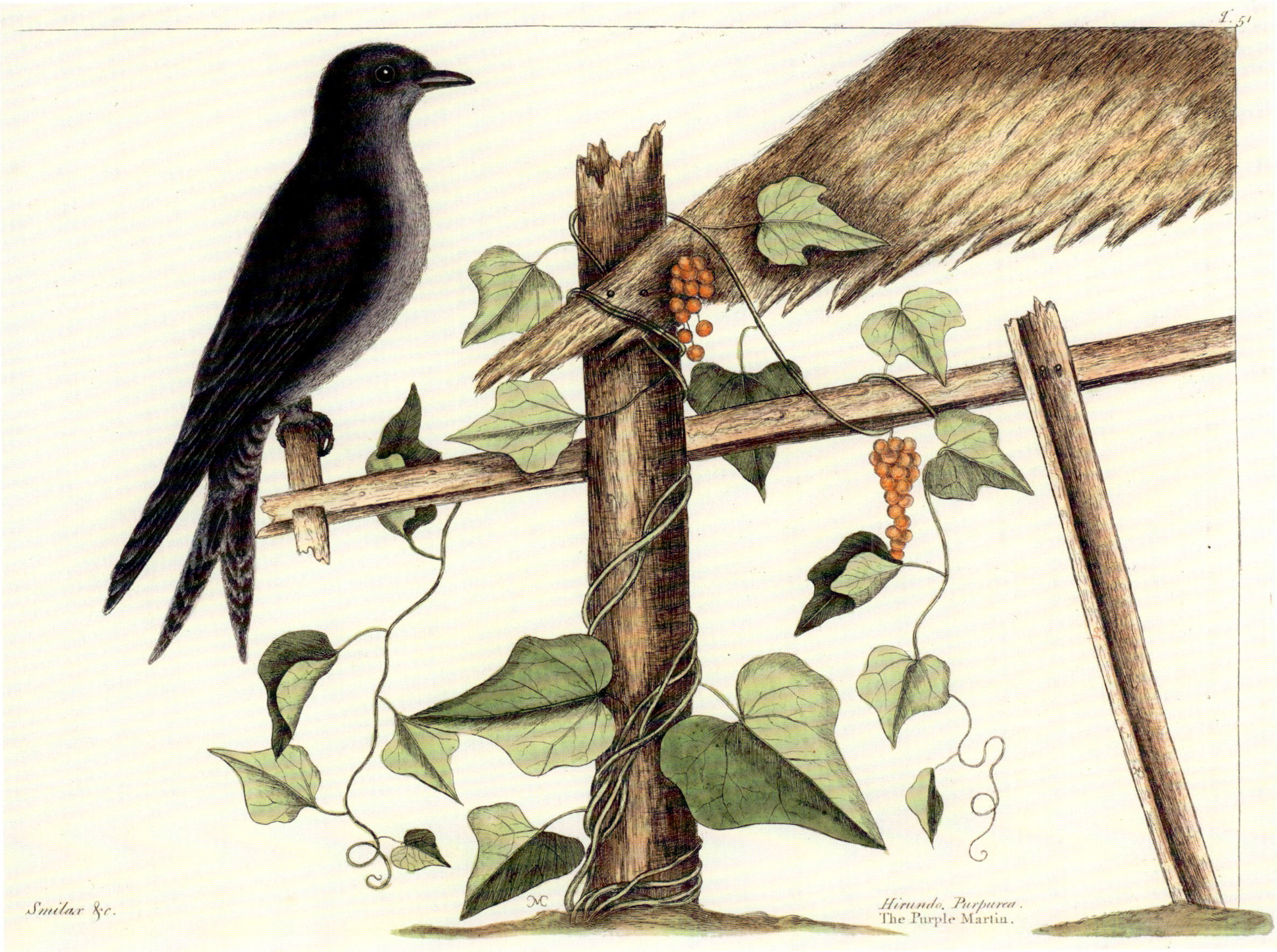

50 *left*
Yellow-breasted chat (*Icterus virens*); spotted wakerobin (*Trillium maculatum*) (vol. 1, t. 50)

51 *above*
Purple martin (*Pogne subis*); Carolina moonseed (*Cocculus carolinus*) (vol. 1, t. 51)

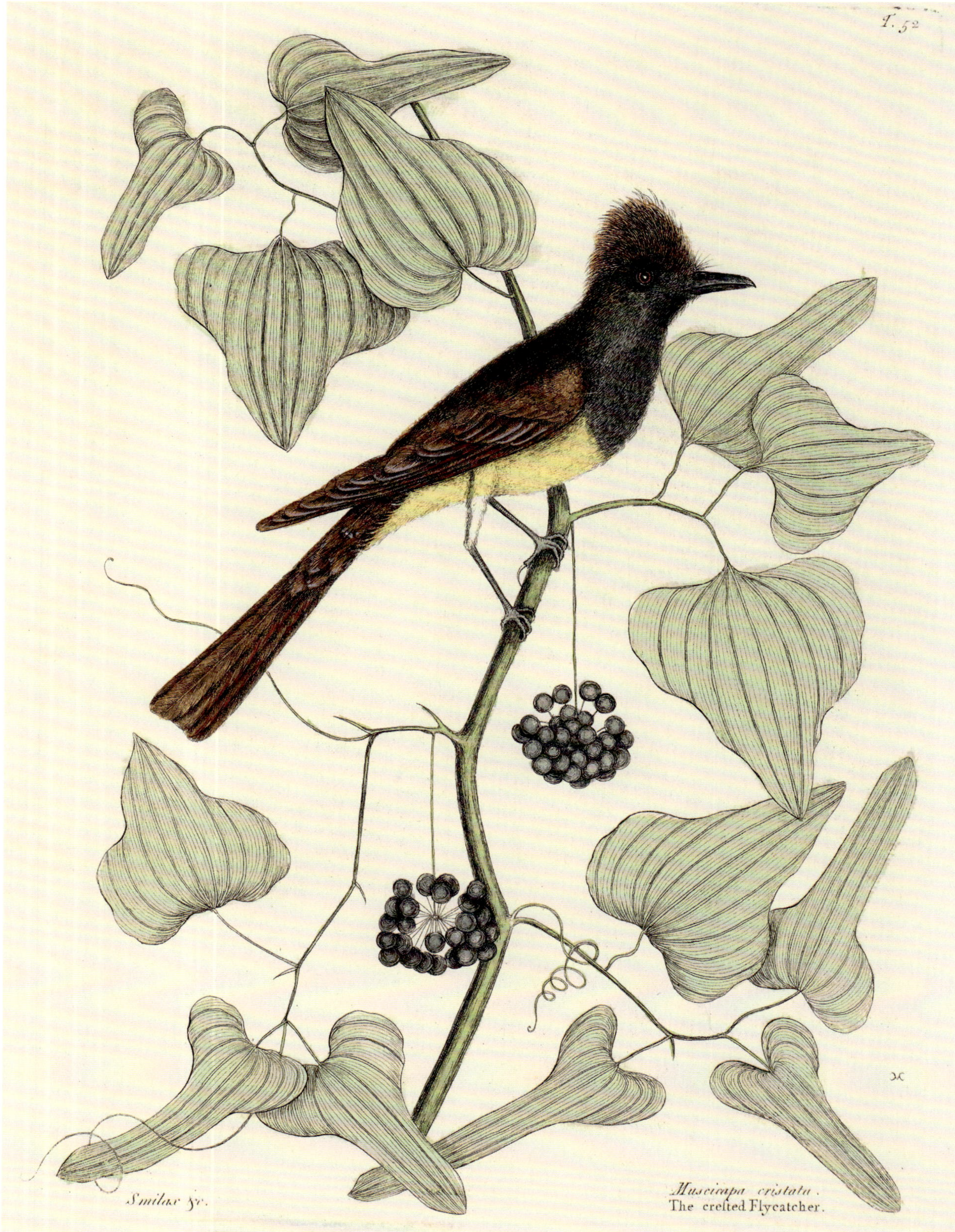

52 *above*

Great crested flycatcher (*Myiarchus crinitus*); bristly greenbrier (*Smilax tamnoides*) (vol. 1, t. 52)

53 *right*

Eastern phoebe (*Sayornis phoebe*); yellow jessamine (*Gelsemium sempervirens*) (vol. 1, t. 53)

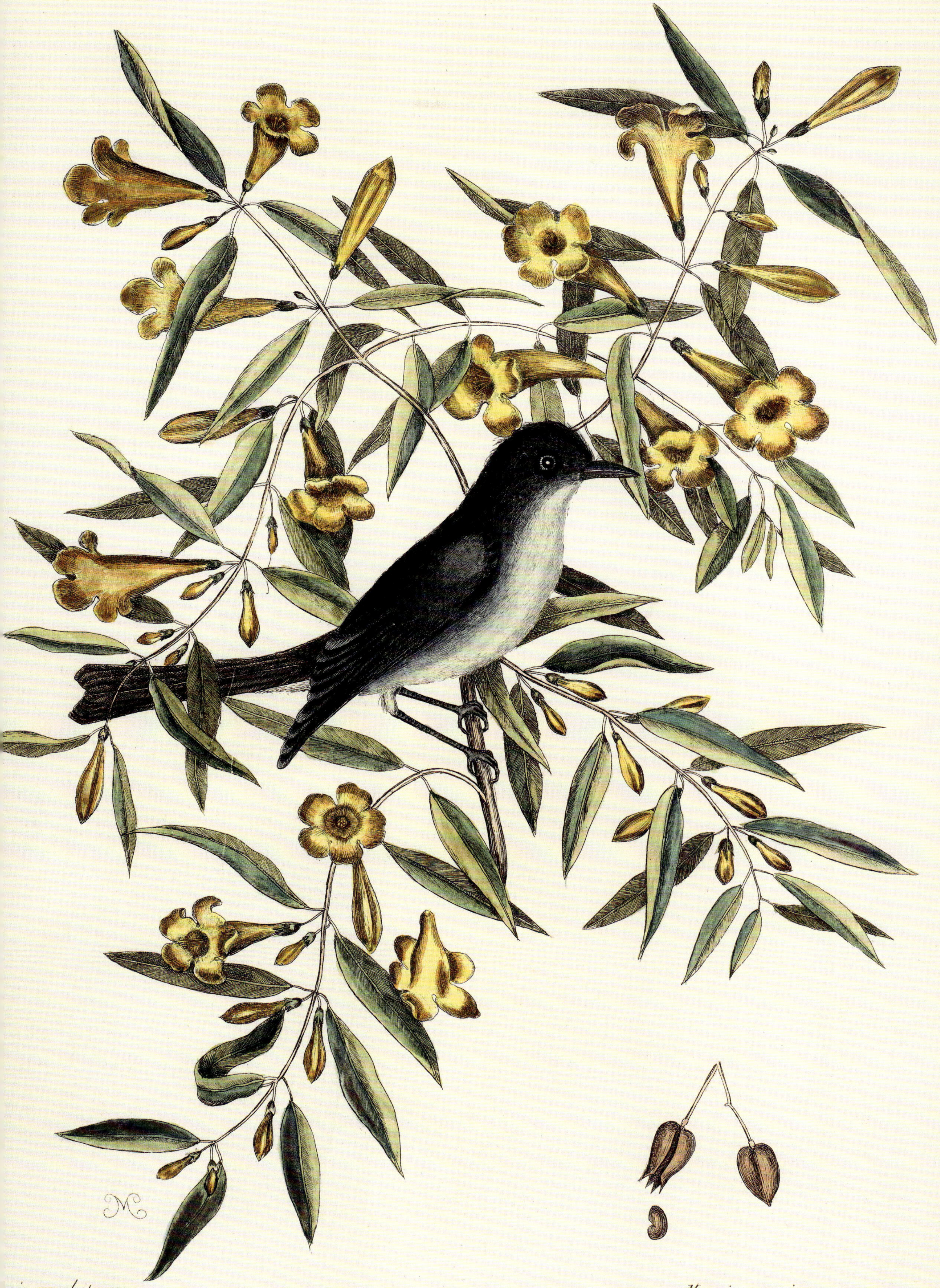

ninum luteum.
yellow Jeſſamy.

Muscicapa nigrescens.
The blackcap Flycatcher.

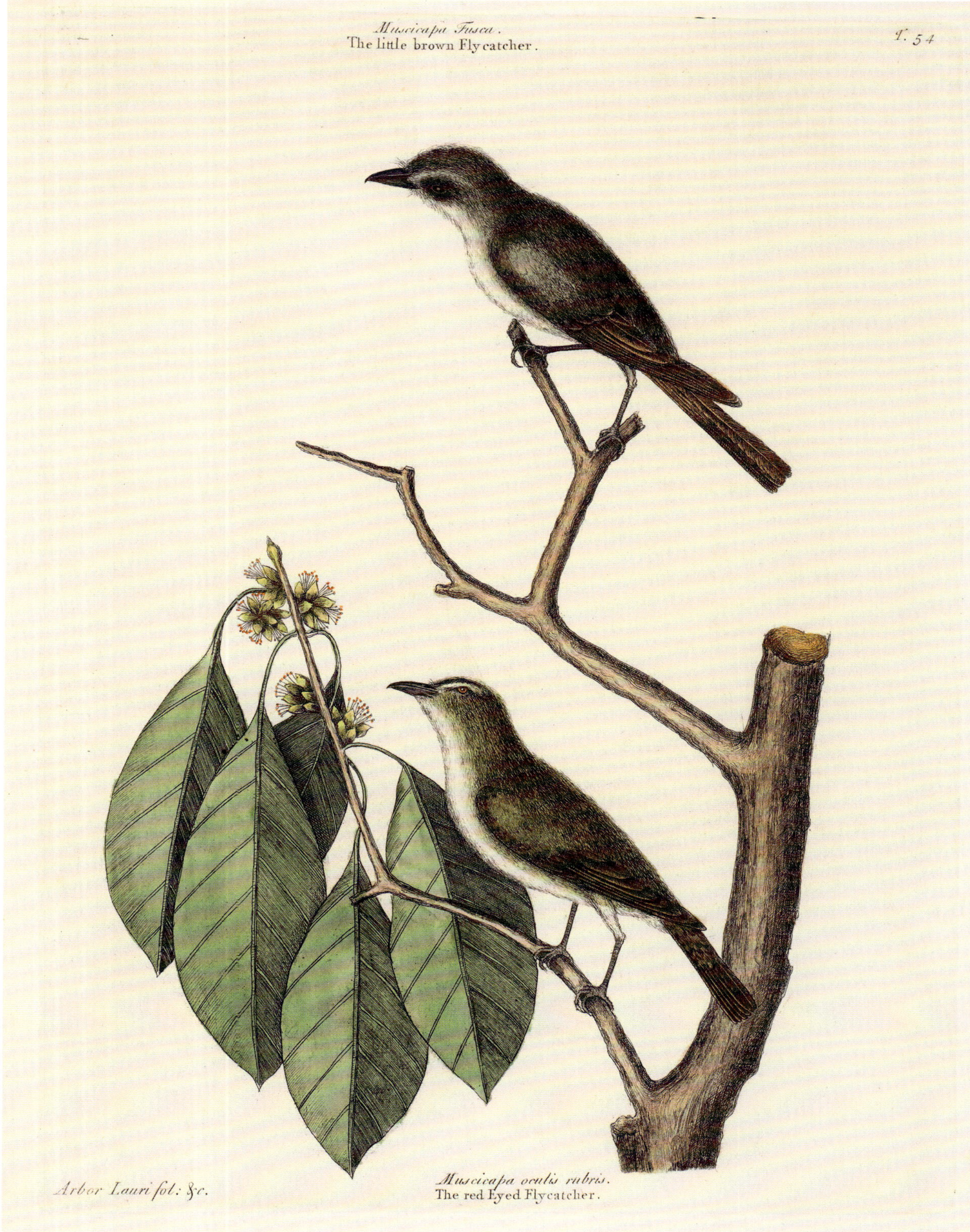

54 *above*

Eastern wood pewee (*Contopus virens*; lower); red-eyed vireo (*Vireo olivaceus*; upper); common sweetleaf (*Symplocos tinctoria*) (vol. 1, t. 54)

55 *right*

Eastern kingbird (*Tyrannus tyrannus*); unidentified insect; sassafras (*Sassafras albidum*) (vol. 1, t. 55)

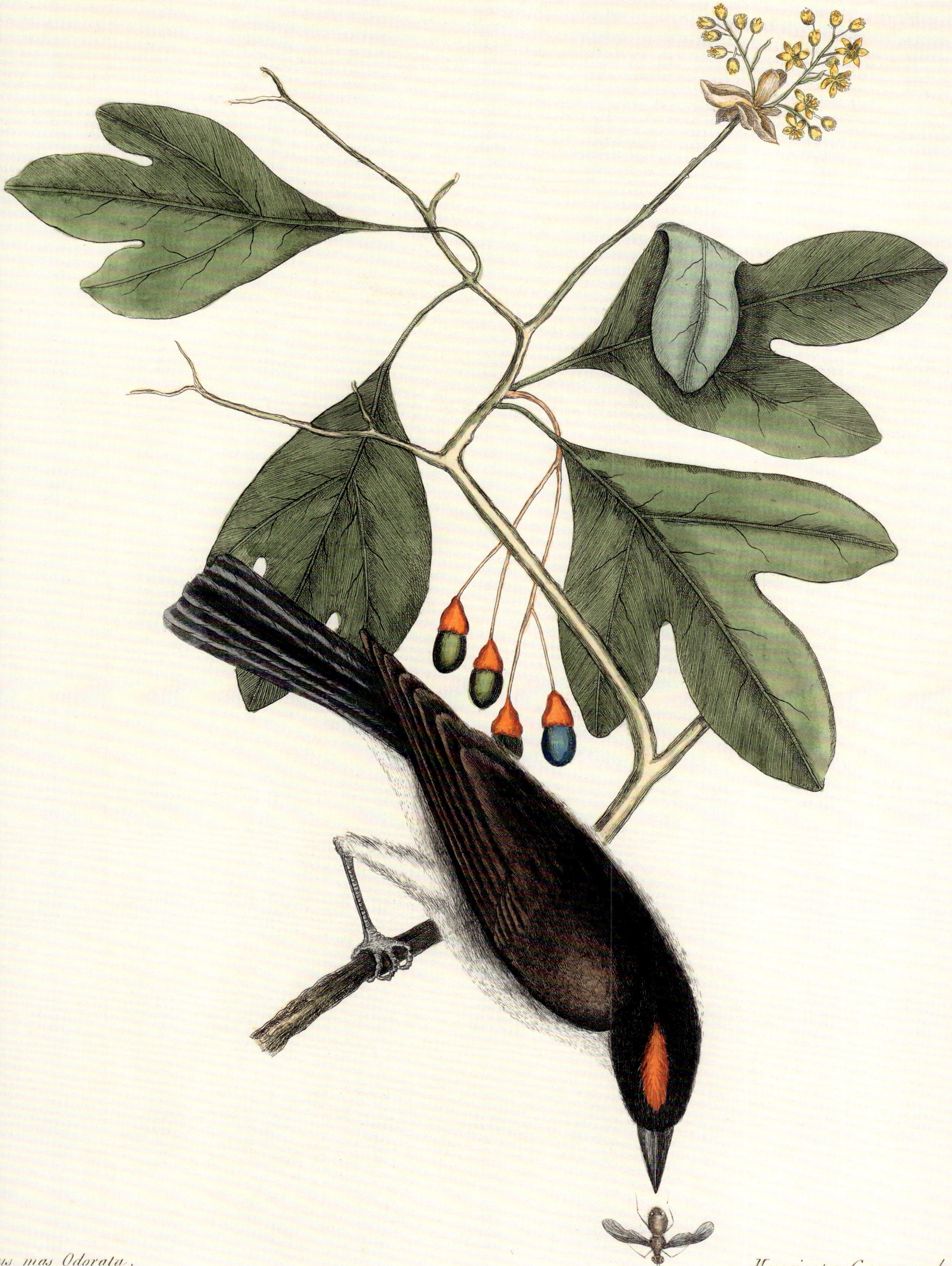

...ous mas Odorata.
Safsafras.

Muscicapa Corona rubra.
The Tyrant.

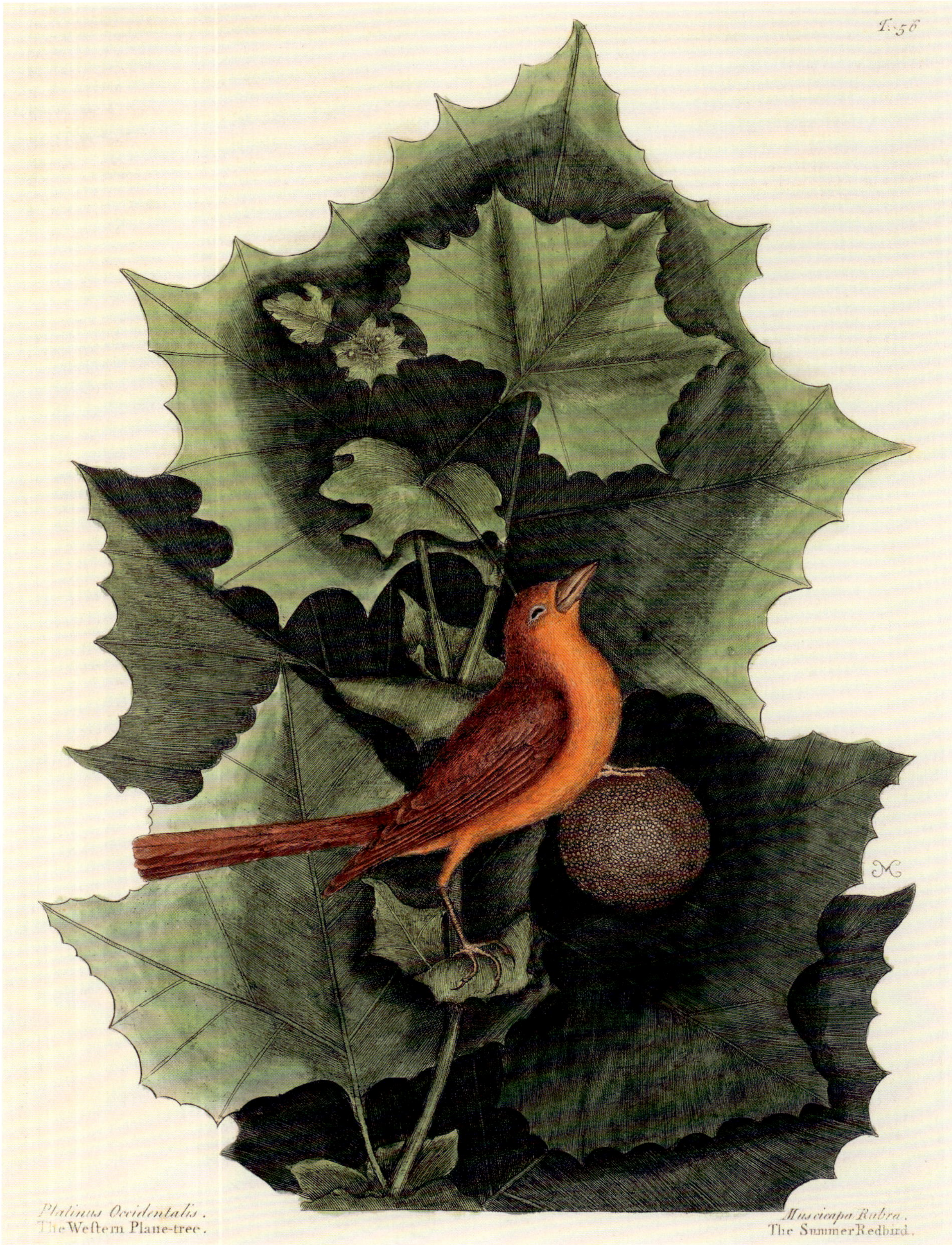

56 *above*

Summer tanager (*Piranga rubra*); American sycamore (*Platanus occidentalis*) (vol. 1, t. 56)

57 *right*

Tufted titmouse (*Bacolophus bicolor*); mountain azalea (*Rhododendron canescens*) (vol. 1, t. 57)

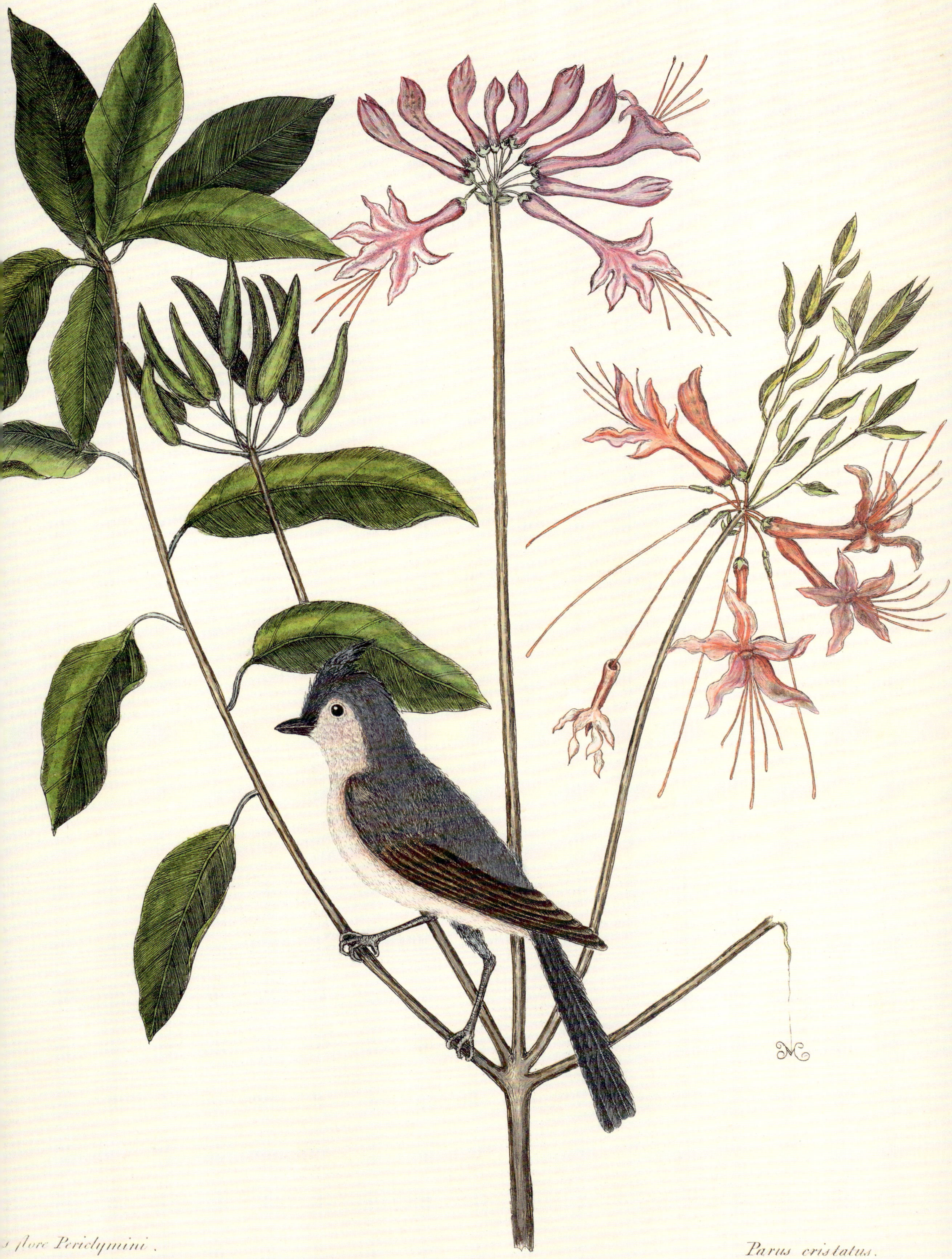

...s flore Periclymini.

Parus cristatus.
The crested Titmous.

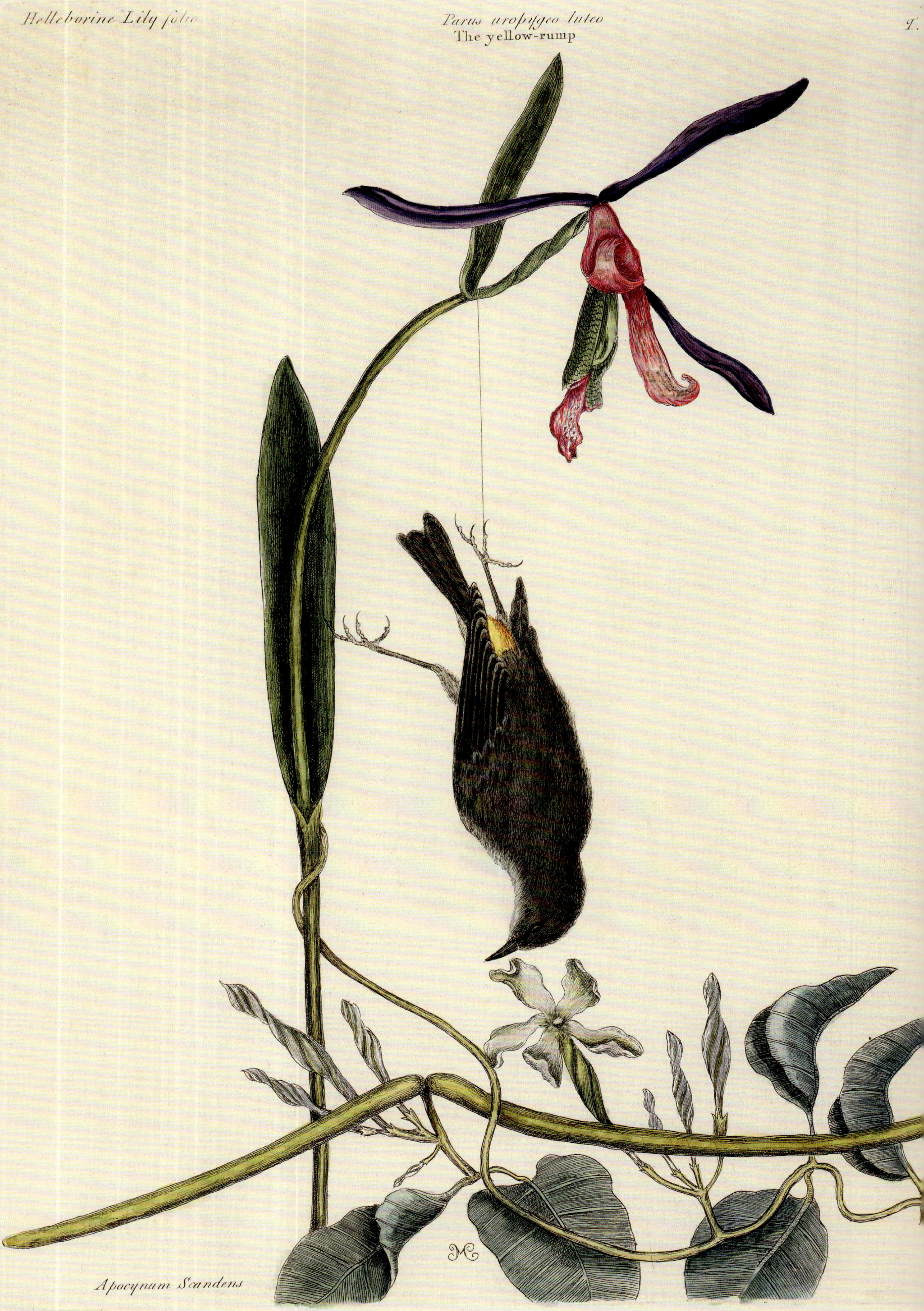
Helleborine Lily folio
Parus uropygeo luteo
The yellow-rump
Apocynum Scandens

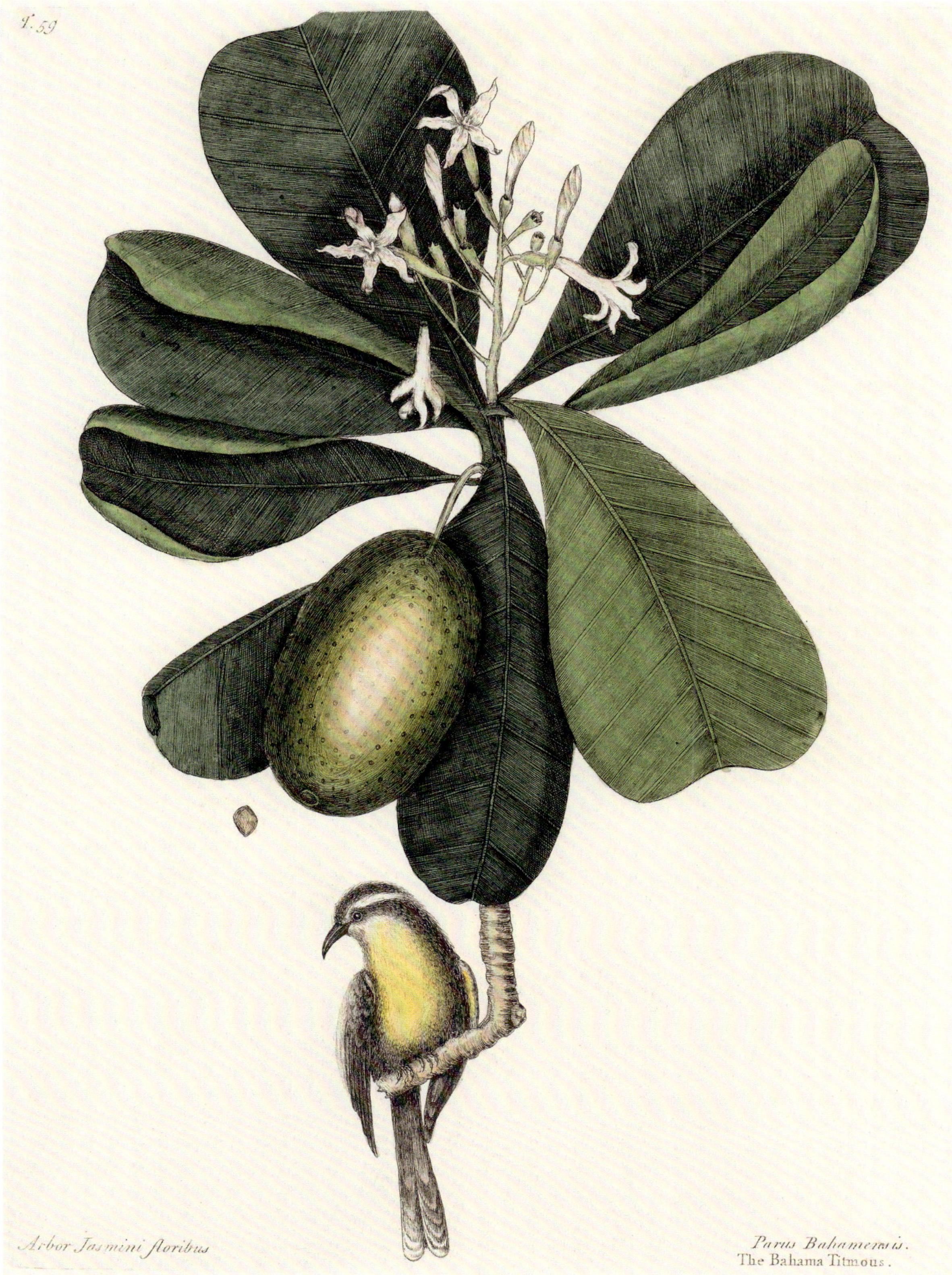

58 *left*
yellow-rumped warbler (*Setophaga coronata*); spreading pogonia (*Cleistesiopsis divaricata*; upper); devil's potato (*Echites umbellatus*; lower) (vol. 1, t. 58)

59 *above*
Bahama bananaquit (*Coereba flaveola*); sevenyear apple (*Casasia clusiifolia*) (vol. 1, t. 59)

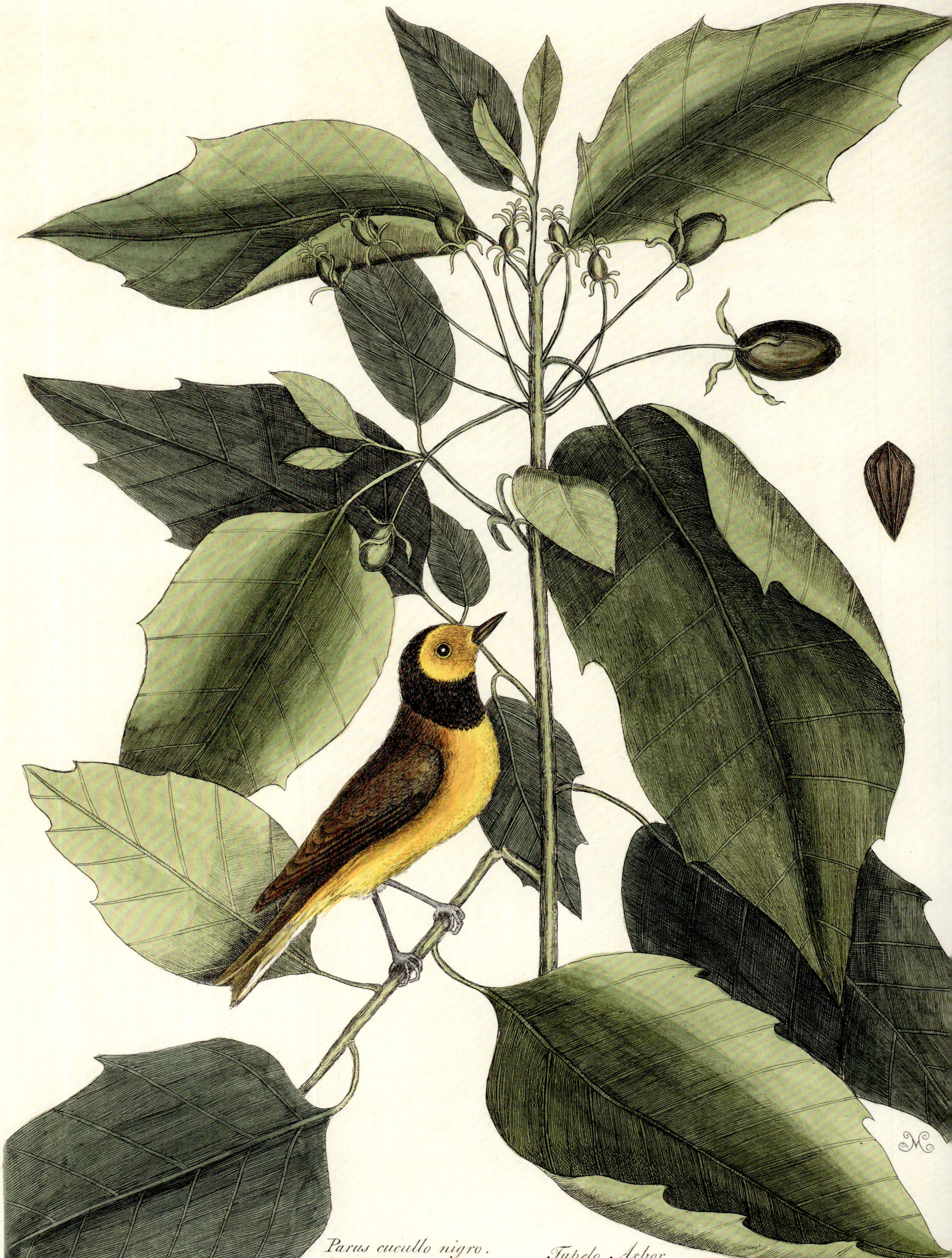

Parus cucullo nigro.
The Hooded Titmous.

Tupelo Arbor.
The water Tupelo.

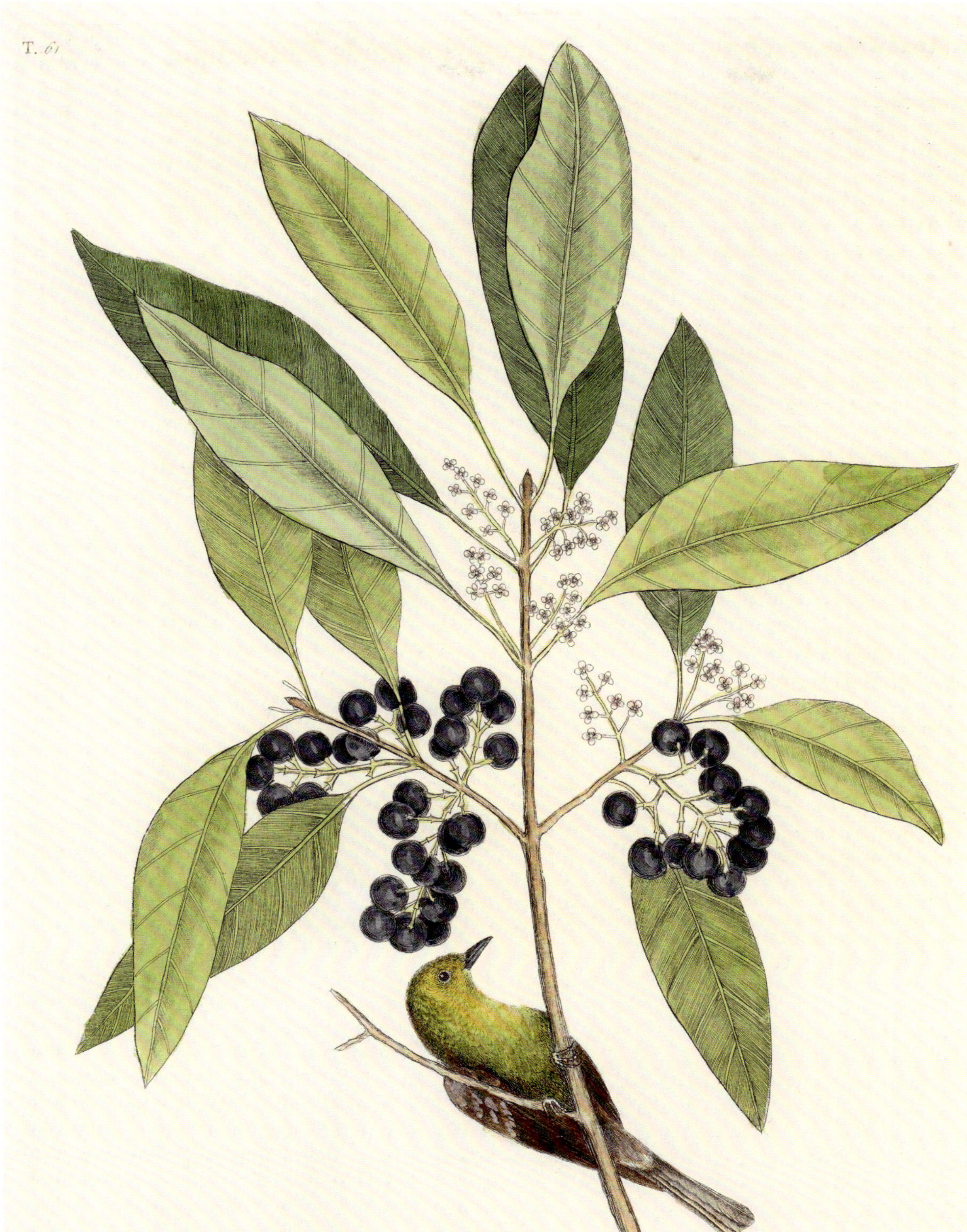

60 *left*
Hooded warbler (*Setophaga citrina*); water tupelo (*Nyssa aquatica*) (vol. 1, t. 60)

61 *above*
Pine warbler (*Setophaga pinus*); American devilwood (*Cartrema americana*) (vol. 1, t. 61)

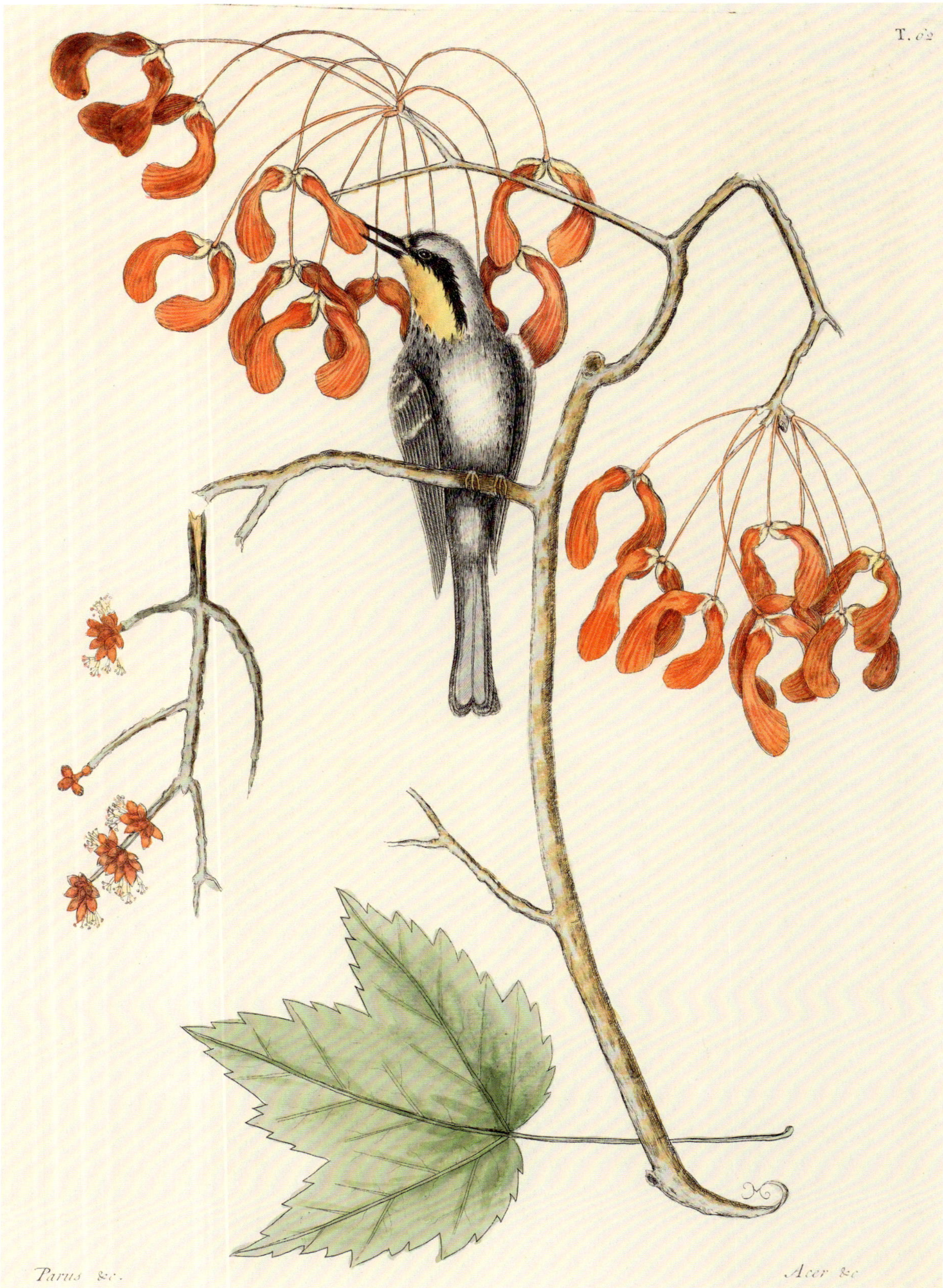

62 *above*
Yellow-throated warbler (*Setophaga dominica*); red maple (*Acer rubrum*) (vol. 1, t. 62)

63 *right*
American yellow warbler (*Setophaga petechia*); swamp redbay (*Persea palustris*) (vol. 1, t. 63)

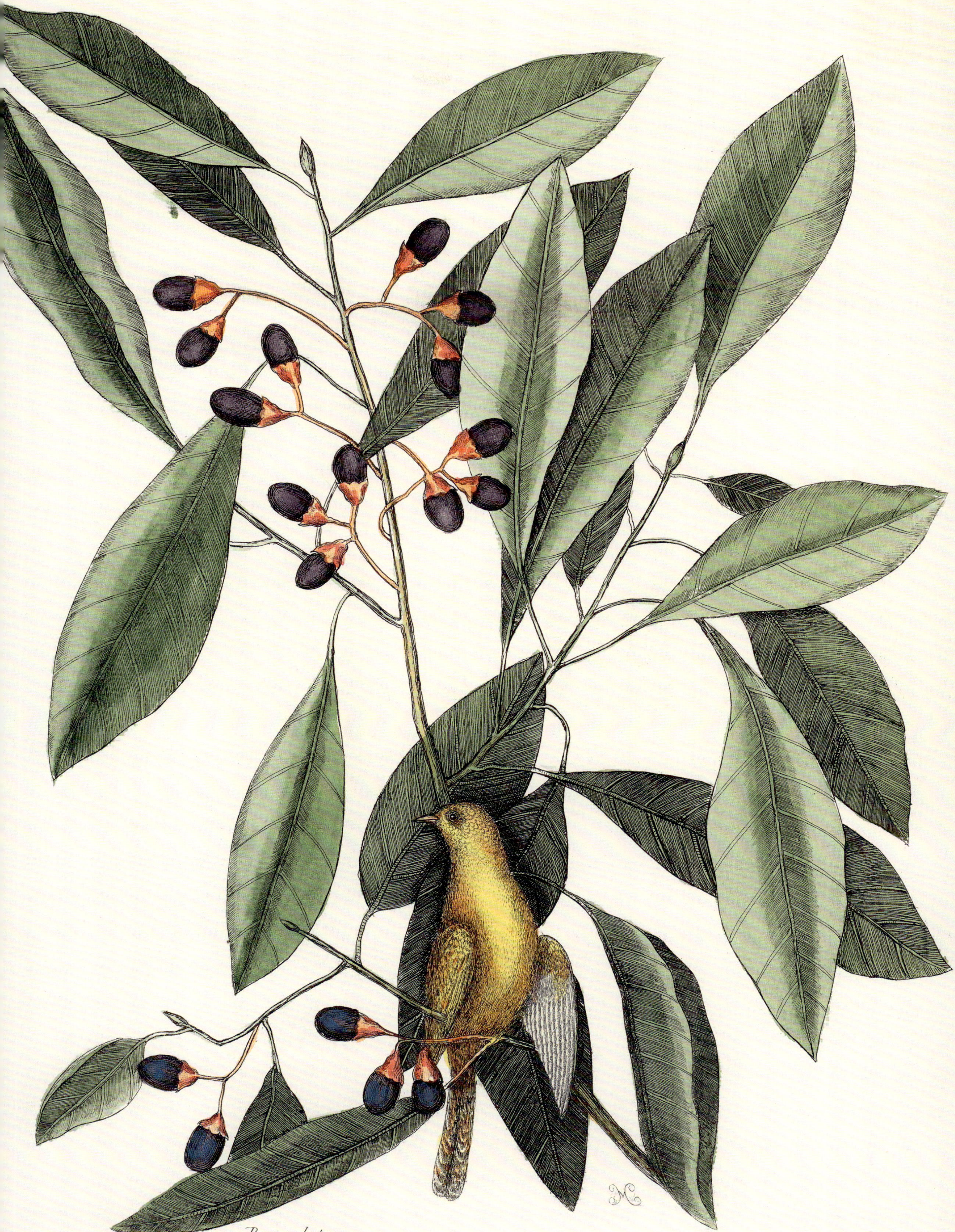

Parus luteus *Laurus Carolinensis*

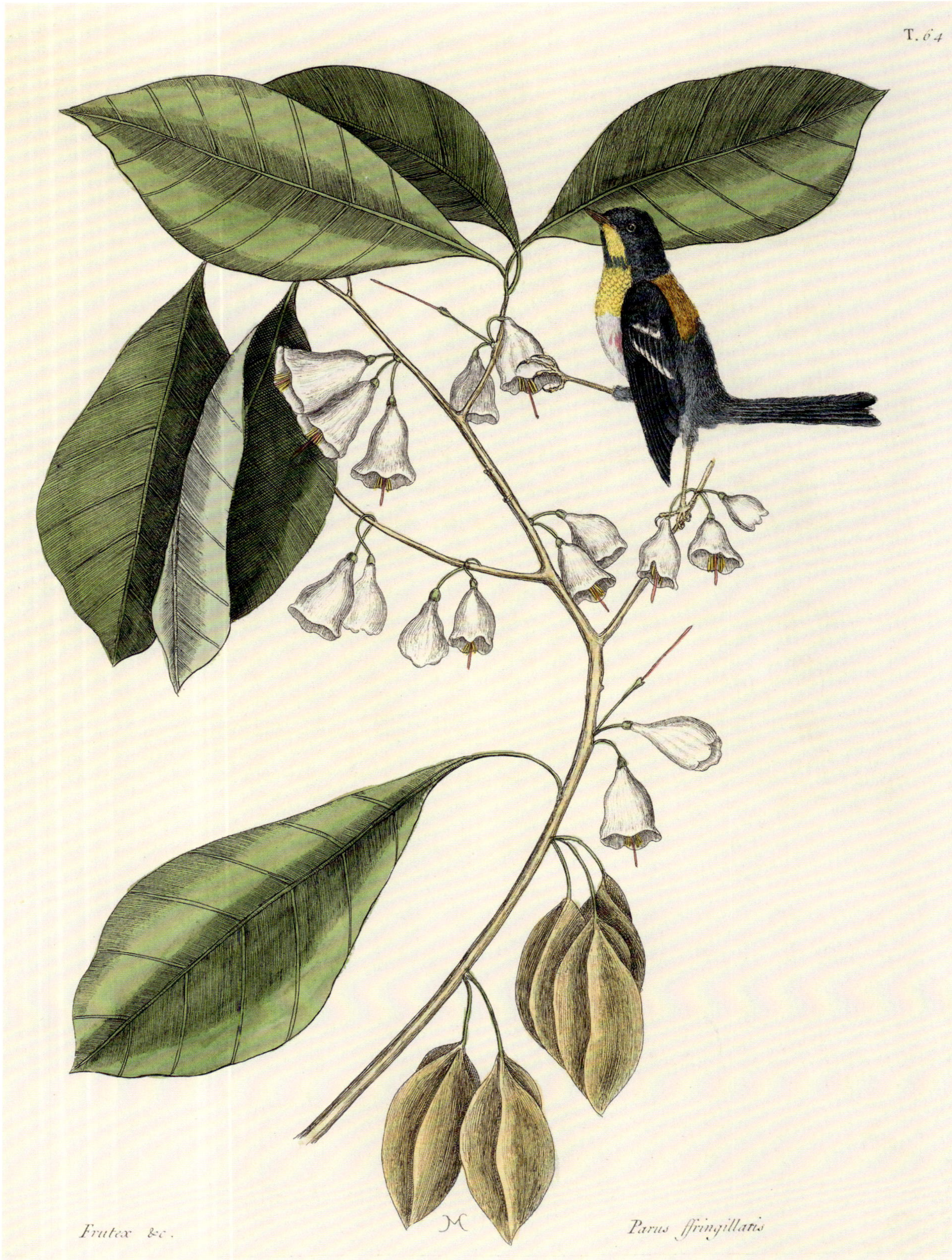

64 *above*
Northern parula (*Setophaga americana*); mountain silverbell (*Halesia tetraptera*) (vol. 1, t. 64)

65 *right*
Ruby-throated hummingbird (*Archilochus colubris*); trumpetcreeper (*Campsis radicans*) (vol. 1, t. 65)

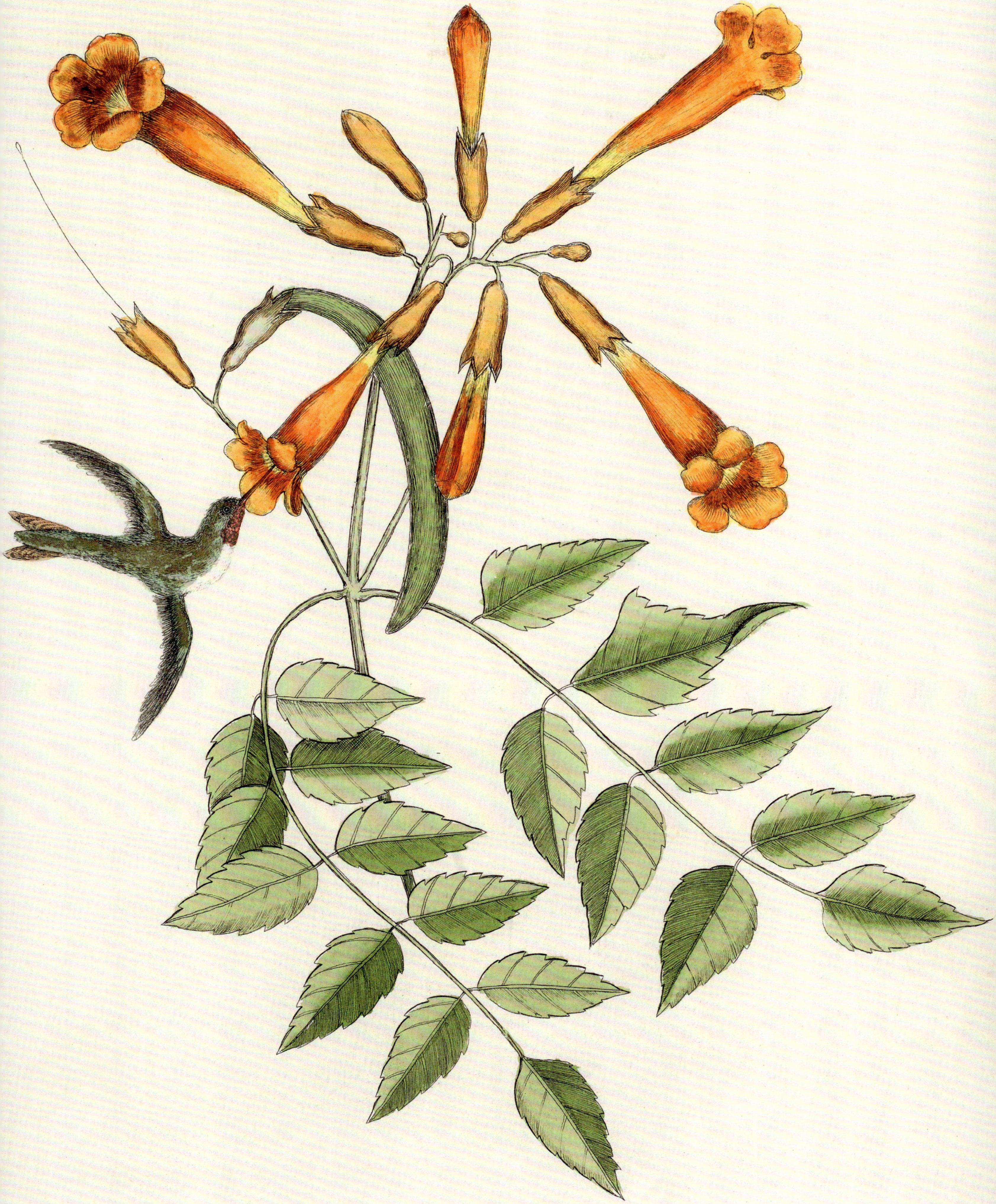

livora &c. M *Bignonia &c.*

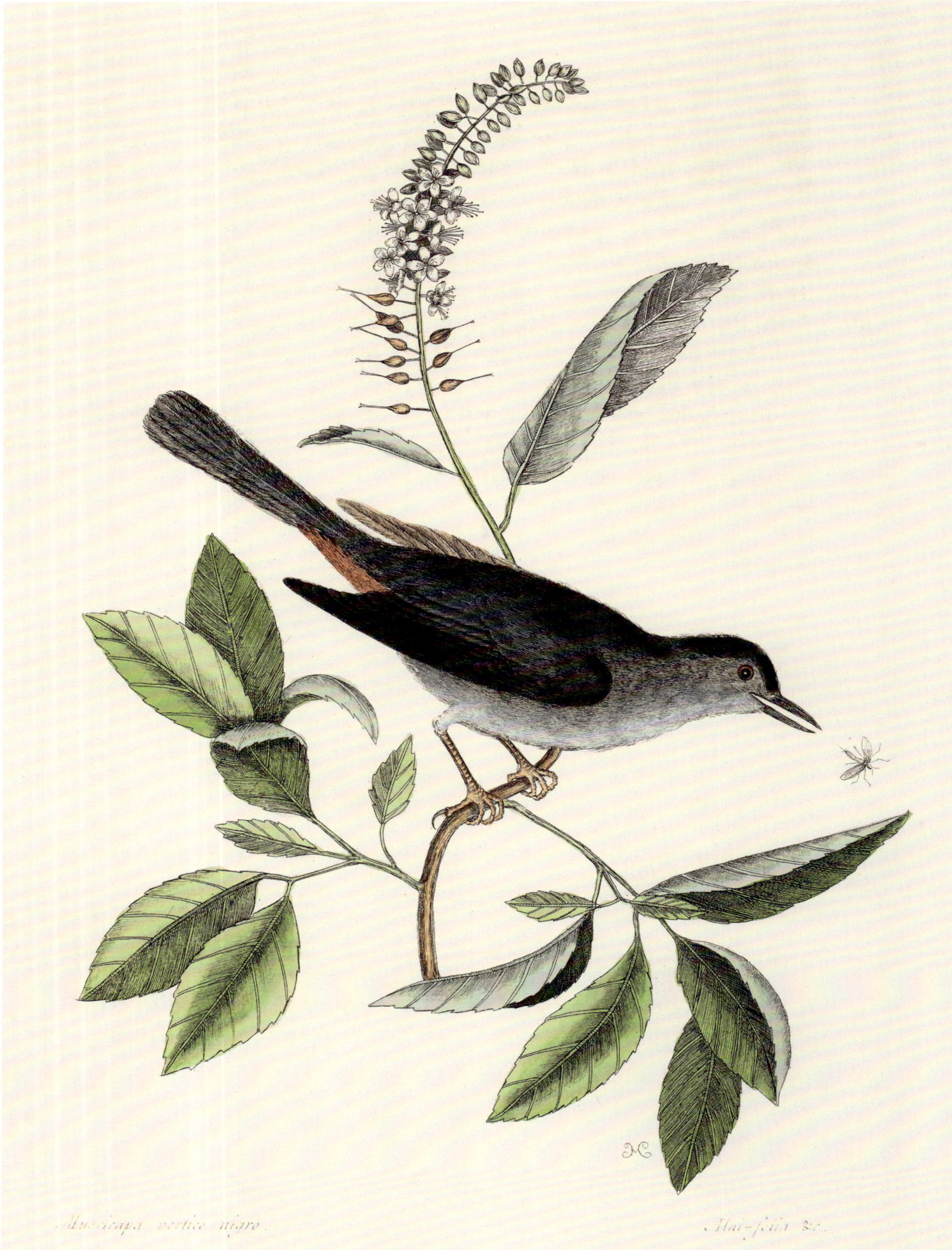

66 *above*

Grey catbird (*Dumetella carolinensis*); unidentified insect; sweet pepper bush (*Clethra alnifolia*) (vol. 1, t. 66)

67 *right*

American redstart (*Setophaga ruticilla*); black walnut (*Juglans nigra*) (vol. 1, t. 67)

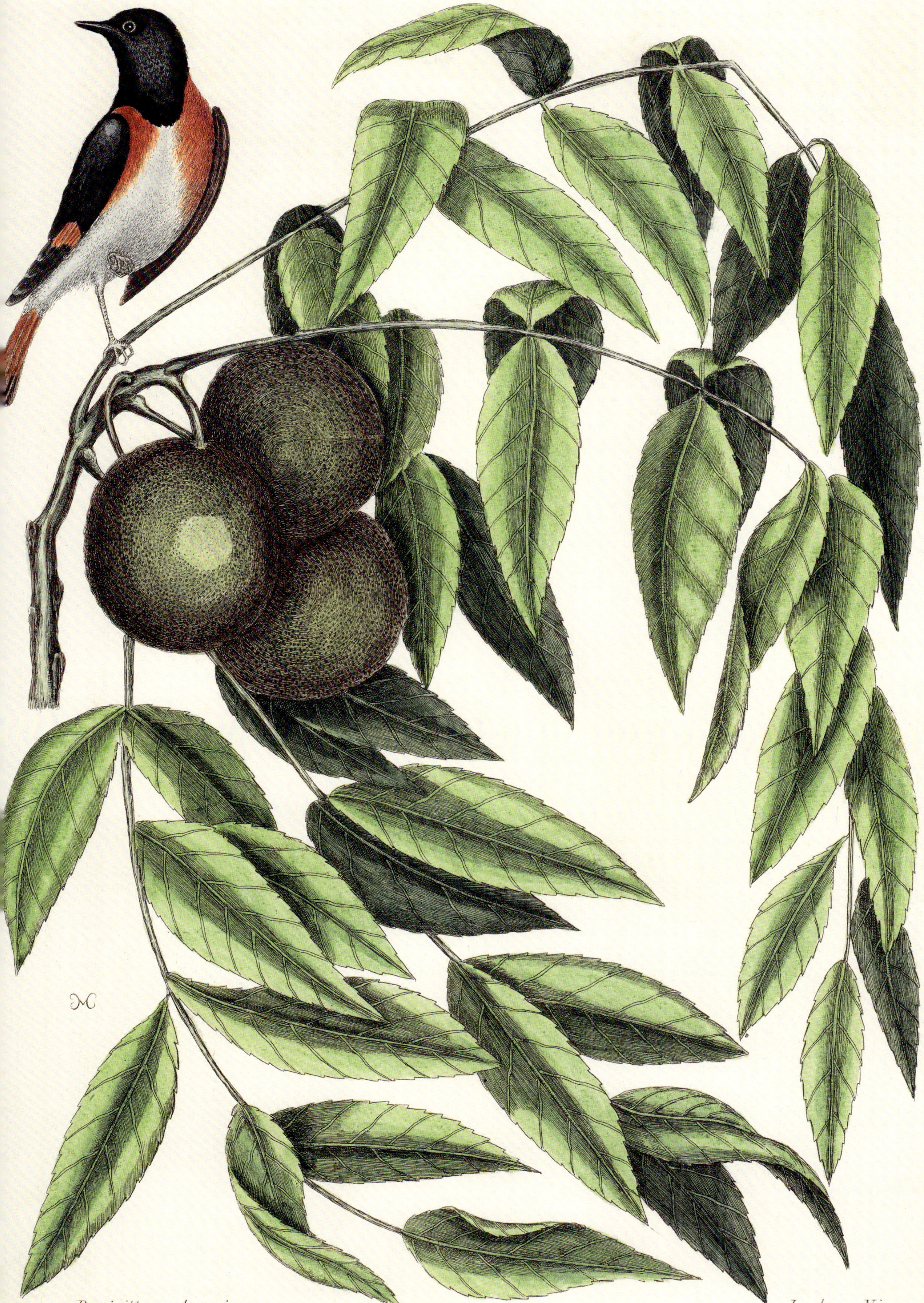

Ruticilla Americana *Juglans Nigra*

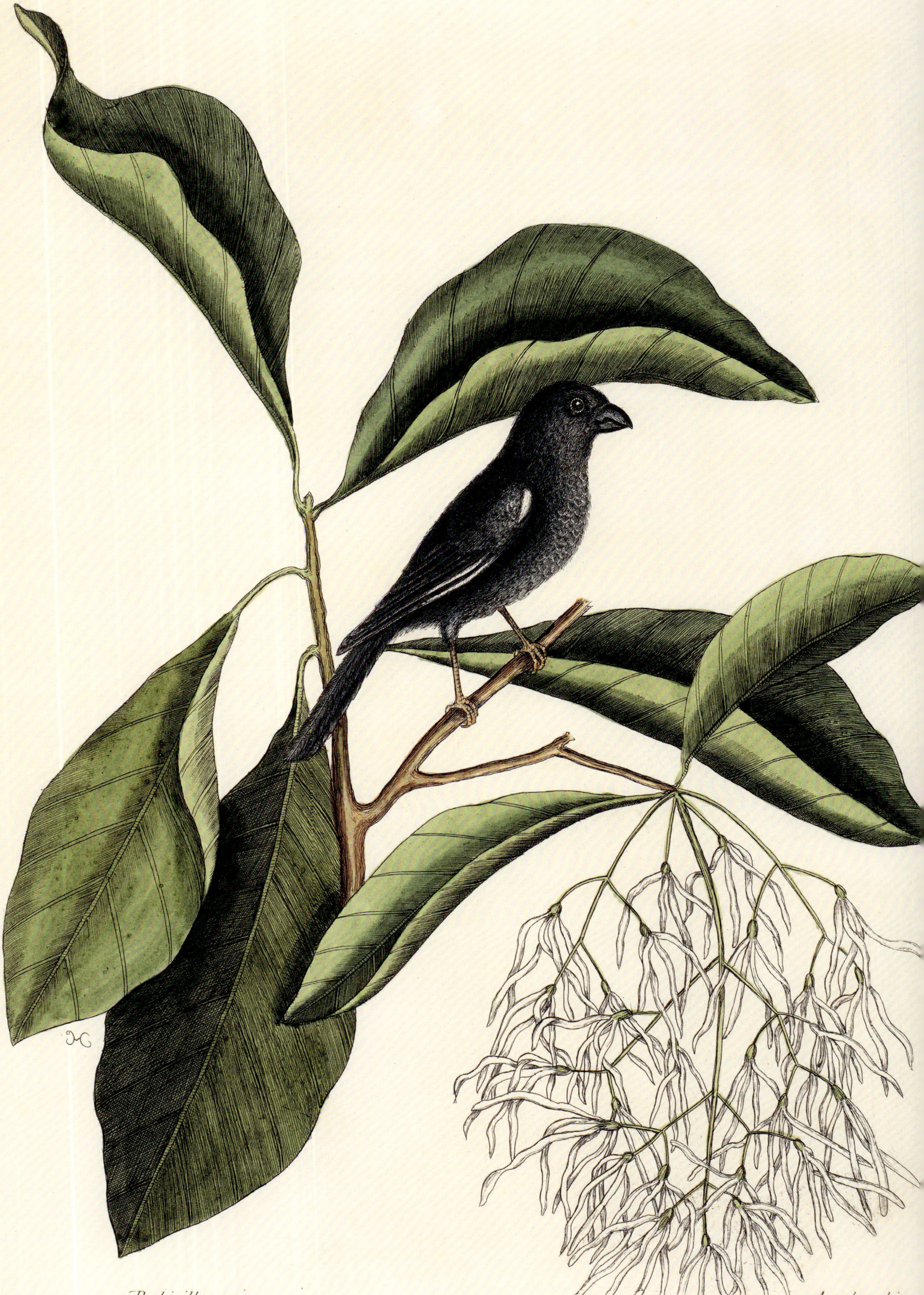

Rubicilla minor nigra.

Amelanchier

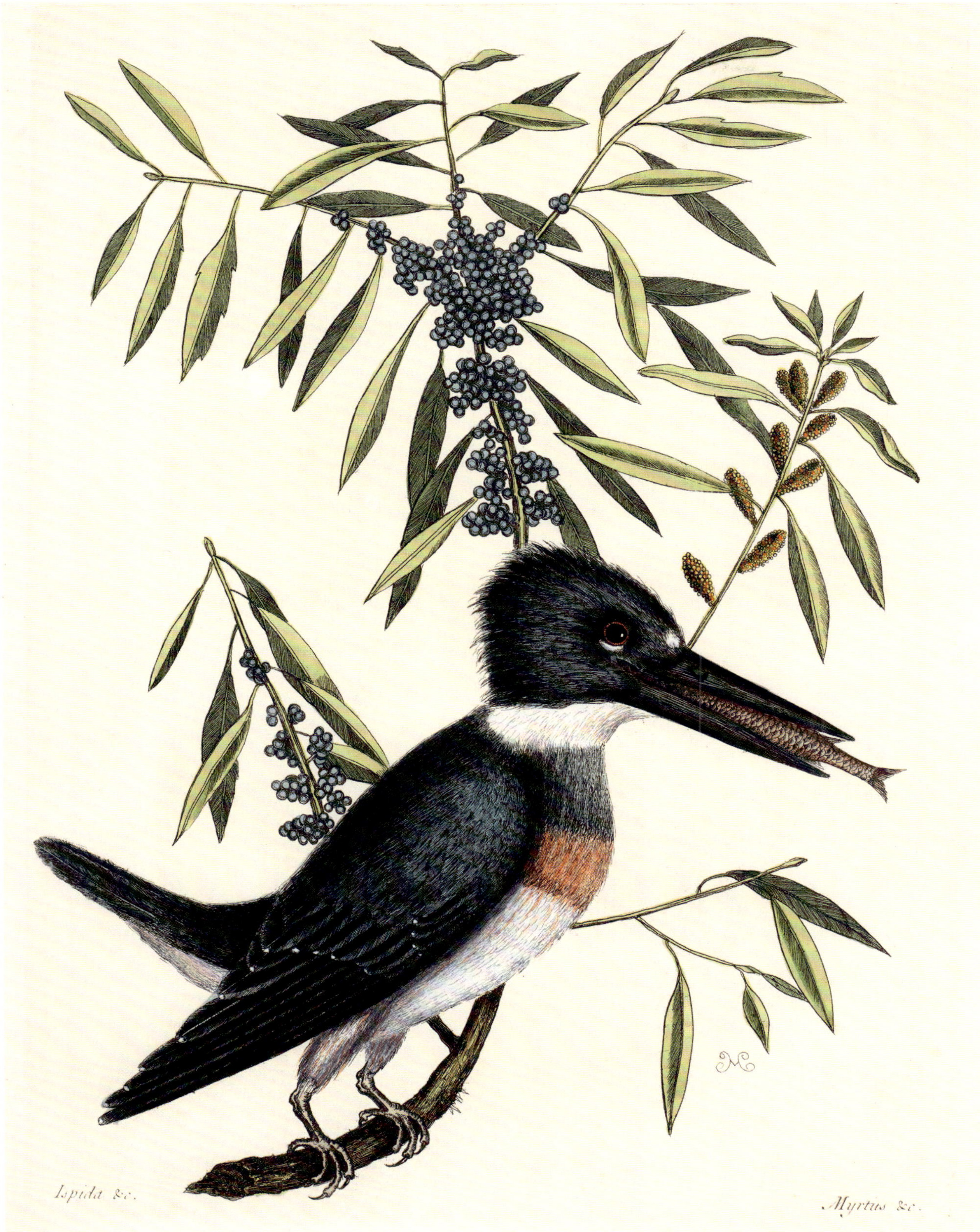

68 *left*
Cuban bullfinch (*Melopyrrha nigra*); fringetree (*Chionanthus virginicus*) (vol. 1, t. 68)

69 *above*
Belted kingfisher (*Megaceryle alcyon*); unidentified fish; wax myrtle (*Myrica cerifera*) (vol. 1, t. 69)

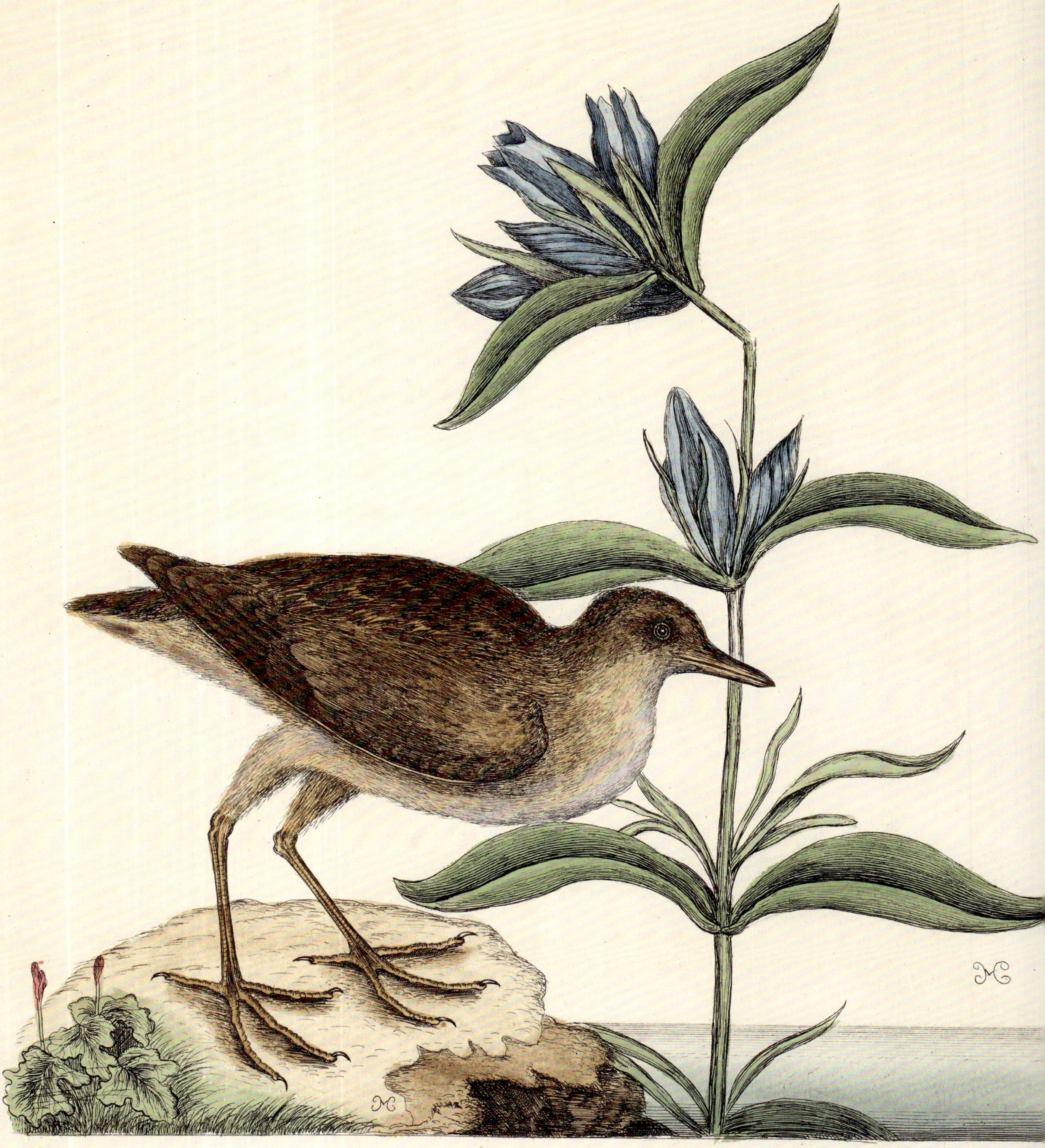

Gallinula Americana *Gentiana &c.*

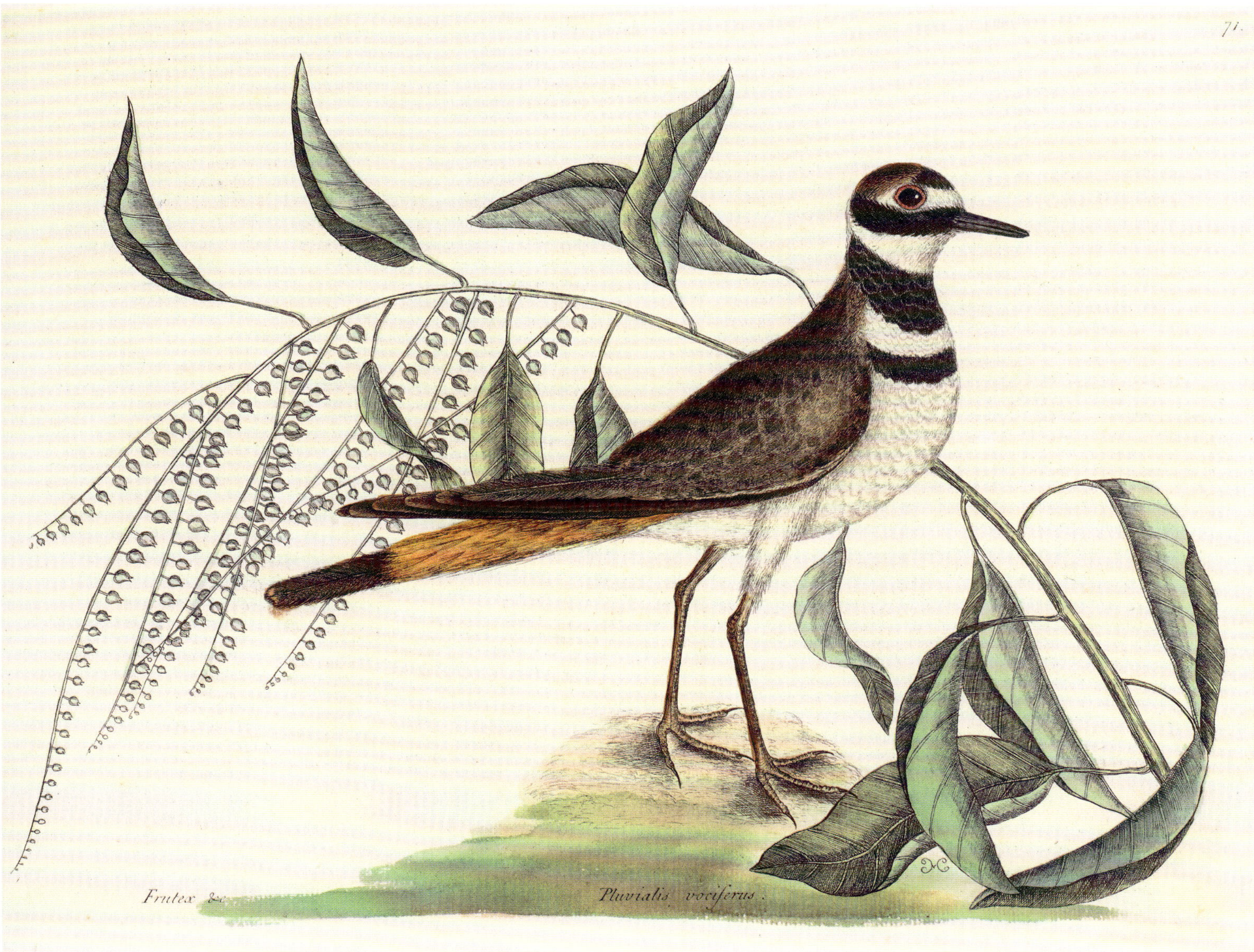

70 *left*
Sora (*Porzana carolina*); Elliott's gentian (*Gentiana catesbaei*) (vol. 1, t. 70)

71 *above*
Killdeer (*Charadrius vociferus*); sorrel tree (*Oxydendrum arboreum*) (vol. 1, t. 71)

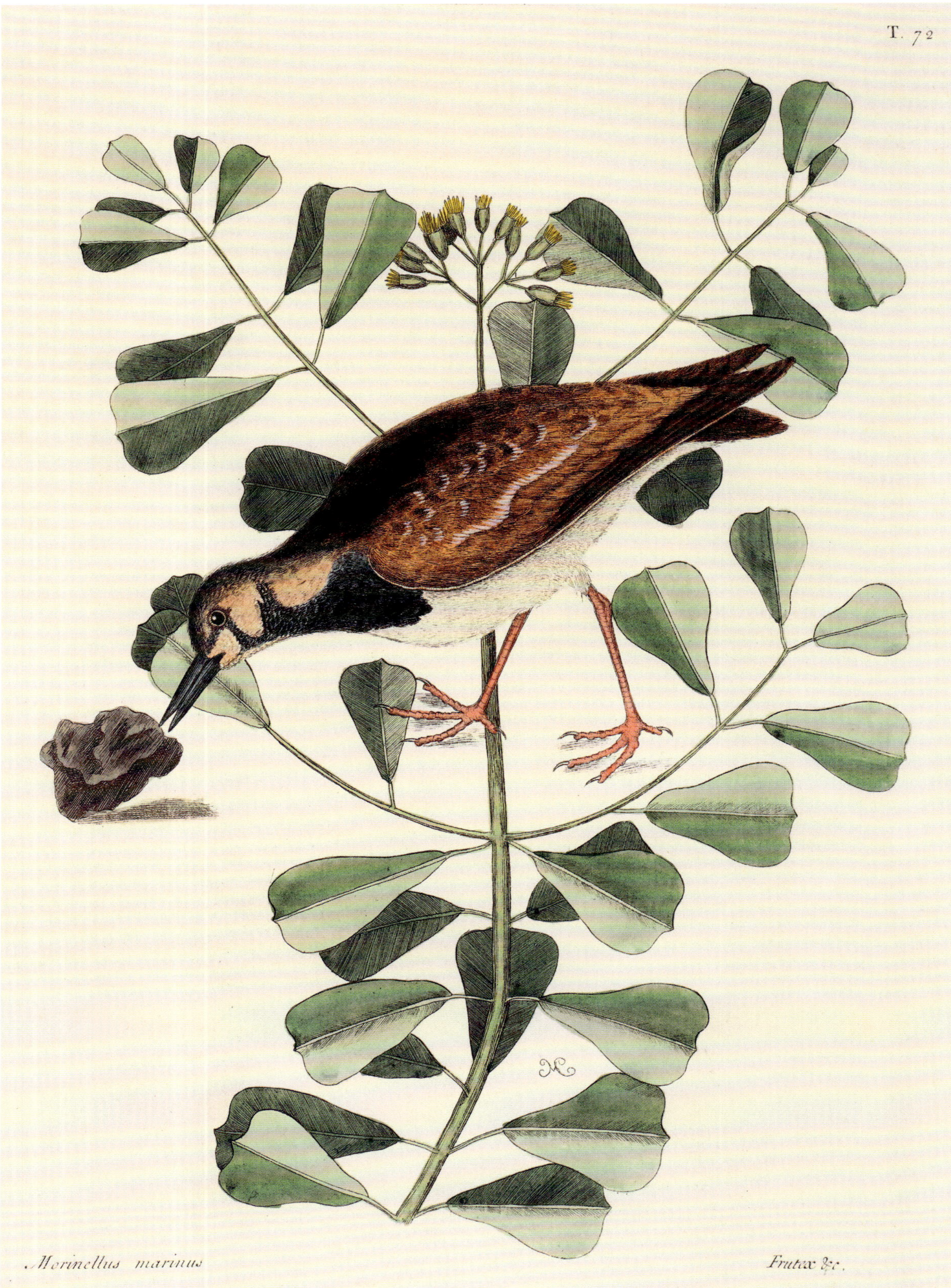

72 *above*

Ruddy turnstone (*Arenaria interpres*); shans (*Salmea petrochioides*) (vol. 1, t. 72)

73 *right*

American flamingo (*Phoenicopterus ruber*); double-forked plexaurella (*Plexaurella dichotoma*) (vol. 1, t. 73)

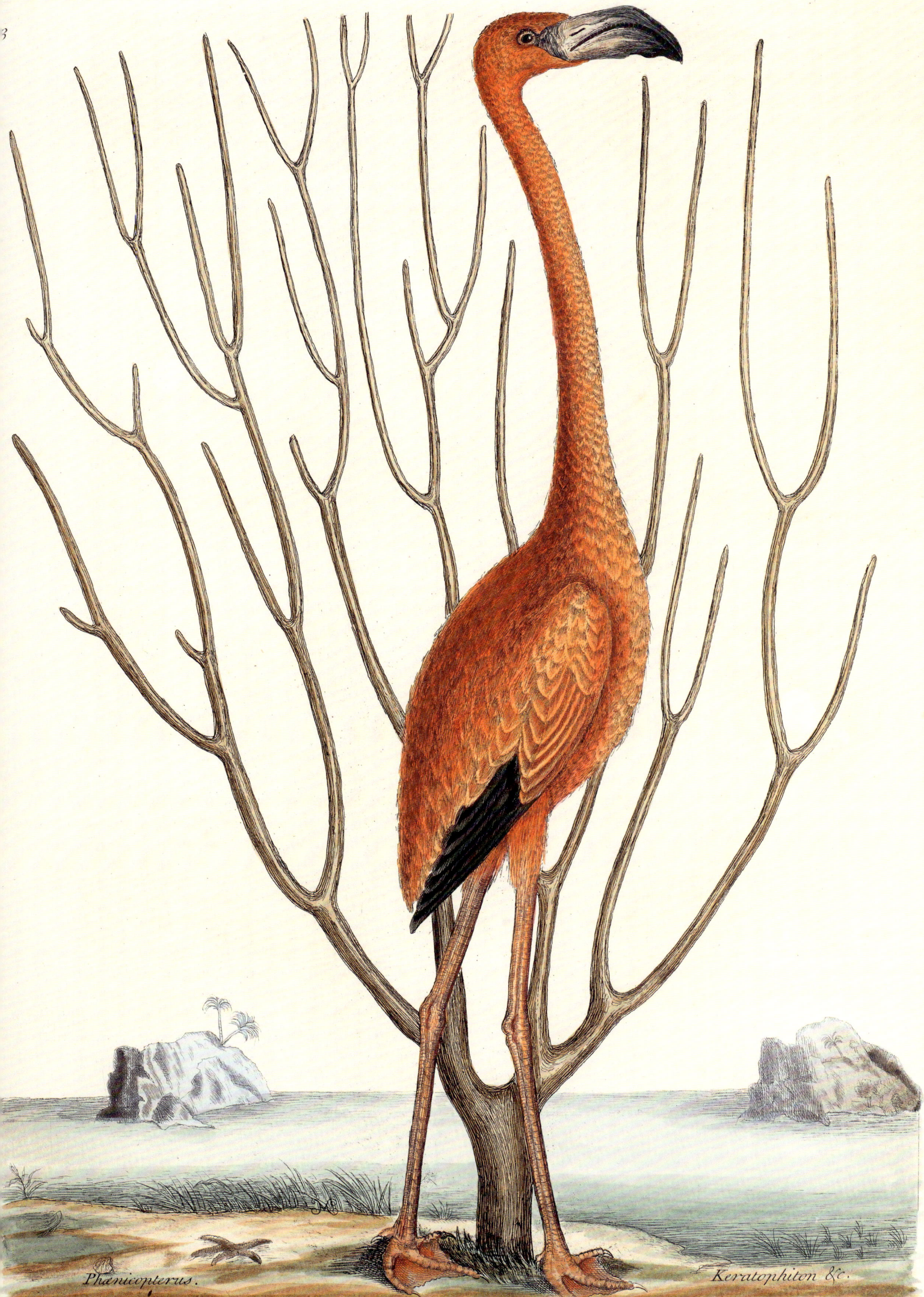
Phænicopterus.
Keratophiton &c.

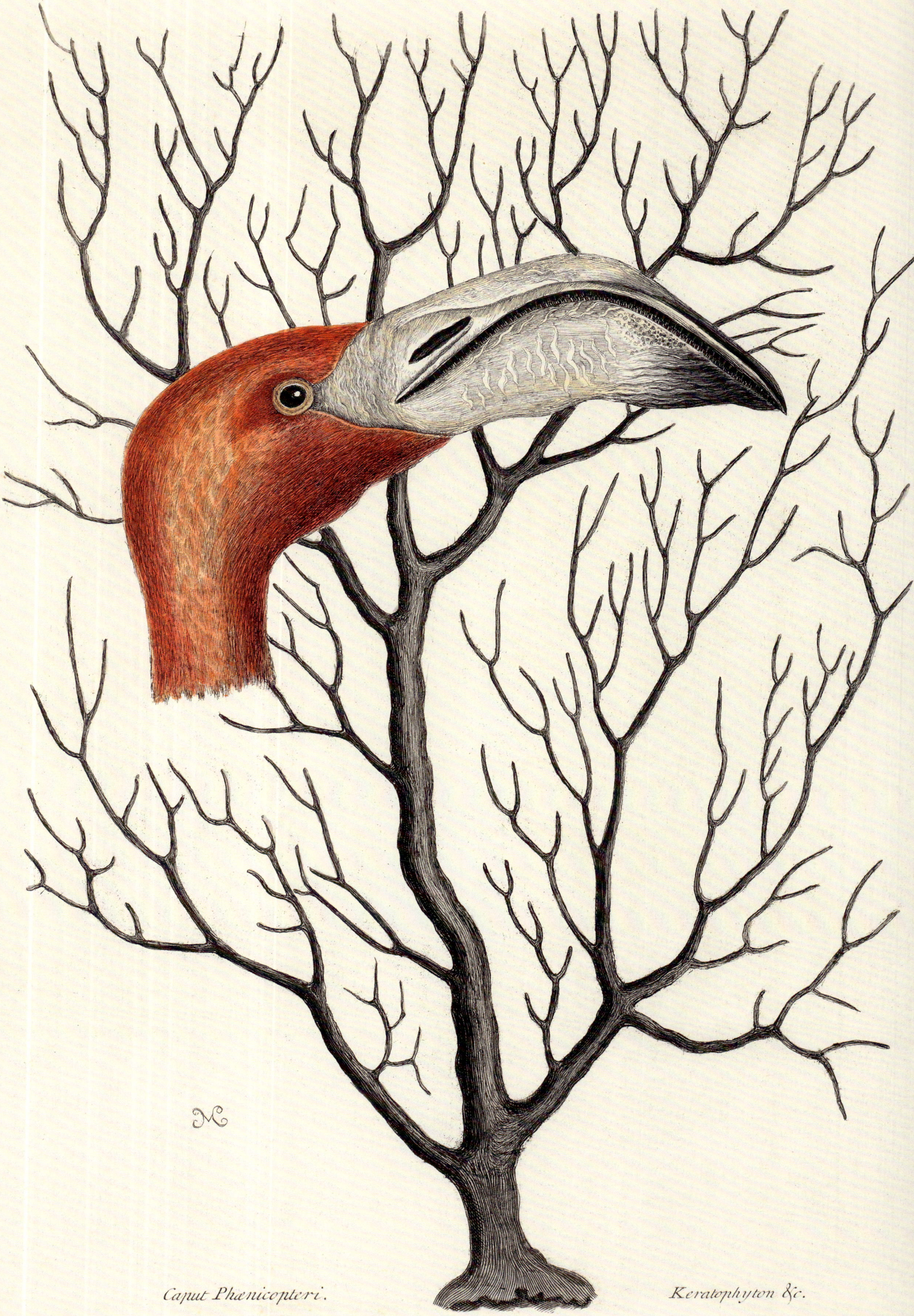

Caput Phœnicopteri.　　*Keratophyton &c.*

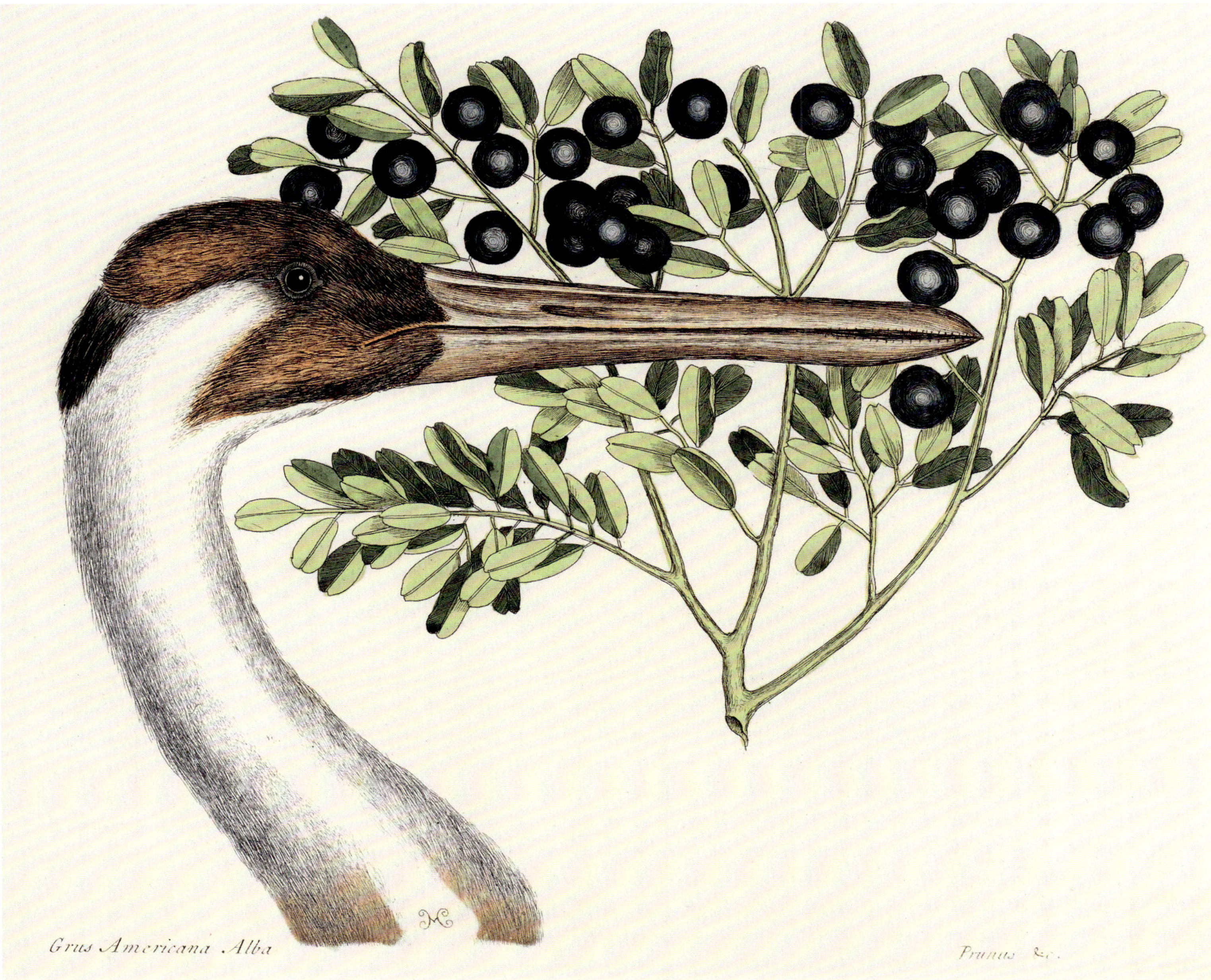

74 *left*
American flamingo (*Phoenicopterus ruber*); bent sea rod (*Plexaurella flexuosa*) (vol. 1, t. 74)

75 *above*
Whooping crane (*Grus americana*); darlingplum (*Reynosia septentrionalis*) (vol. 1, t. 75)

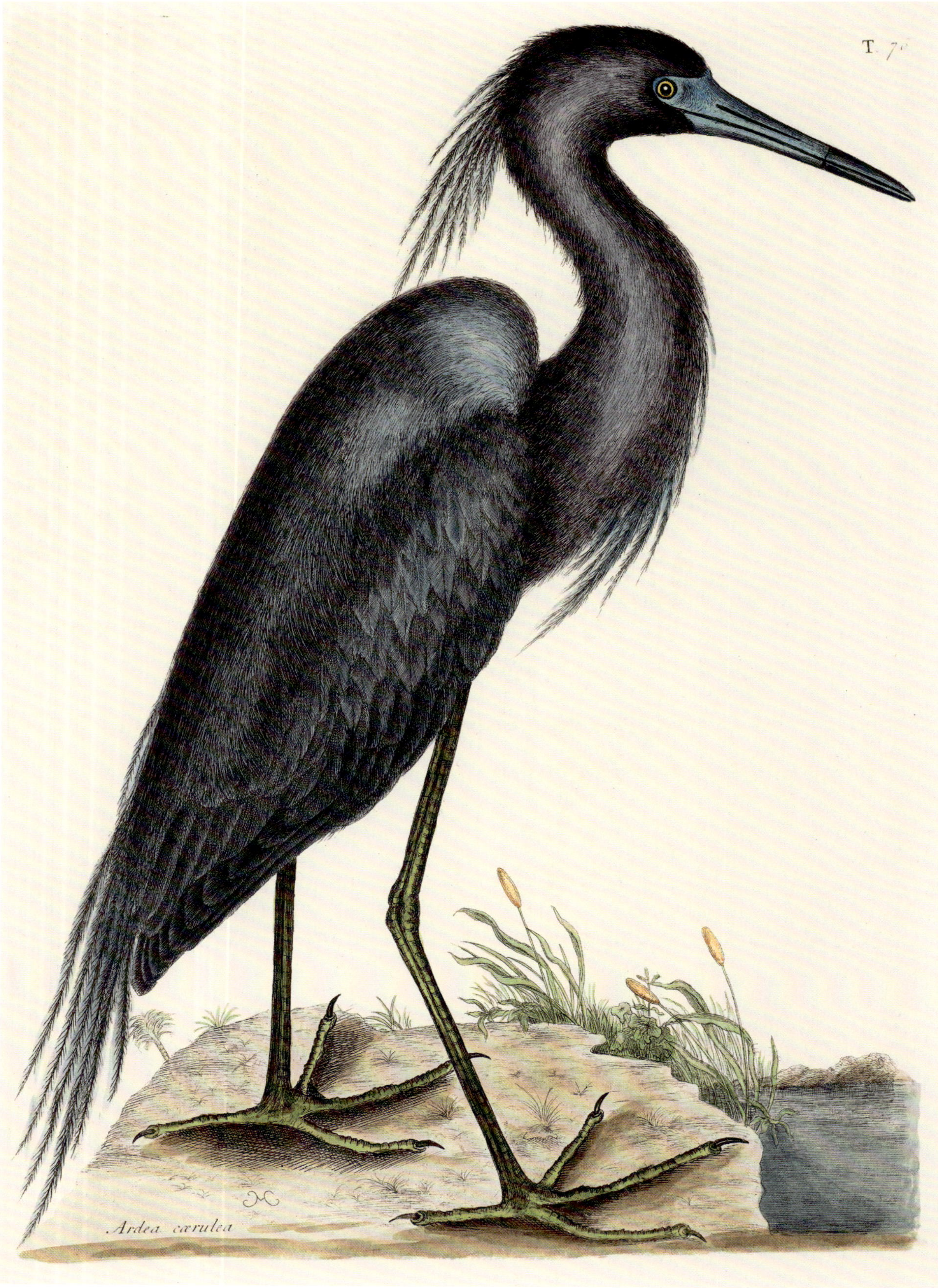

76 *above*
Little blue heron (*Egretta caerulea*) (vol. 1, t. 76)

77 *right*
Little blue heron (*Egretta caerulea*); Bahamian phymosia (*Phymosia abutiloides*) (vol. 1, t. 77)

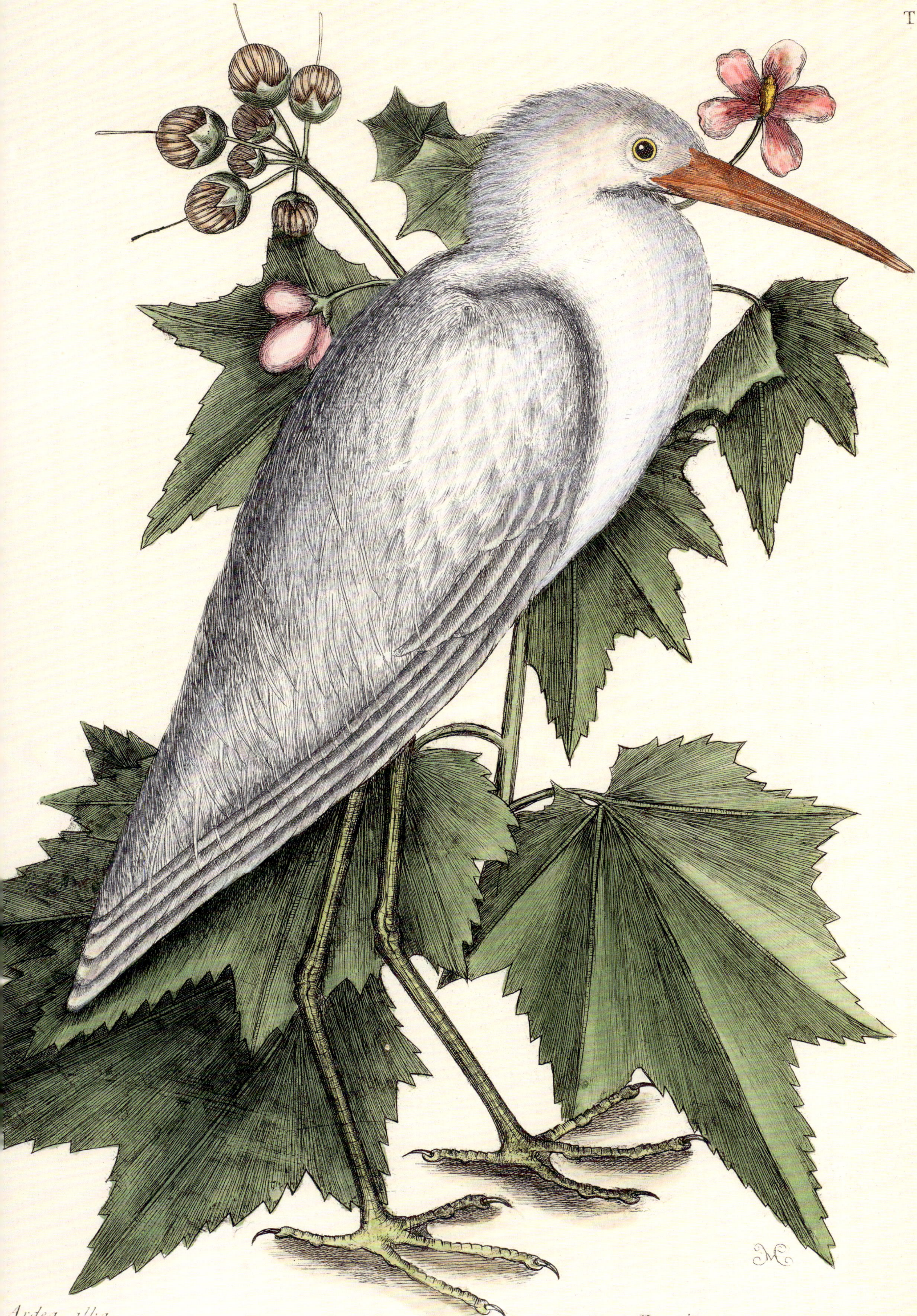

Ardea alba *Ketmia &c.*

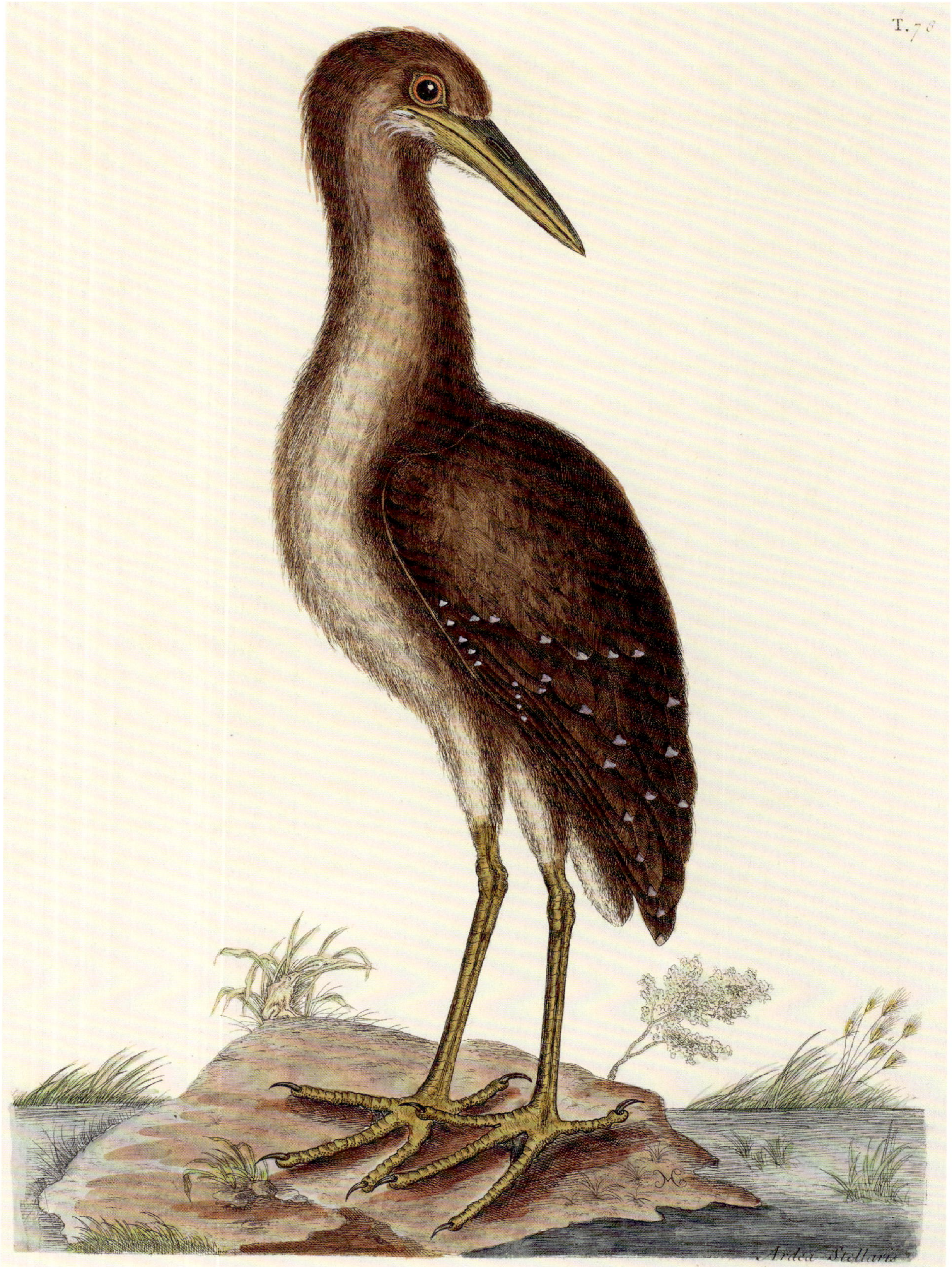

78 *above*

American bittern (*Botaurus lentiginosus*) (vol. 1, t. 78)

79 *right*

Yellow-crowned night-heron (*Nyctanassa violacea*); gullfeed (*Scaevola plumieri*) (vol. 1, t. 79)

T. 7

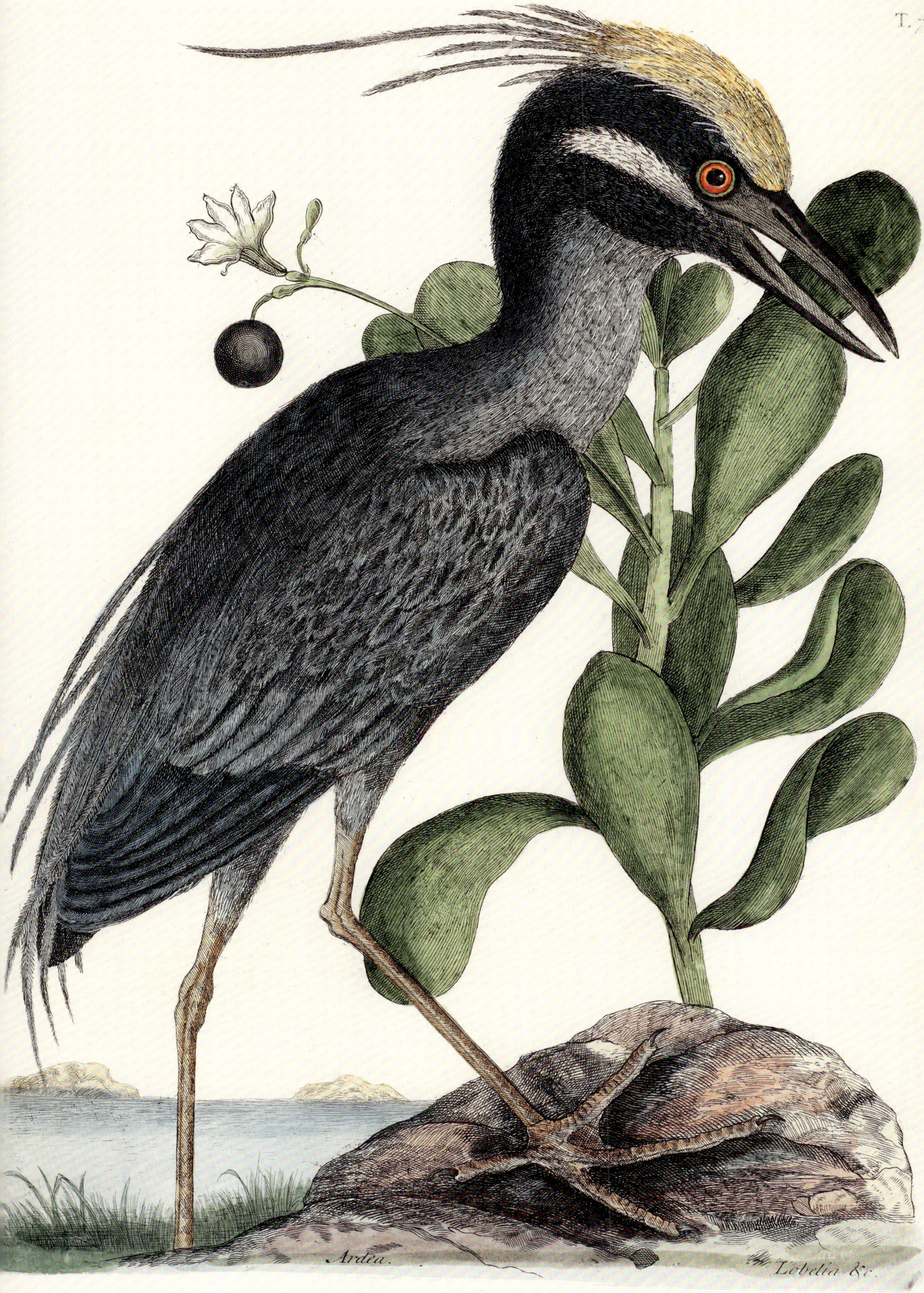

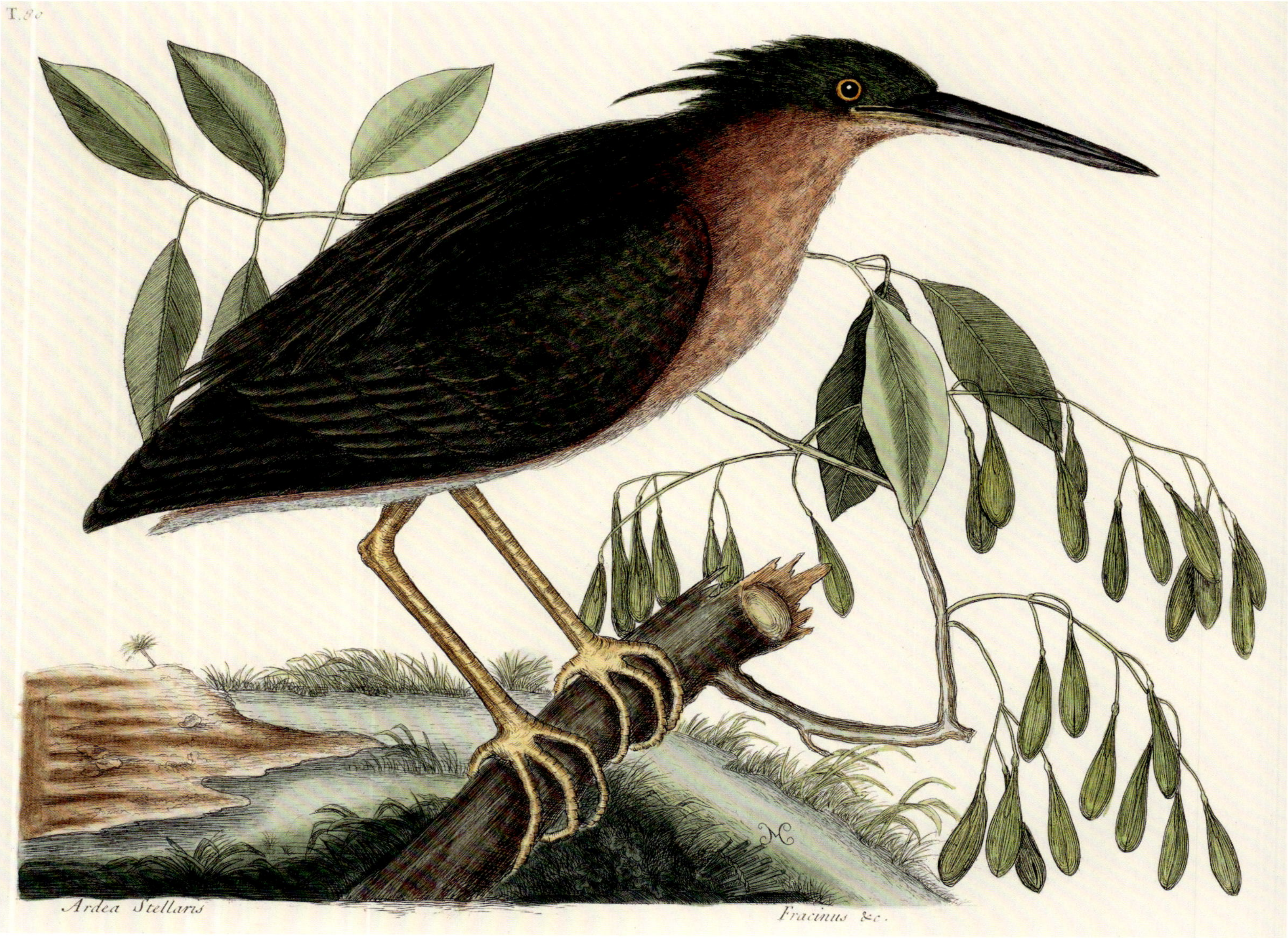

80 *above*

Green heron (*Butorides virescens*); Carolina ash (*Fraxinus caroliniana*) (vol. 1, t. 80)

81 *right*

Wood stork (*Mycteria americana*) (vol. 1, t. 81)

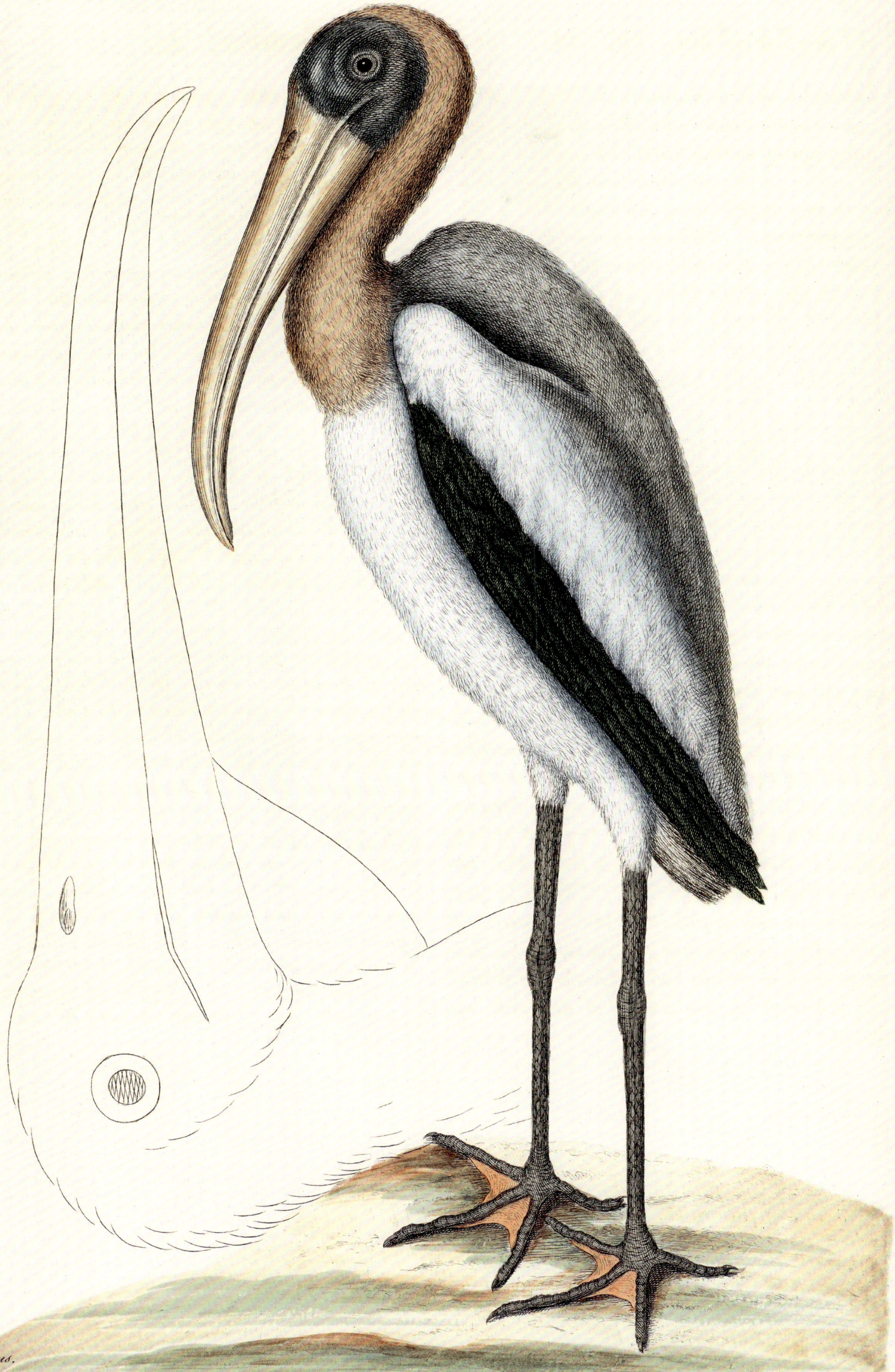

[...]anus Sylvaticus.
[...]e Wood Pelican.

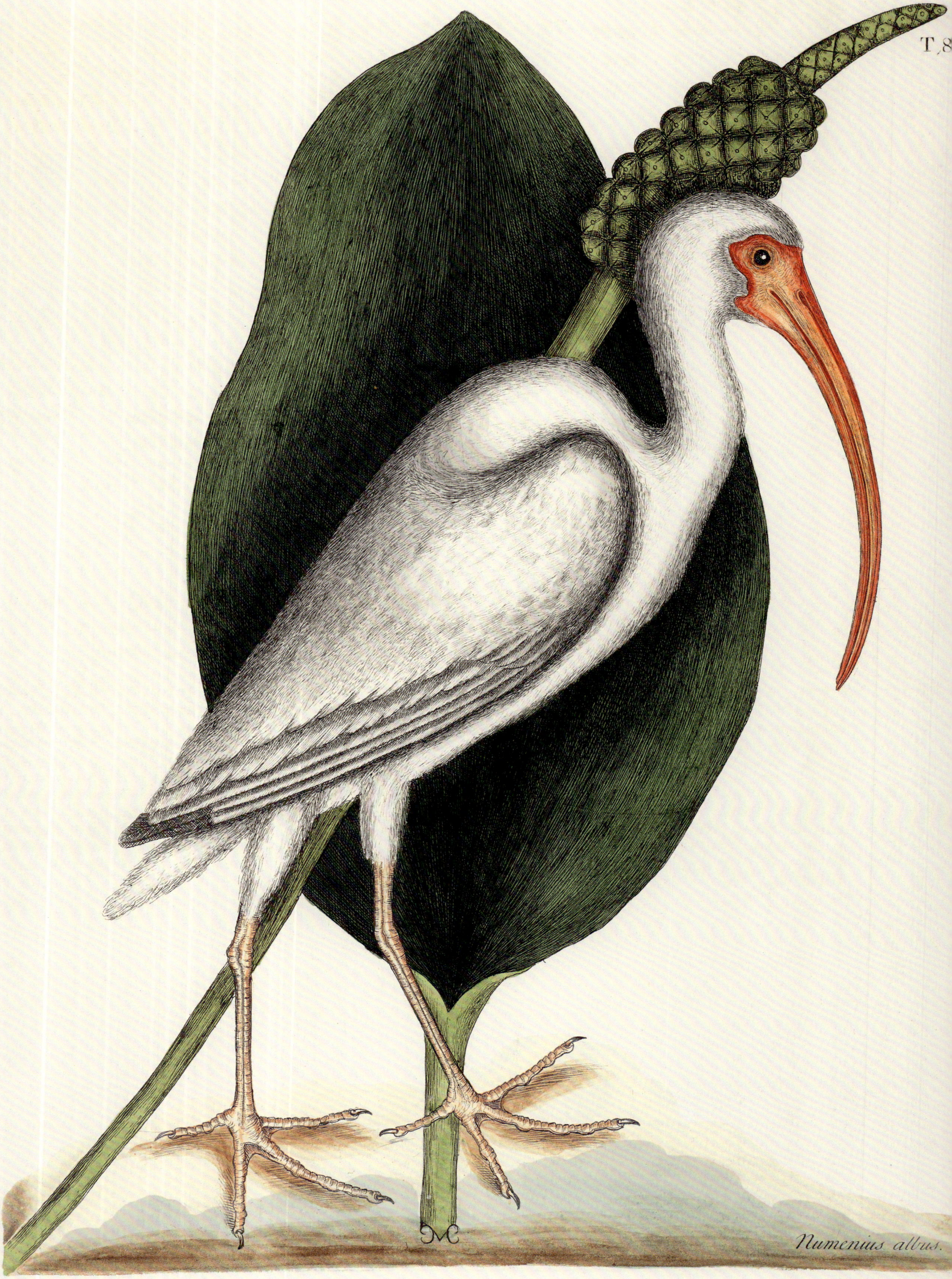
T. 8
Numenius albus.

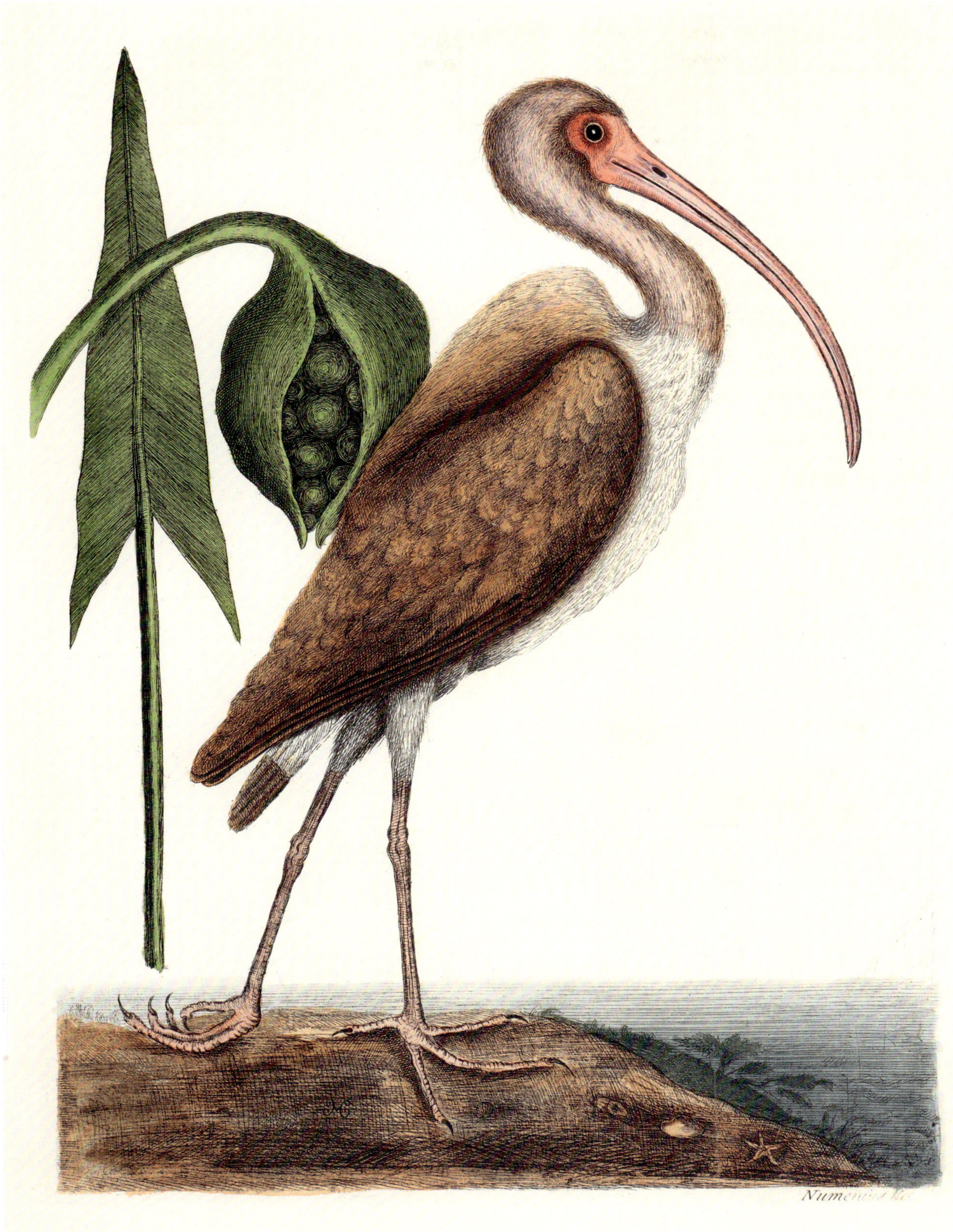

82 *left*
White ibis (*Eudocimus albus*); goldenclub (*Orontium aquaticum*) (vol. 1, t. 82)

83 *above*
White ibis (*Eudocimus albus*); green arrow arum (*Peltandra virginica*) (vol. 1, t. 83)

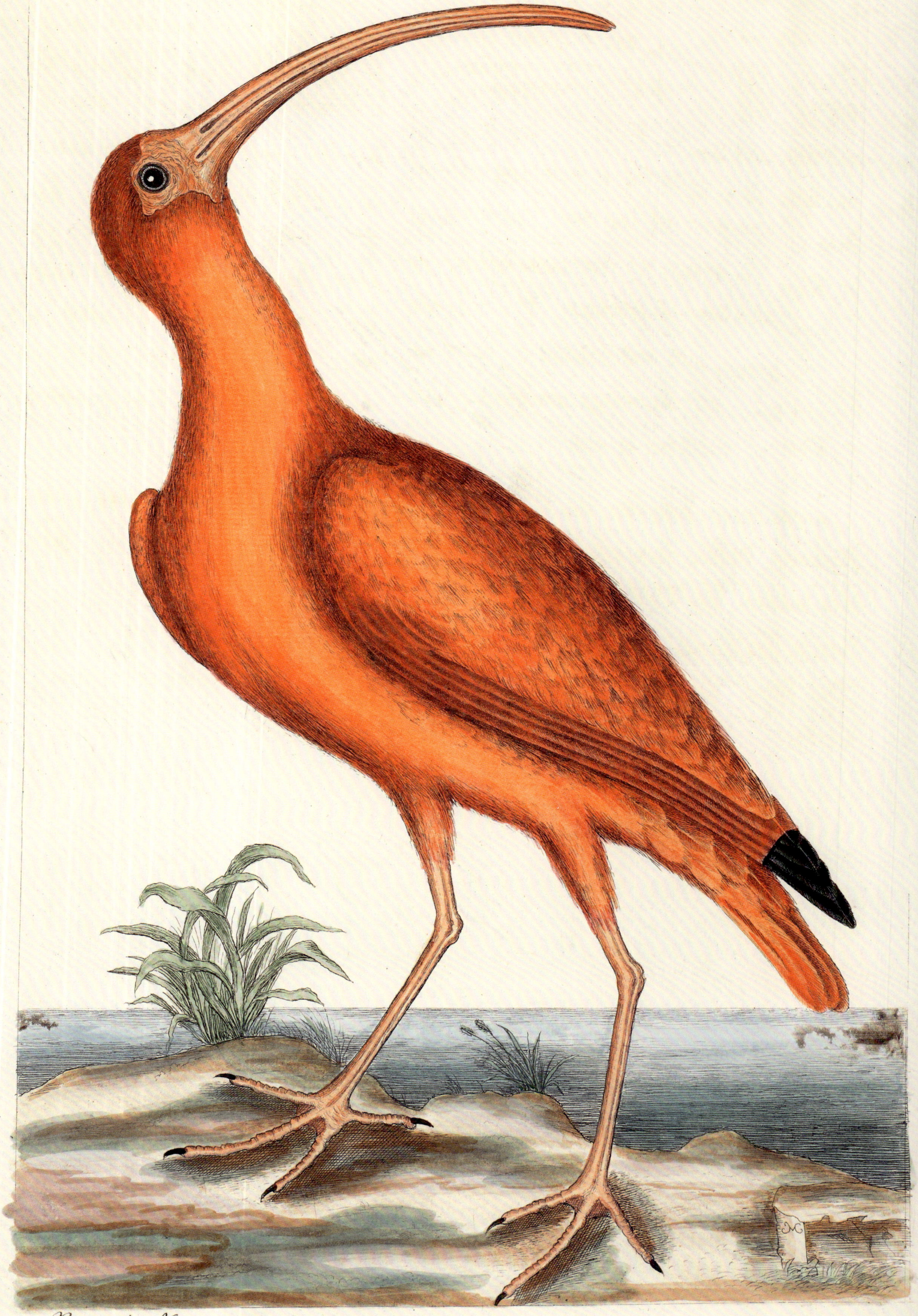

Numenius &c

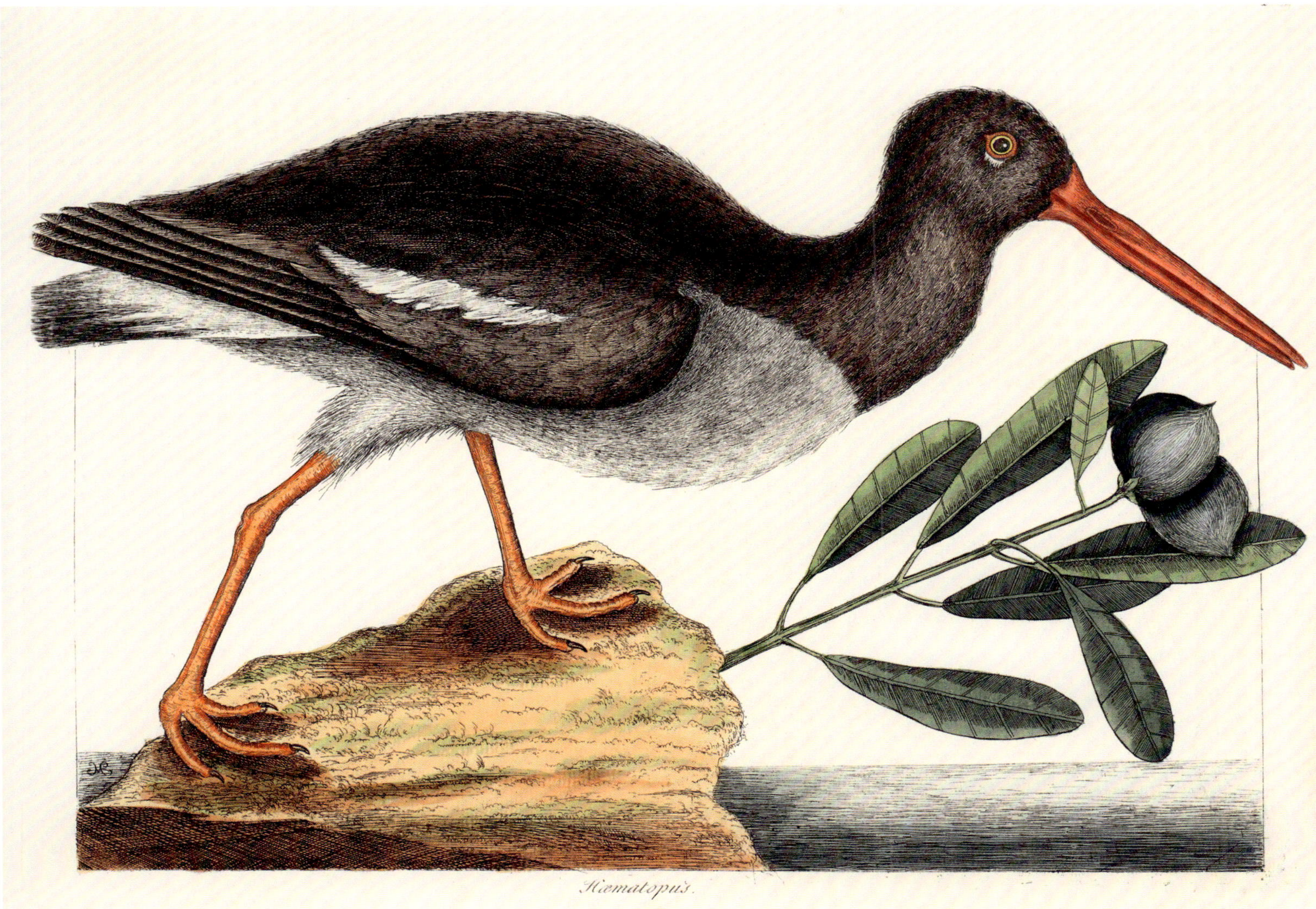

84 *left*
Scarlet ibis (*Eudocimus ruber*) (vol. 1, t. 84)

85 *above*
American oystercatcher (*Haematopus palliatus*); black mangrove (*Avicennia germinans*) (vol. 1, t. 85)

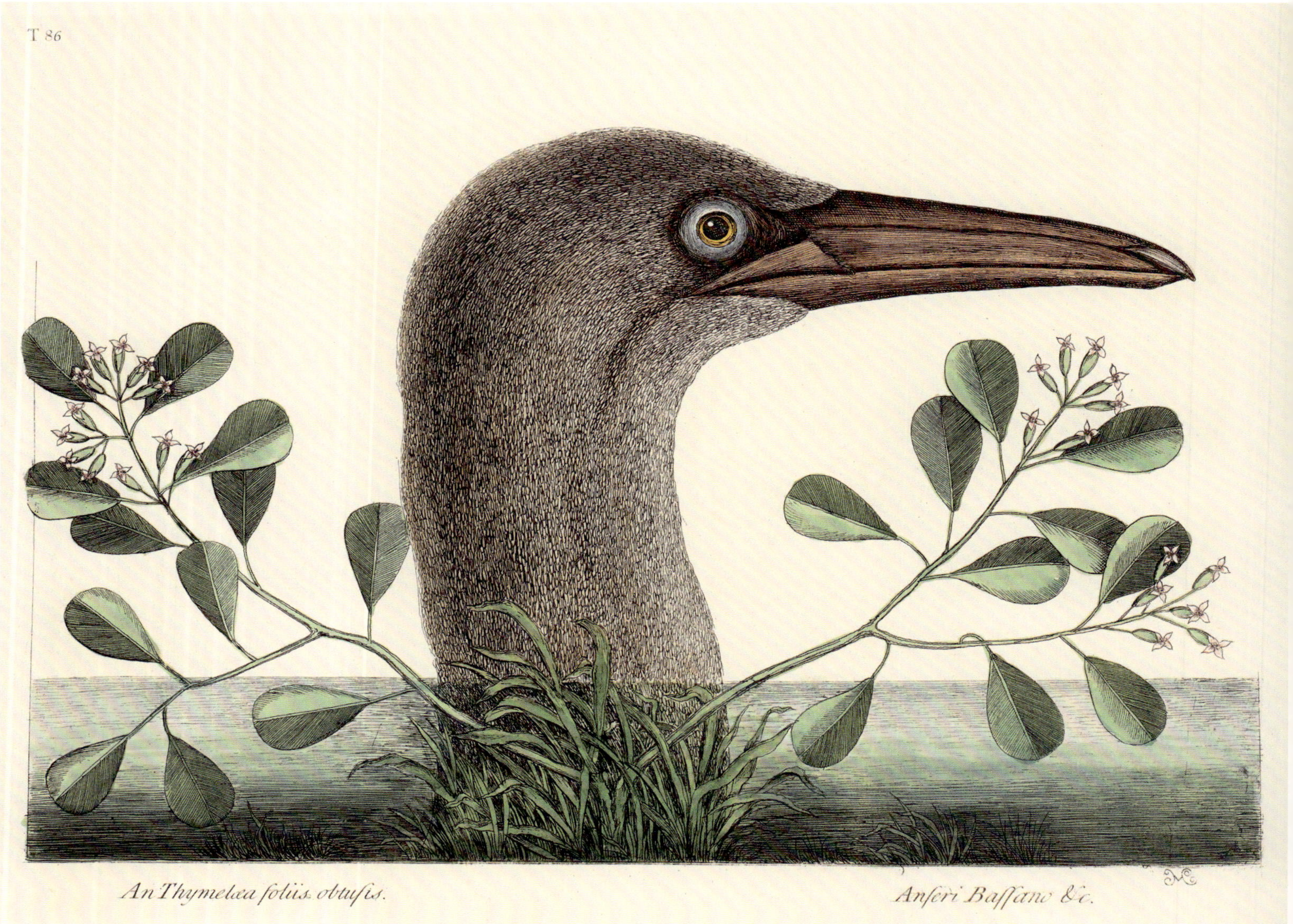

86 *above*
Northern gannet (*Morus bassanus*); white mangrove (*Laguncularia racemosa*) (vol. 1, t. 86)

87 *right*
Brown booby (*Sula leucogaster*) (vol. 1, t. 87)

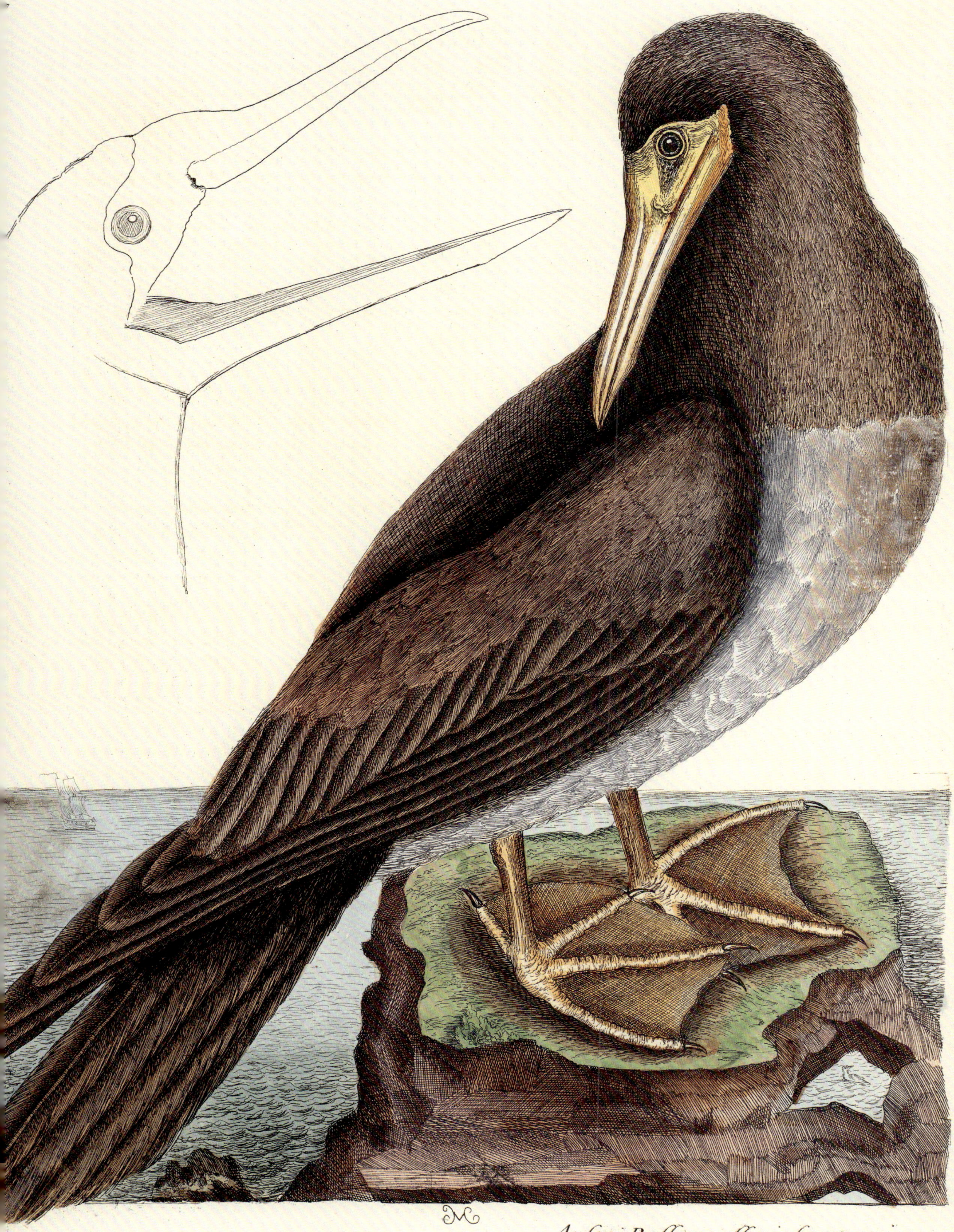

Anseri Bassano affinis fusca avis.

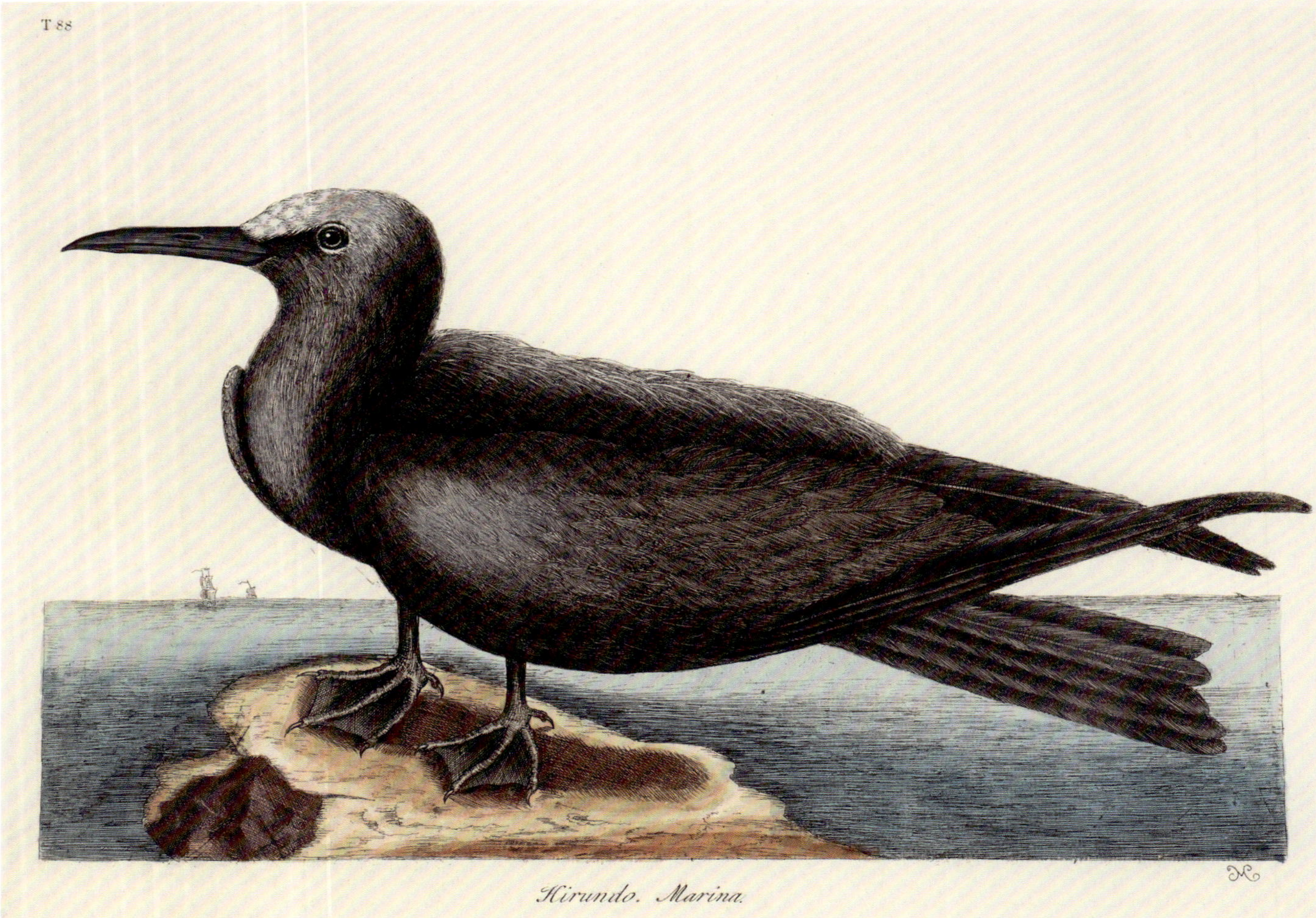
T 88
Hirundo. Marina.

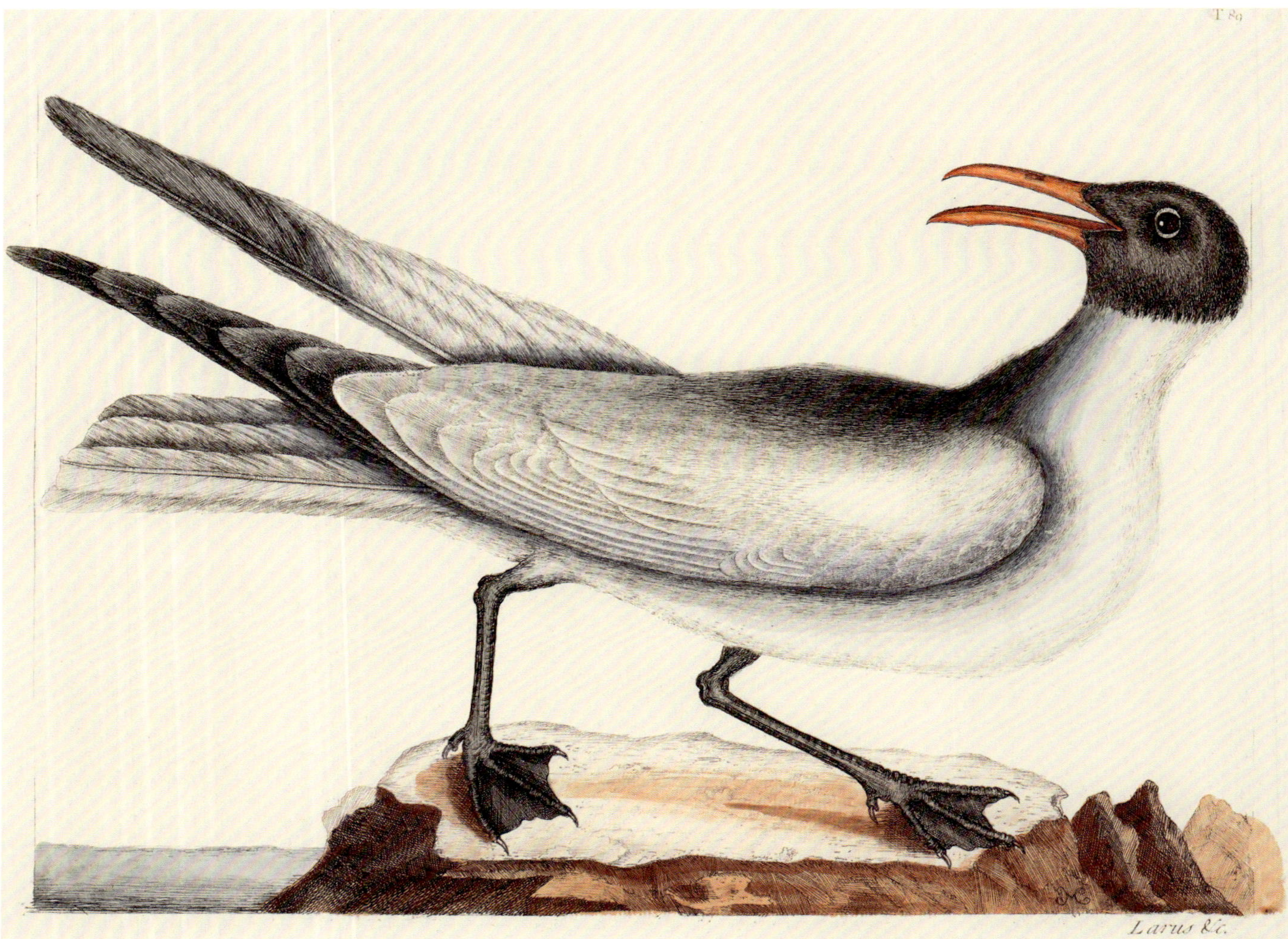
T 89
Larus &c.

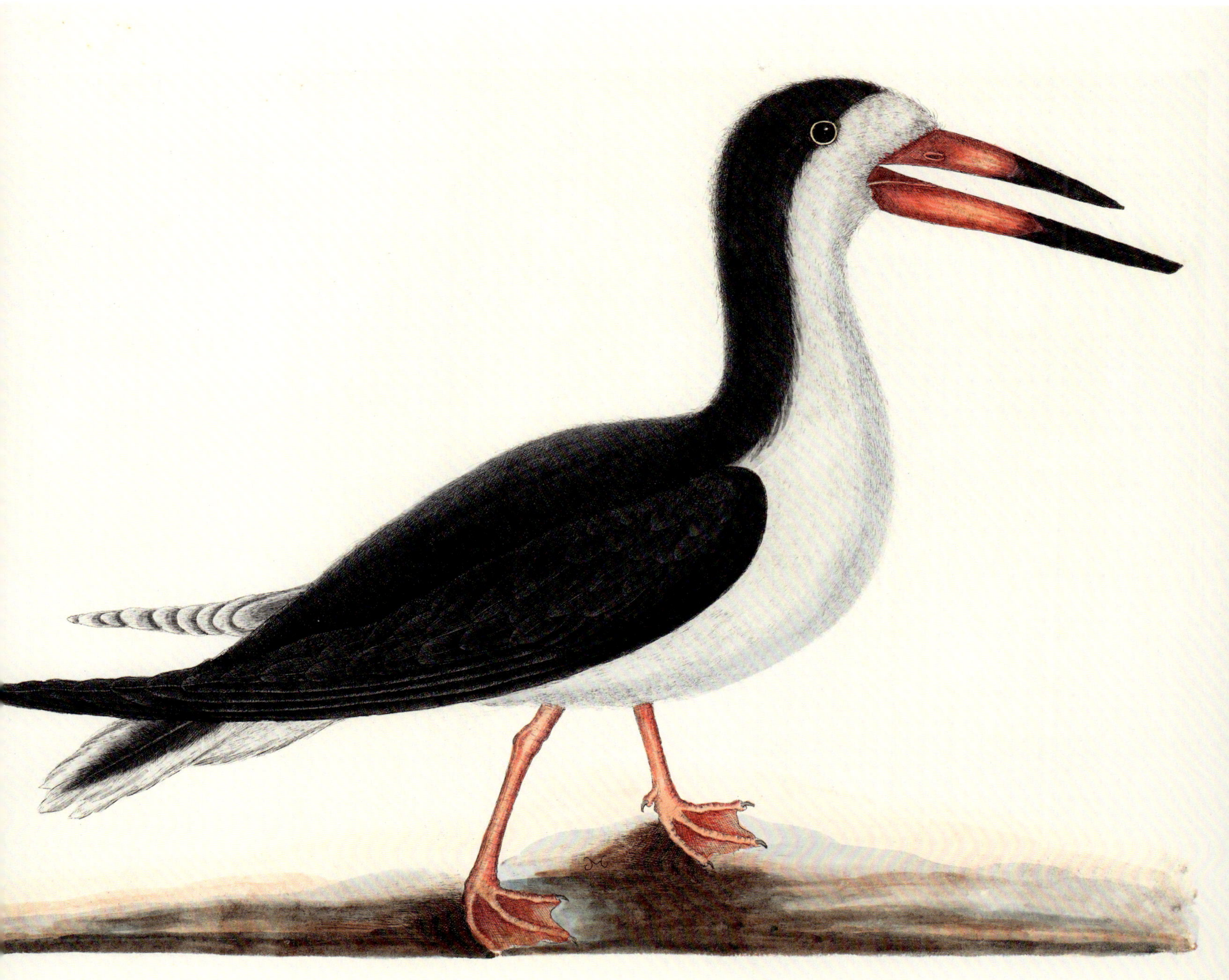

88 *left top*
Brown noddy (*Ancus stolidus*) (vol. 1, t. 88)

89 *left bottom*
Laughing gull (*Leucophaeus atricilla*) (vol. 1, t. 89)

90 *above*
Black skimmer (*Rynchops niger*) (vol. 1, t. 90)

T 91

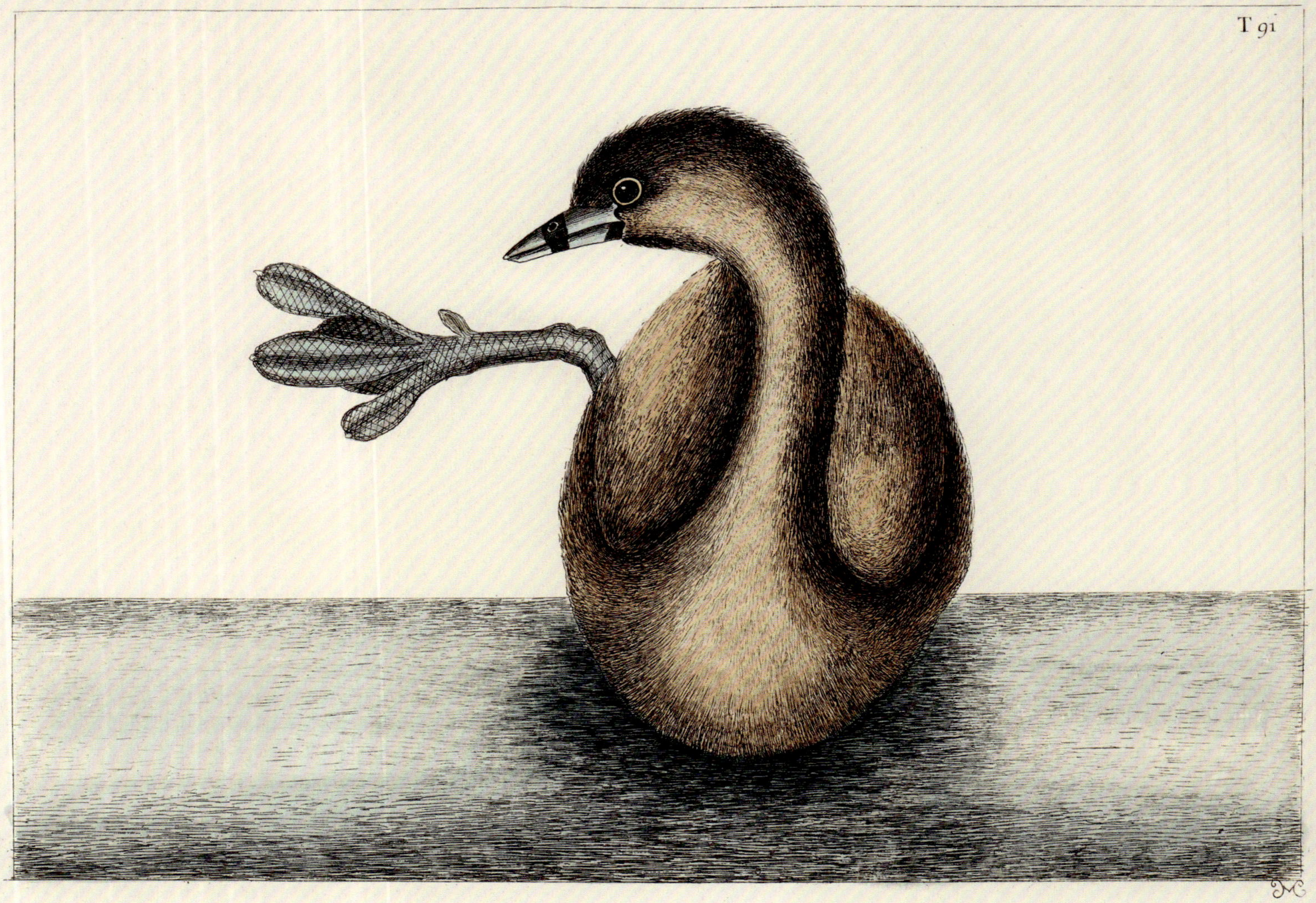

Podicipes &c.

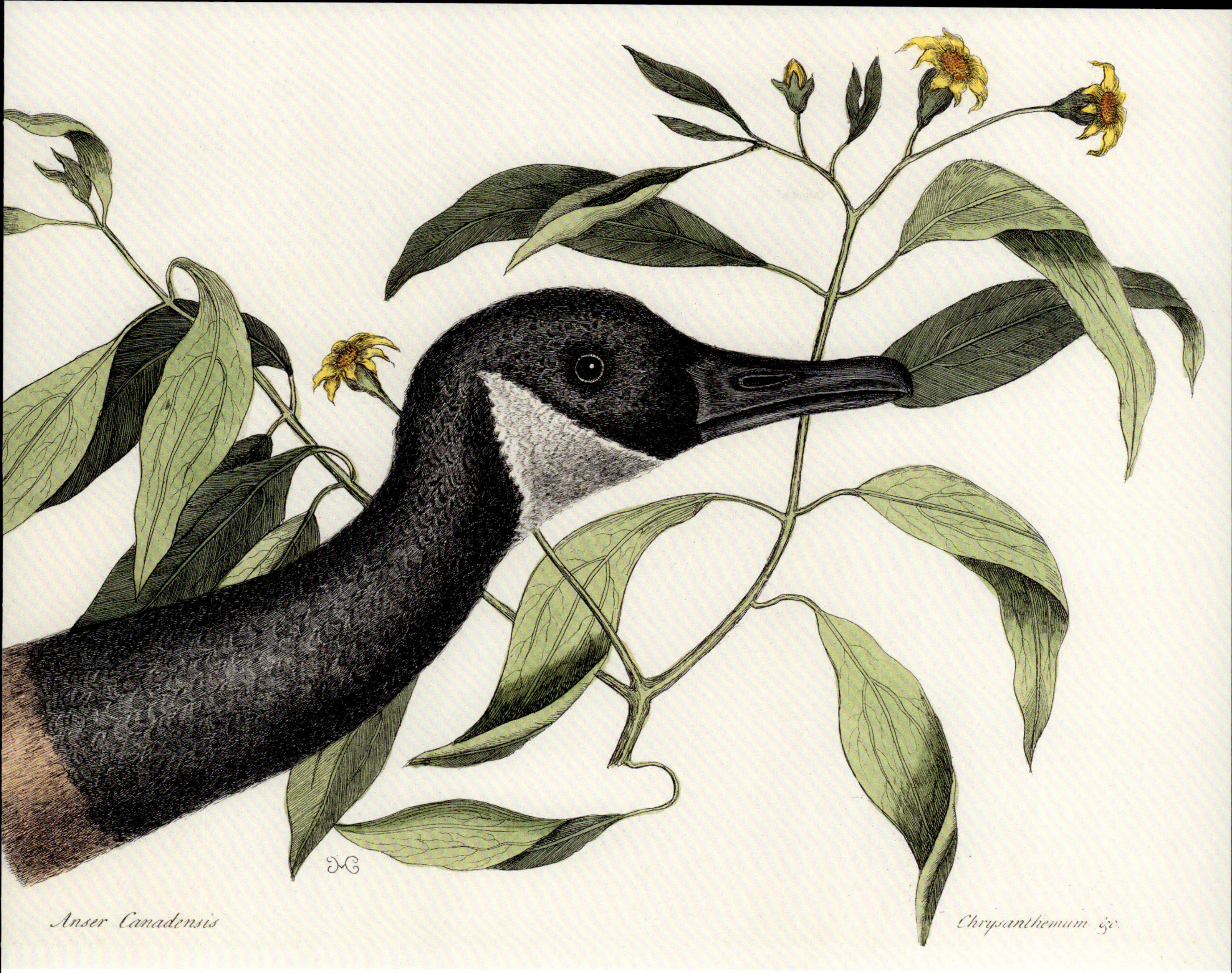

91 *above left*
Pied-billed grebe (*Podilymbus podiceps*) (vol. 1, t. 91)

92 *above*
Canadian goose (*Branta canadensis*); rong bush (*Wedelia bahamensis*) (vol. 1, t. 92)

following pages

93
White-cheeked pintail (*Anas bahamensis*); sea ox-eye (*Borrichia arborescens*) (vol. 1, t. 93)

94 *left top*
Hooded merganser (*Lophodytes cucullatus*) (vol. 1, t. 94)

95 *left bottom*
Bufflehead (*Bucephala albeola*) (vol. 1, t. 95)

96 *right*
Northern shoveler (*Anas clypeata*) (vol. 1, t. 96)

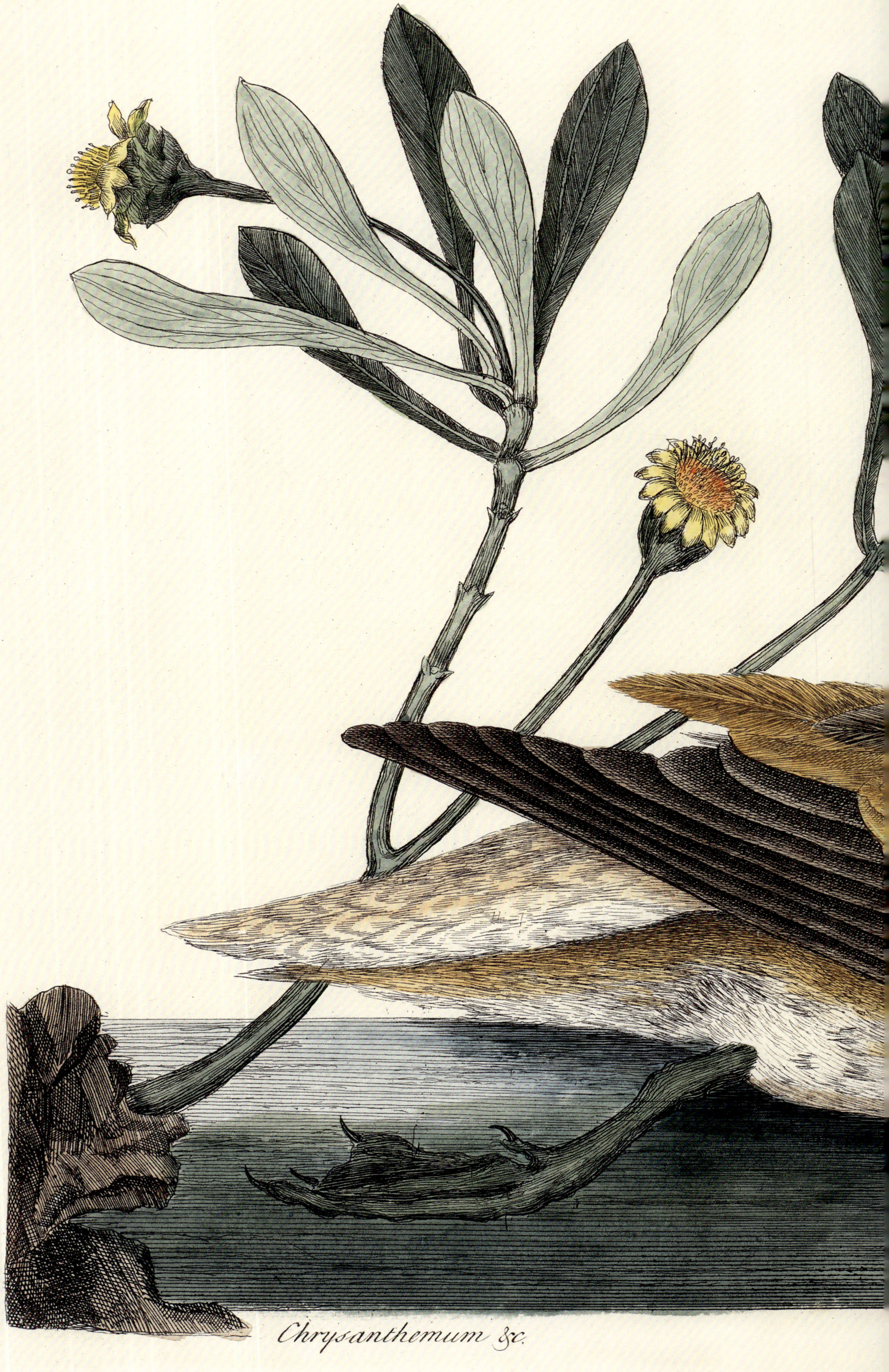

Chrysanthemum &c.

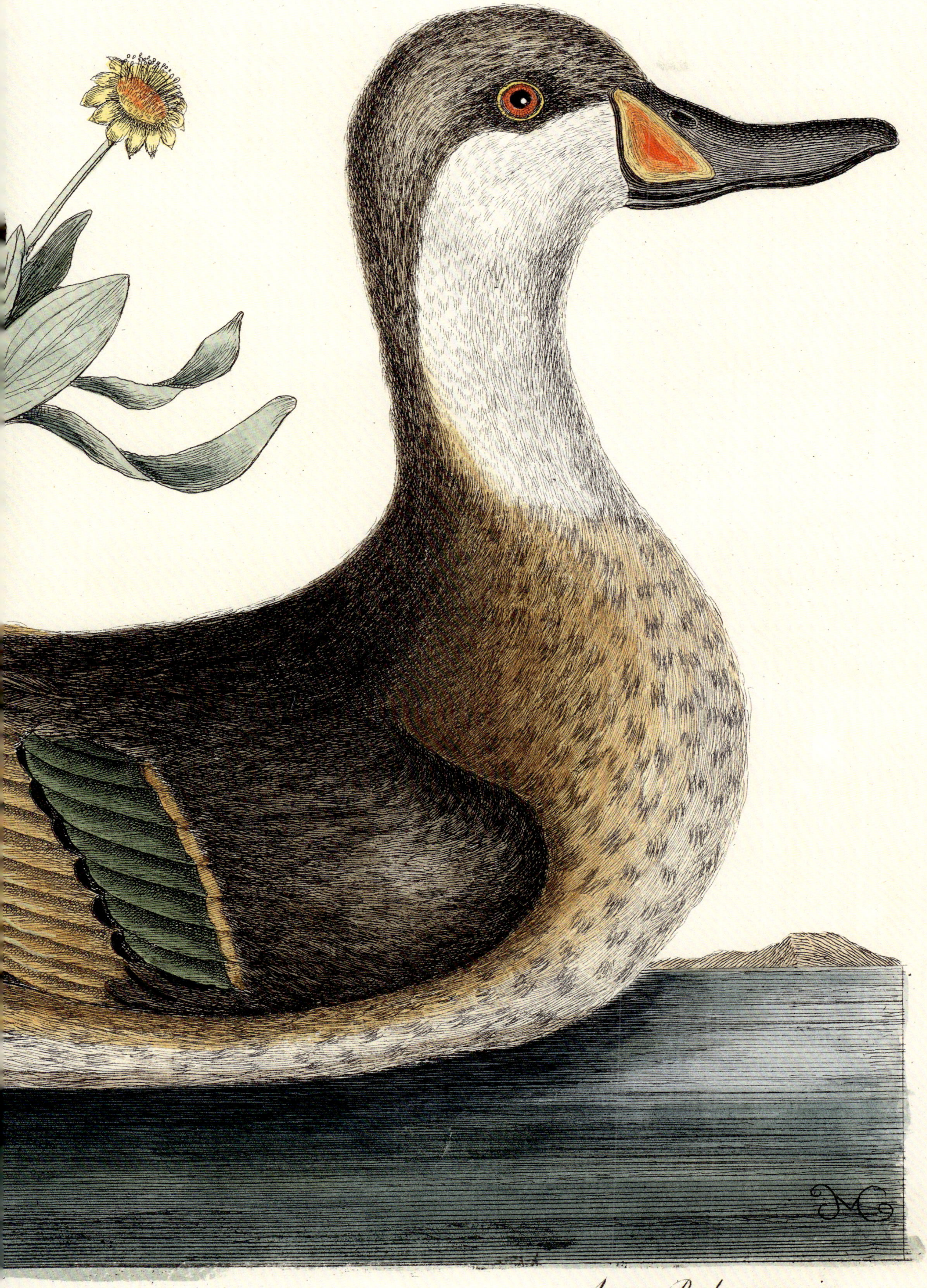

Anas Bahamensis.

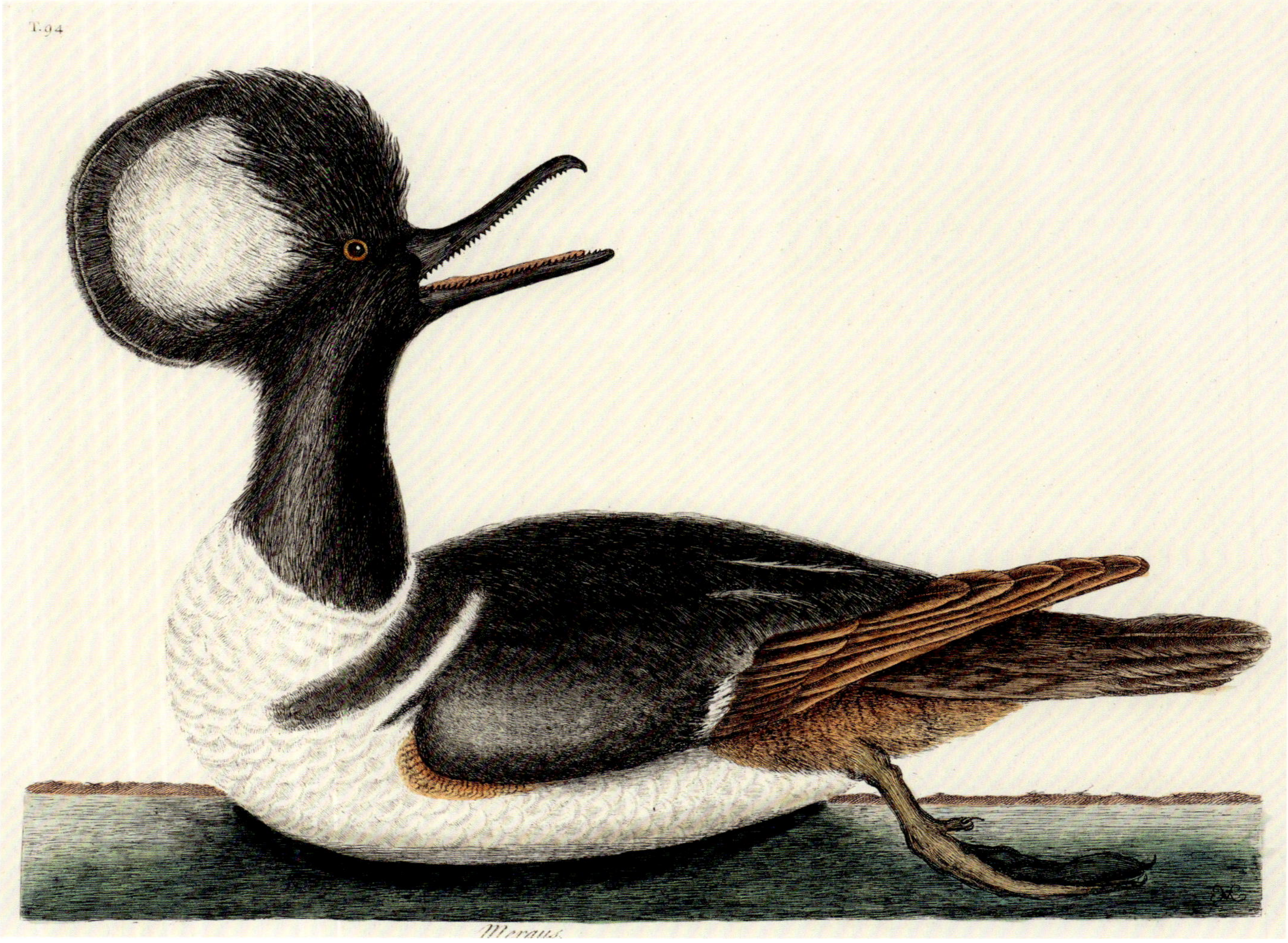
T.94
Mergus.

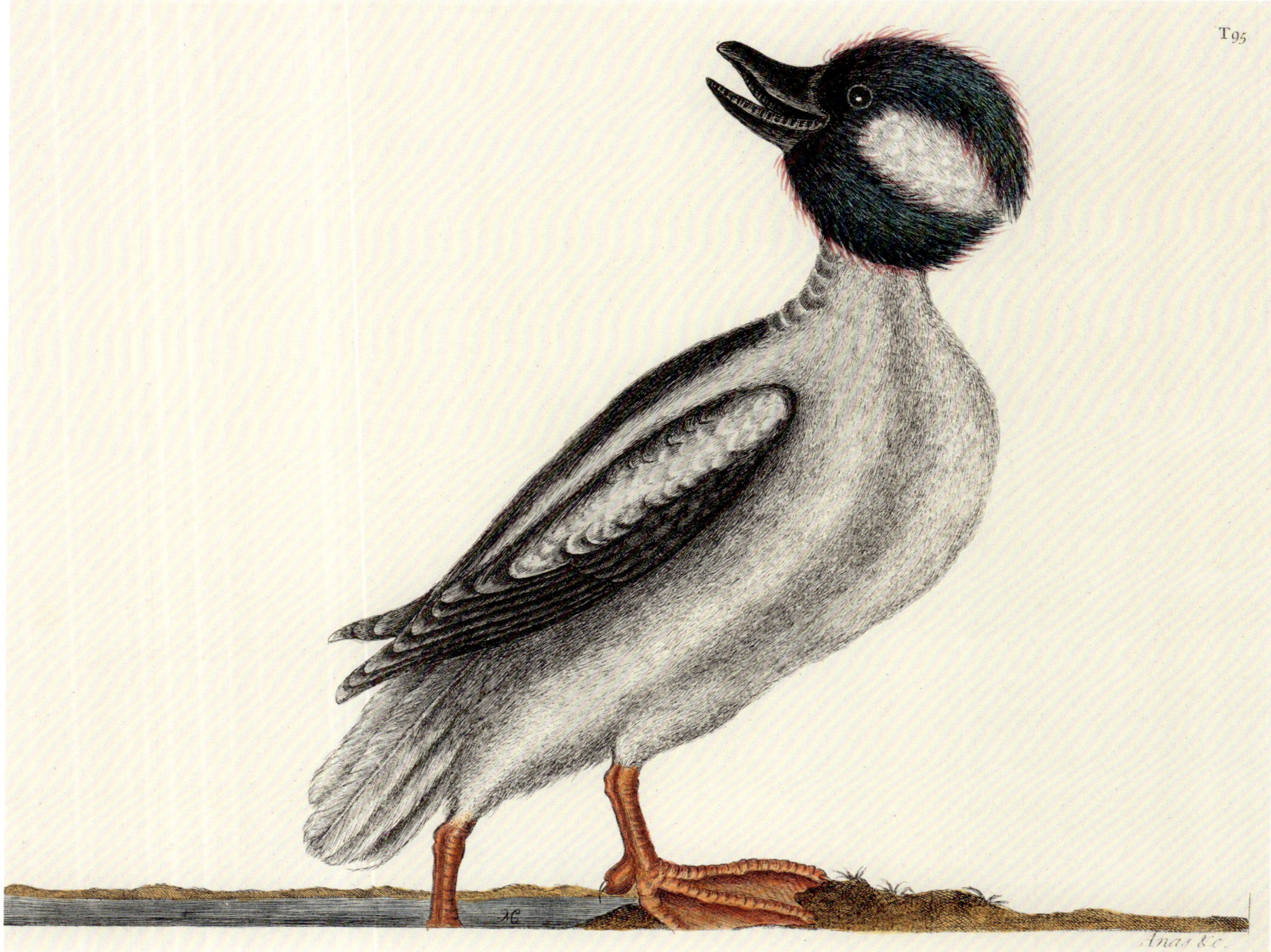
T95
Anas &c.

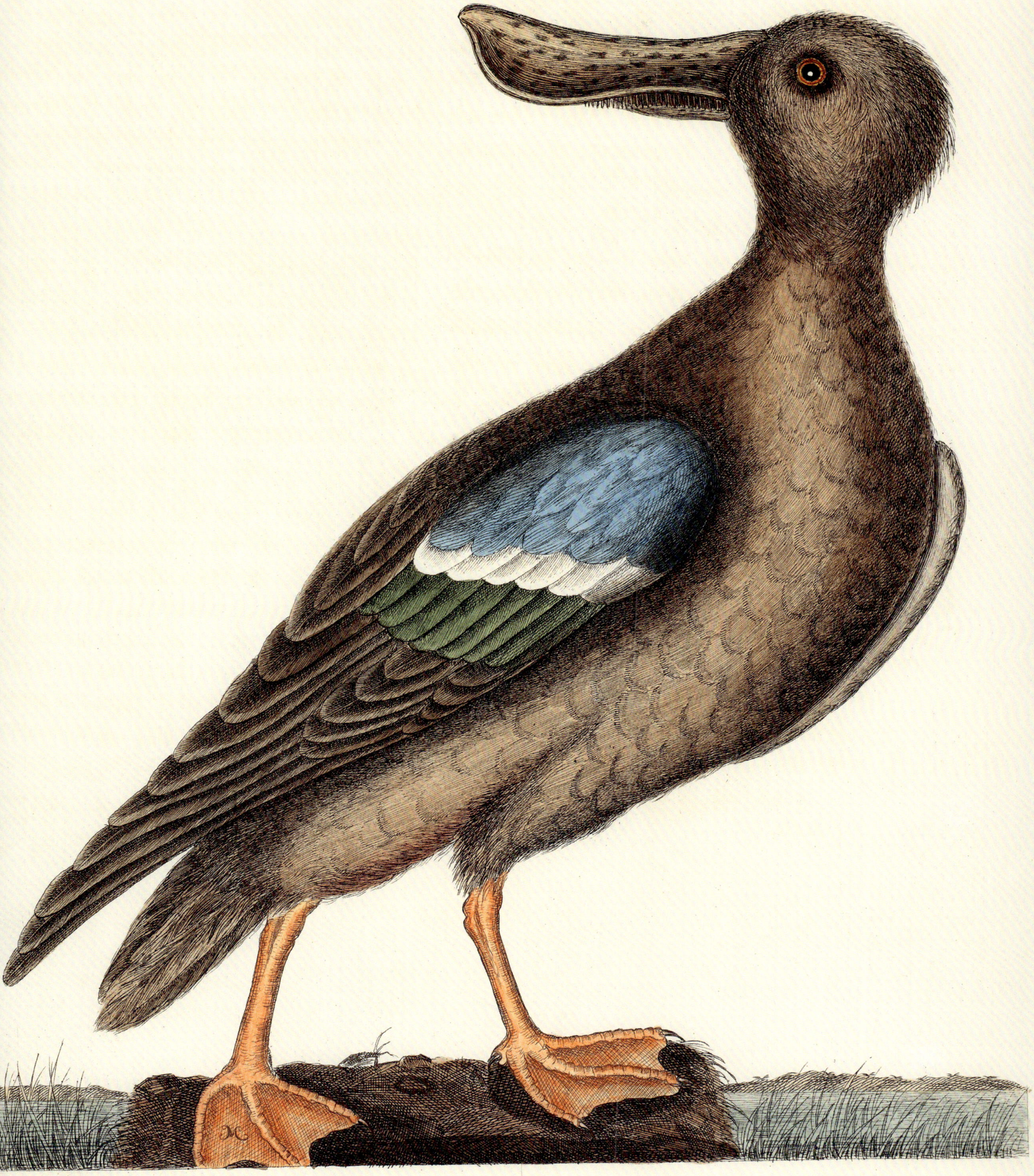

Anas &c.

Anas cristata elegans.

The Summer Duck.

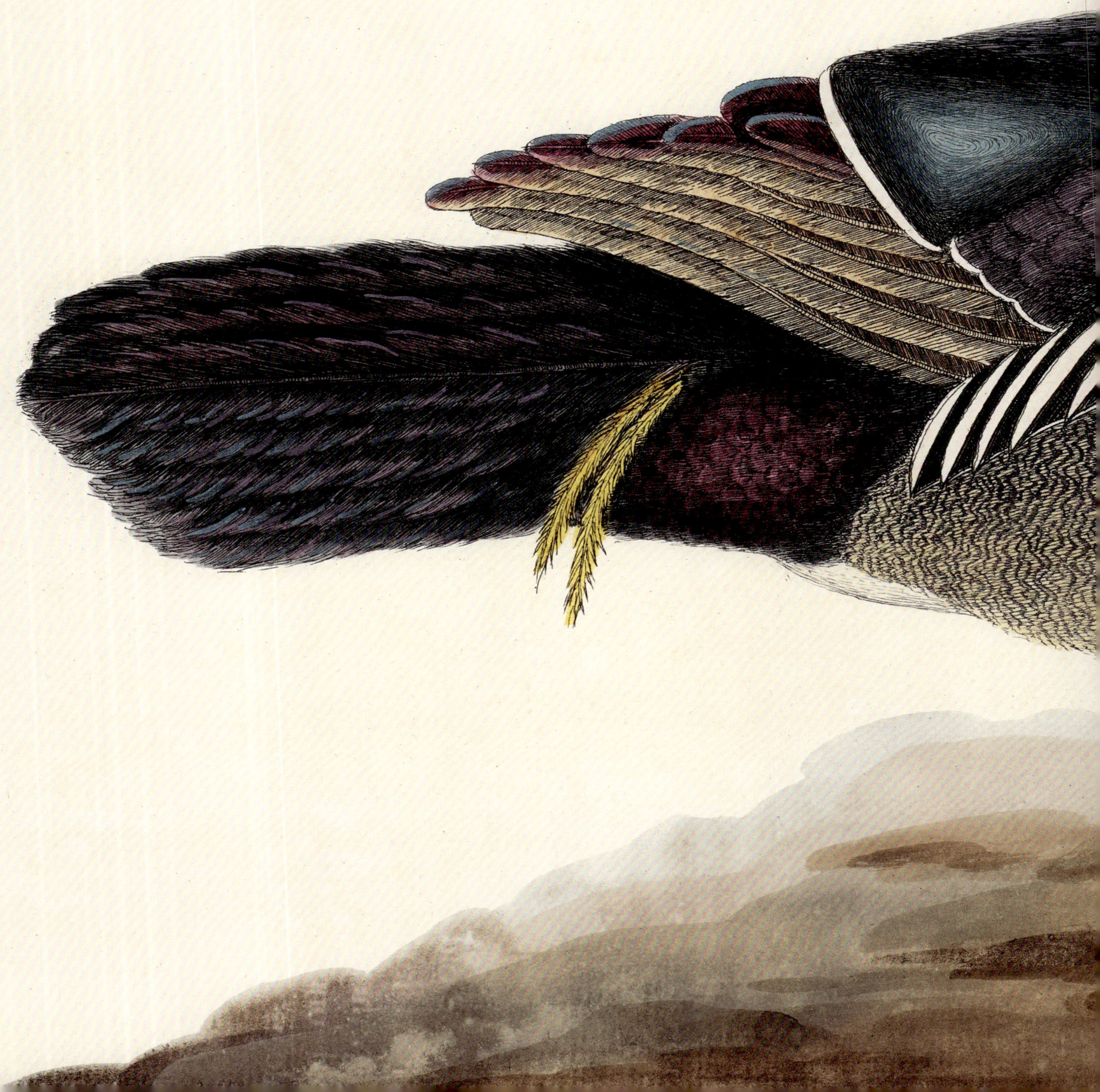

T 97

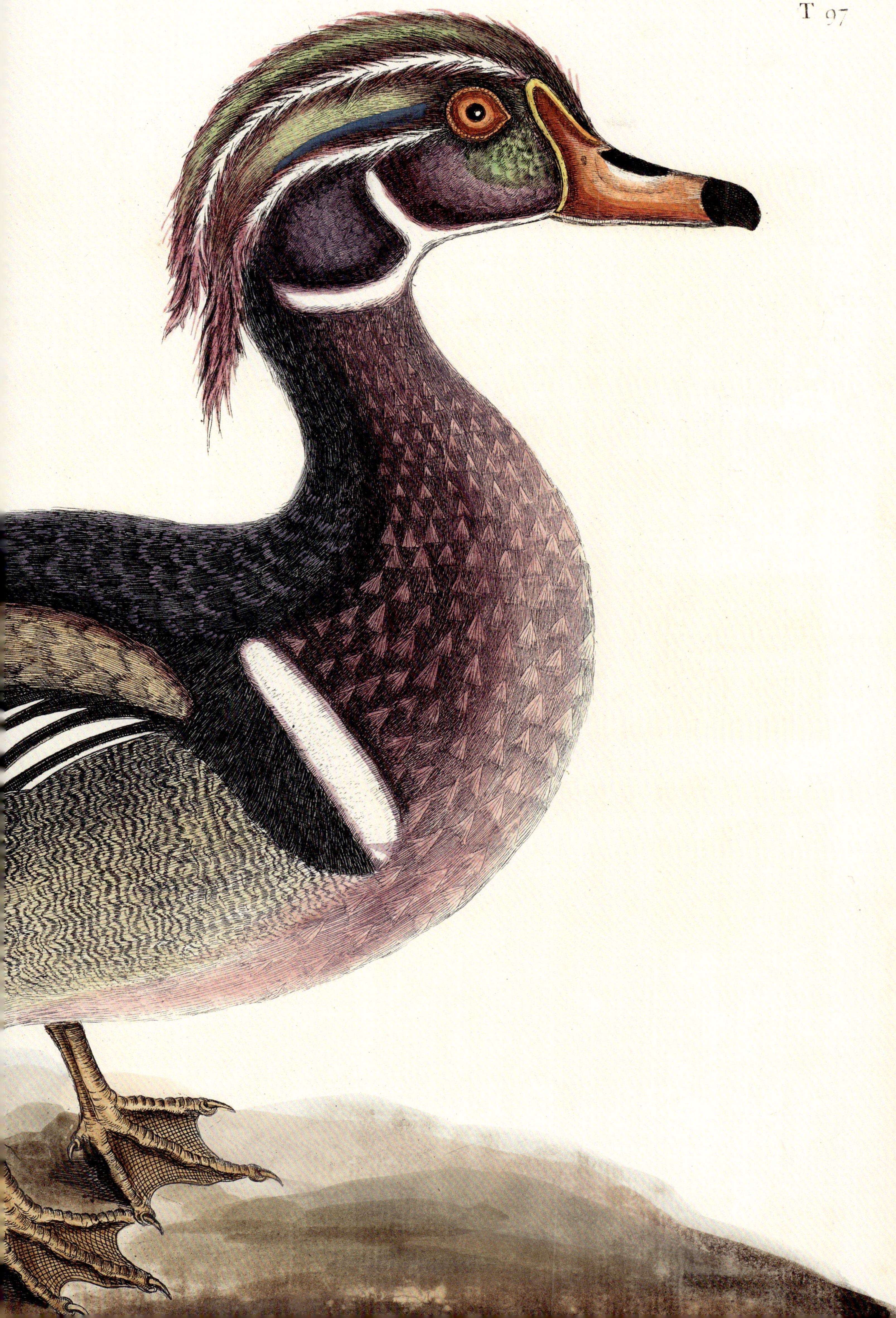

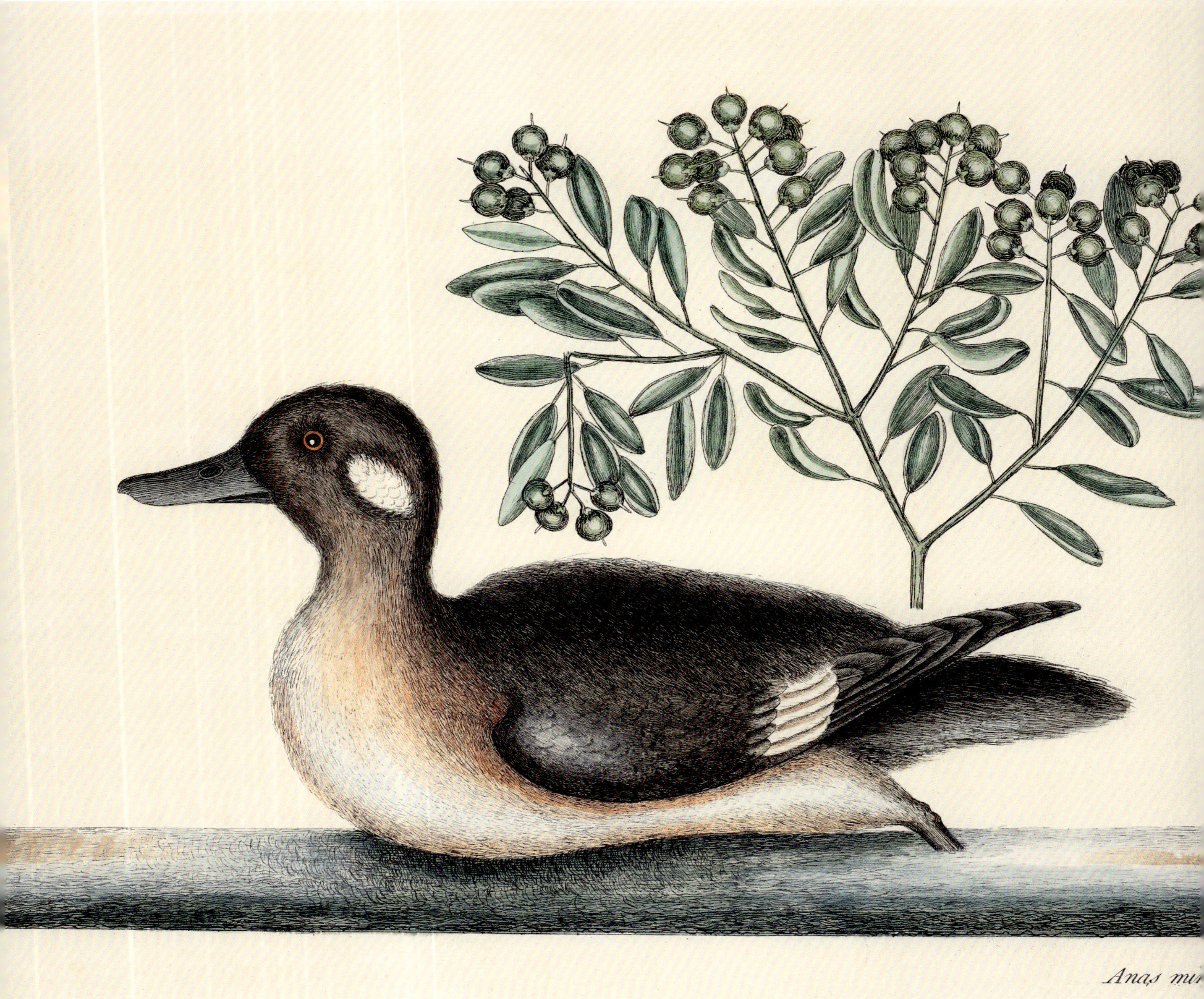

97 *previous pages*
Wood duck (*Aix sponsa*) (vol. 1, t. 97)

98 *above*
Bufflehead (*Bucephala albeola*); joewood (*Jacquinia keyensis*) (vol. 1, t. 98)

99 *right top*
Blue-winged teal (*Anas discors*) (vol. 1, t. 99)

100 *right bottom*
Blue-winged teal (*Anas discors*) (vol. 1, t. 100)

T 99

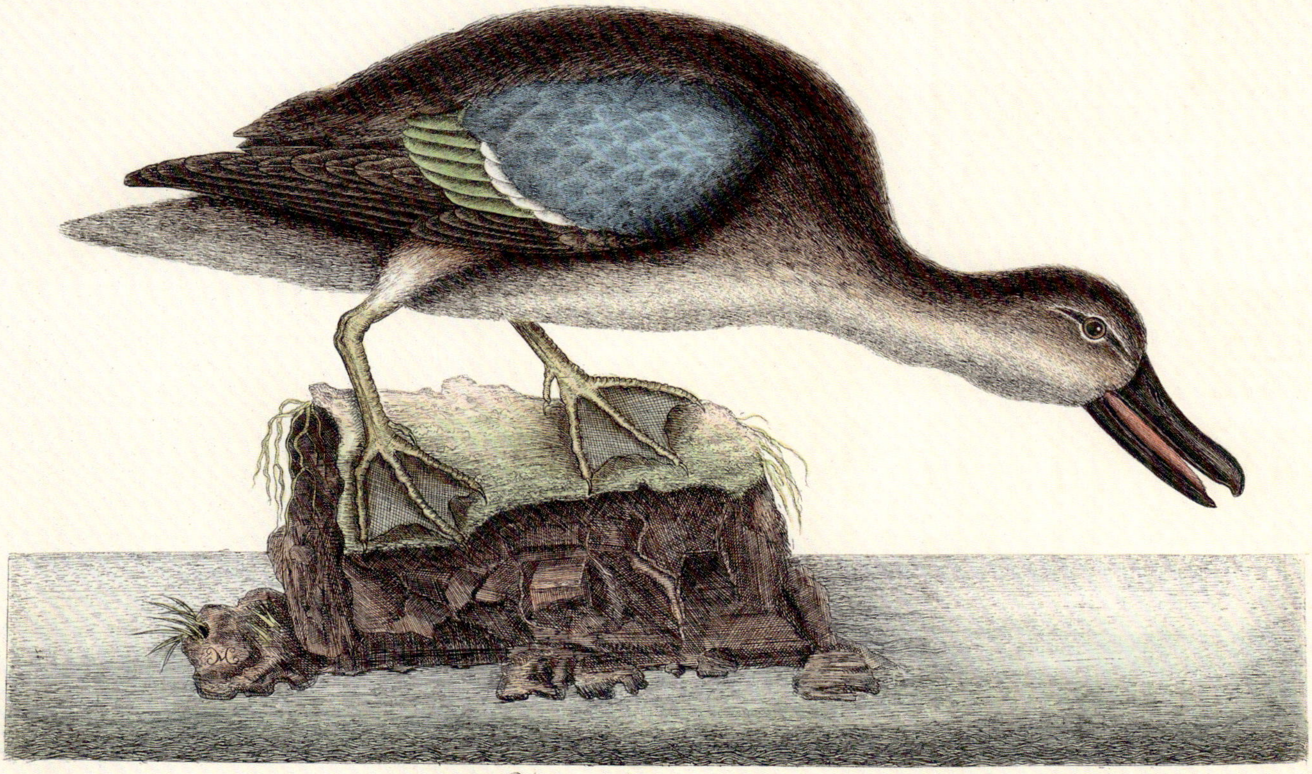

Querquedula &c.

T 100

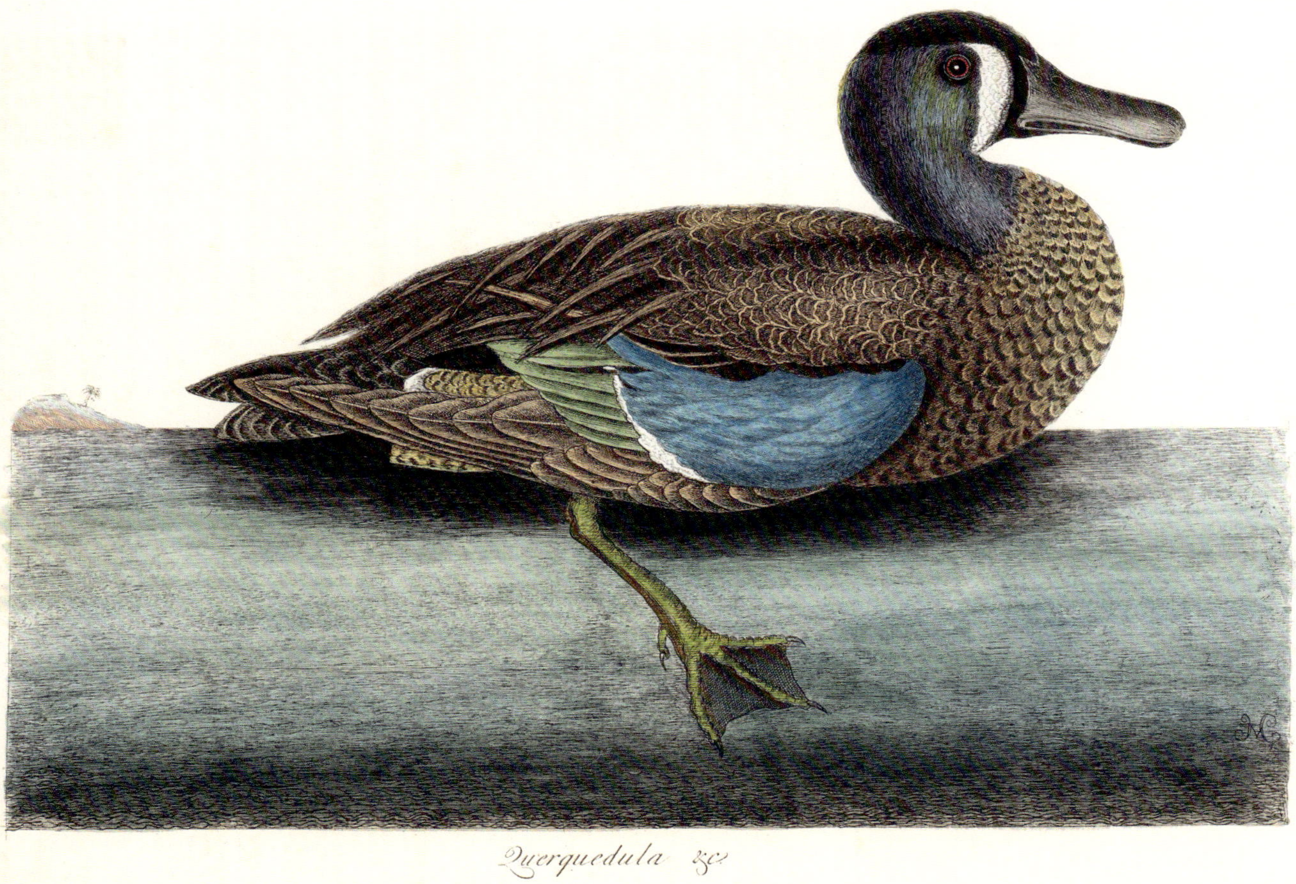

Querquedula &c.

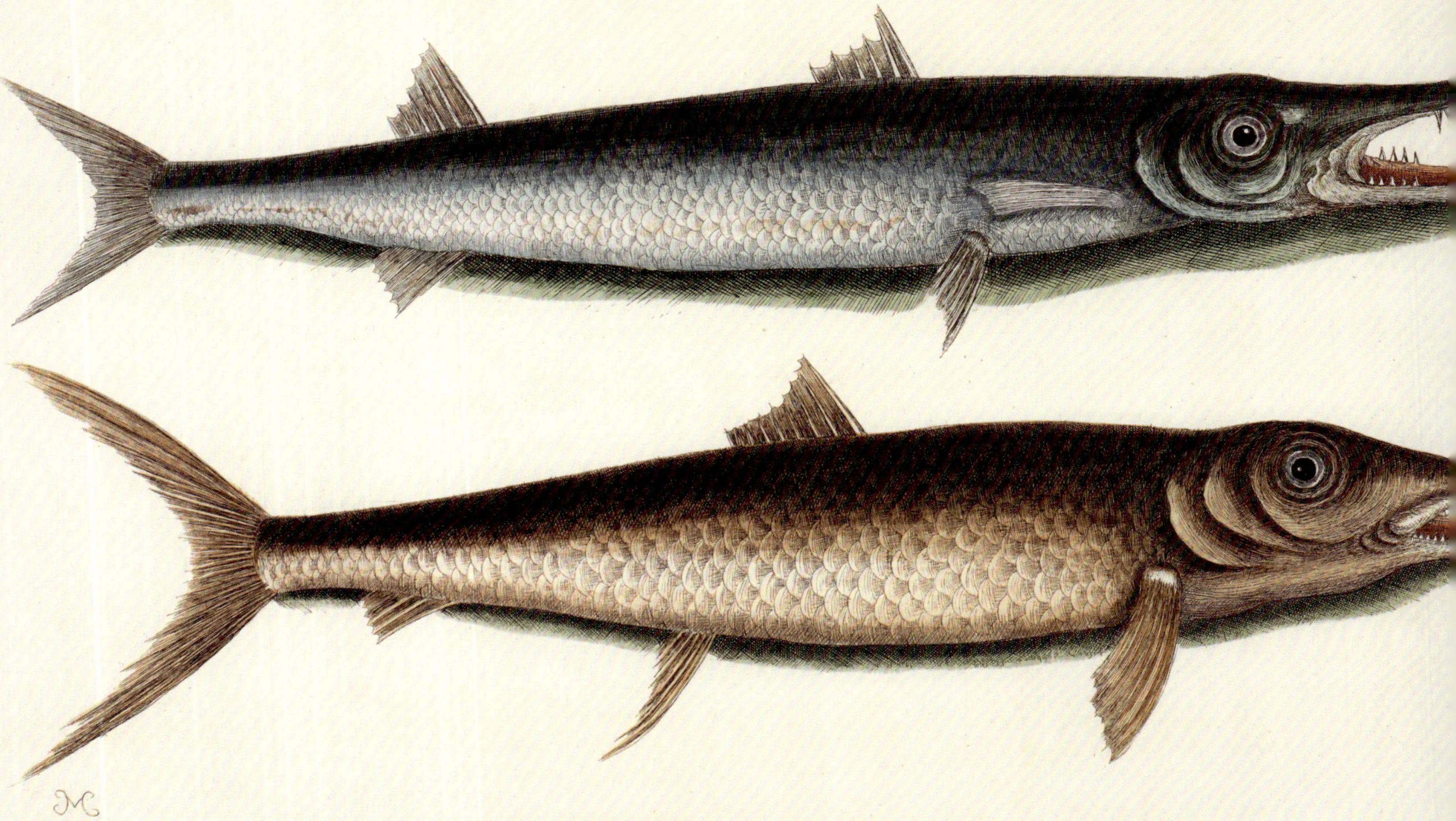

Vulpis Bahamensis.

Perca &c.

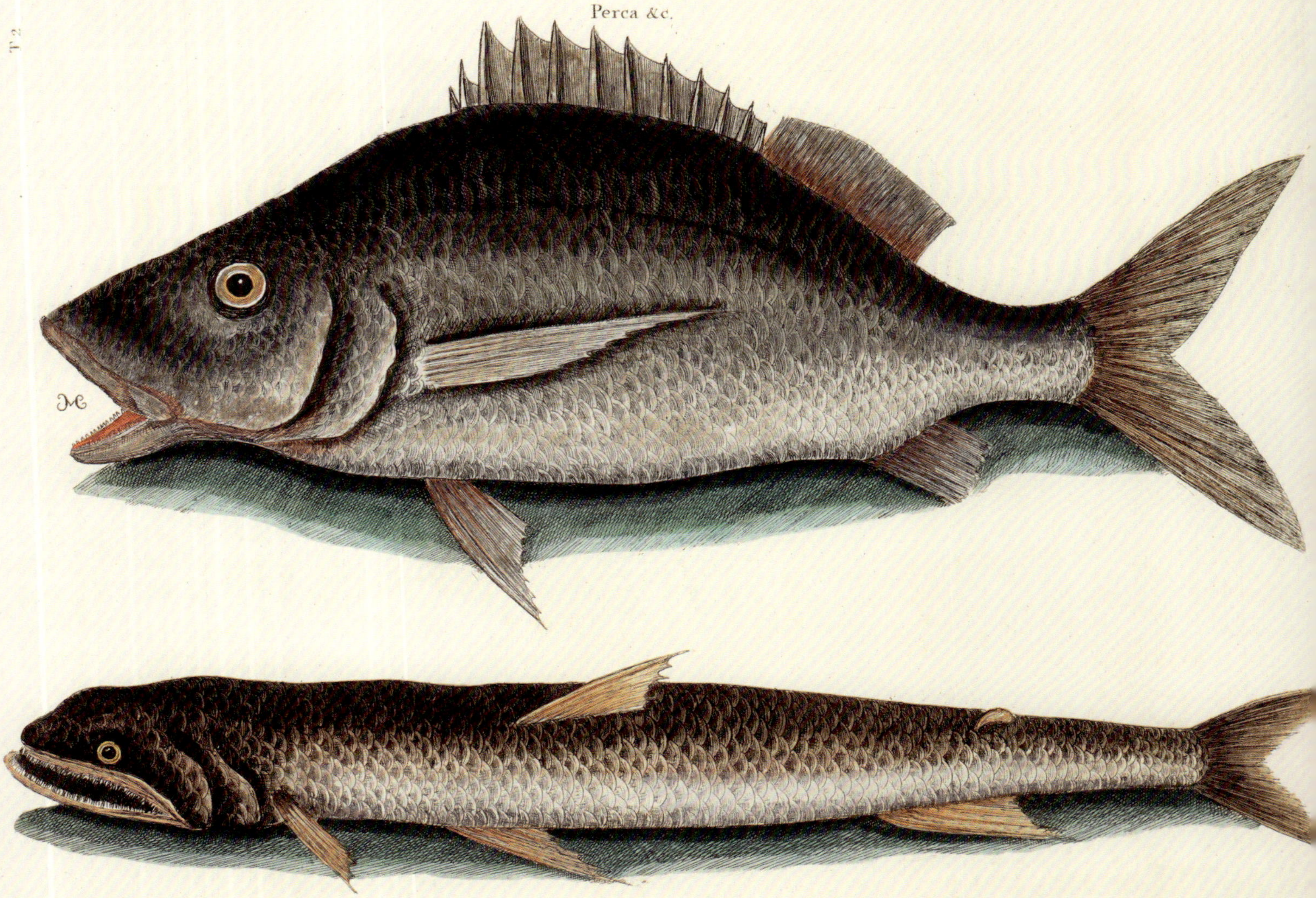

Saurus &c.

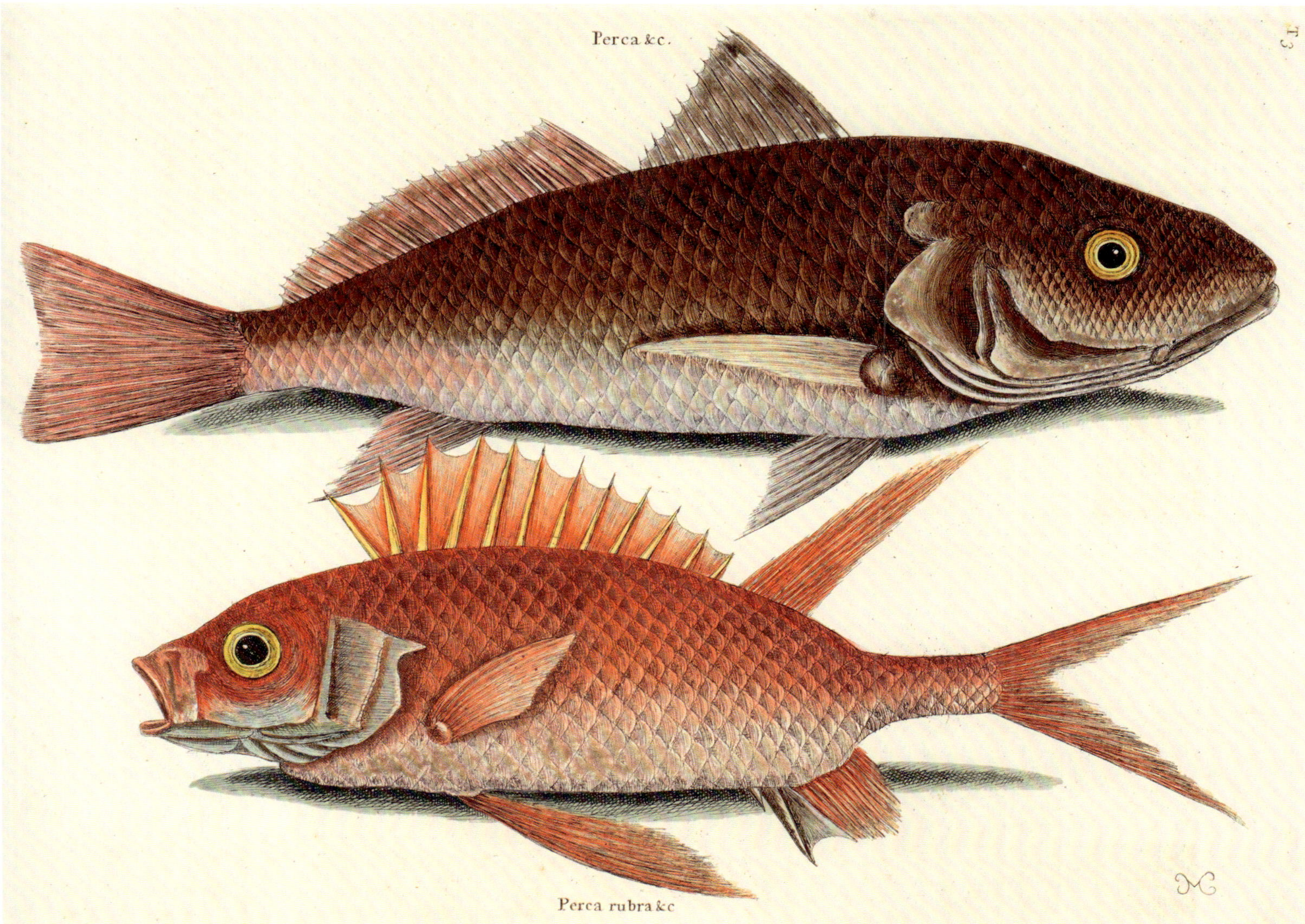

101 *left top*
Great barracuda (*Sphyraena barracuda*; upper); bonefish (*Albula vulpes*; lower) (vol. 2, t. 1)

102 *left bottom*
White margate (*Haemulon album*; upper); inshore lizardfish (*Synodus foetens*; lower) (vol. 2, t. 2)

103 *above*
Altantic croaker (*Micropogonias undulatus*; upper); longspine squirrelfish (*Holocentrus rufus*; lower) (vol. 2, t. 3)

T 4

Perca rhomboidalis &c.

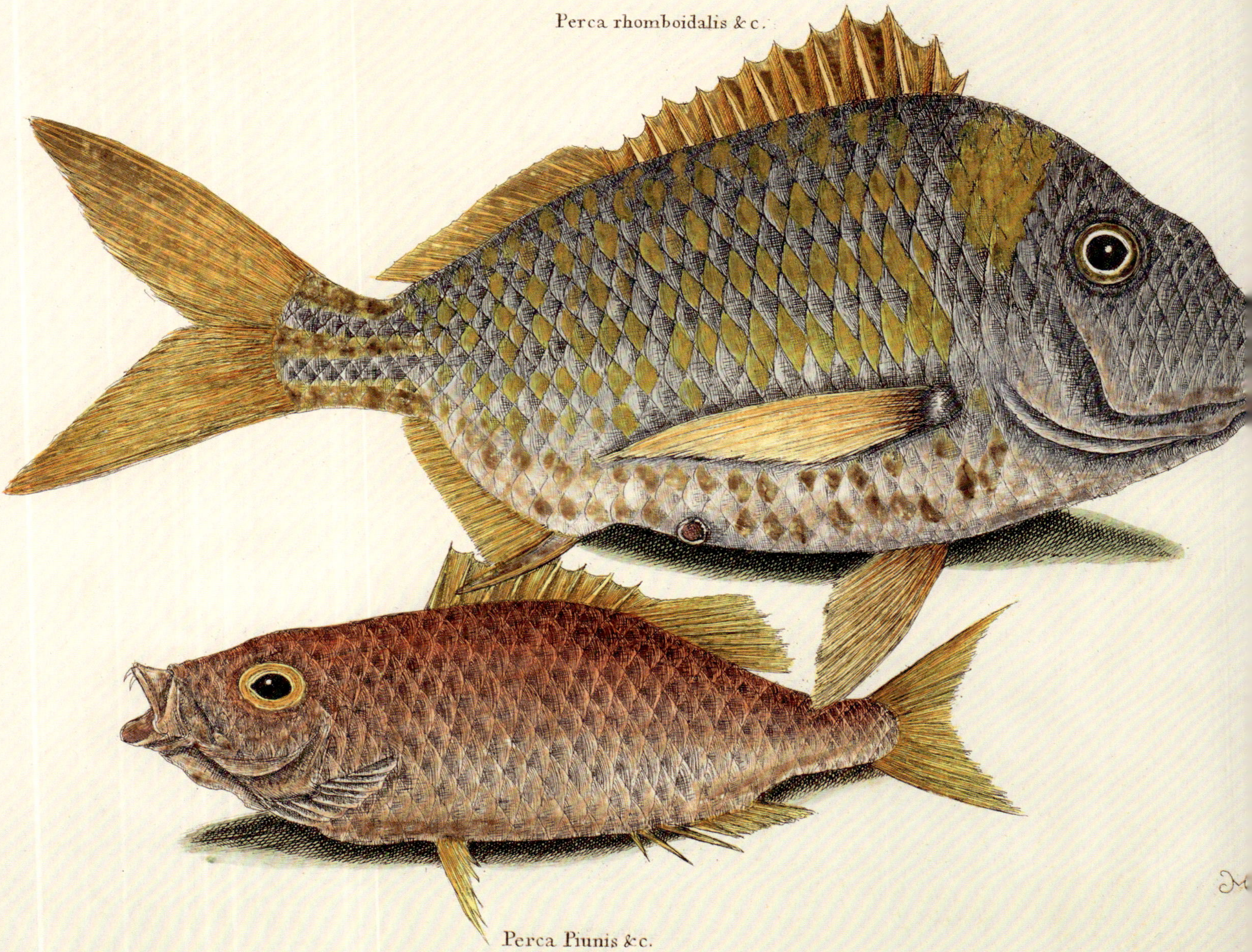

Perca Piunis &c.

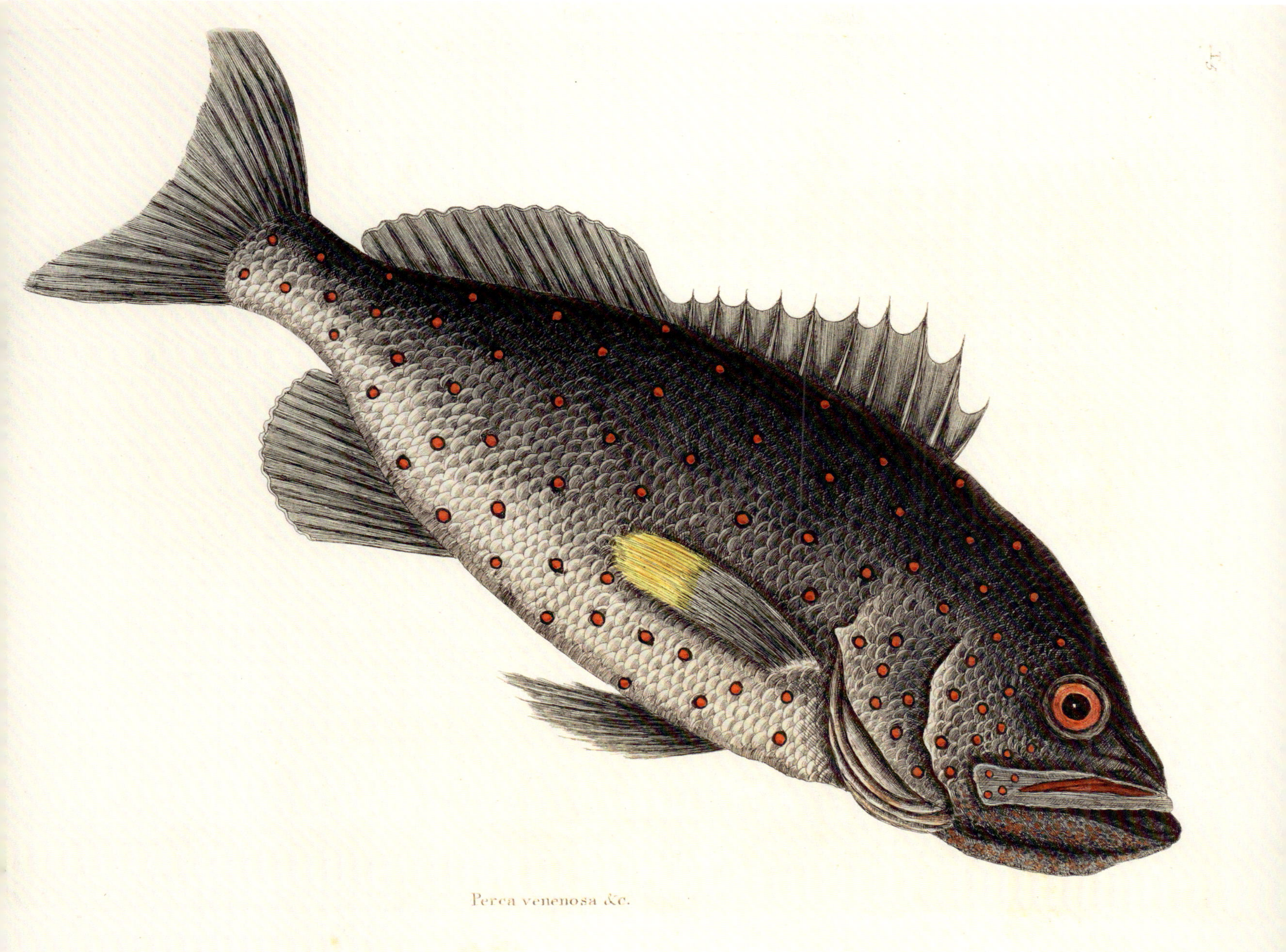

104 *left*

Porkfish (*Anisotremus virginicus*; upper); schoolmaster snapper (*Lutjanus apodus*; lower) (vol. 2, t. 4)

105 *above*

Yellowfin grouper (*Mycteroperca venenosa*) (vol. 2, t. 5)

following pages

106

White grunt (*Haemulon plumeri*; upper); perhaps flathead grey mullet (*Mugil cephalus*) (vol. 2, t. 6)

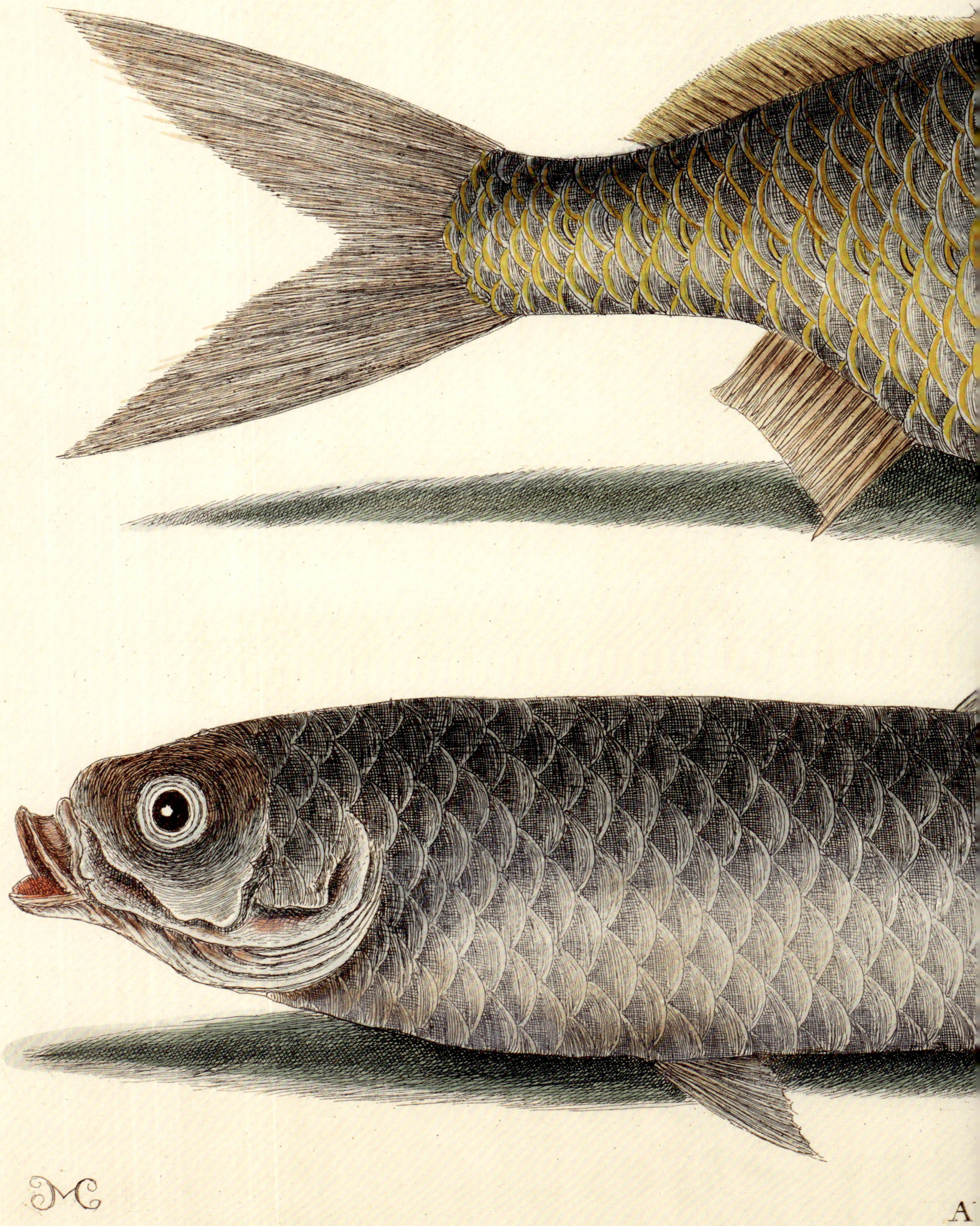

A

c. T 6

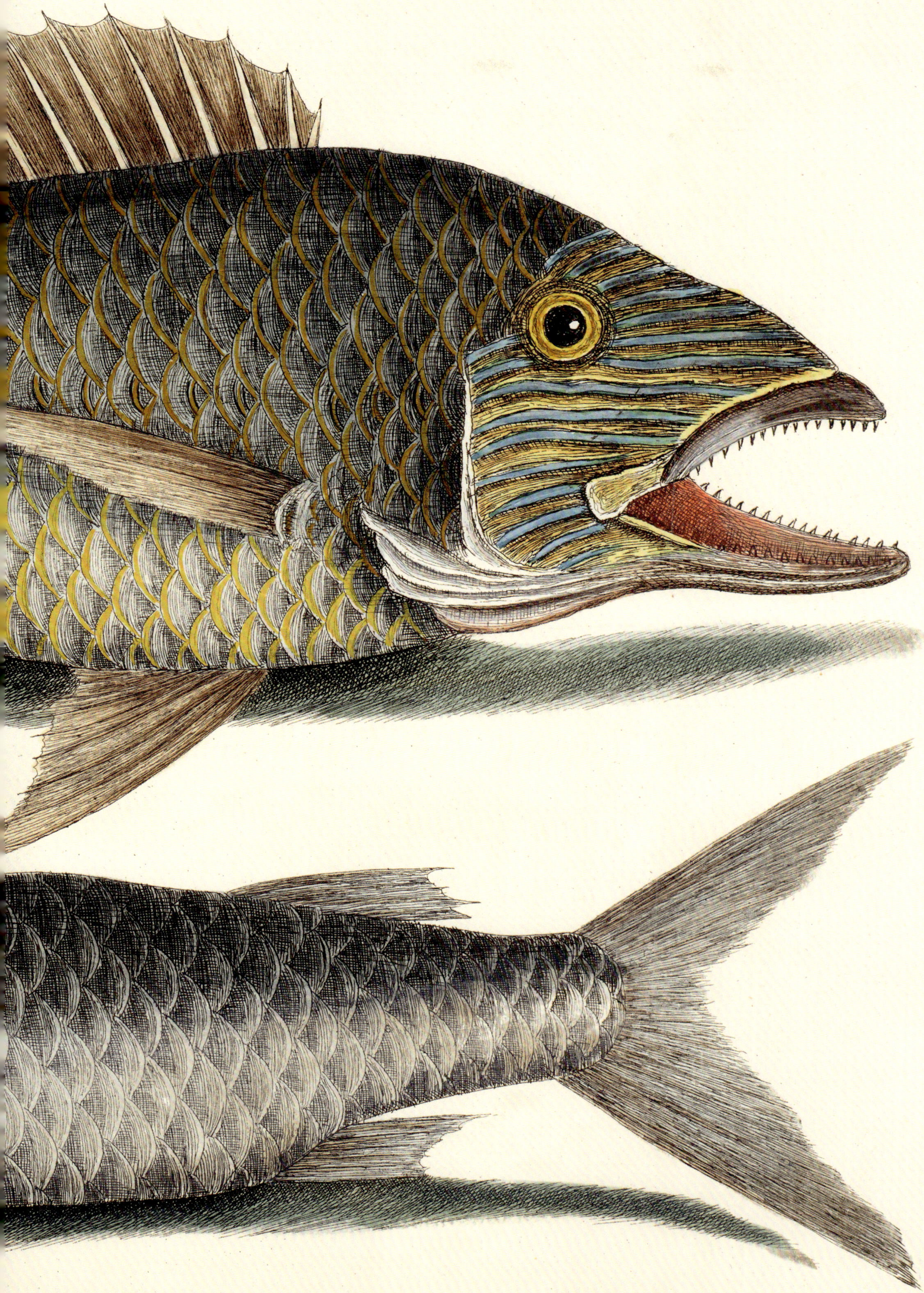

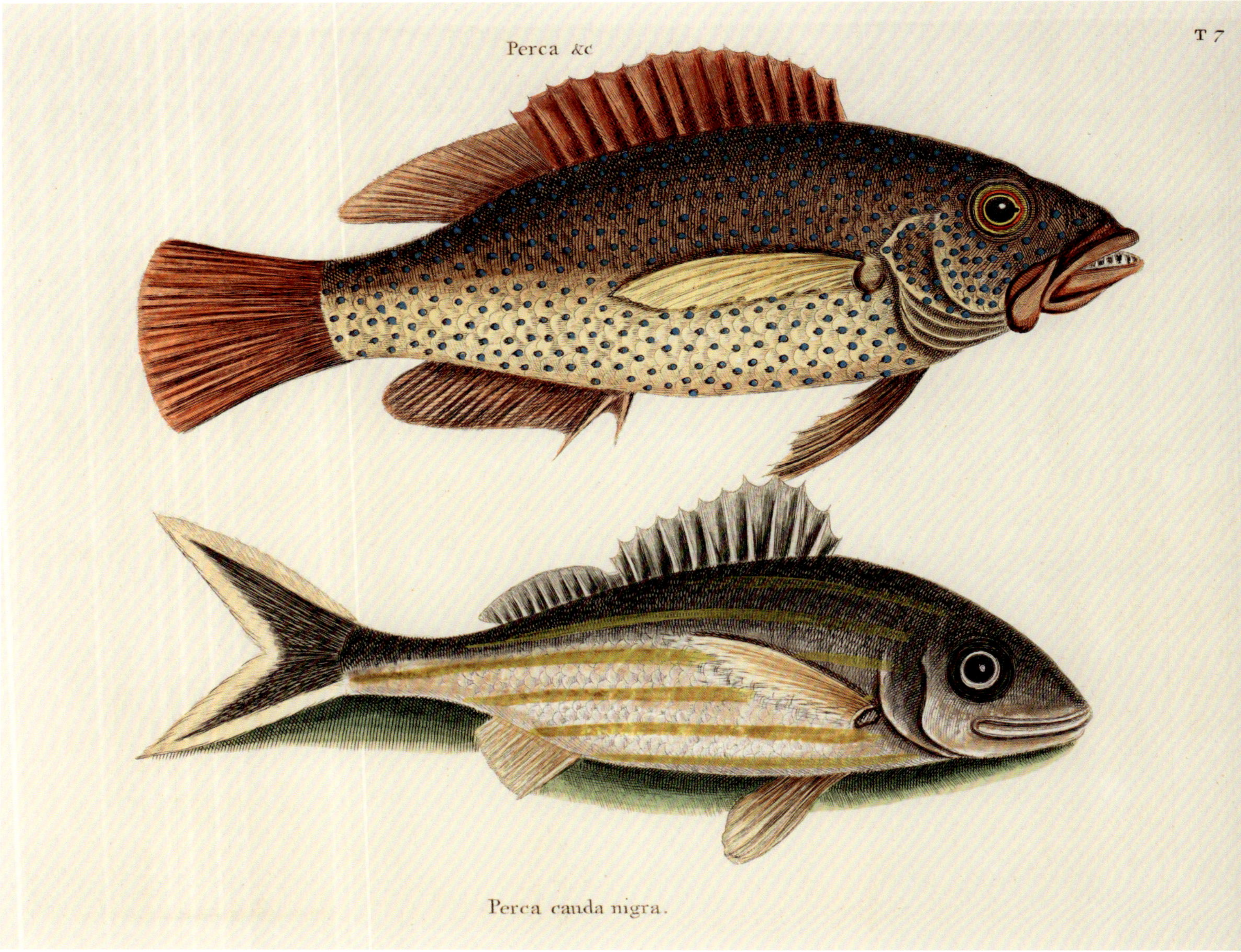

107 *above*
Coney (*Cephalopholis fulva*; upper); cottonwick grunt (*Haemulon melanurum*; lower) (vol. 2, t. 7)

108 *right*
Unidentified flying fish (upper); Bermuda chub (*Kyphosus saltatrix*; centre); pumpkinseed (*Lepomis gibbosus*; lower) (vol. 2, t. 8)

Hirundo

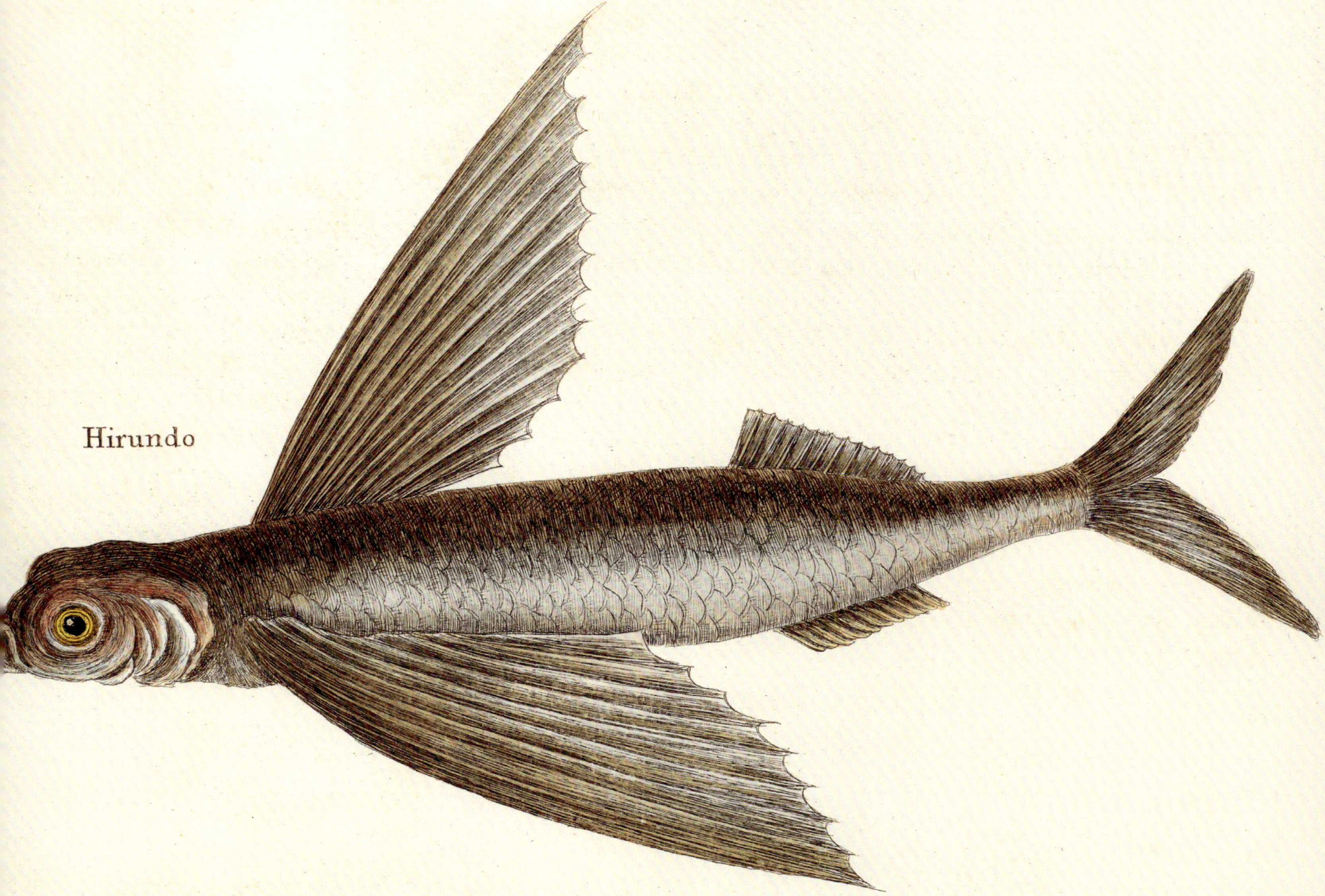

Perca &c.

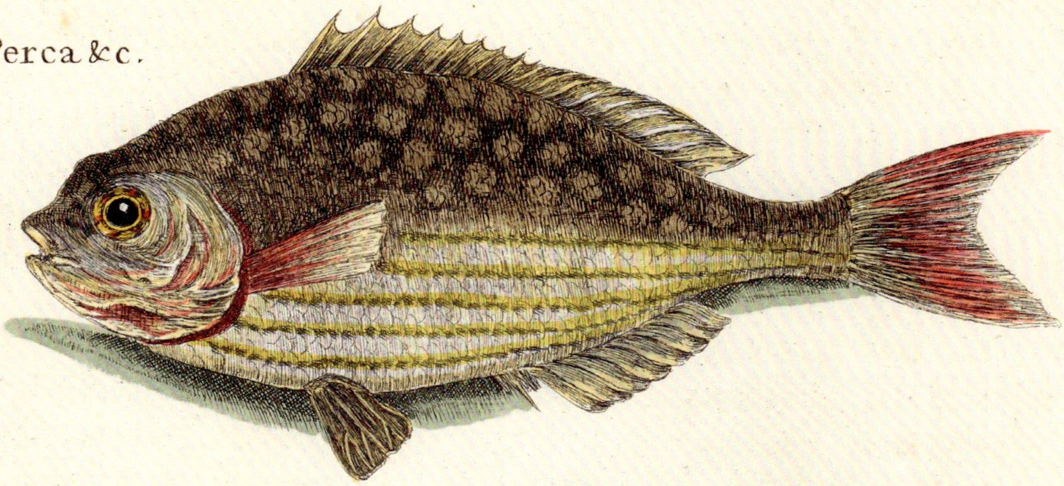

Perca fluviatilis &c.

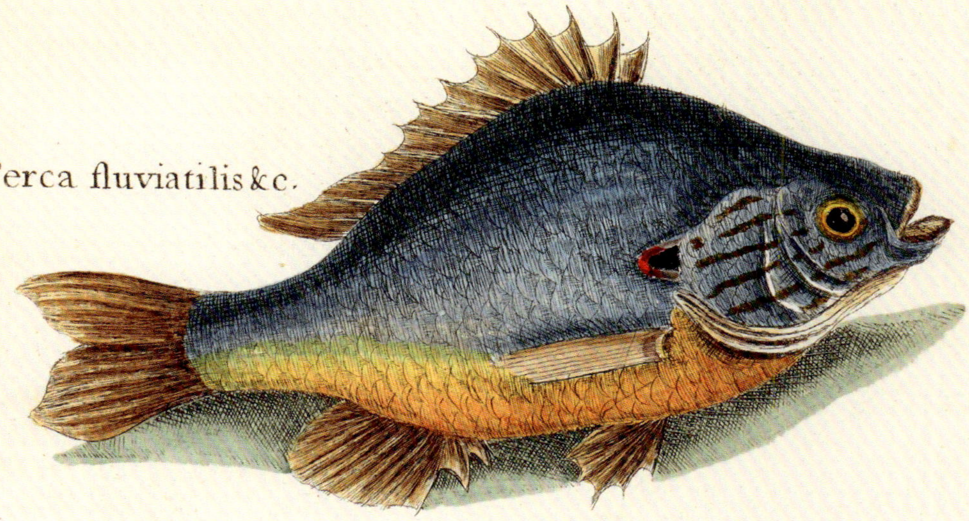

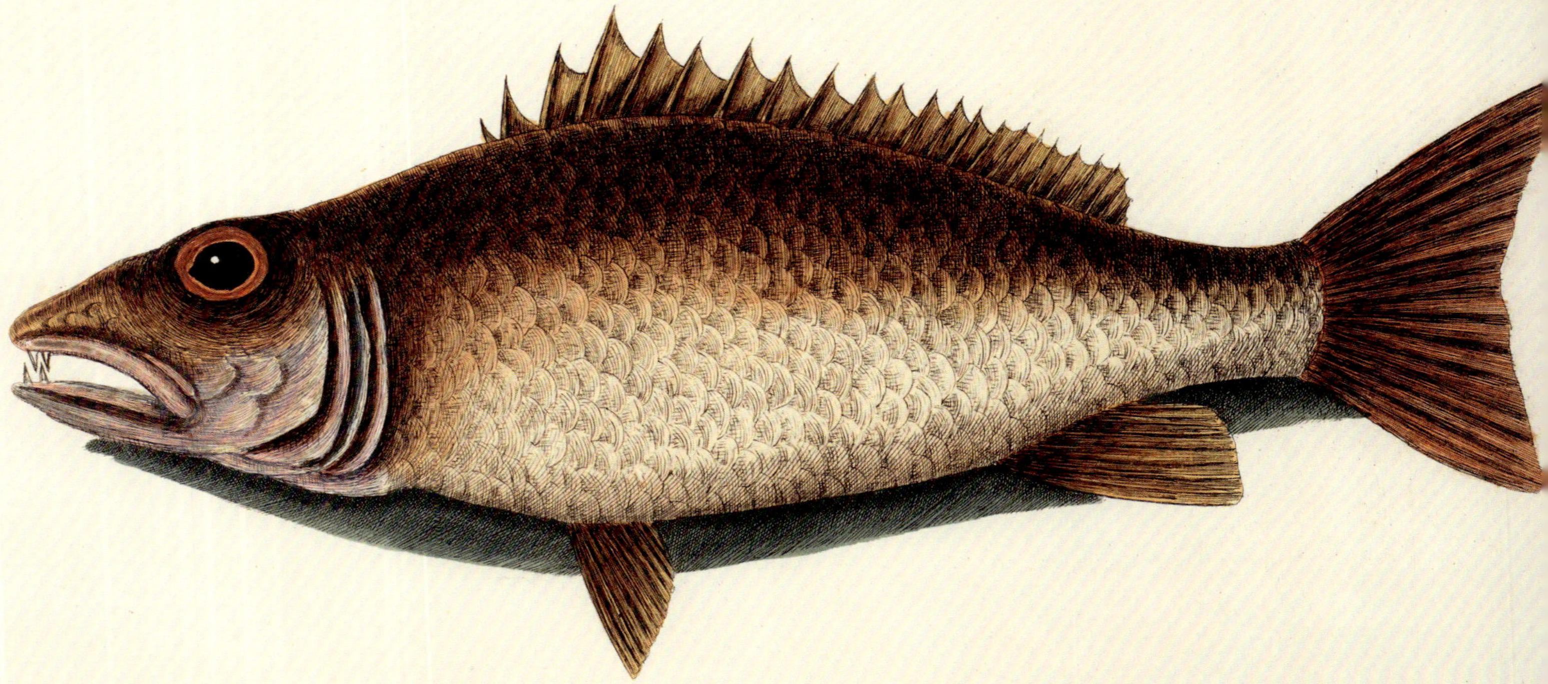

Turdus Rhomboidelis

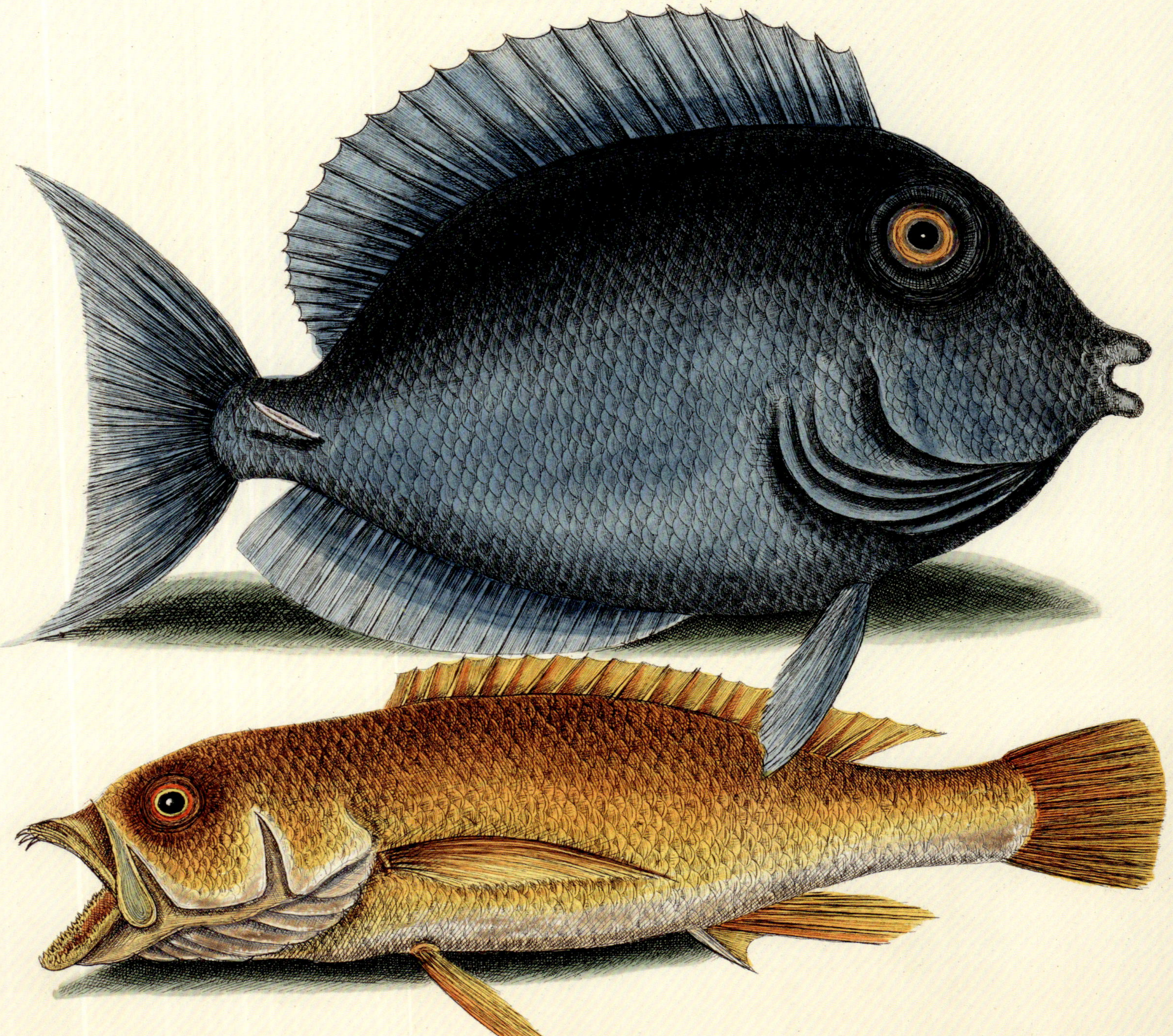

Turdus &c

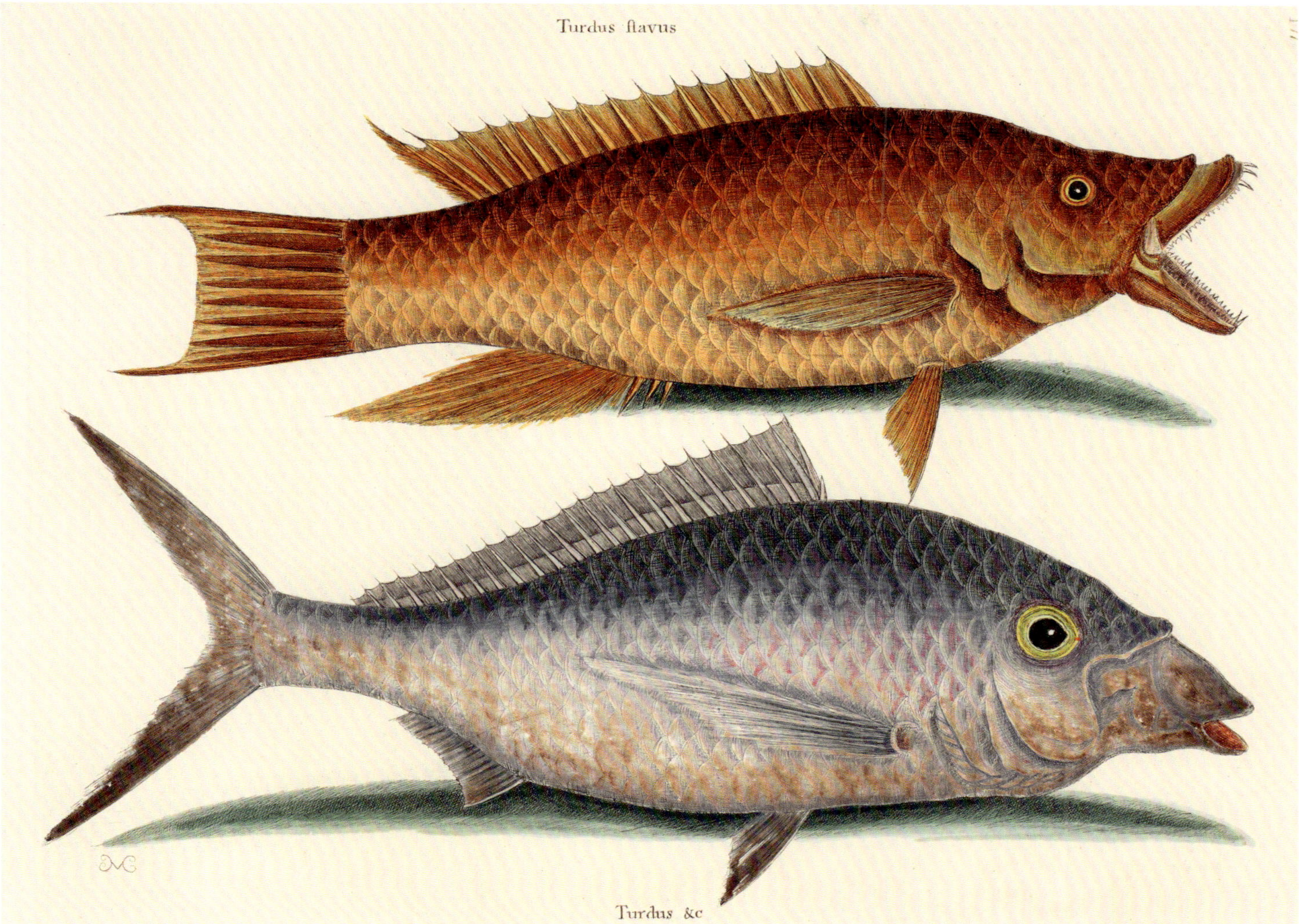

109 *top left*
Grey snapper (*Lutjanus griseus*) (vol. 2, t. 9)

110 *bottom left*
Atlantic blue tang (*Acanthurus coeruleus*; upper); coney (*Cephalopholis fulva*; lower) (vol. 2, t. 10)

111 *above*
Spanish hogfish (*Bodianus rufus*; upper); yellowfin mojarra (*Gerres cinereus*; lower) (vol. 2, t. 11)

following pages

112
Pudding wife wrasse (*Halichoeres radiatus*; upper); southern kingfish (*Menticirrhus americanus*; lower) (vol. 2, t. 12)

T/2

Turdus &

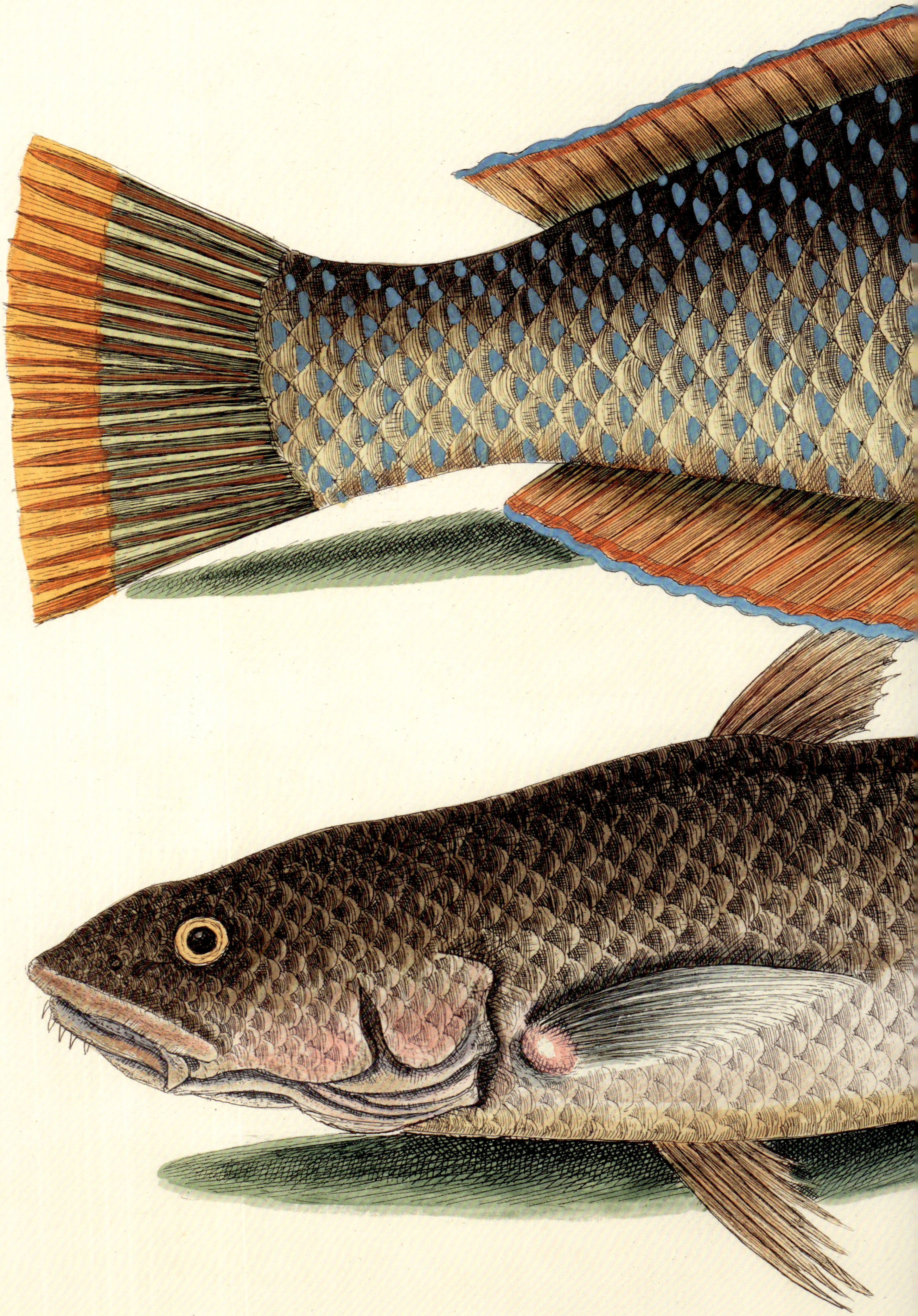

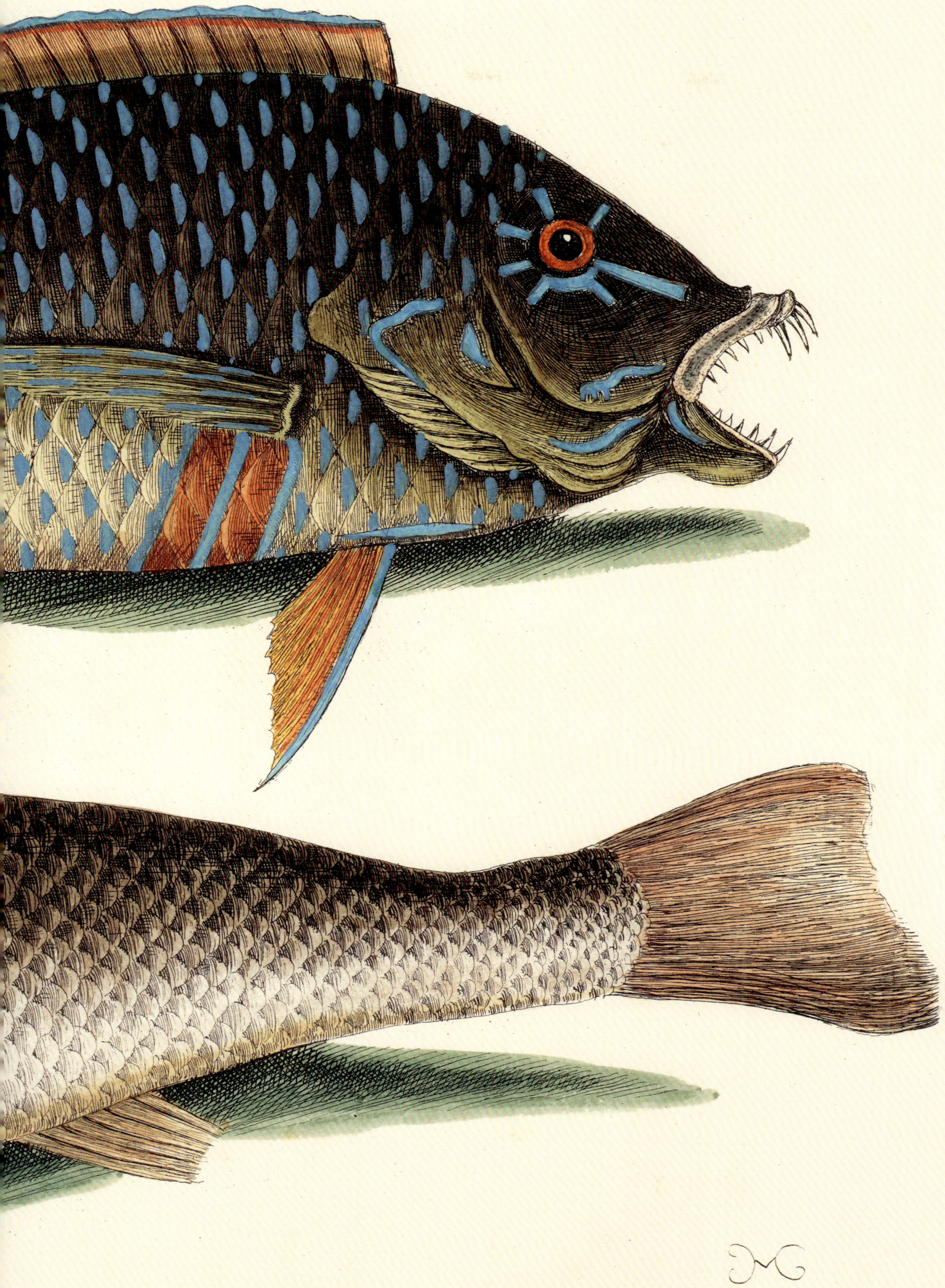

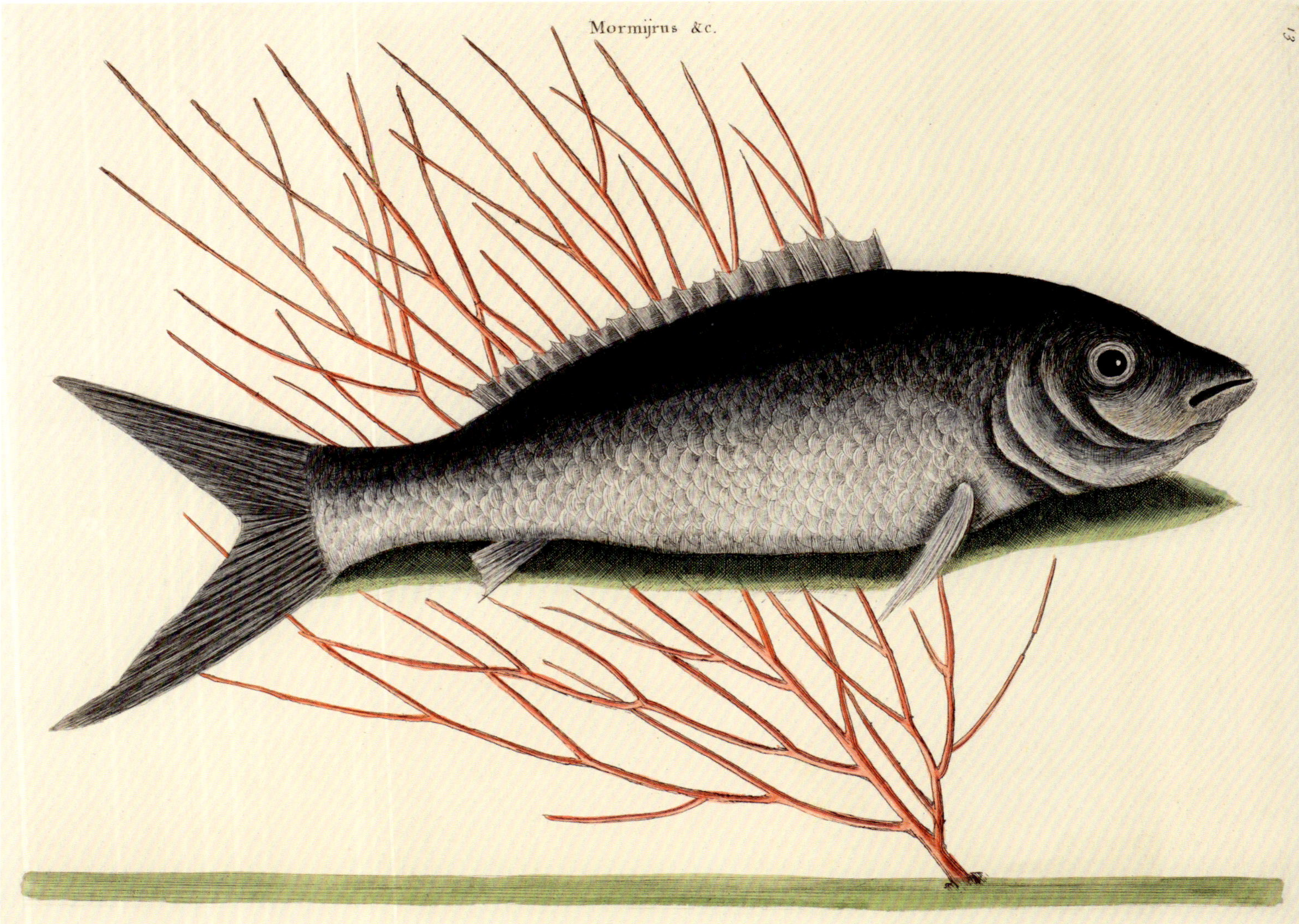

113 *above*
Possibly mottled mojarra (*Ulaema lefroyi*); unidentified gorgonid coral (vol. 2, t. 13)

114 *top right*
Red hind (*Epinephelus guttatus*; upper); bluefish (*Pomatomus saltatrix*; lower) (vol. 2, t. 14)

115 *bottom right*
Great hogfish (*Lachnolaimus maximus*); saucereye porgy (*Calamus calamus*) (vol. 2, t. 15)

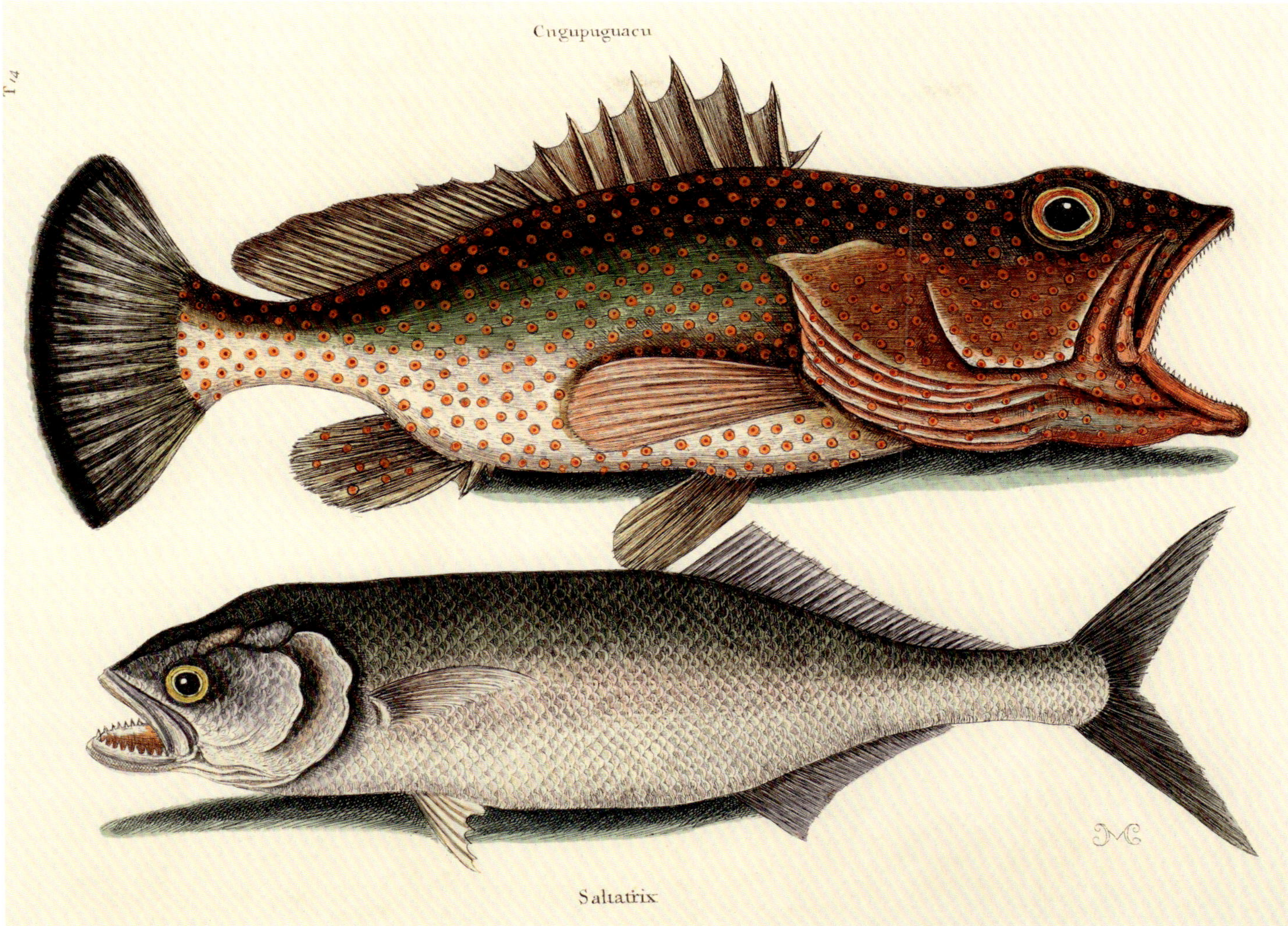
Cugupuguacu
T.4
Saltatrix

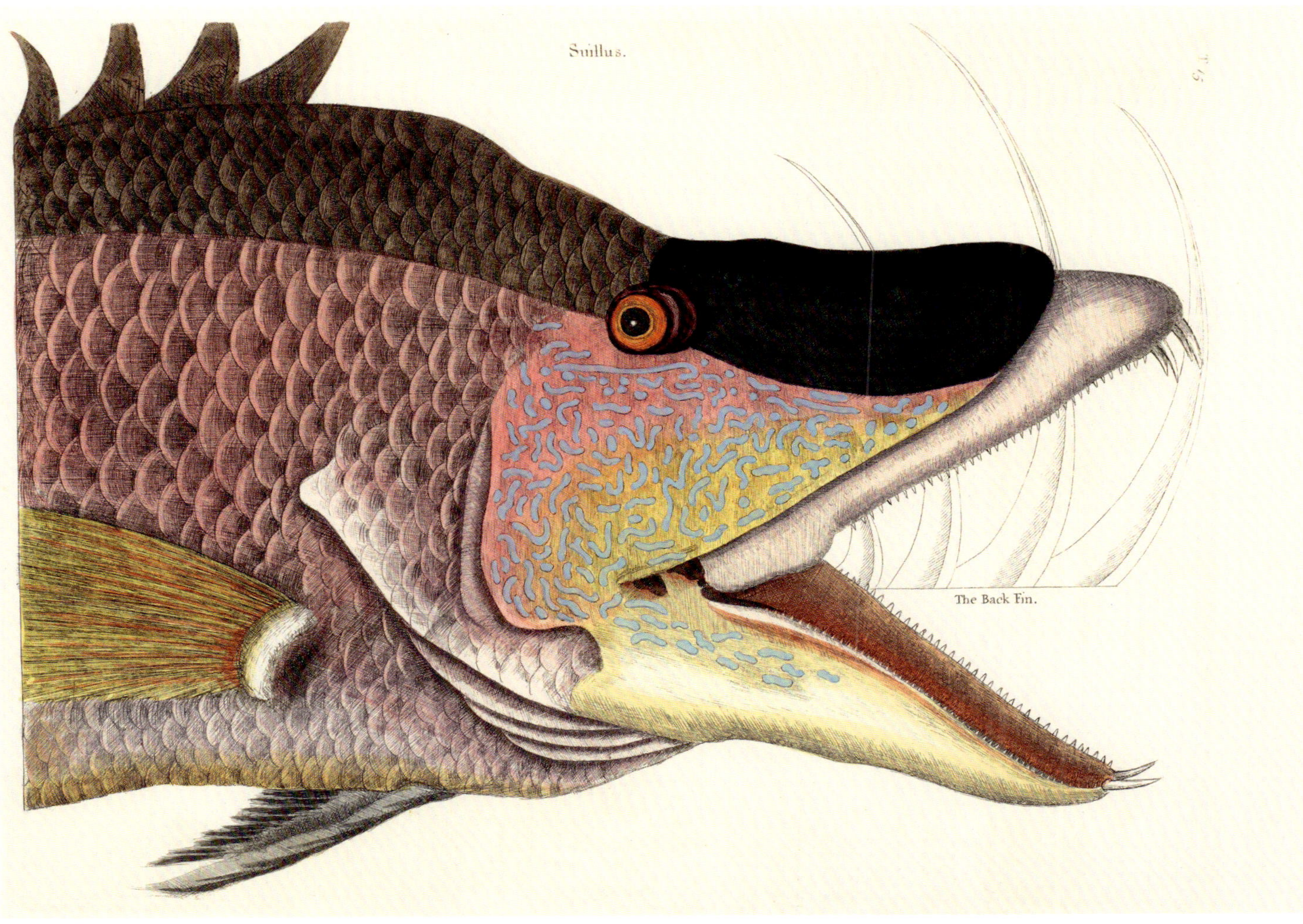
Suillus.
T.5
The Back Fin.

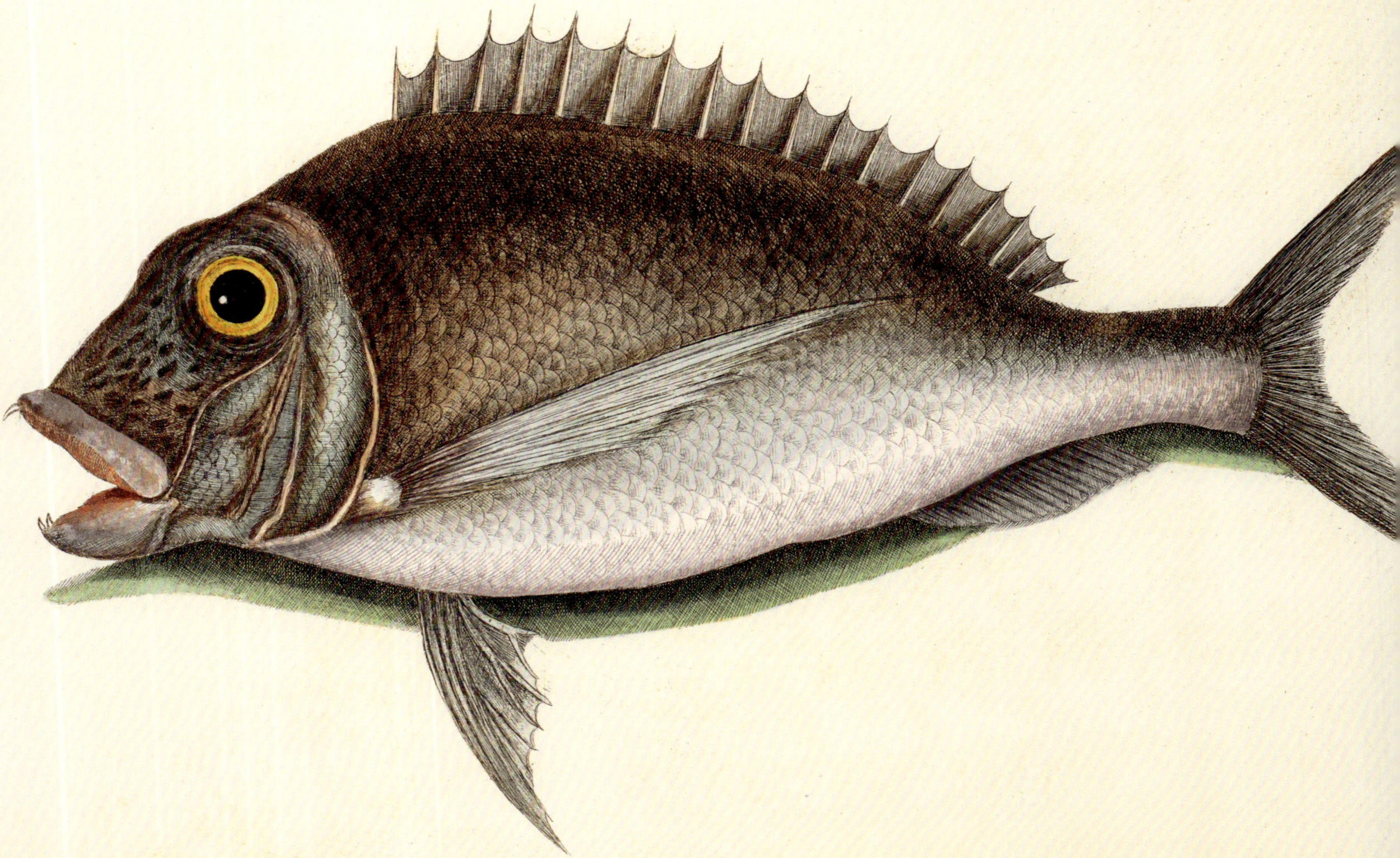

Aurata.

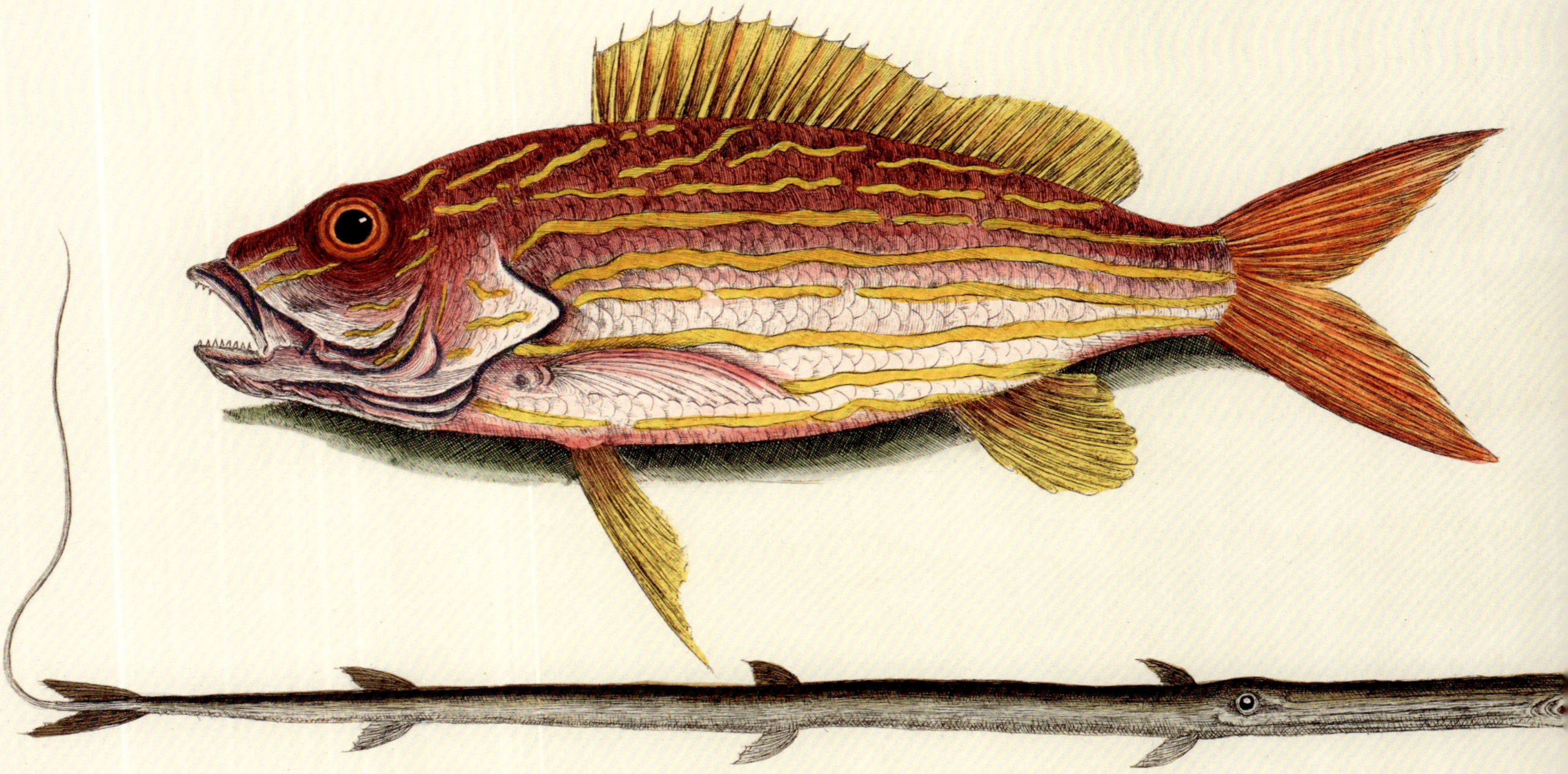

Petimbuabo &c

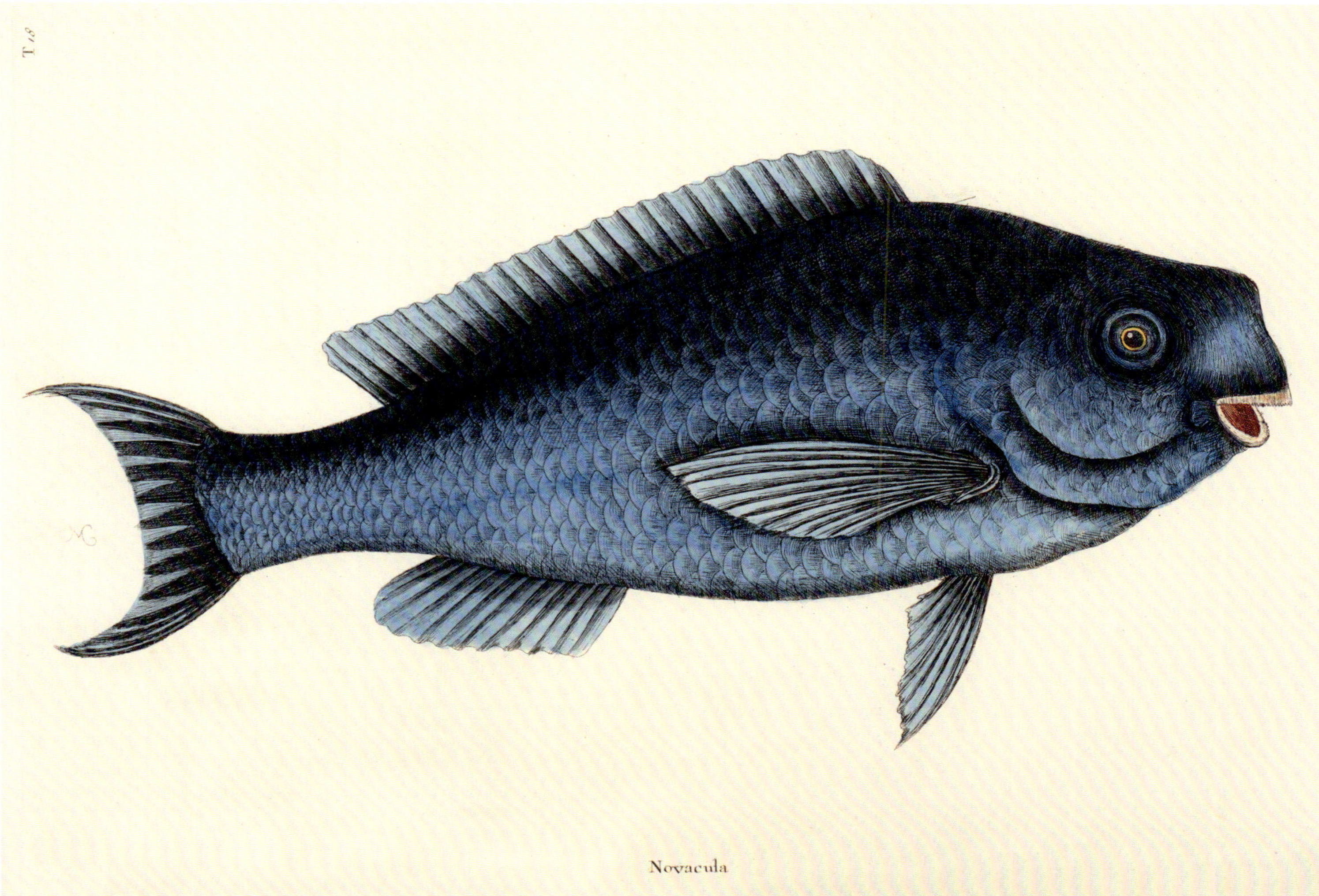

116 *top left*
Saucereye porgy (*Calamus calamus*) (vol. 2, t. 16)

117 *bottom left*
Lane snapper (*Lutjanus synagris*; upper); blue-spotted cornetfish (*Fistularia tabacaria*; lower) (vol. 2, t. 17)

118 *above*
Blue parrotfish (*Scarus coeruleus*) (vol. 2, t. 18)

following pages

119
Scribbled leatherjacket filefish (*Aluterus scriptus*) (vol. 2, t. 19)

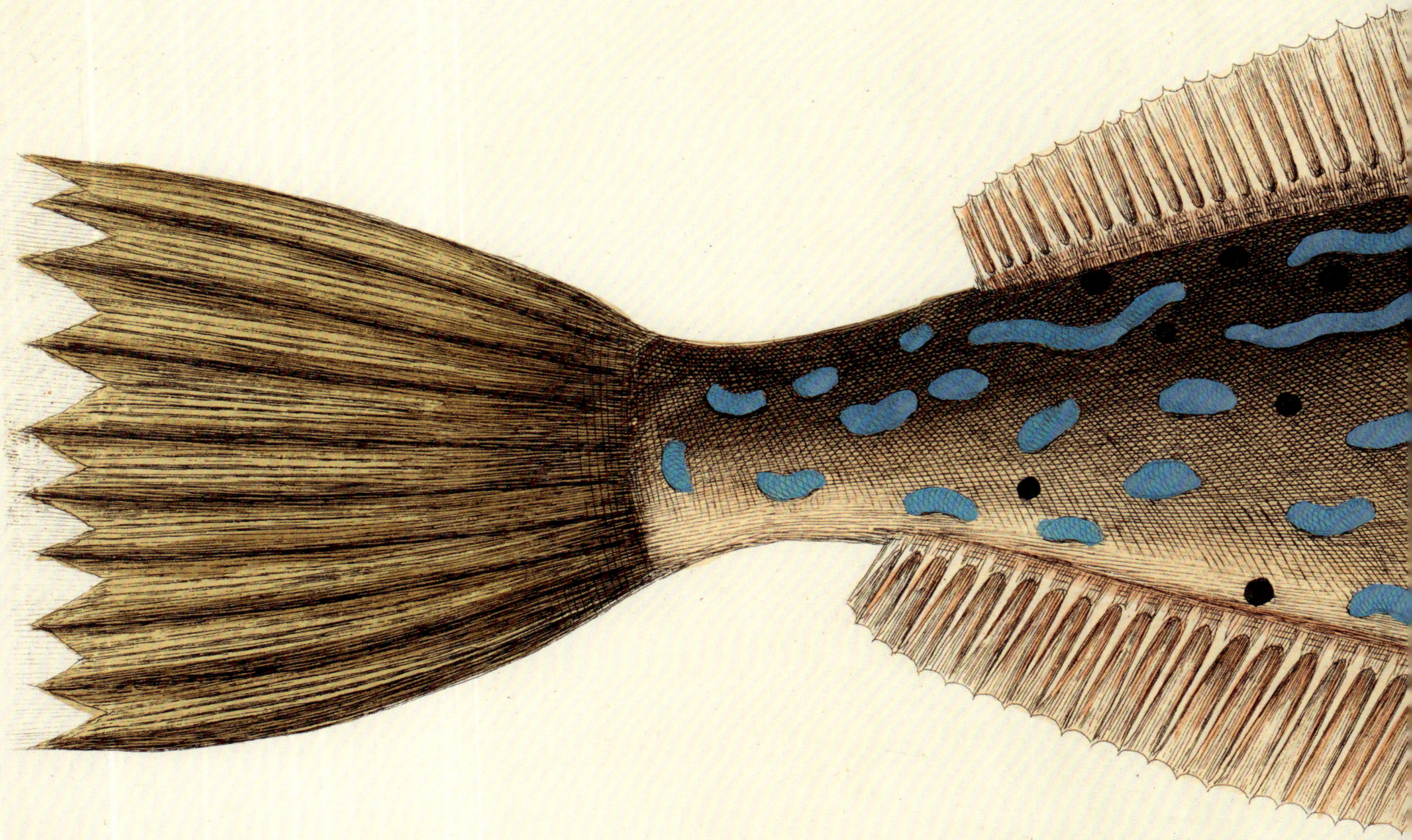

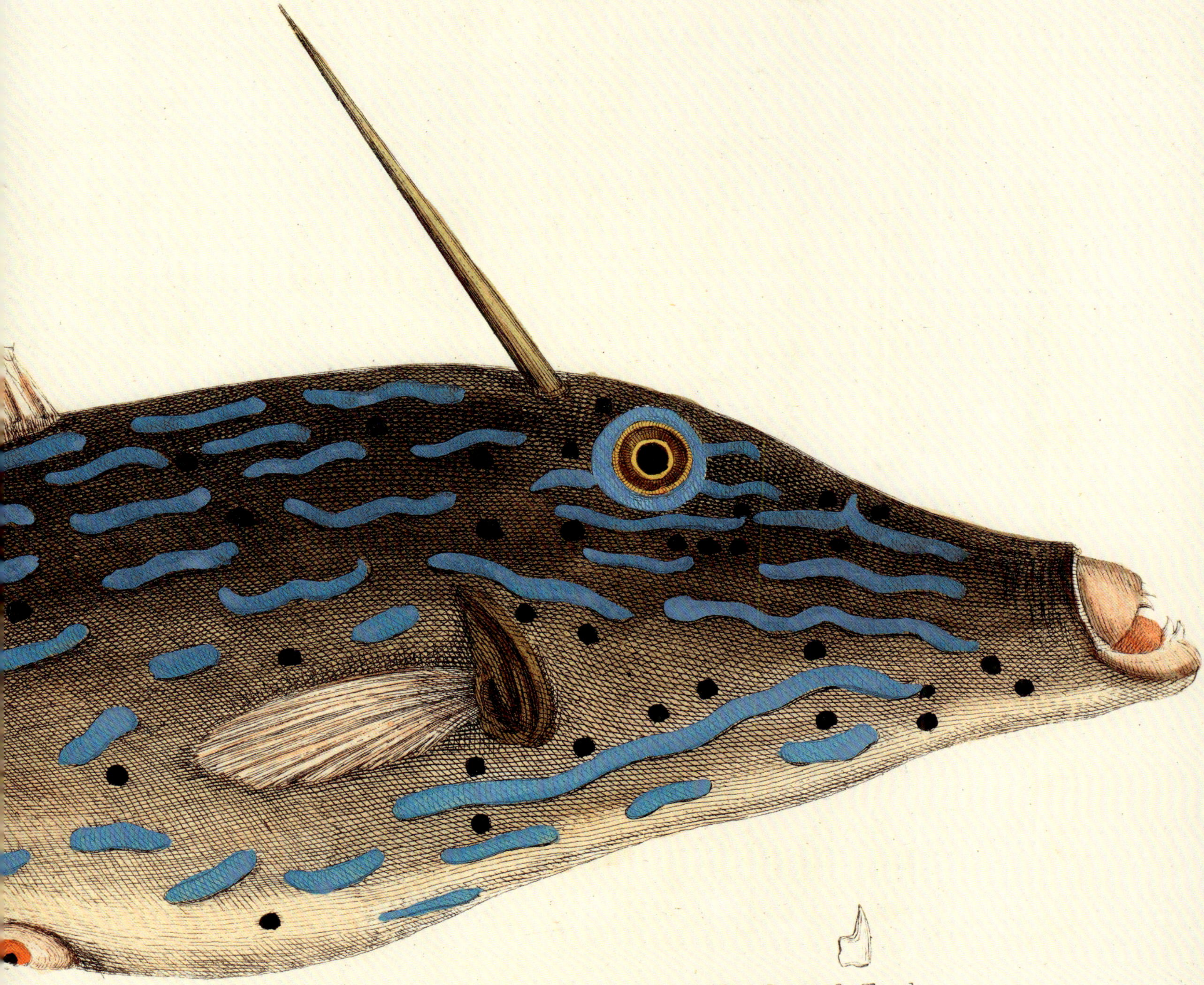

The form of a Tooth

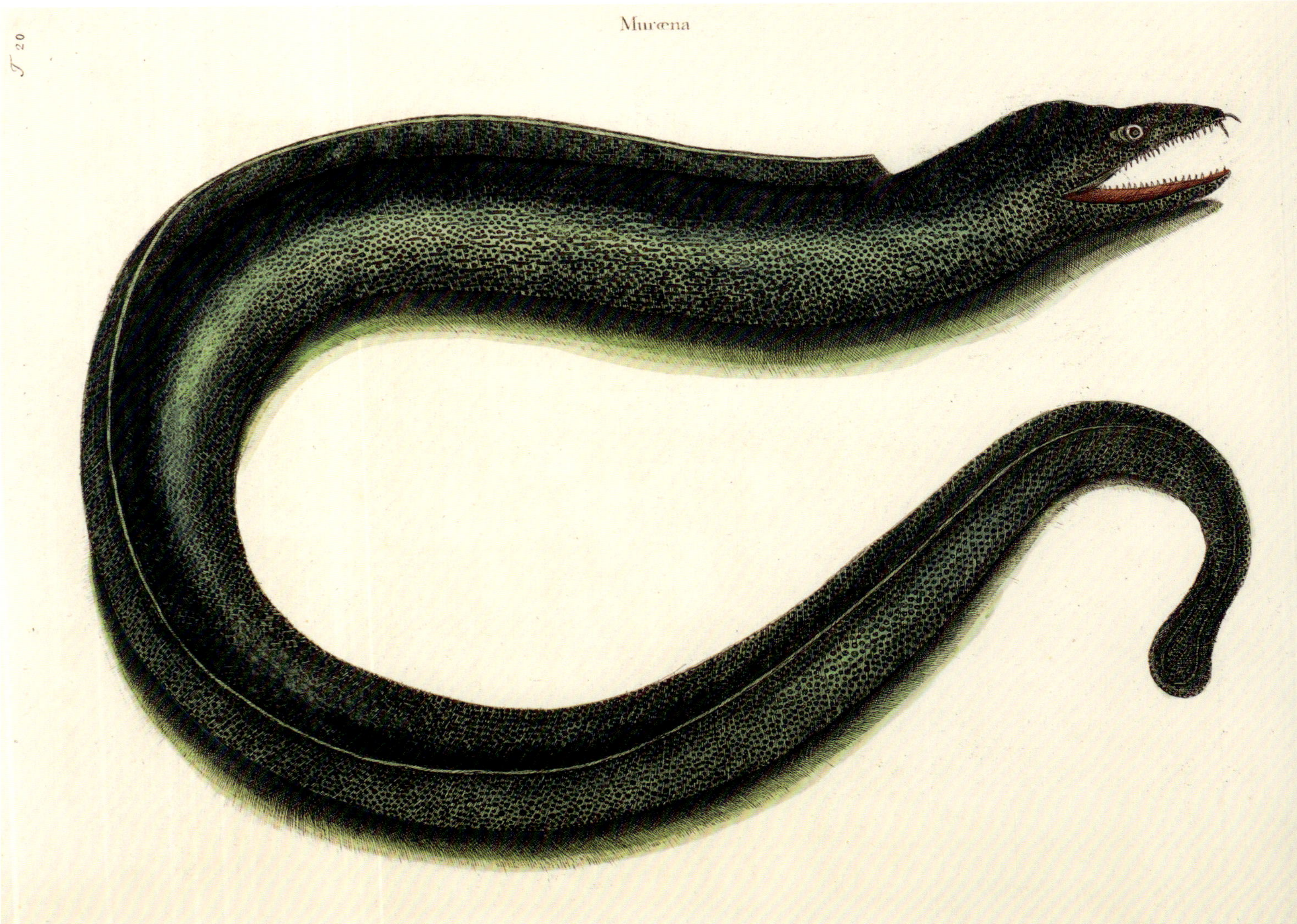

120 *above*

Green moray (*Gymnothorax funebris*) (vol. 2, t. 20)

121 *right*

Spotted moray (*Gymnothorax moringa*); unidentified gorgonid coral (vol. 2, t. 21)

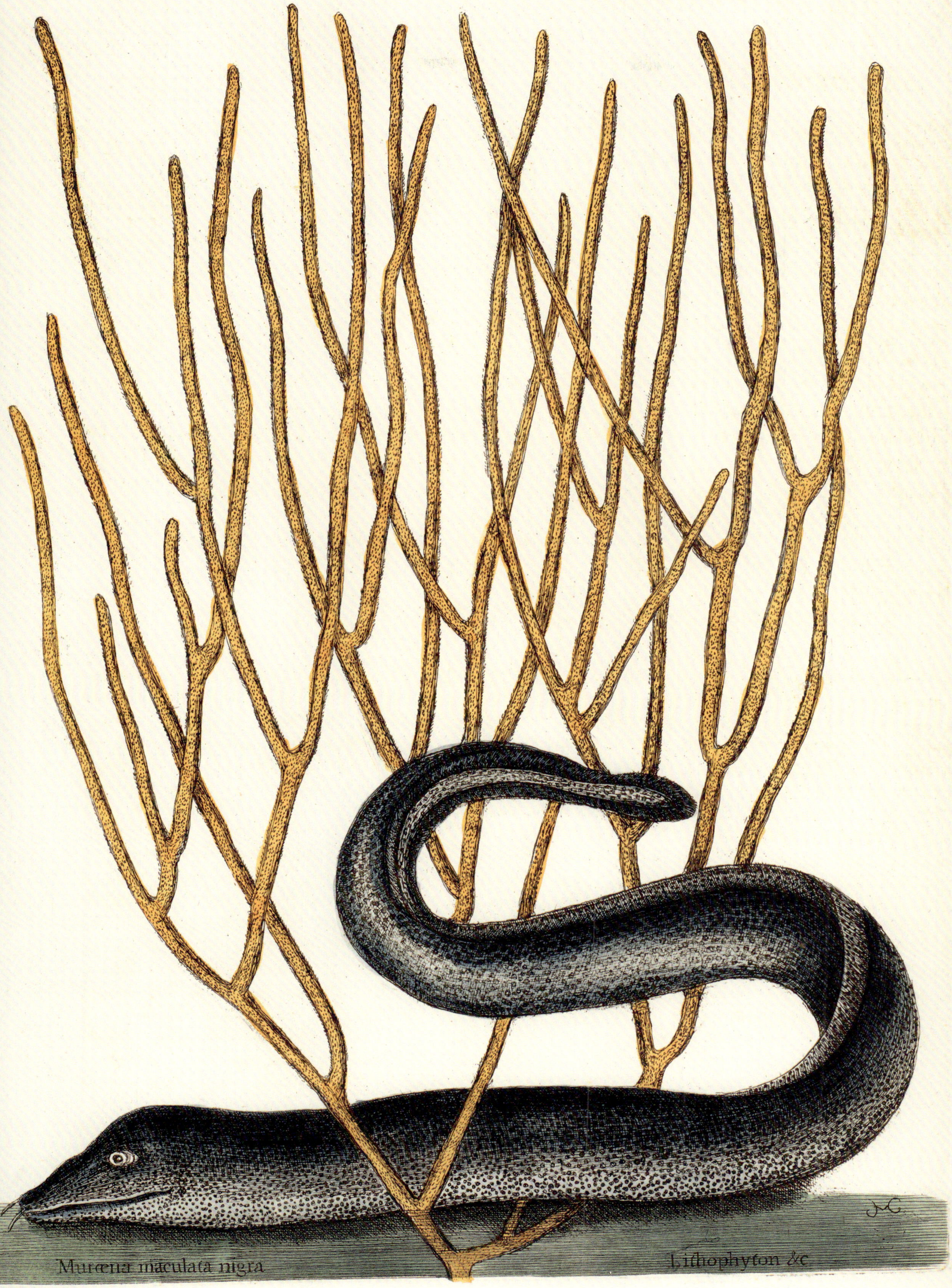
Murœna maculata nigra
Lithophyton &c.

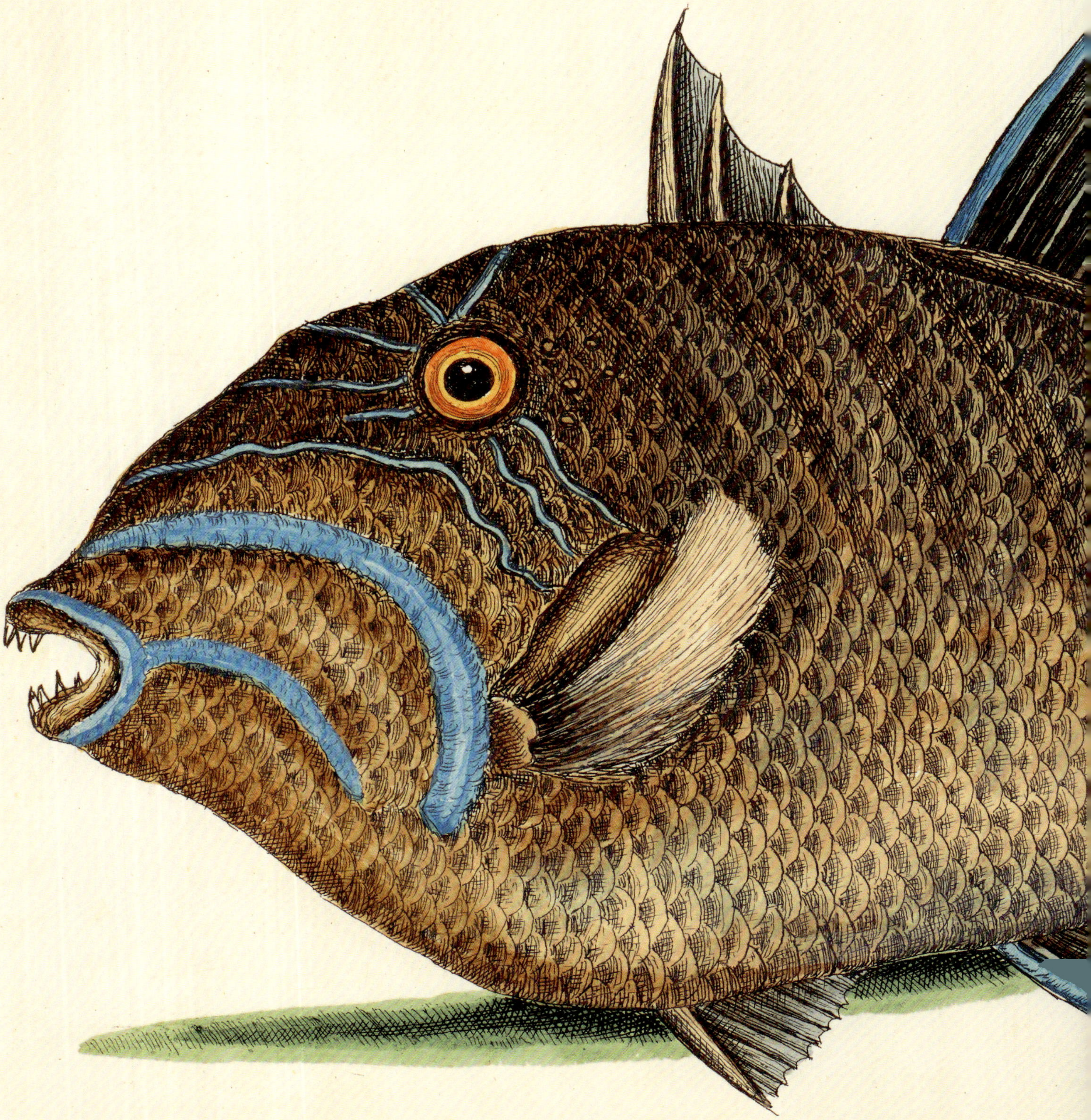

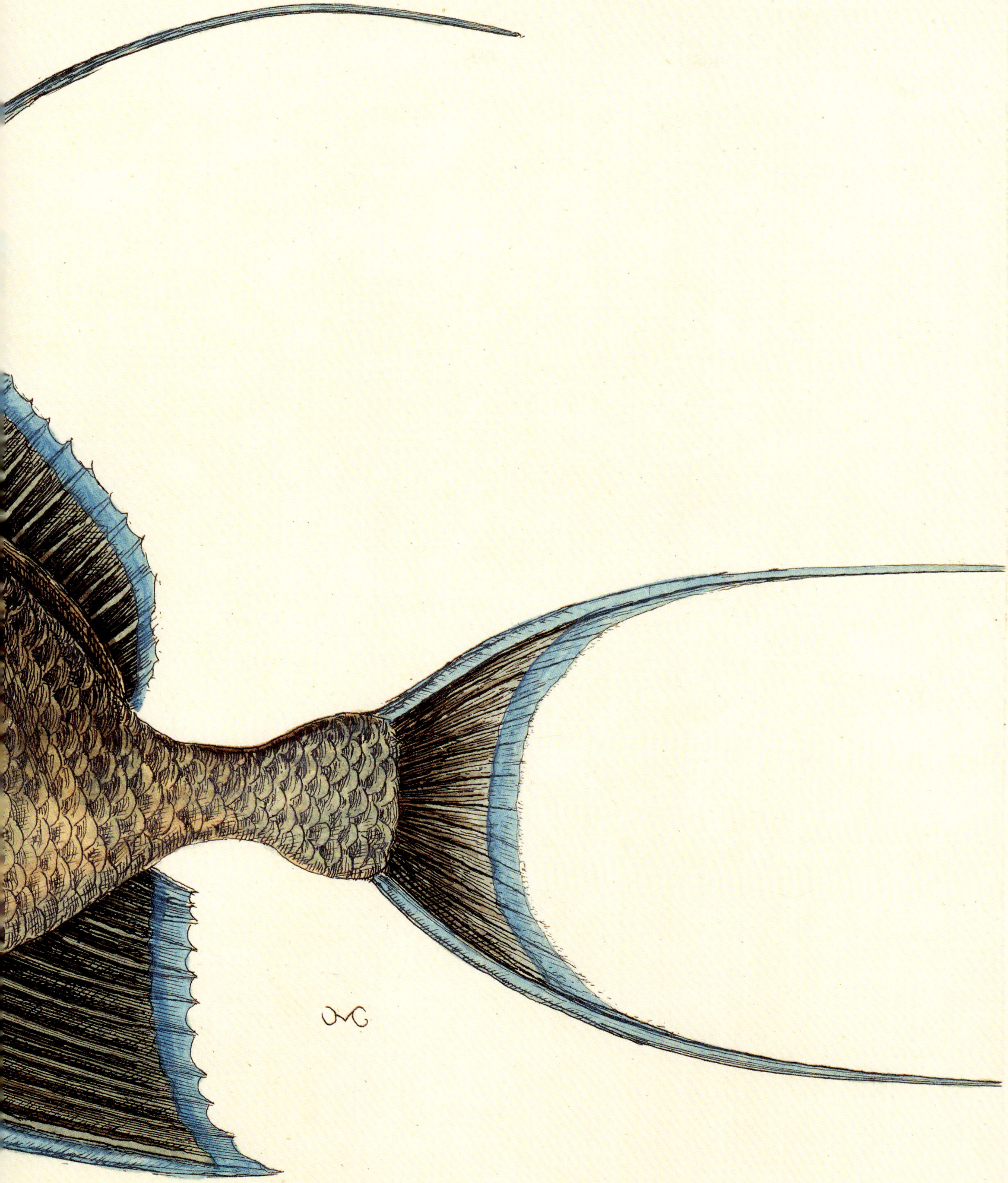

Guaperva &c

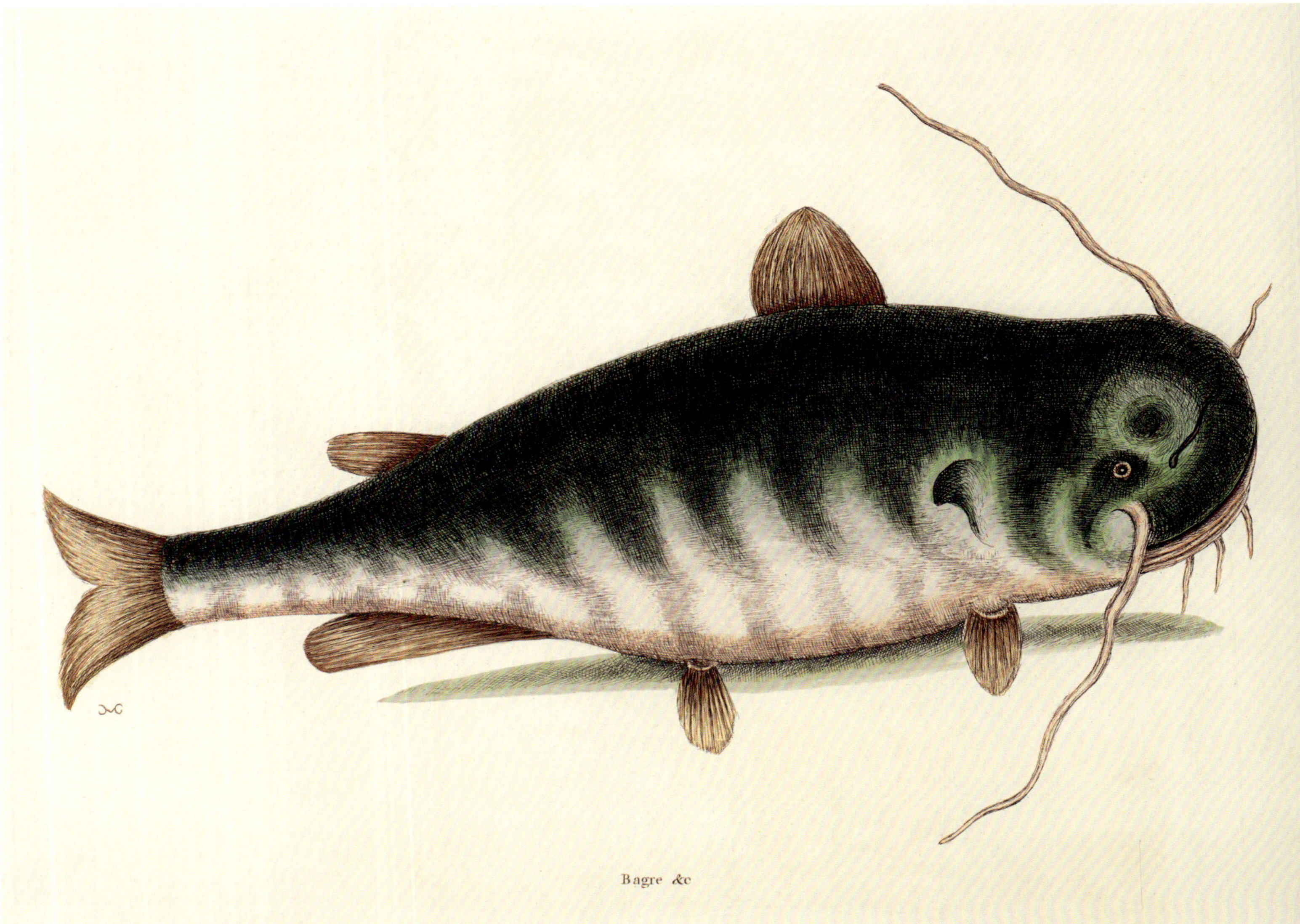

122 *previous pages*
Queen triggerfish (*Balistes vetula*) (vol. 2, t. 22)

123 *above*
White bullhead (*Ameiurus catus*) (vol. 2, t. 23)

124 *right*
Perhaps ladyfish (*Elops saurus*); coinvine (*Dalbergia ecastaphyllum*) (vol. 2, t. 24)

Arbor Populi &c

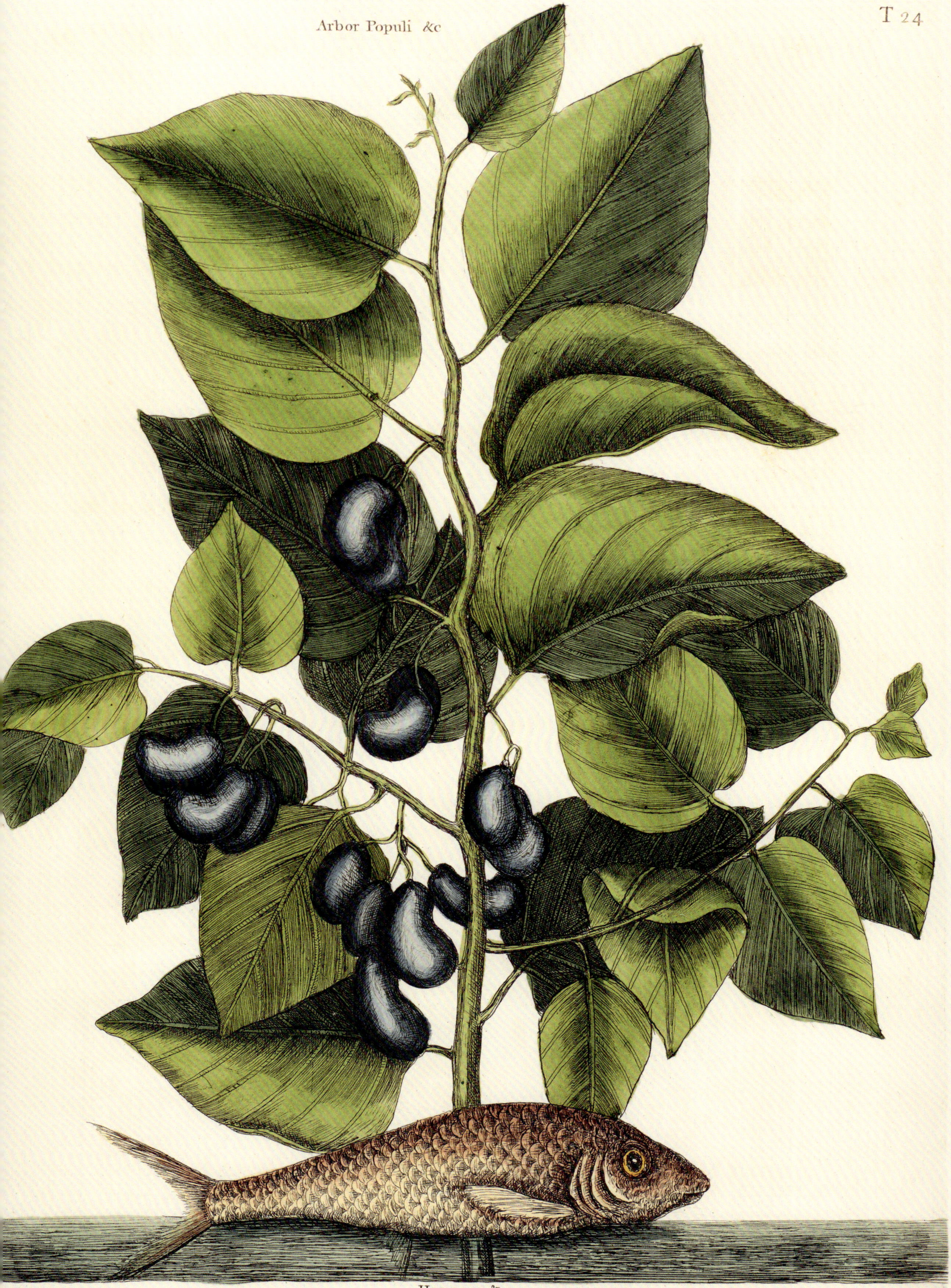

Harengus &c

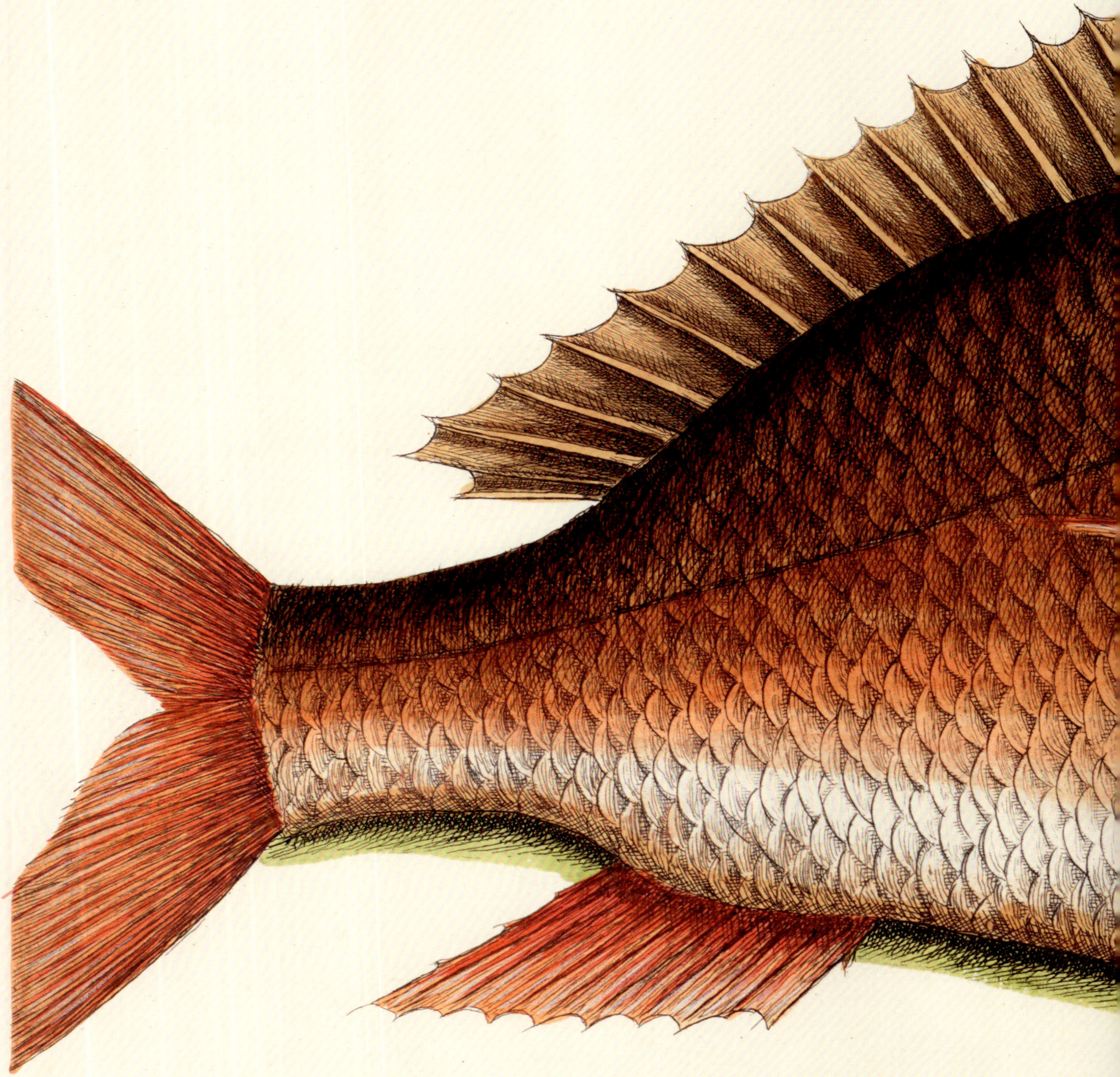

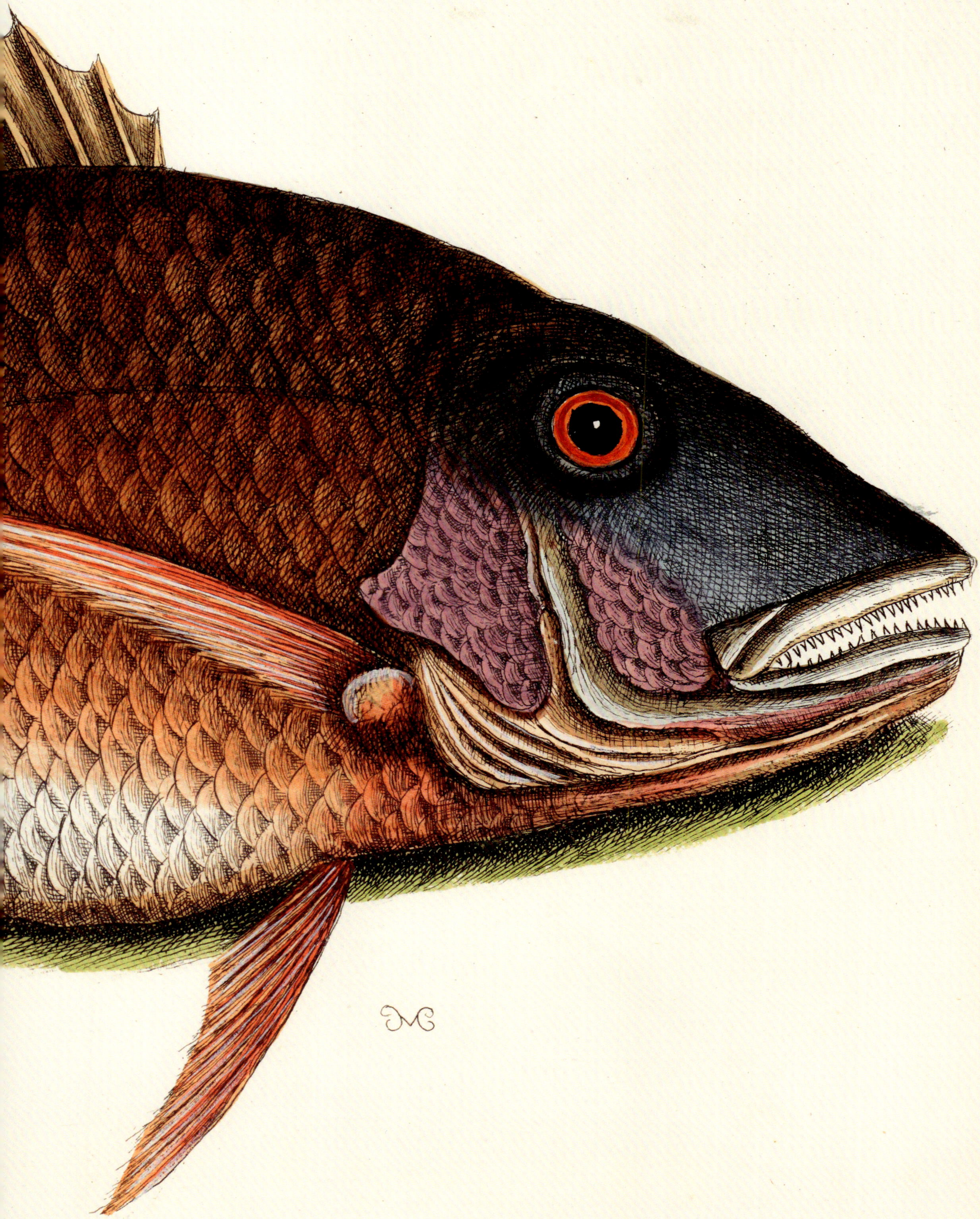

Phylanthos &c

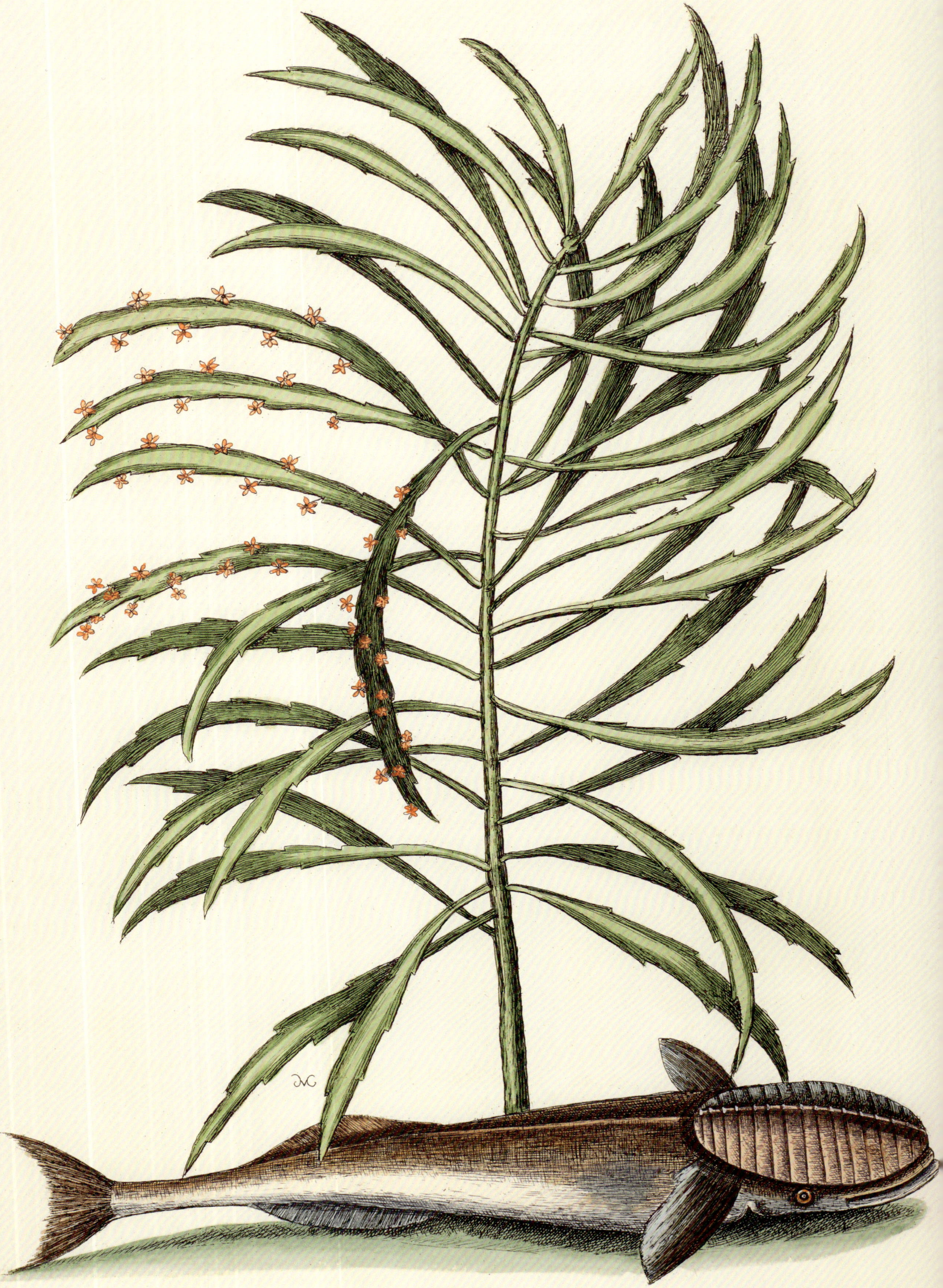

Remora.

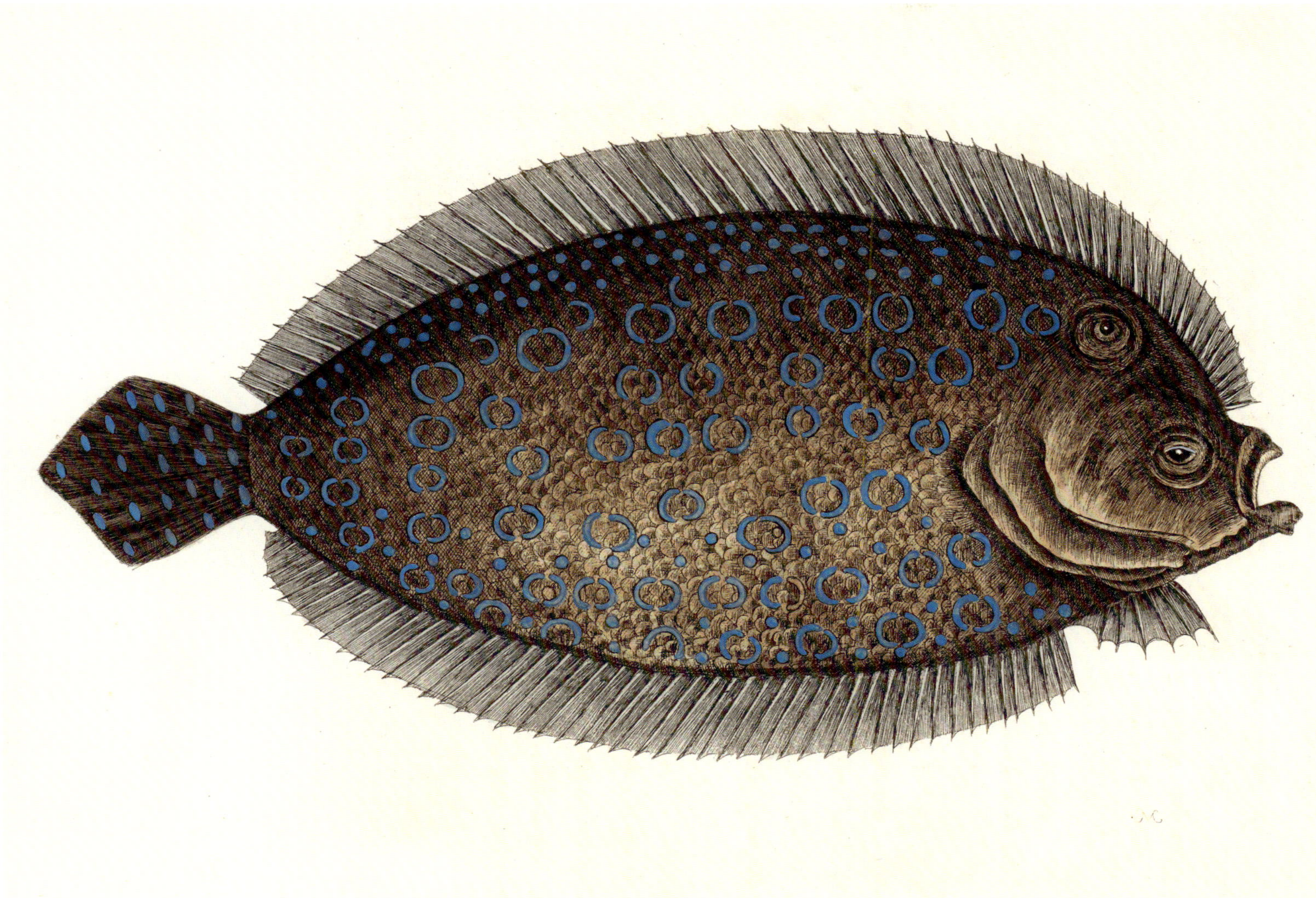

previous pages

125

Mutton snapper (*Lutjanus analis*) (vol. 2, t. 25)

these pages

126 *left*

Remora (*Remora remora*); Abraham bush (*Phyllanthus epiphyllanthus*) (vol. 2, t. 26)

127 *above*

Plate fish (*Bothus lunatus*) (vol. 2, t. 27)

following pages

128 *left*

Checkered puffer (*Sphoeroides testudineus*); bastard torch (*Nectandra coriacea*; upper); red milk pea (*Galactia rudolphioides*; lower) (vol. 2, t. 28)

129 *right top*

Stoplight parrotfish (*Sparisoma viride*) (vol. 2, t. 29)

130 *right bottom*

Florida gar (*Lepisosteus platyrhincus*); fruiting branch of coastal doghobble (*Leucothoe axillaris*); flowering branch unidentified (vol. 2, t. 30)

Cornus &c

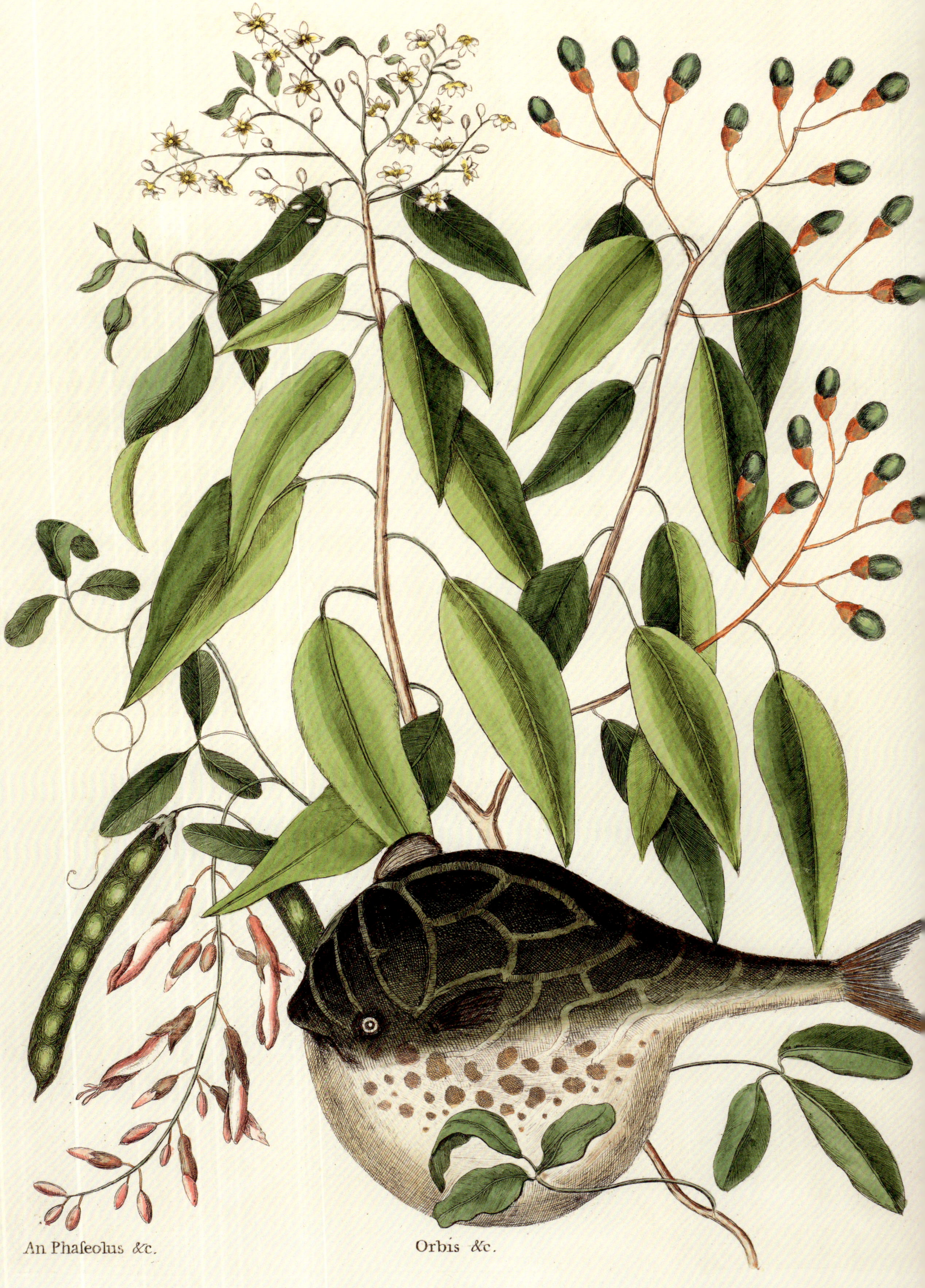

An Phaseolus &c.

Orbis &c.

T 29

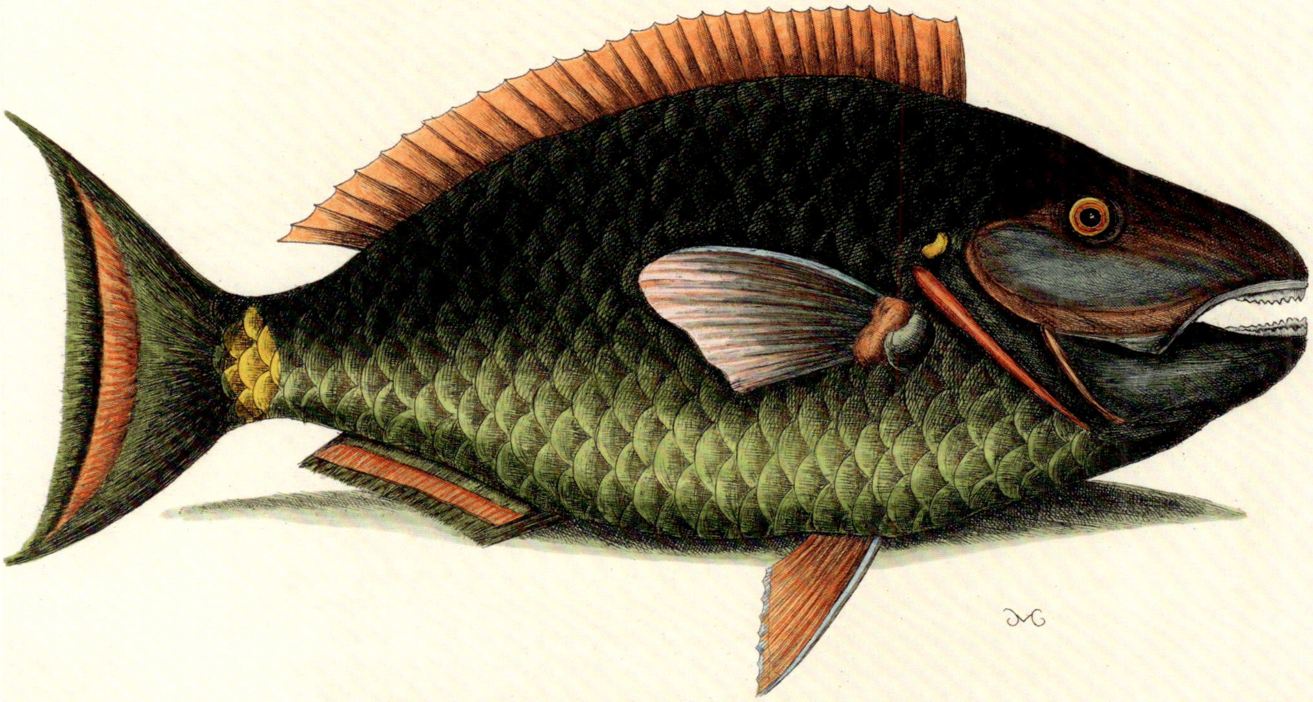

Pſittacus &c.

T 30

Frutex aquaticus &c

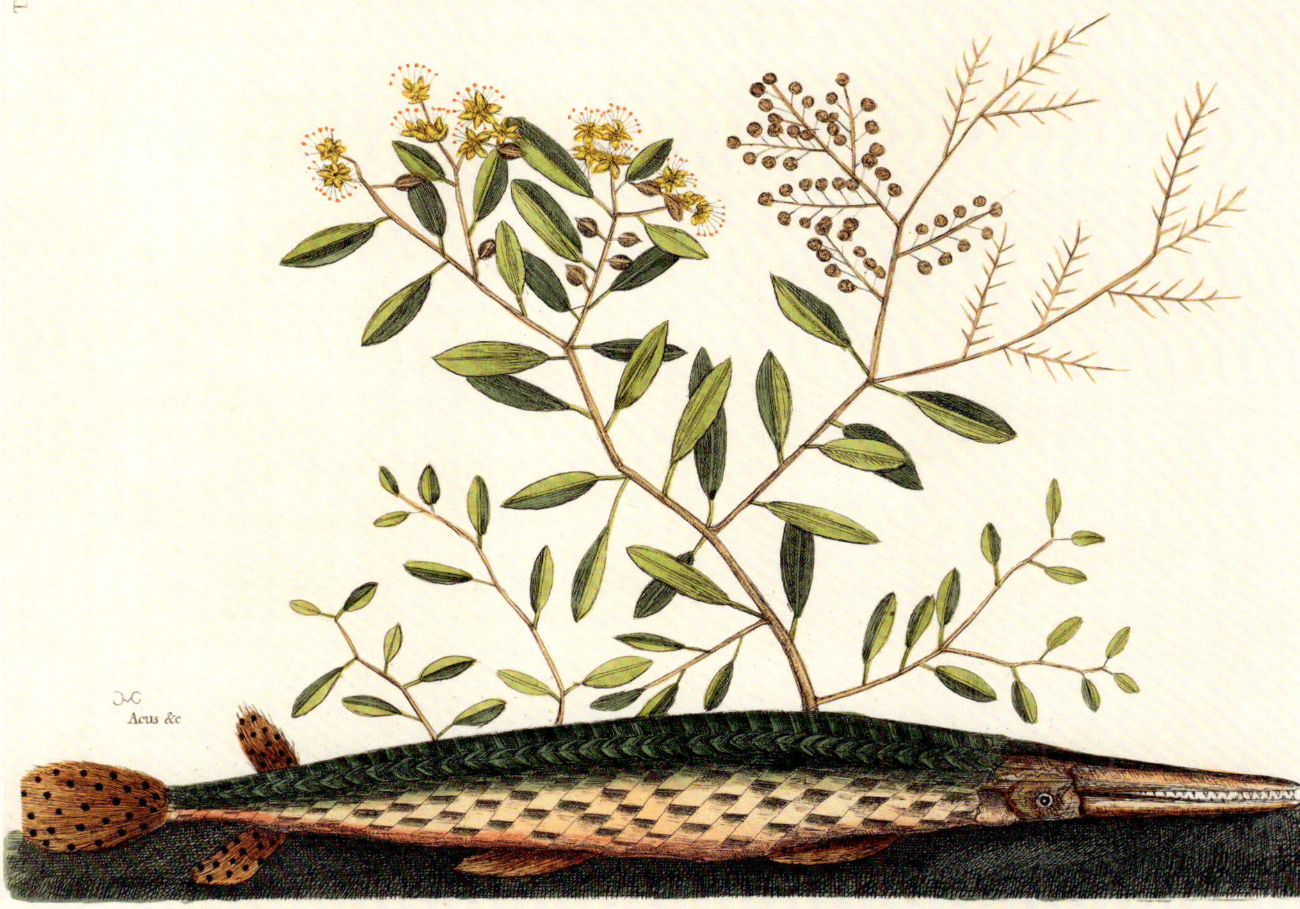

Acus &c

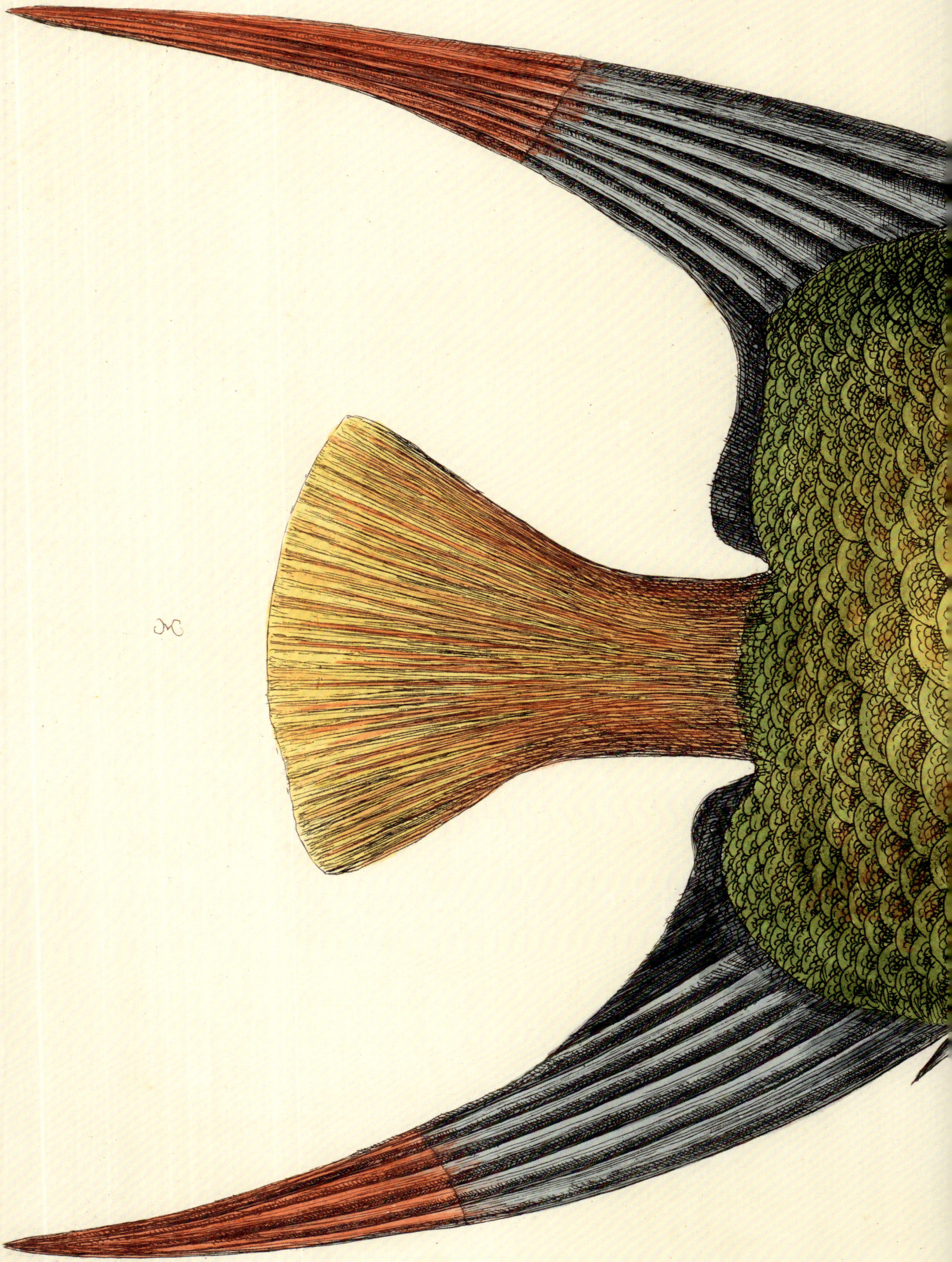

Ar

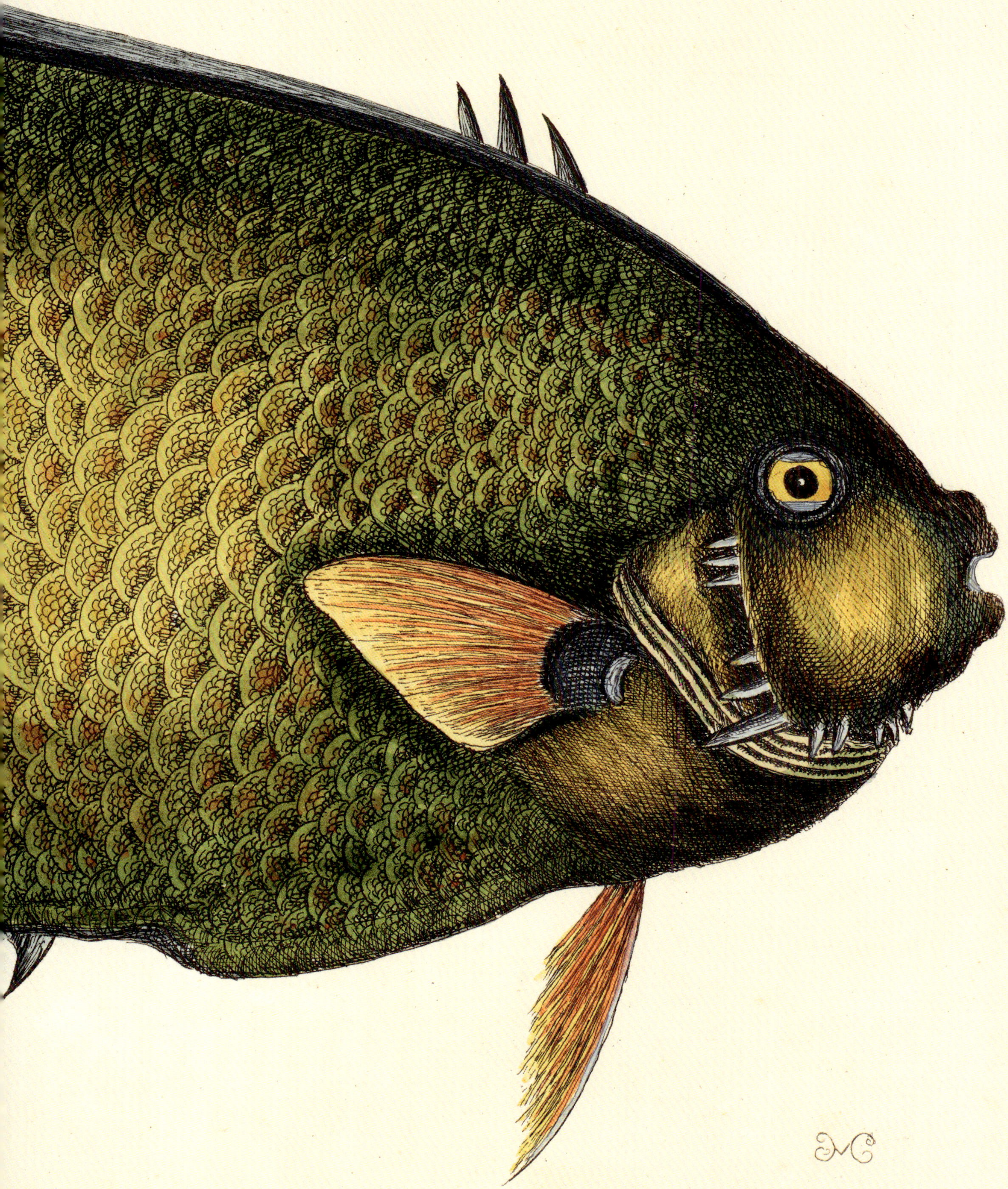

&c.

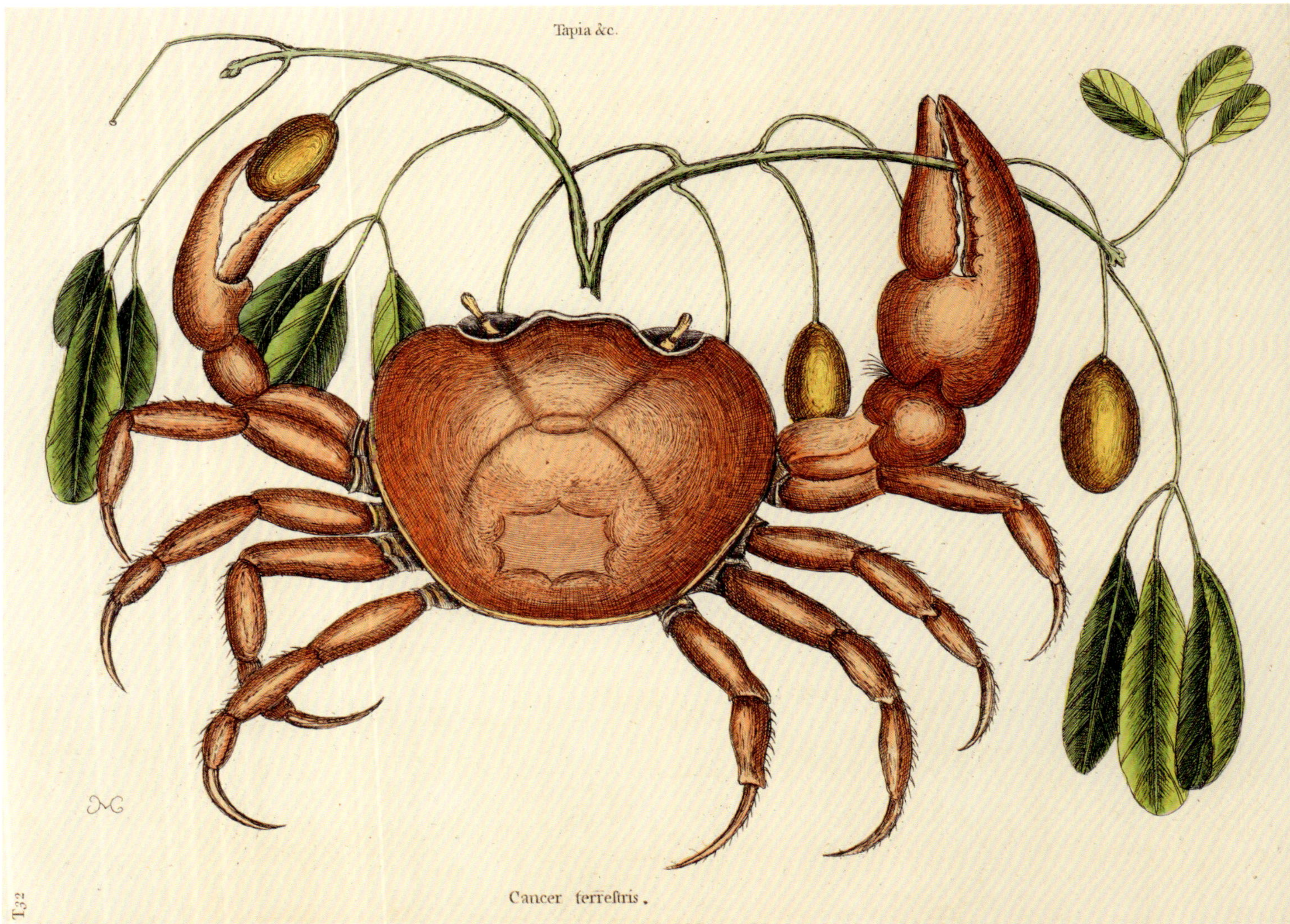

previous pages

131
Queen angelfish (*Holacanthus ciliaris*) (vol. 2, t. 31)

132 *above*
Purple land crab (*Gecarcinus ruricola*); blackwood (*Picrodendron baccatum*) (vol. 2, t. 32)

133 *right*
Caribbean hermit crab (*Coenobita clypeatus*); West Indian topsnail (*Cittarium pica*); button mangrove (*Conocarpus erectus*; upper); sea torchwood (*Amyris elemifera*; lower) (vol. 2, t. 33)

Manghala &c. T.33

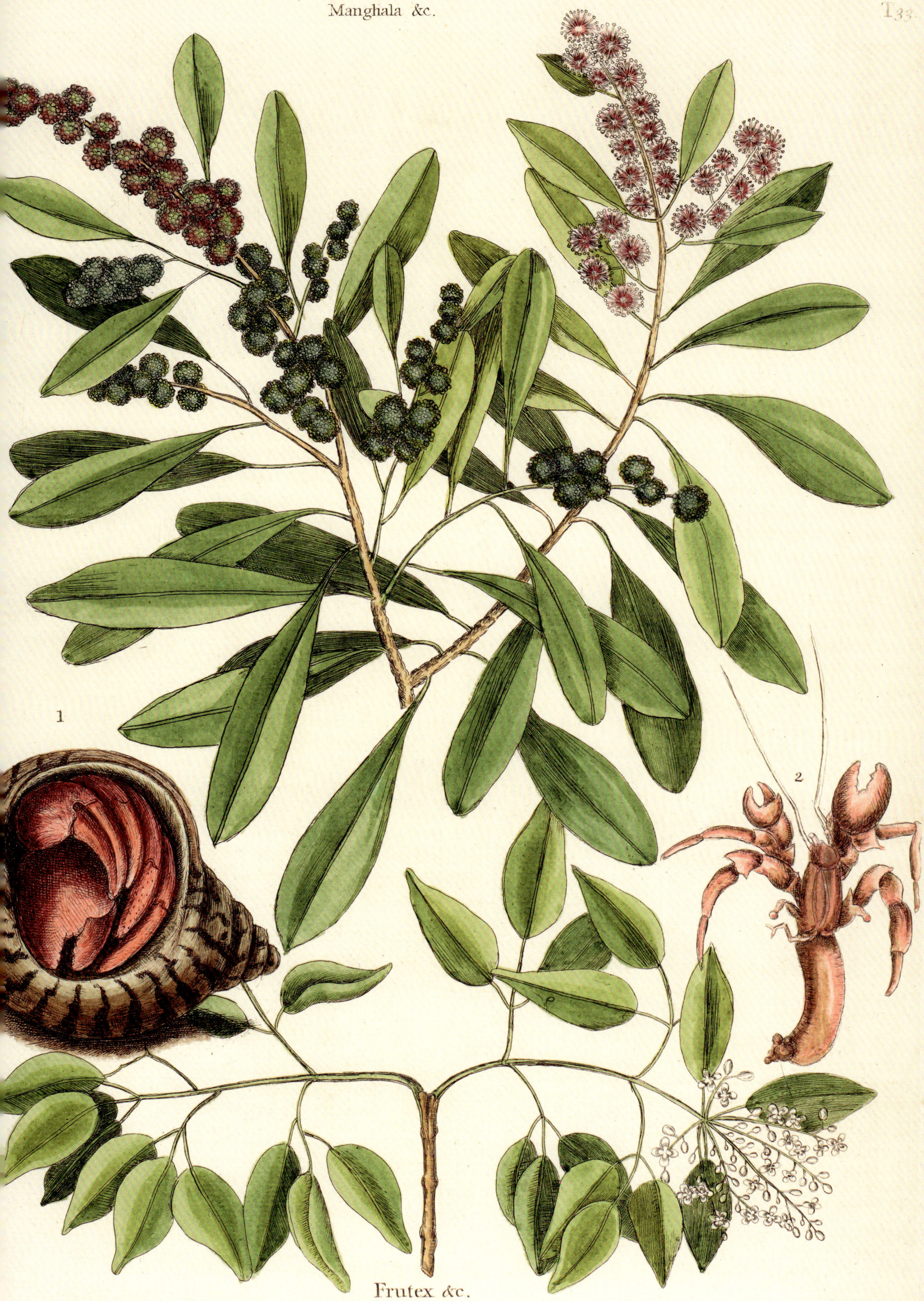

Frutex &c.

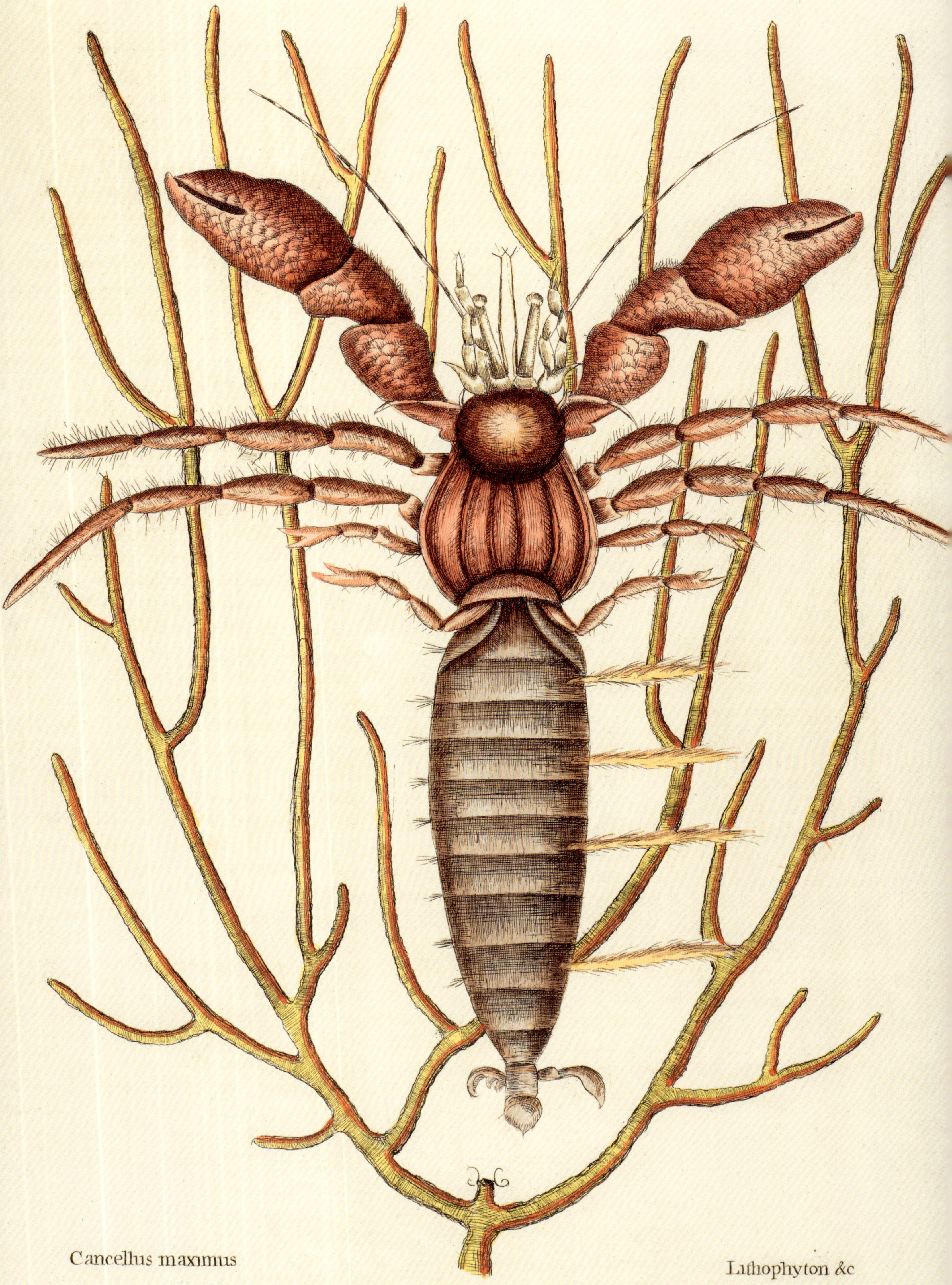

Cancellus maximus

Lithophyton &c

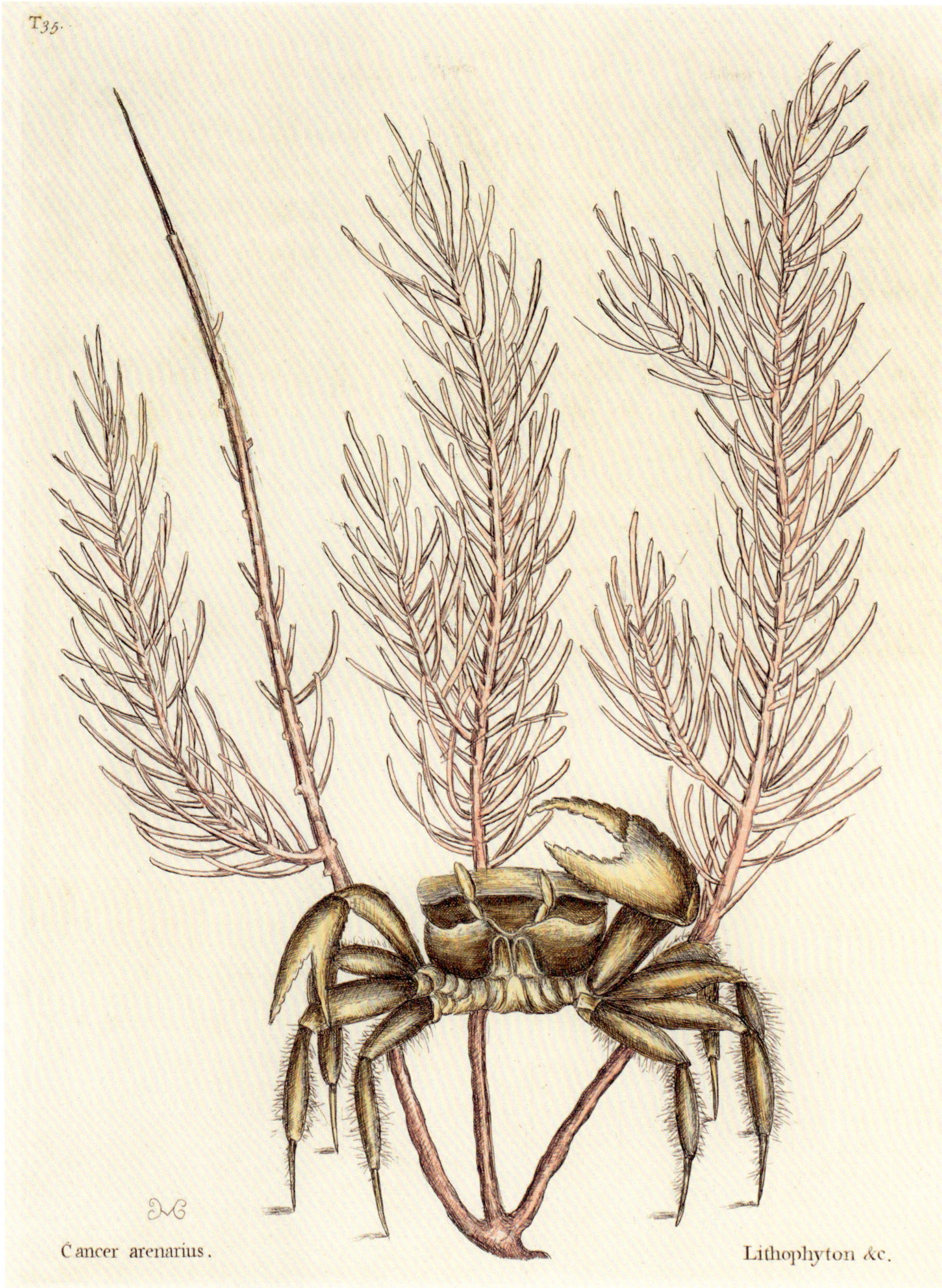

134 *left*

Giant hermit crab (*Petrochirus diogenes*); probably angular sea whip (*Pterogorgia anceps*) (vol. 2, t. 34)

135 *above*

Atlantic ghost crab (*Ocypode quadrata*); probably purple sea plume (*Pseudopterogorgia acerosa*) (vol. 2, t. 35)

Pagurus Maculatus.

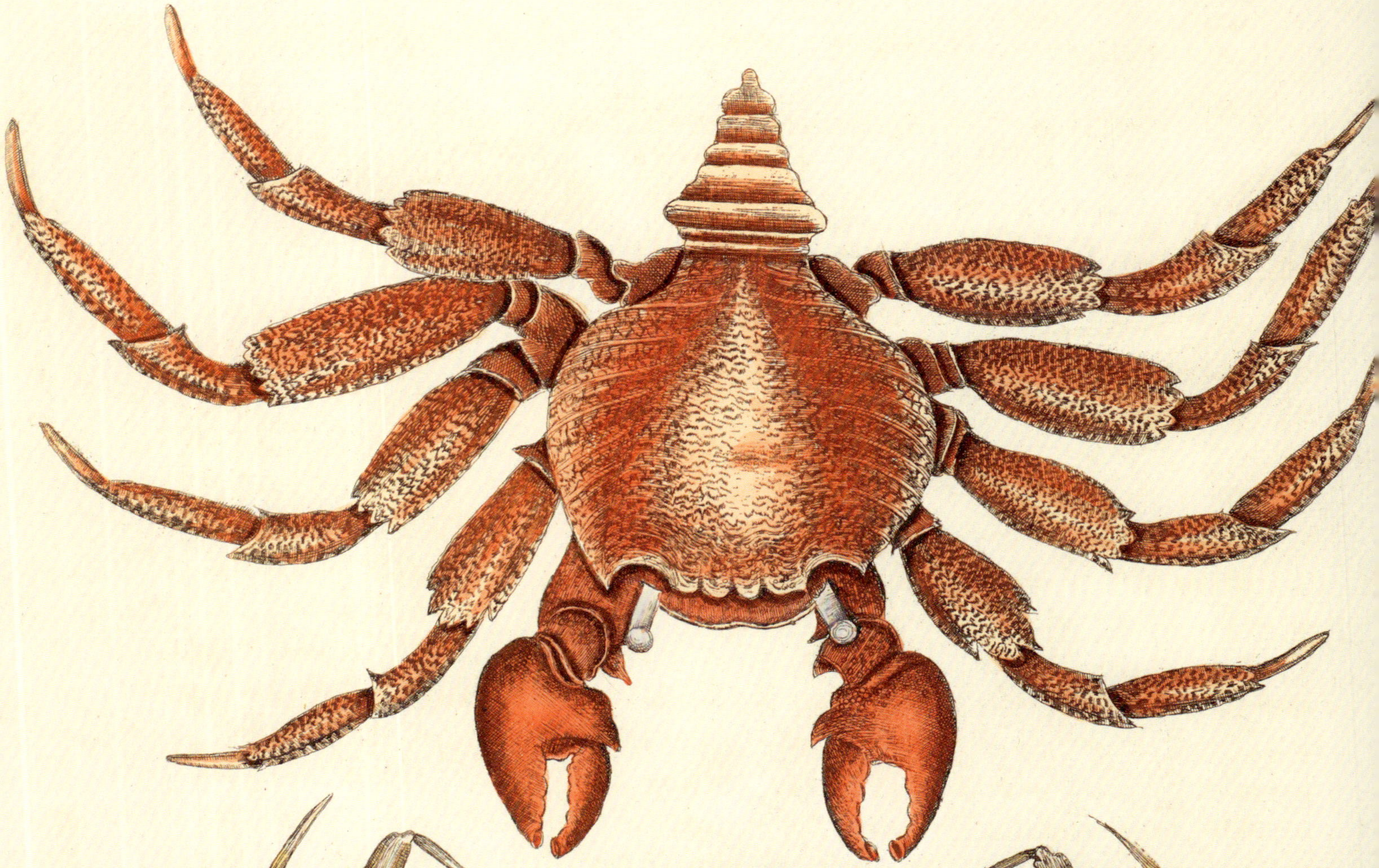

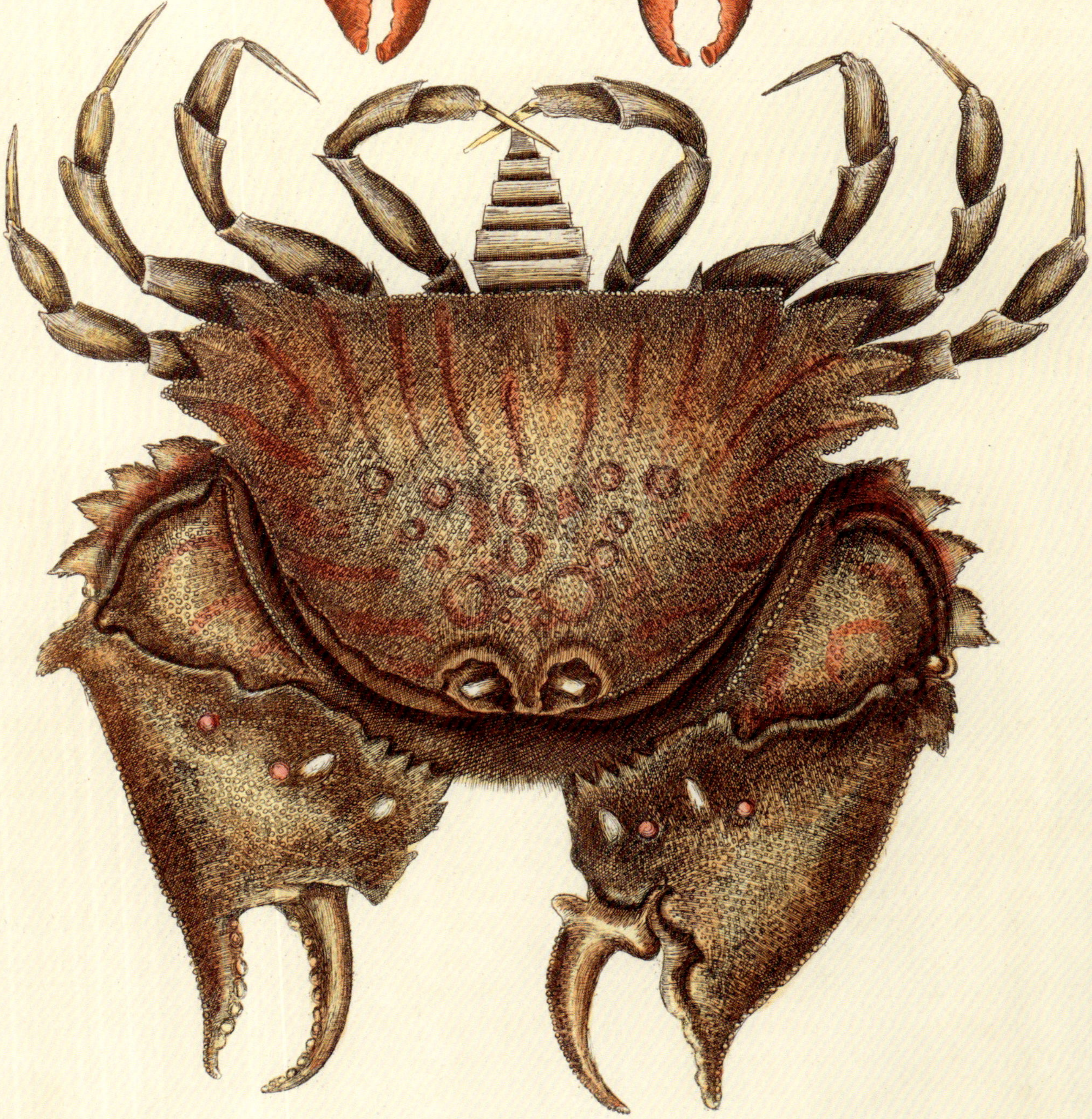

Cancer chelis craſsiſsimis.

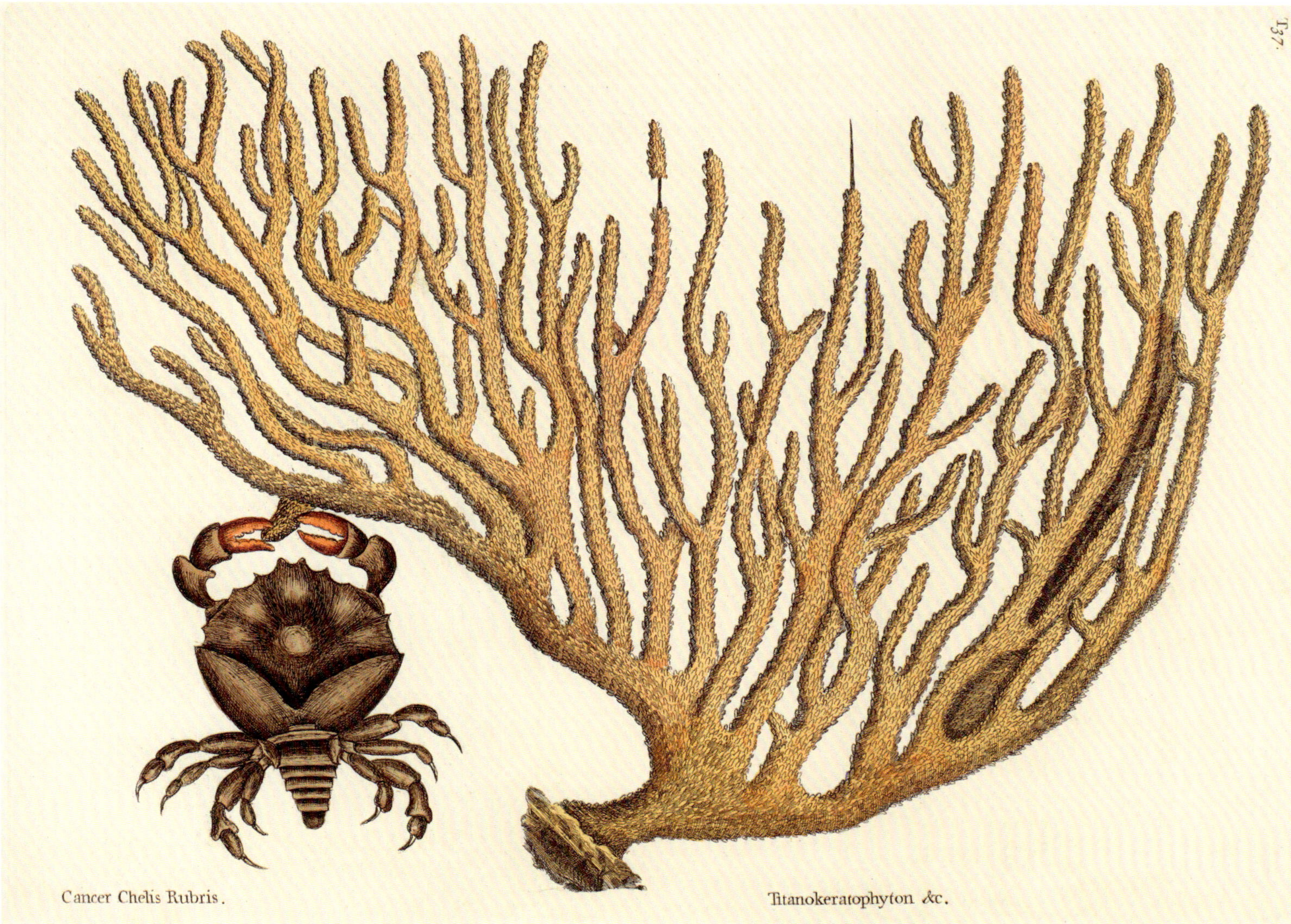

136 *left*
Sally Lightfoot crab (*Grapsus grapsus*; upper); flame box crab (*Calappa flammea*; lower) (vol. 2, t. 36)

137 *above*
Redeye sponge crab (*Dromia erythropus*); spiny sea fan (*Muricea muricata*) (vol. 2, t. 37)

T 38

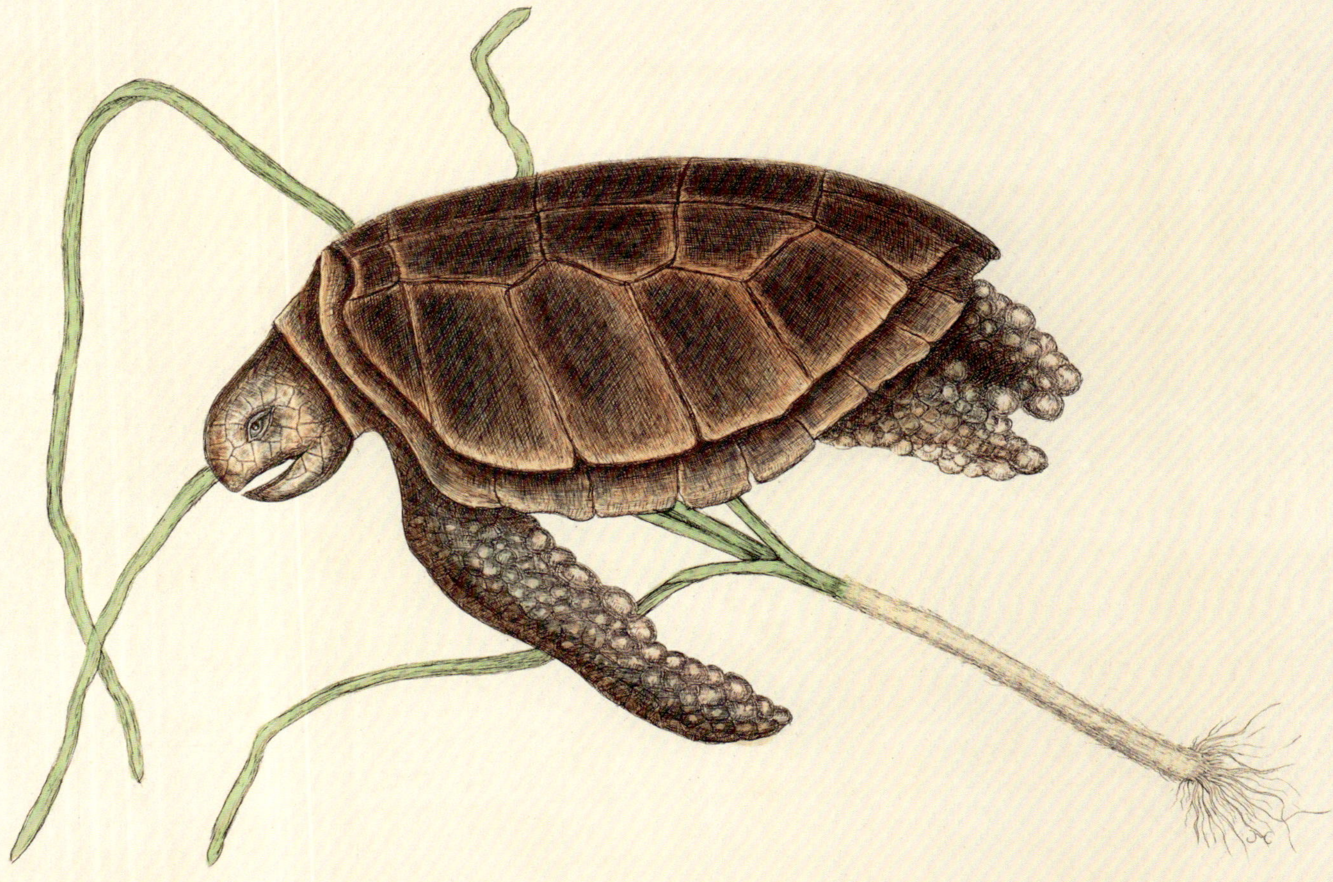

Teſtudo &c.

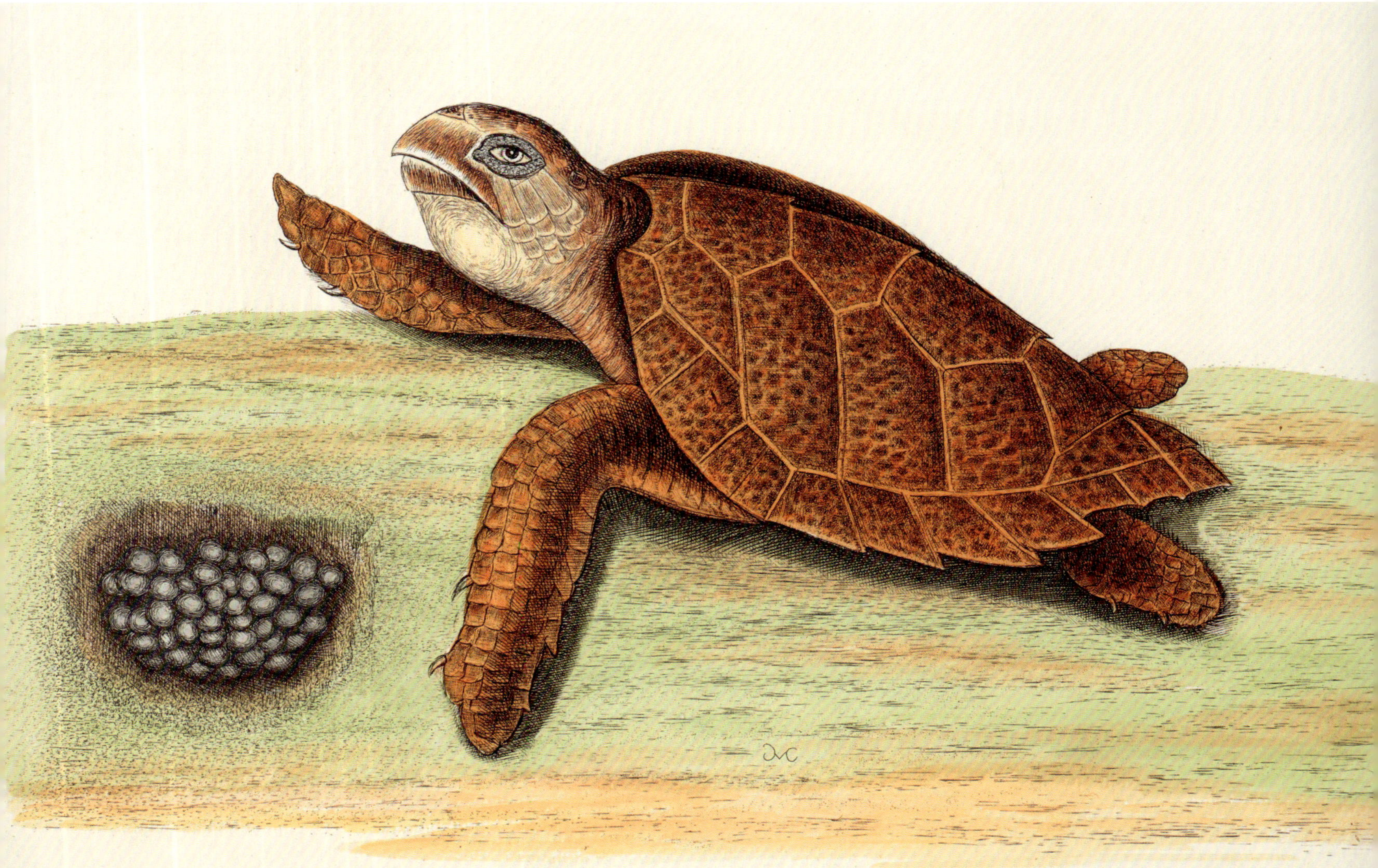

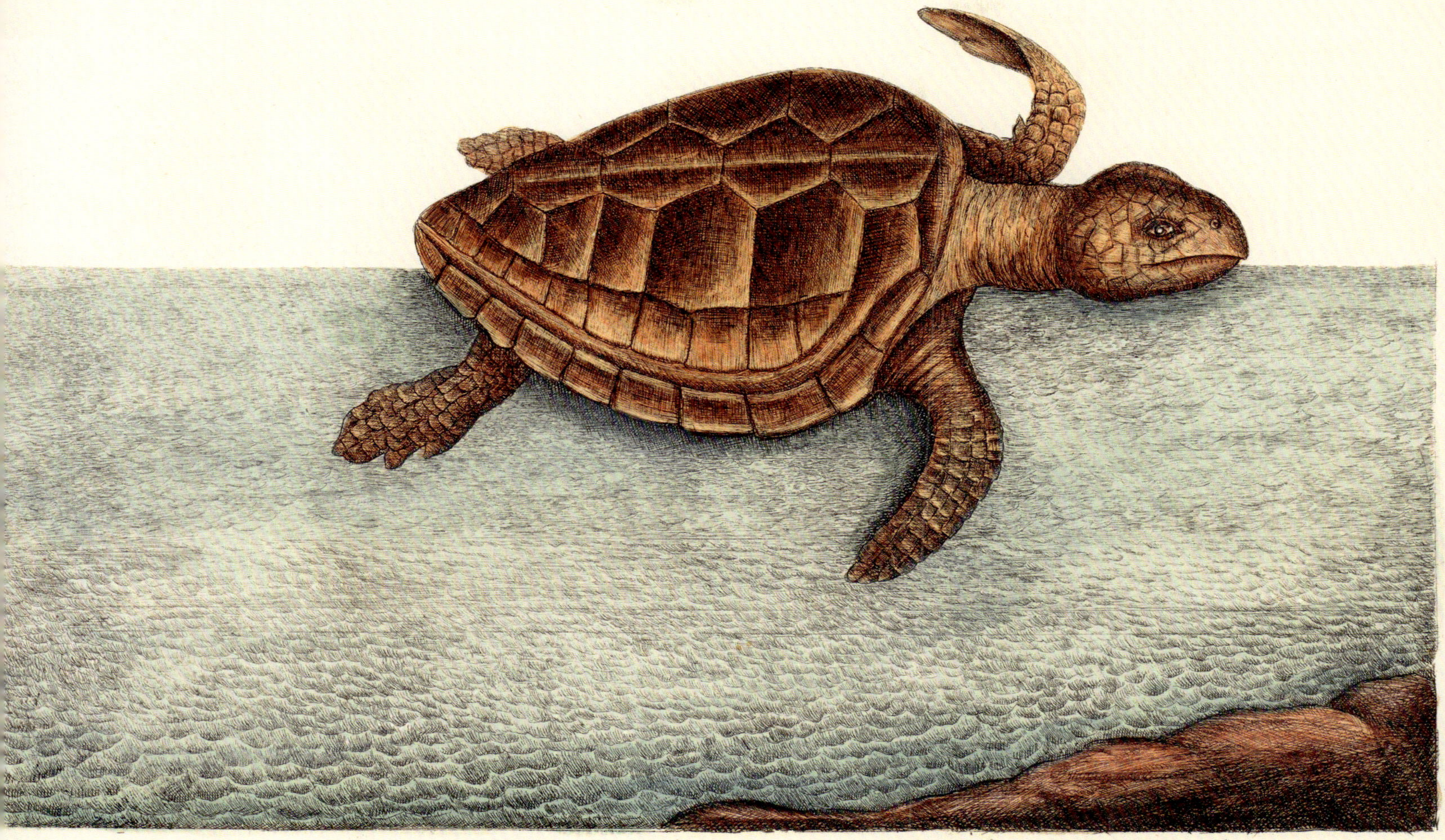

Testudo &c.

138 *top left*
Green turtle (*Chelonia mydas*); turtlegrass (*Thalassia testudinum*) (vol. 2, t. 38)

139 *bottom left*
Hawksbill turtle (*Eretmochelys imbricata*) (vol. 2, t. 39)

140 *above*
Loggerhead turtle (*Caretta caretta*) (vol. 2, t. 40)

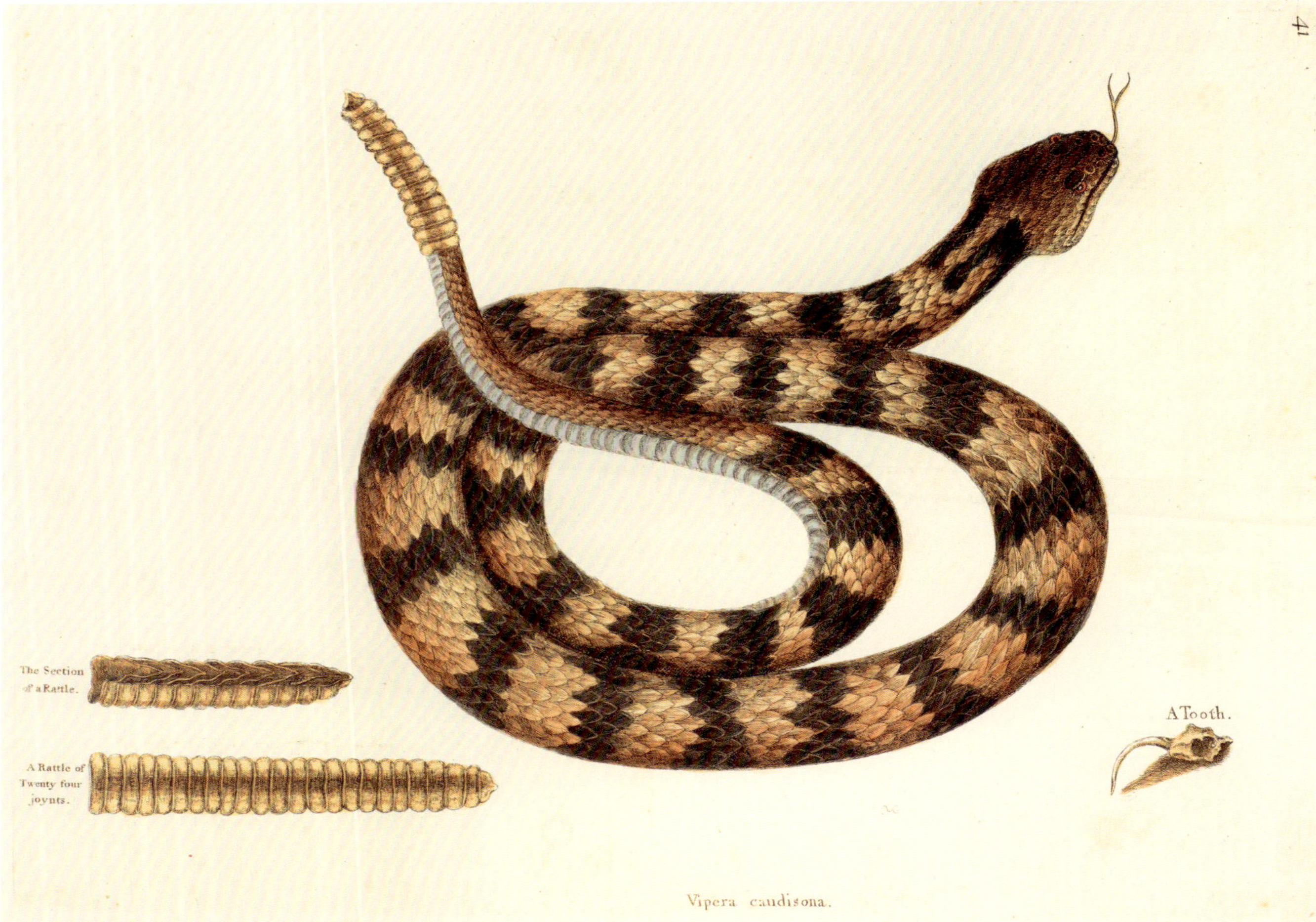
41
The Section of a Rattle.
A Rattle of Twenty four joynts.
A Tooth.
Vipera caudisona.

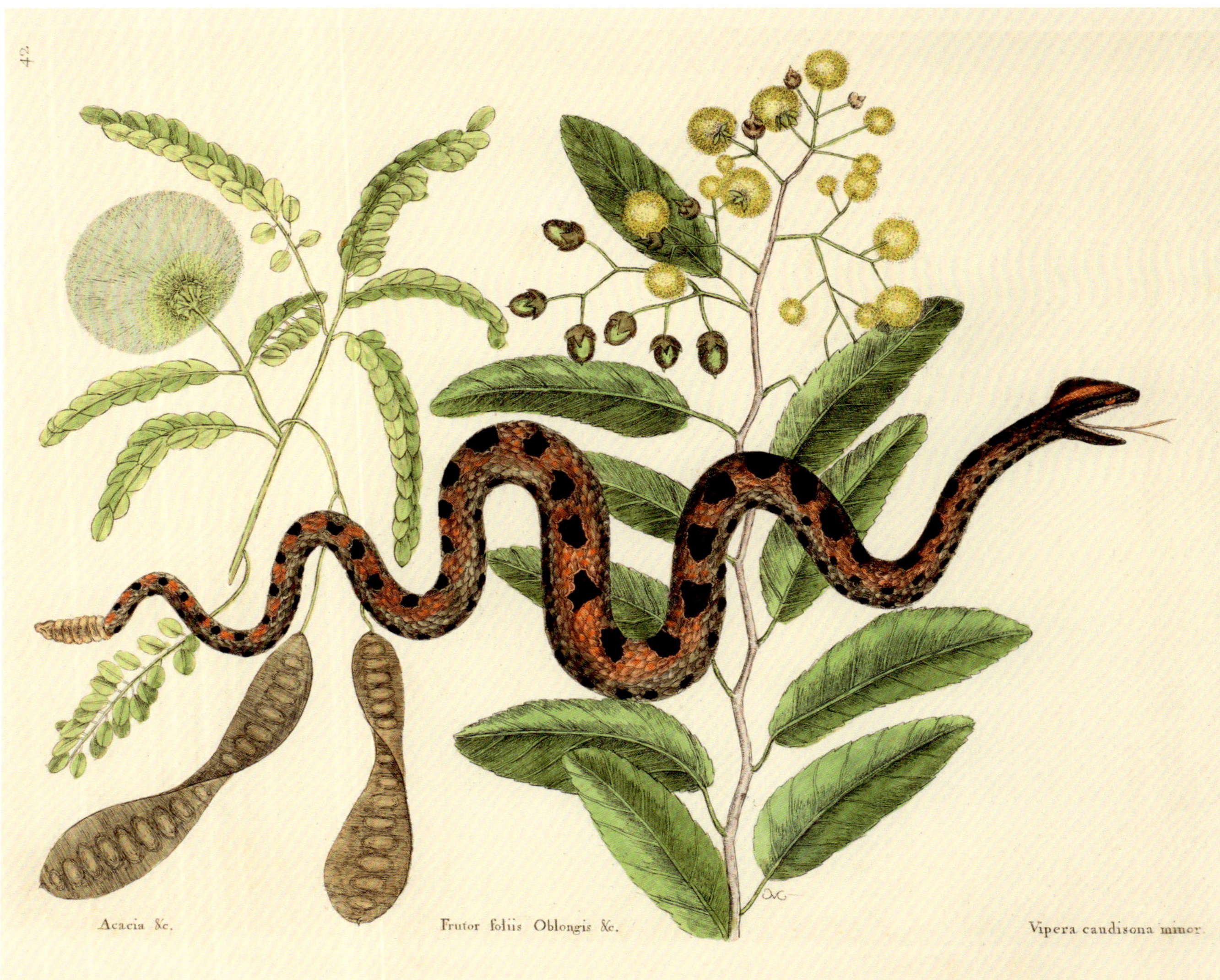
42
Acacia &c.
Frutor foliis Oblongis &c.
Vipera caudisona minor.

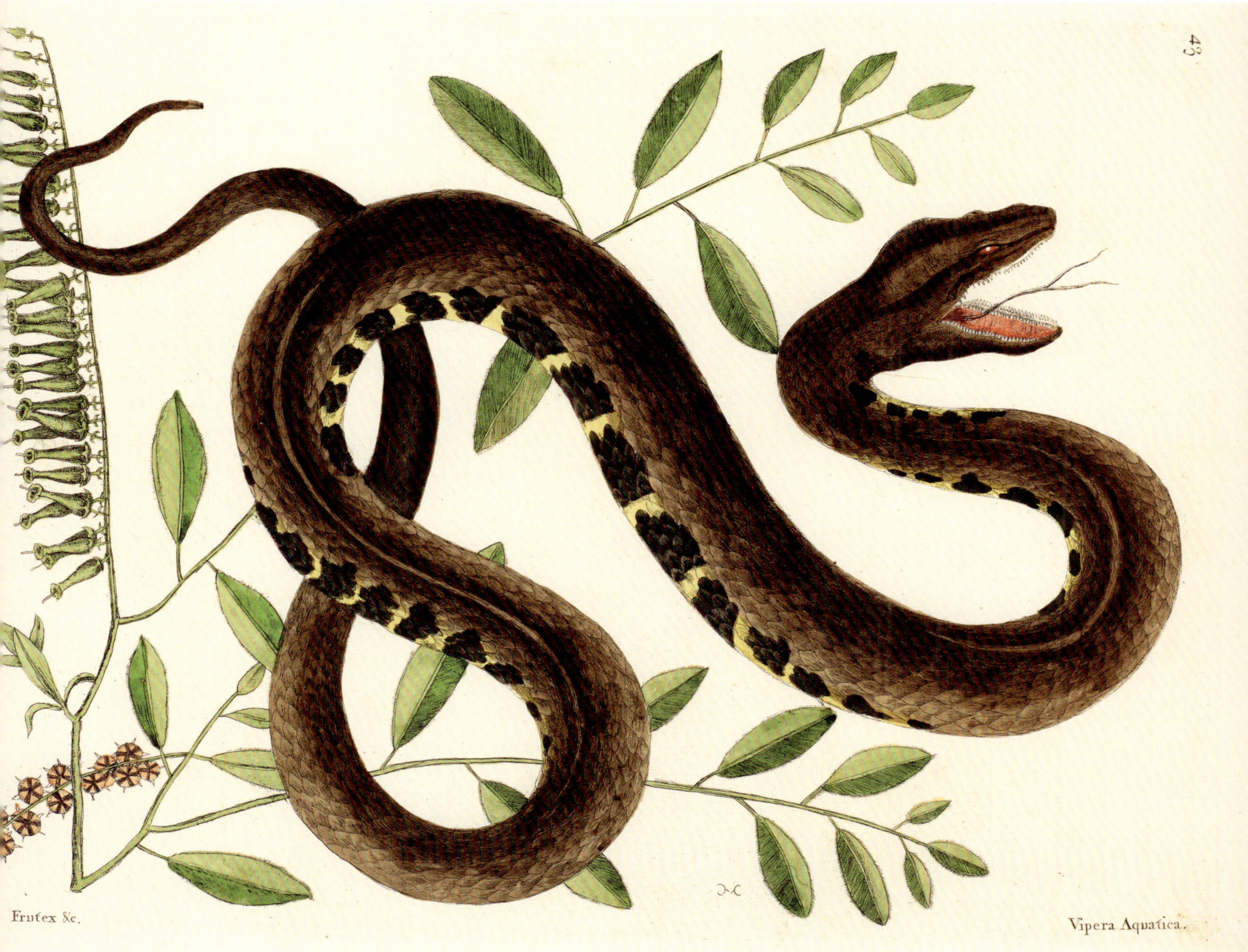

141 *top left*
Timber rattlesnake (*Crotalus horridus*) (vol. 2, t. 41)

142 *bottom left*
Pygmy rattlesnake (*Sistrurus miliarus*); horseflesh mahogany (*Lysiloma sabicu*; left); Banara (*Banara minutiflora*; right) (vol. 2, t. 42)

143 *above*
Northern cottonmouth (*Agkistrodon piscivorus*); swamp doghobble (*Eubotrys racemosa*) (vol. 2, t. 43)

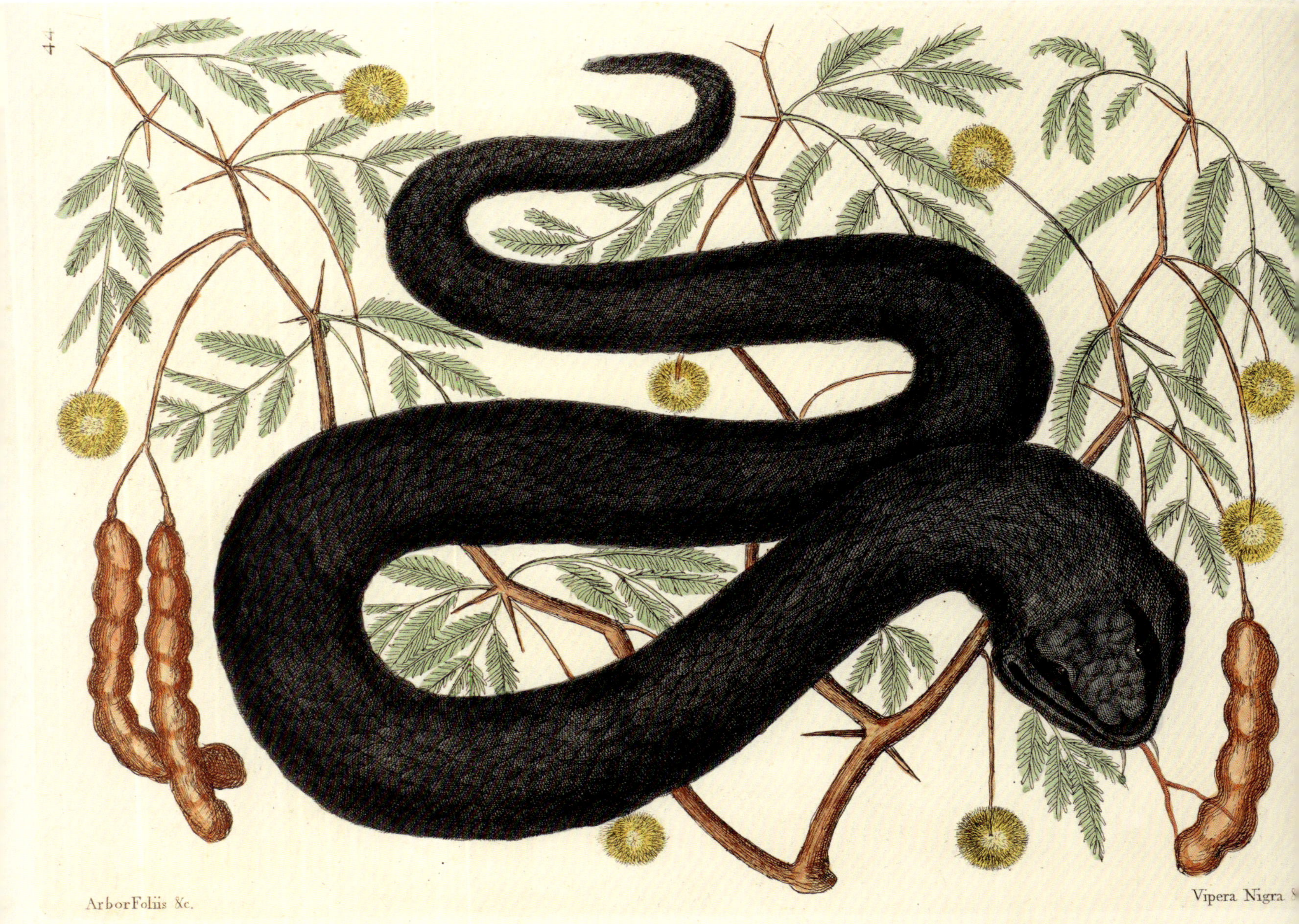
44
Arbor Foliis &c.
Vipera Nigra

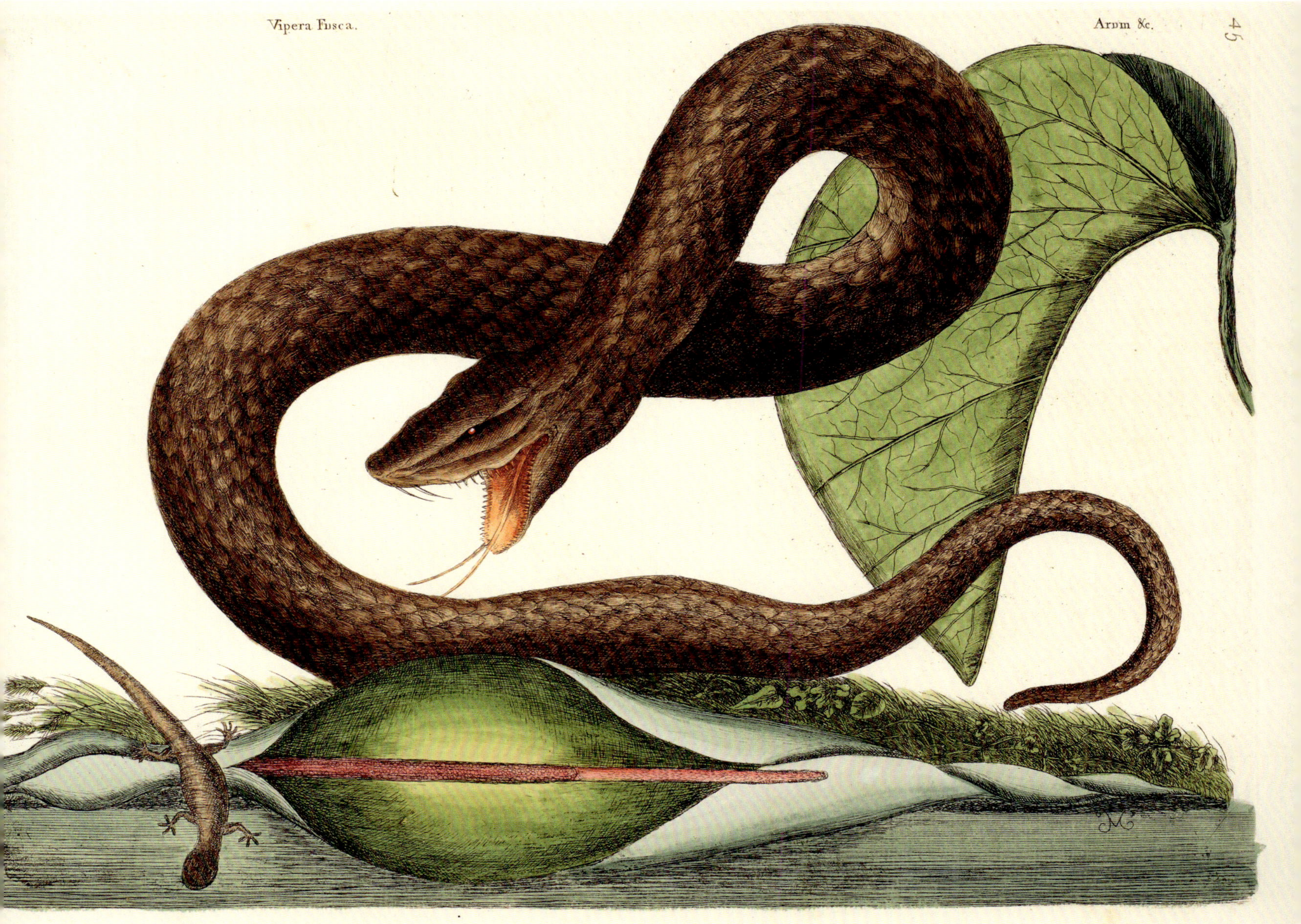

144 *left*
Eastern hog-nosed snake (*Heterodon platirhinos*); poponax (*Vachellia tortuosa*) (vol. 2, t. 44)

145 *above*
Eastern hog-nosed snake (*Heterodon platirhinos*); probably mole salamander (*Ambystoma talpoideum*); tannia (*Xanthosoma sagittifolium*) (vol. 2, t. 45)

following pages

146
Plain-bellied watersnake (*Nerodia erythrogaster*); cascarilla (*Croton eluteria*) (vol. 2, t. 46)

46

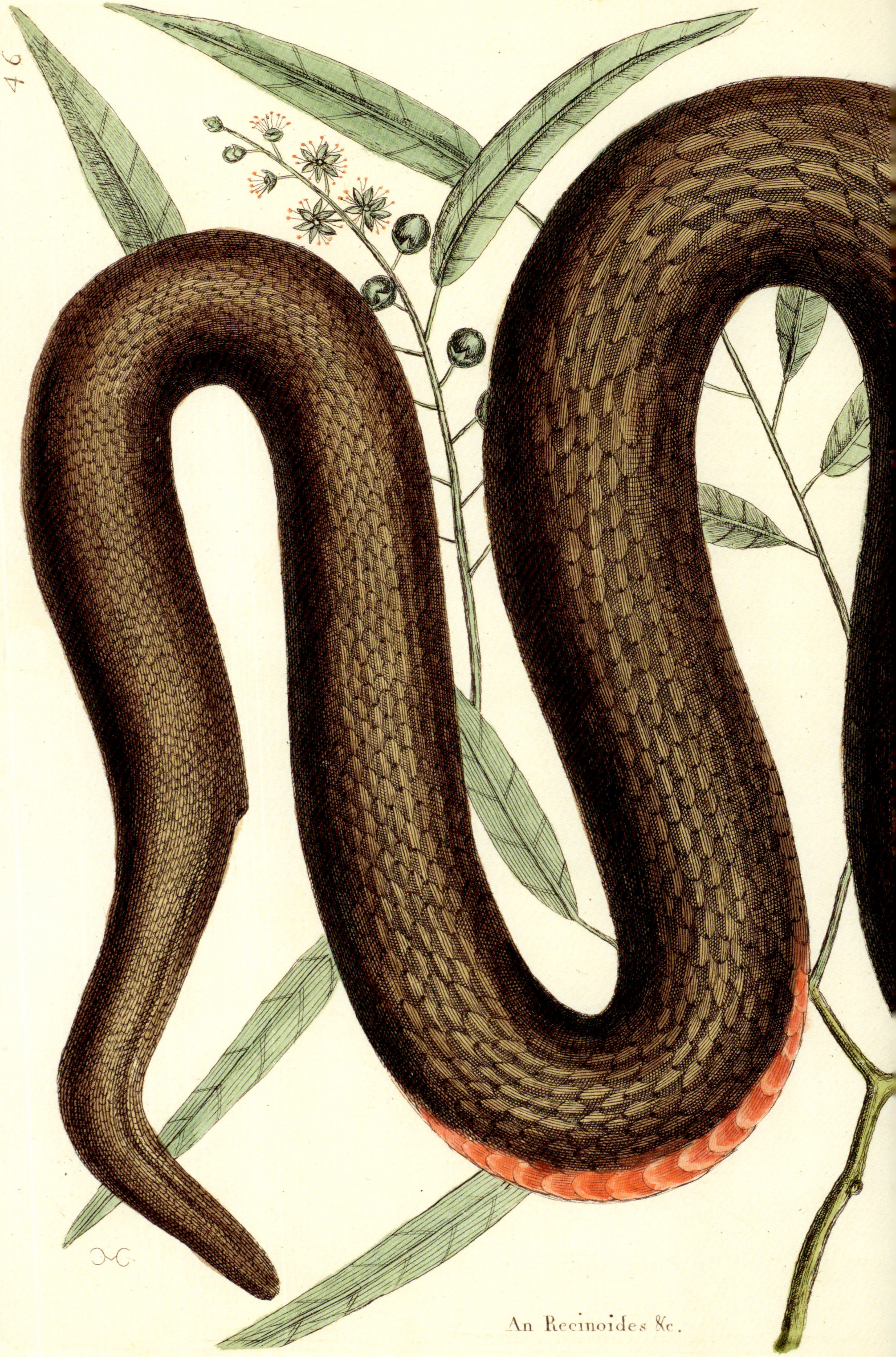

An Recinoides &c.

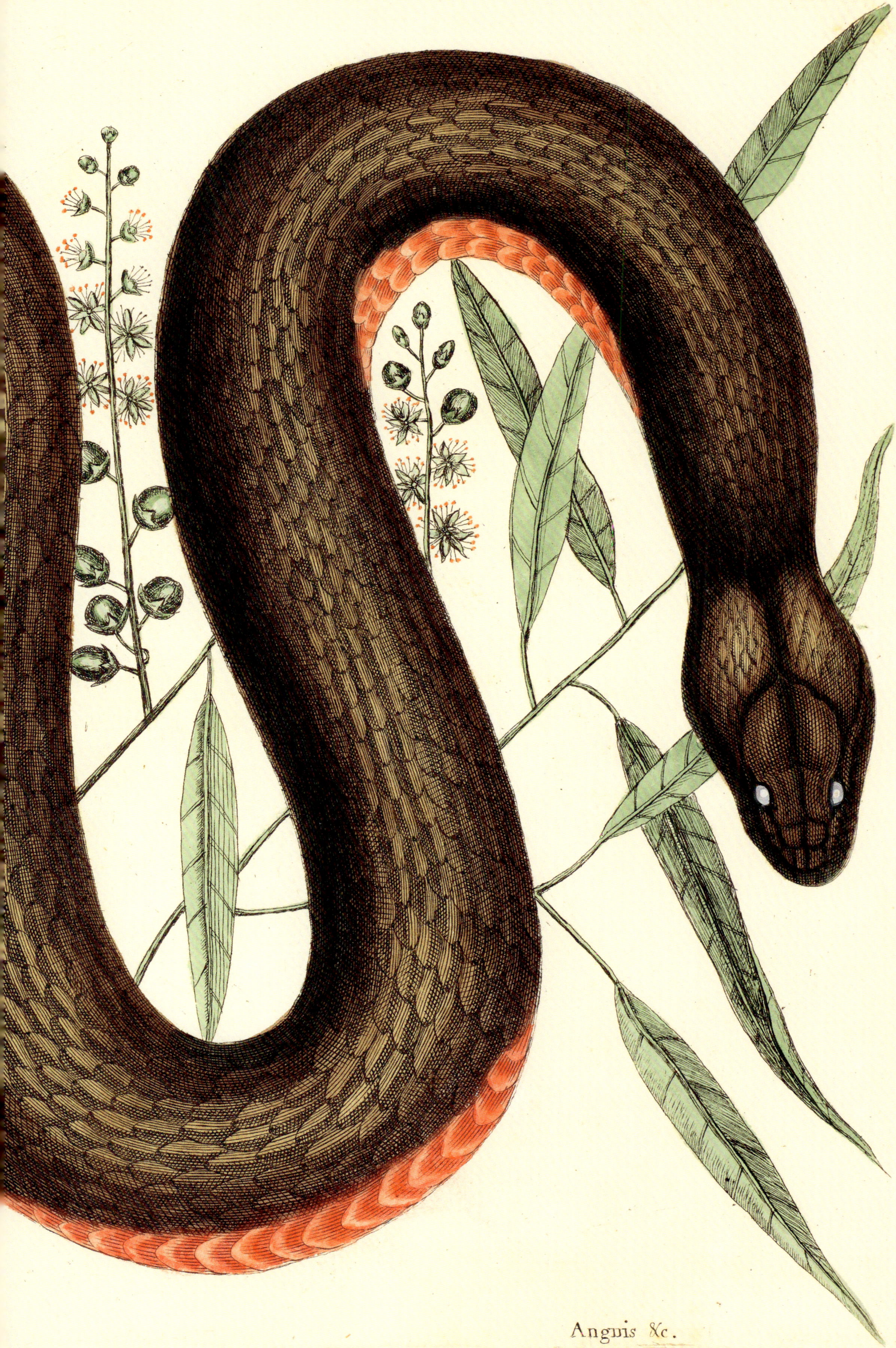

Anguis &c.

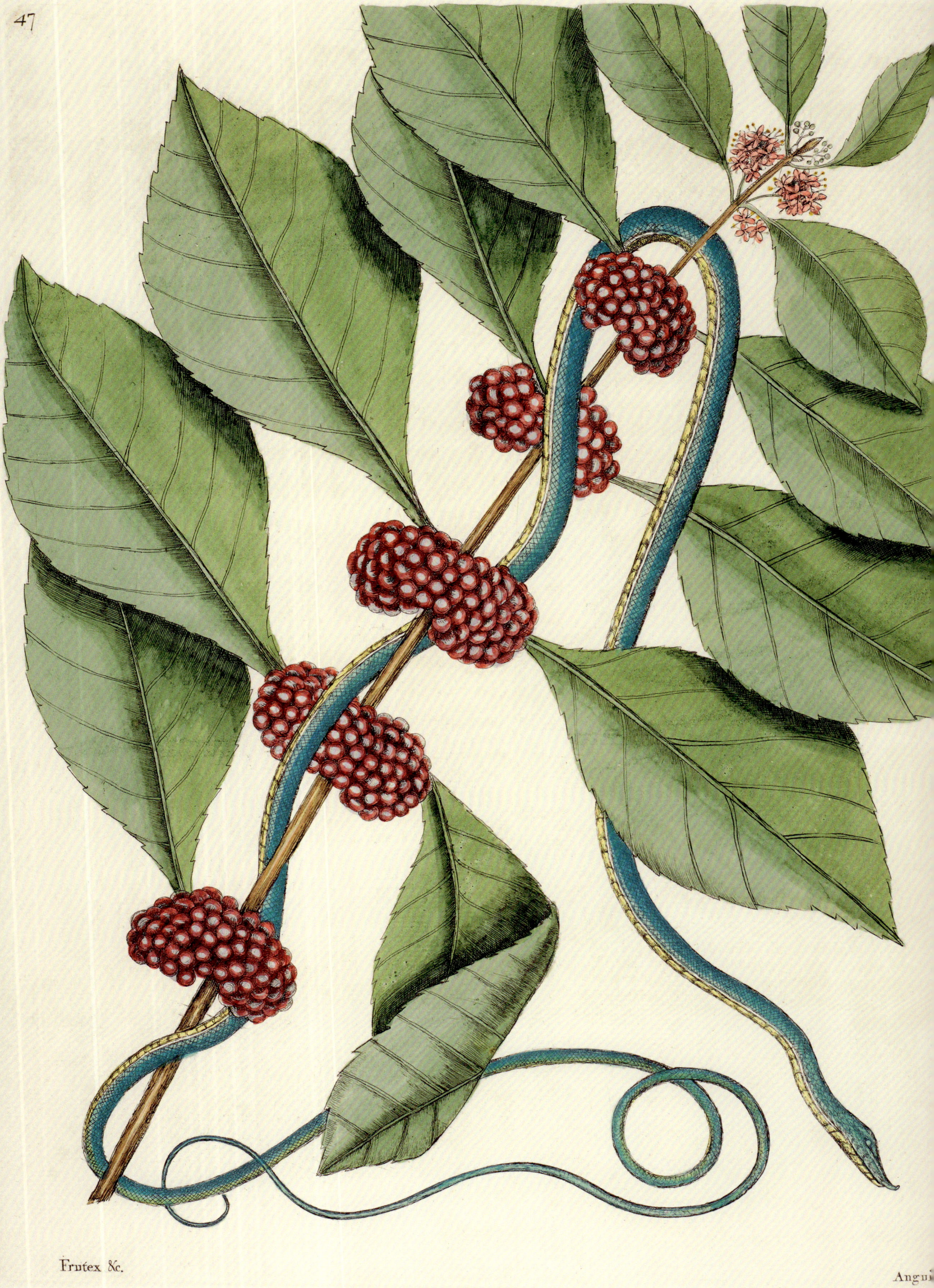

Frutex &c.

Angui

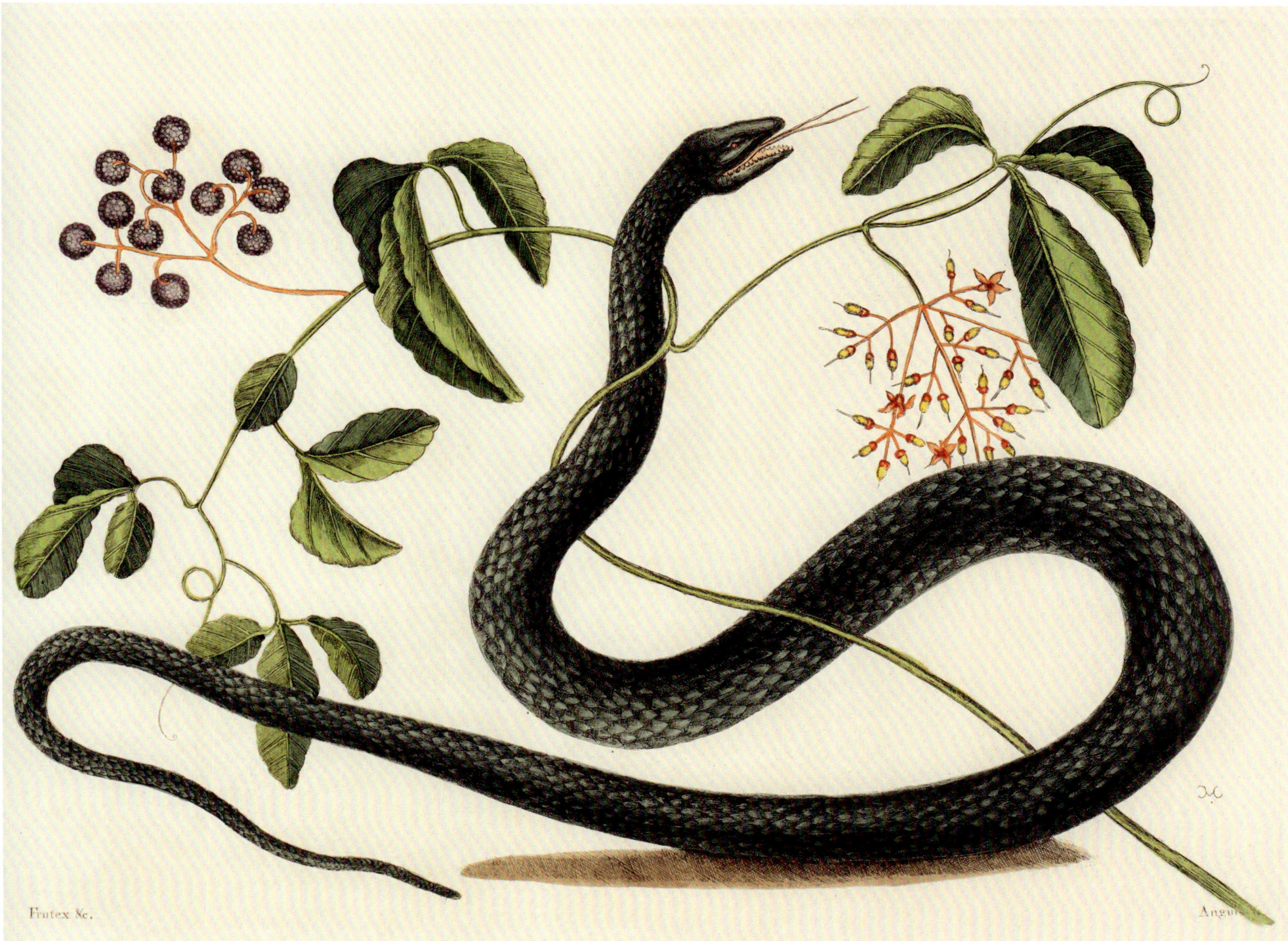

147 *left*

Rough greensnake (*Opheodrys aestivus*); American beautyberry (*Callicarpa americana*) (vol. 2, t. 47)

148 *above*

Eastern ratsnake (*Pantherophis alleghaniensis*); spoonleaf treebine (*Cissus obovata*) (vol. 2, t. 48)

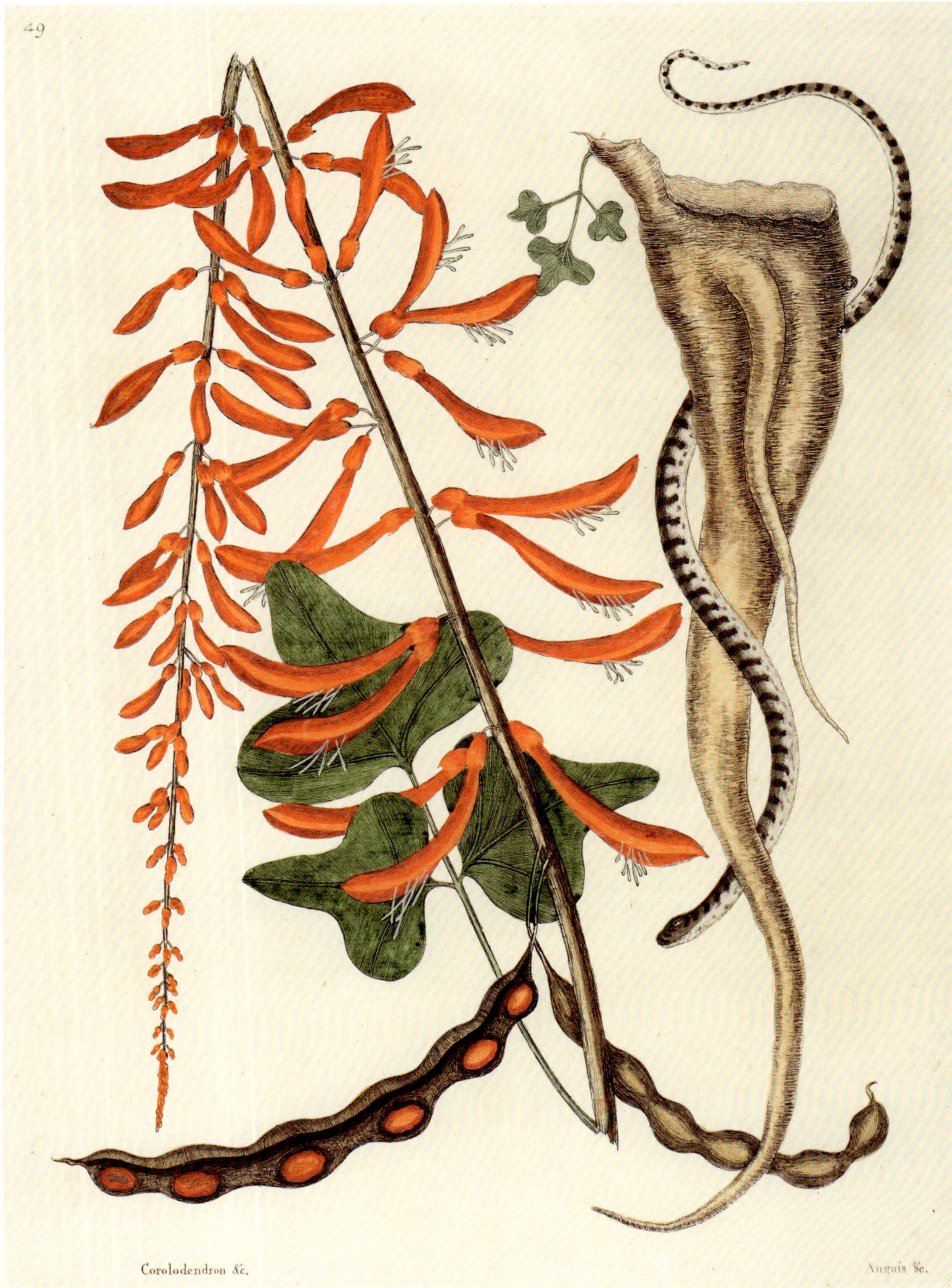

149 *above*
Dekay's brownsnake (*Storeria dekayi*); coral bean (*Erythrina herbacea*) (vol. 2, t. 49)

150 *right*
Eastern ribbon snake (*Thamnophis saurita*); wild cinnamon (*Canella winterana*) (vol. 2, t. 50)

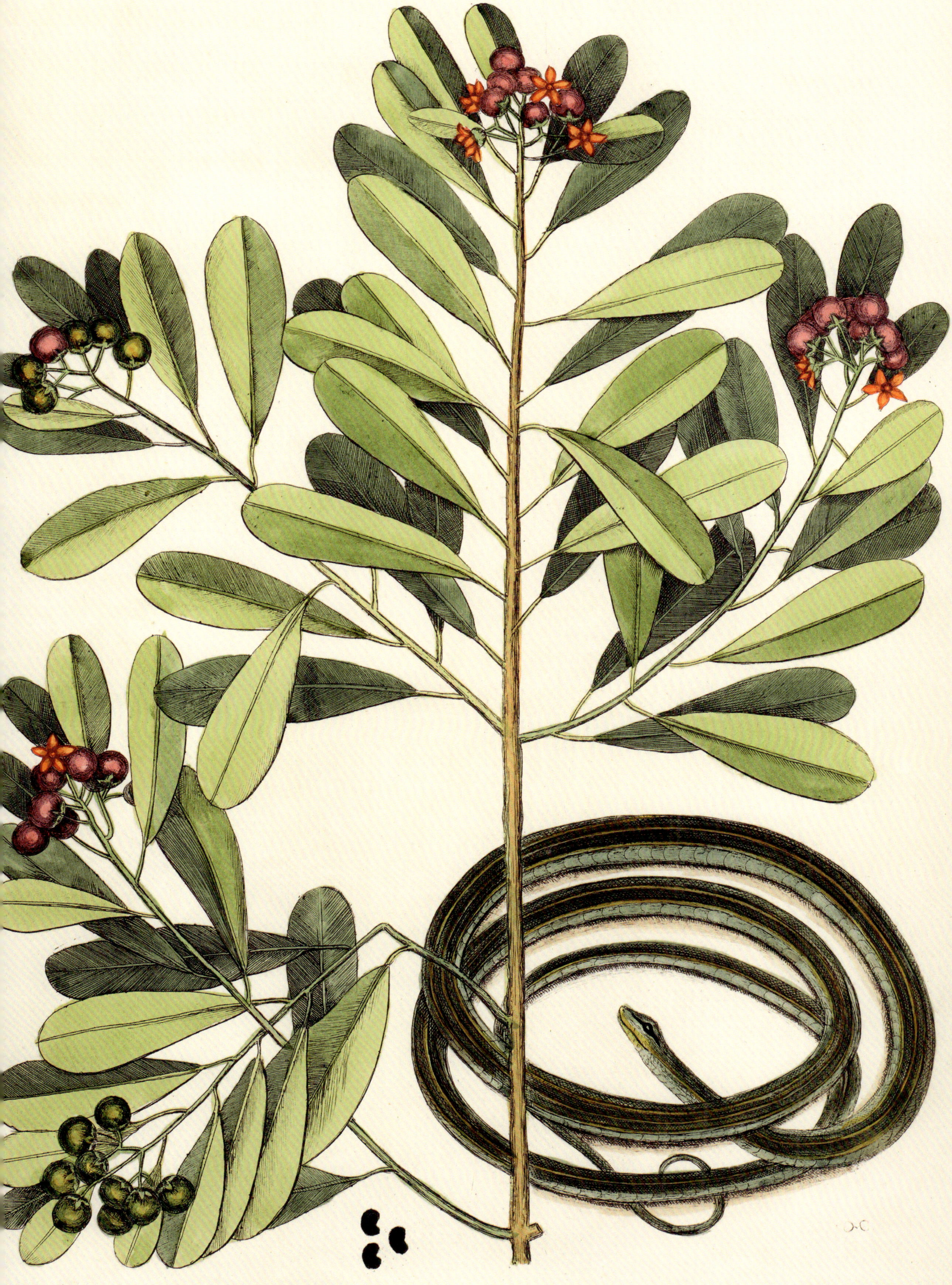

Calyculato. &c.

Anguis &c.

51

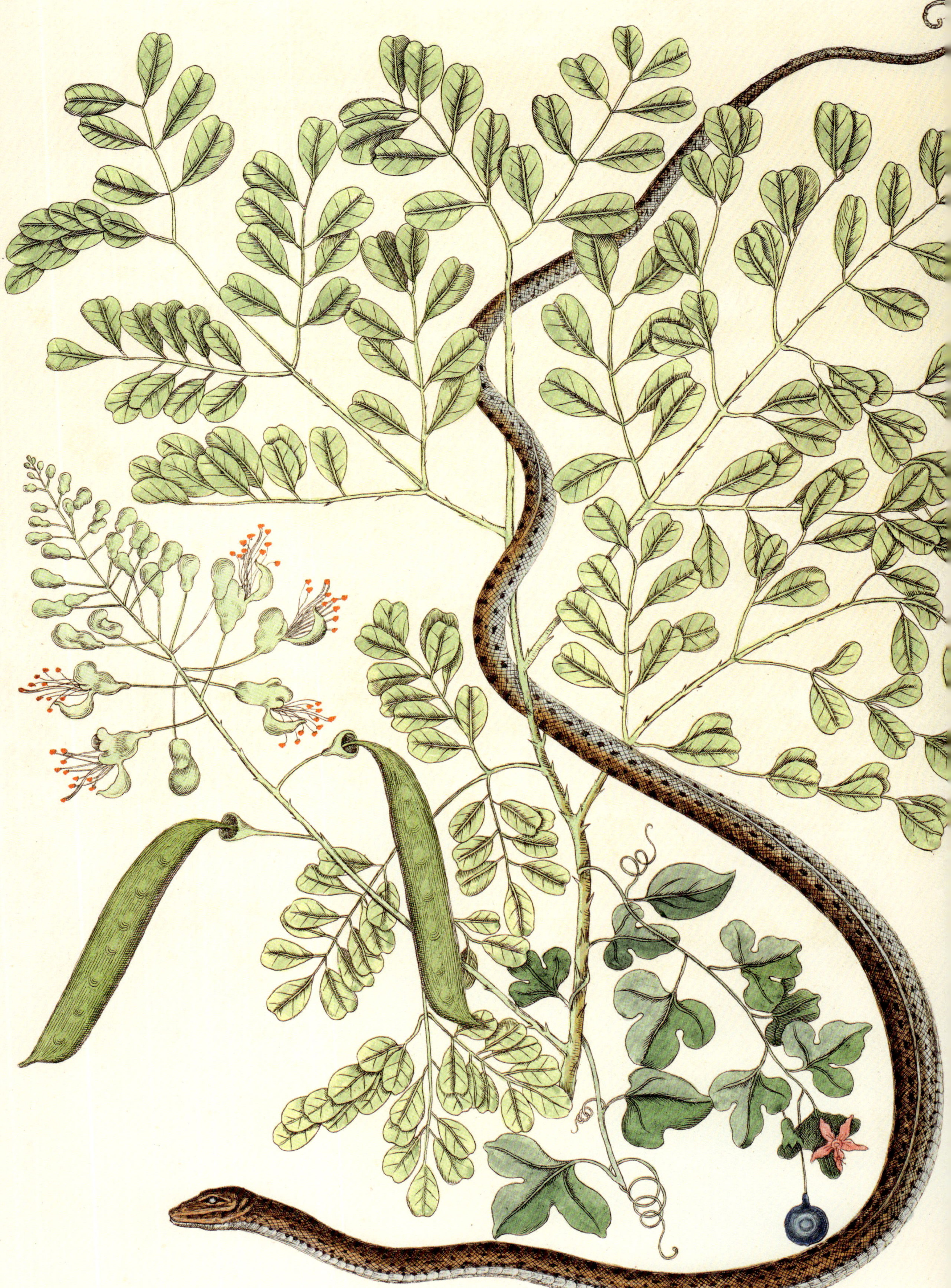

Pseudo &c. Flos. Passioims &c. Ang

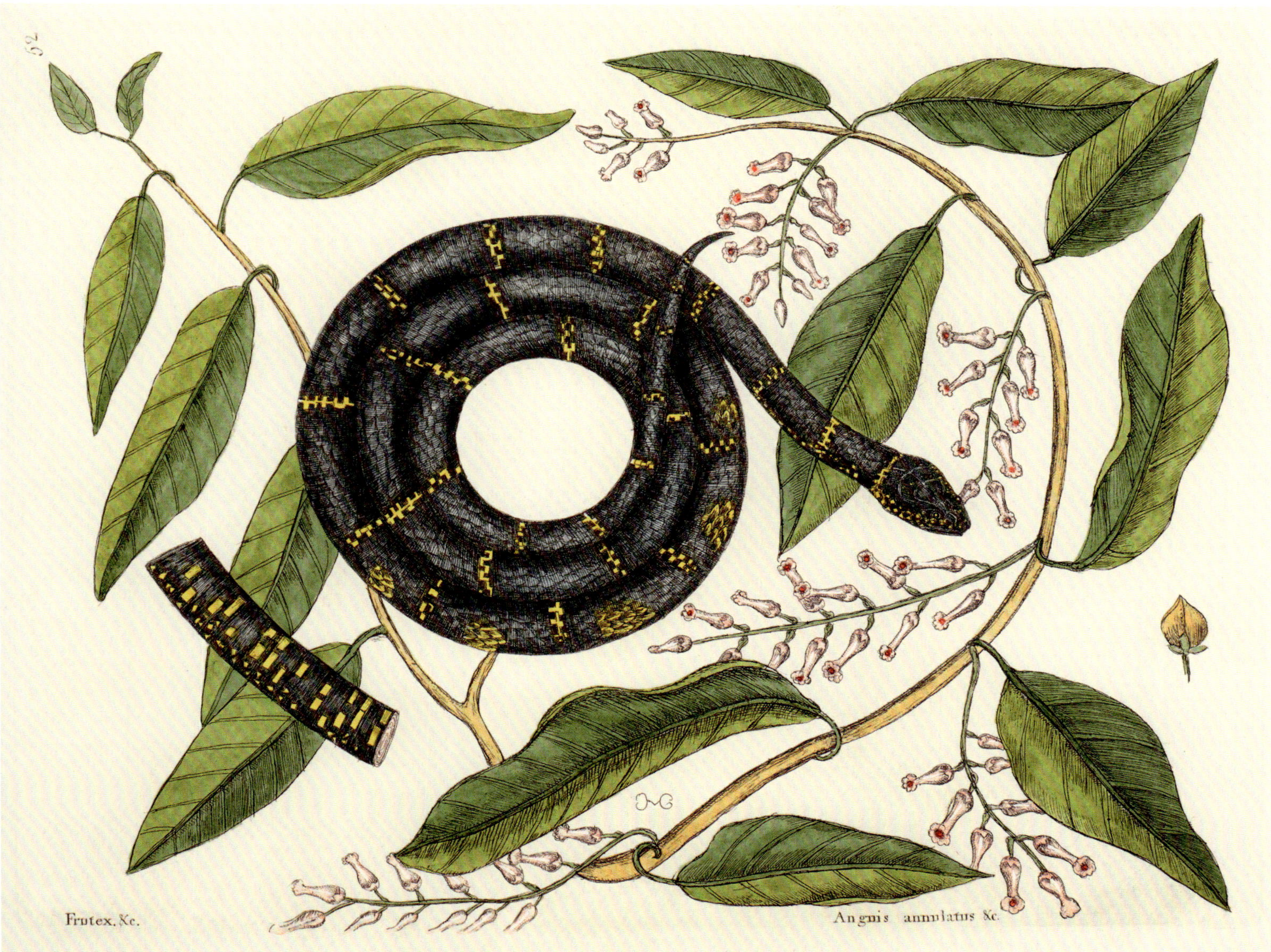

151 *left*

Common garter snake (*Thamnophis sirtalis*); Bahama caesalpinia (*Caesalpinia bahamensis*); corky-stemmed passion flower (*Passiflora suberosa*) (vol. 2, t. 51)

152 *above*

Eastern kingsnake (*Lampropeltis getula*); coastal doghobble (*Leucothoe axillaris*) (vol. 2, t. 52)

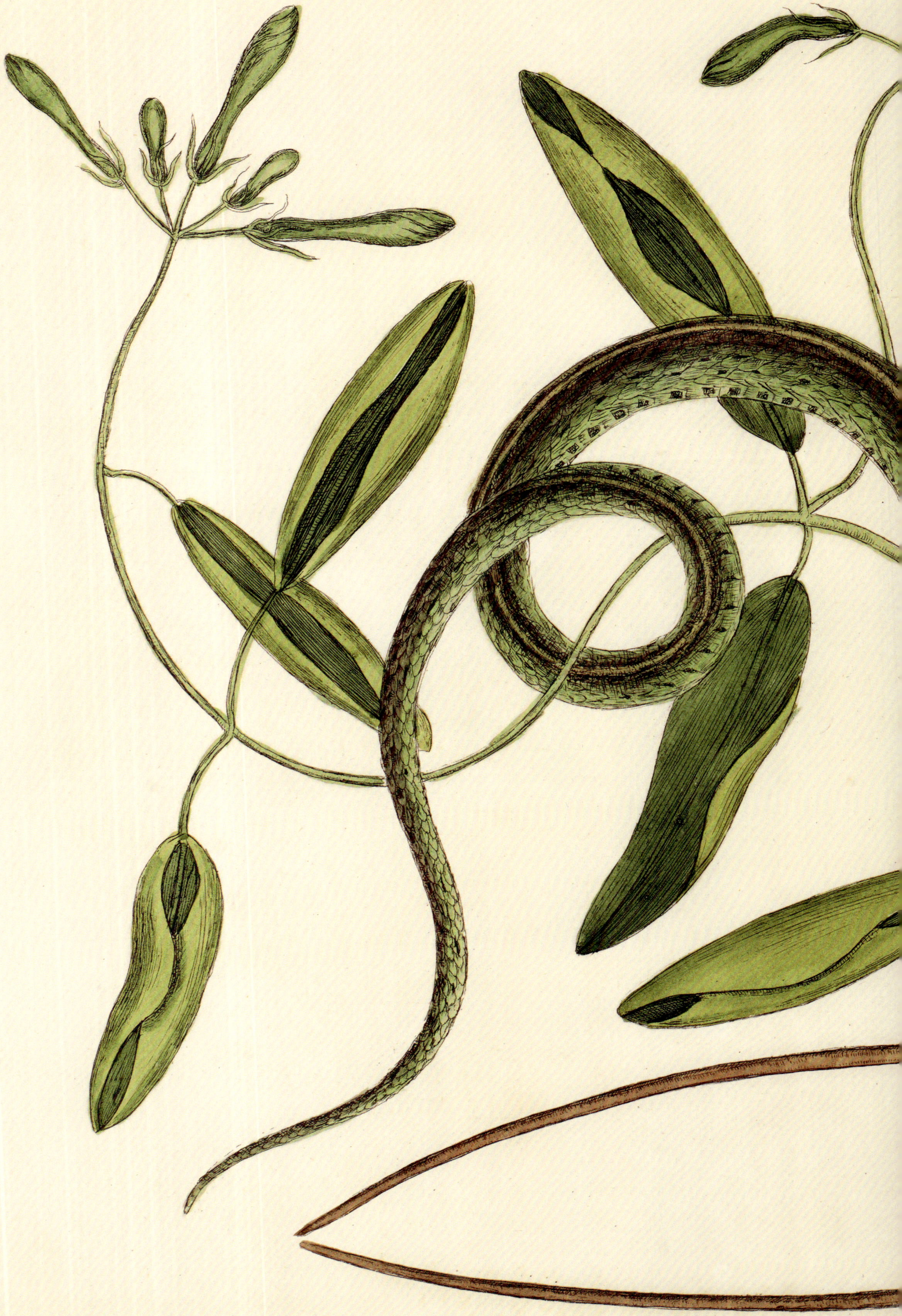

Apocynum &c.

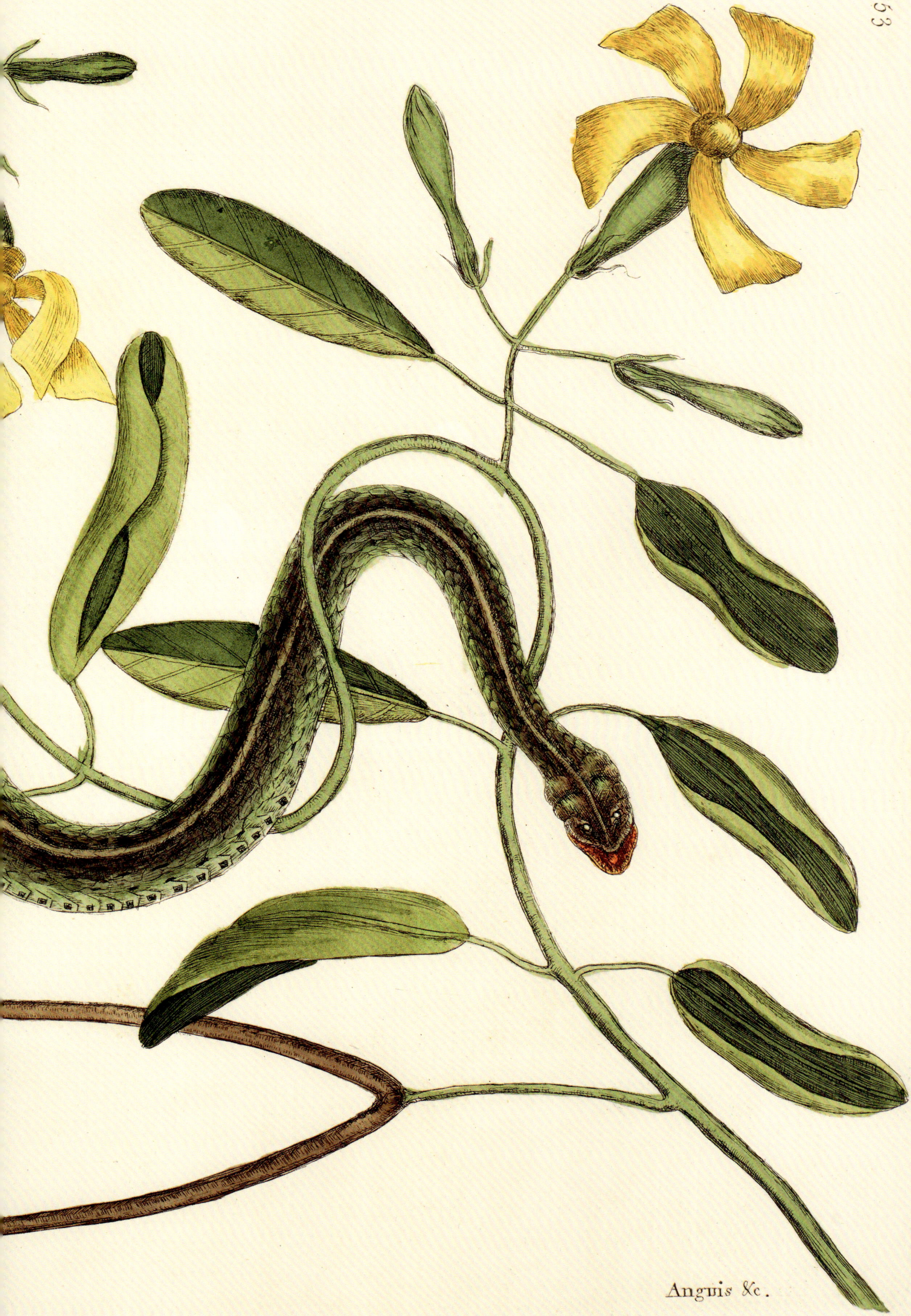

Anguis &c.

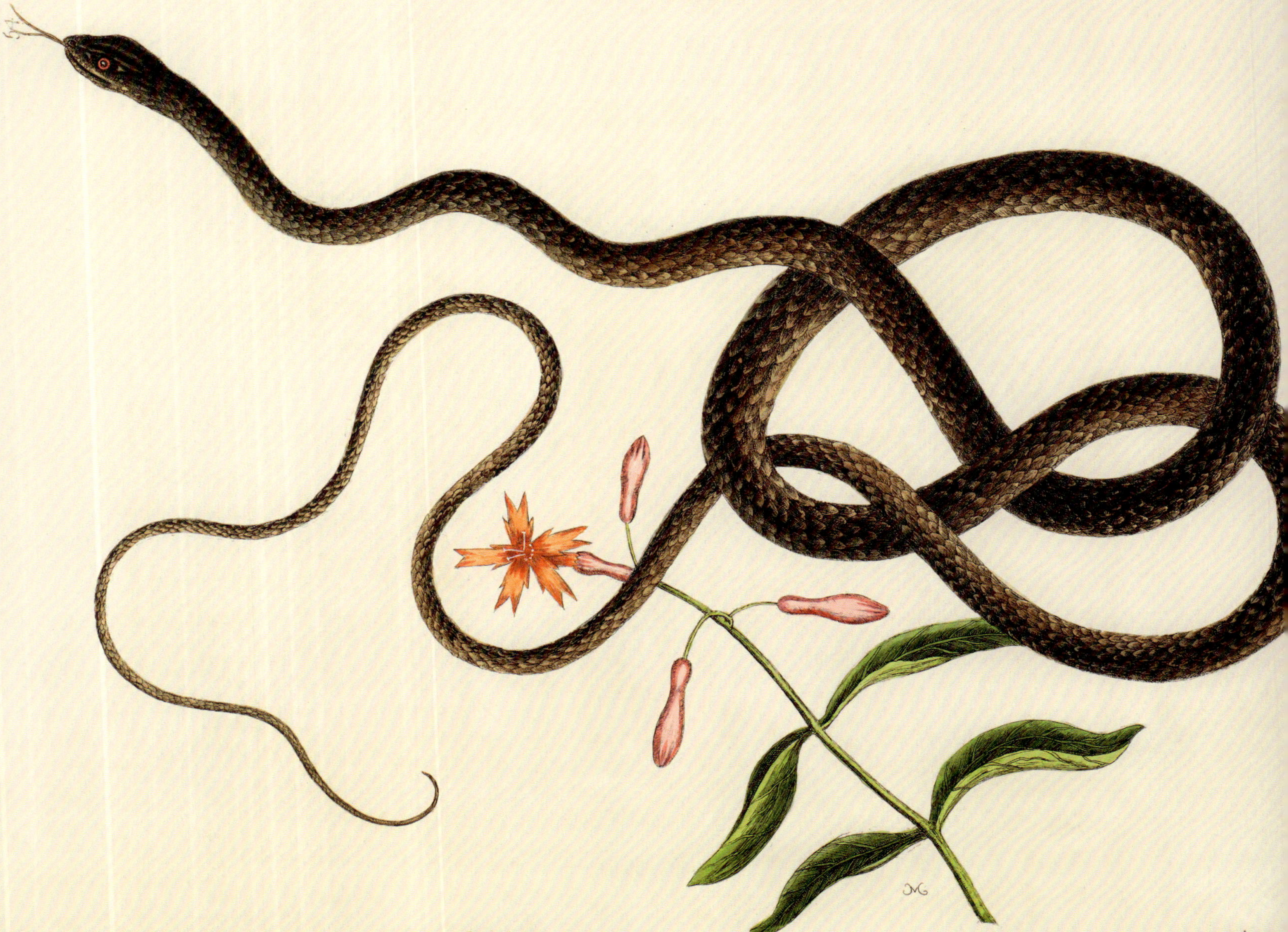

Lychnis &c. Angnis &c.

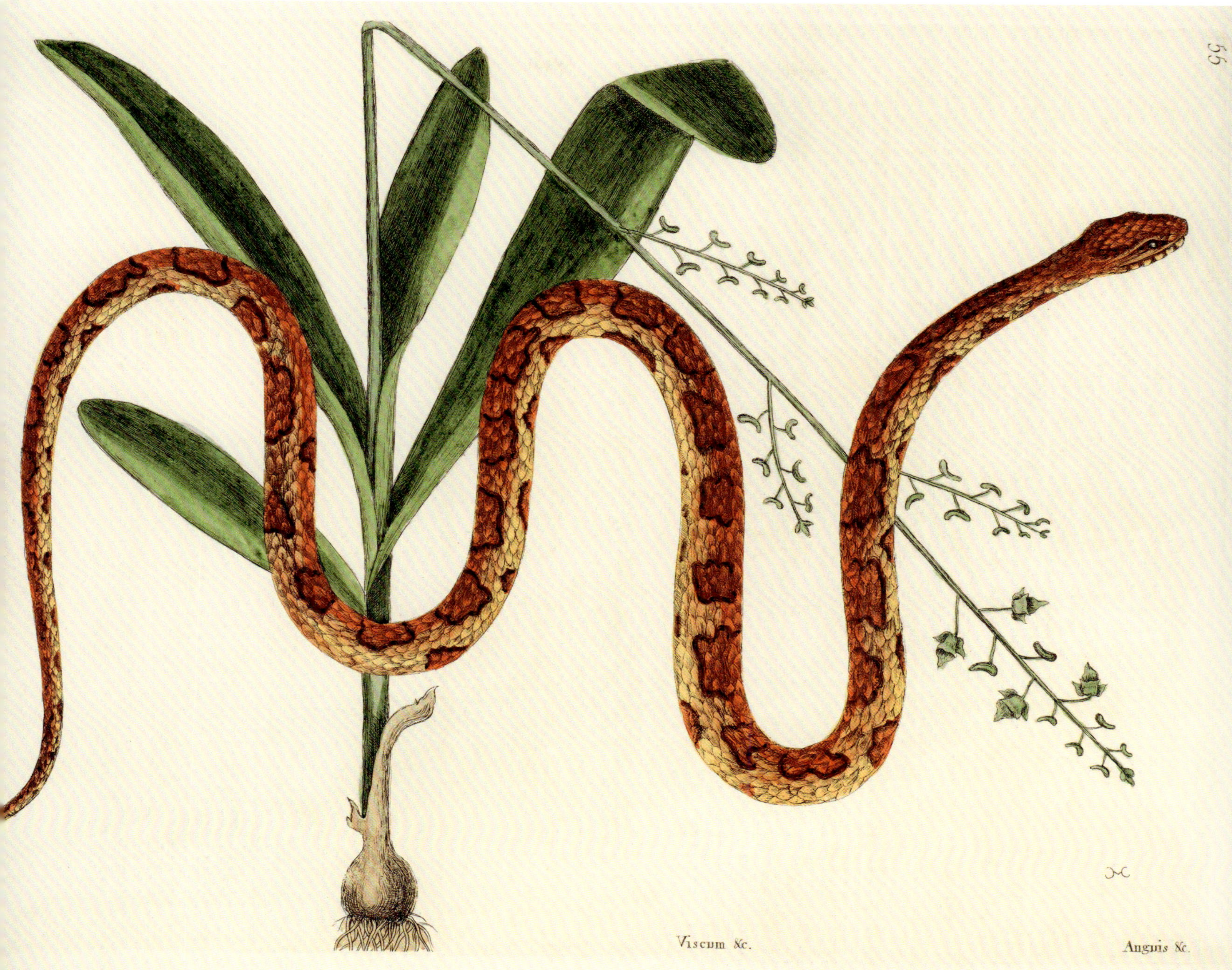

previous pages

153
Common garter snake (*Thamnophis sirtalis*); hammock viper's-tail (*Pentalinon luteum*) (vol. 2, t. 53)

154 *left*
Eastern coachwhip (*Masticophis flagellum*); fire pink (*Silene virginica*) (vol. 2, t. 54)

155 *above*
Red corn snake (*Pantherophis guttatus*); greater yellowspike orchid (*Polystachya concreta*) (vol. 2, t. 55)

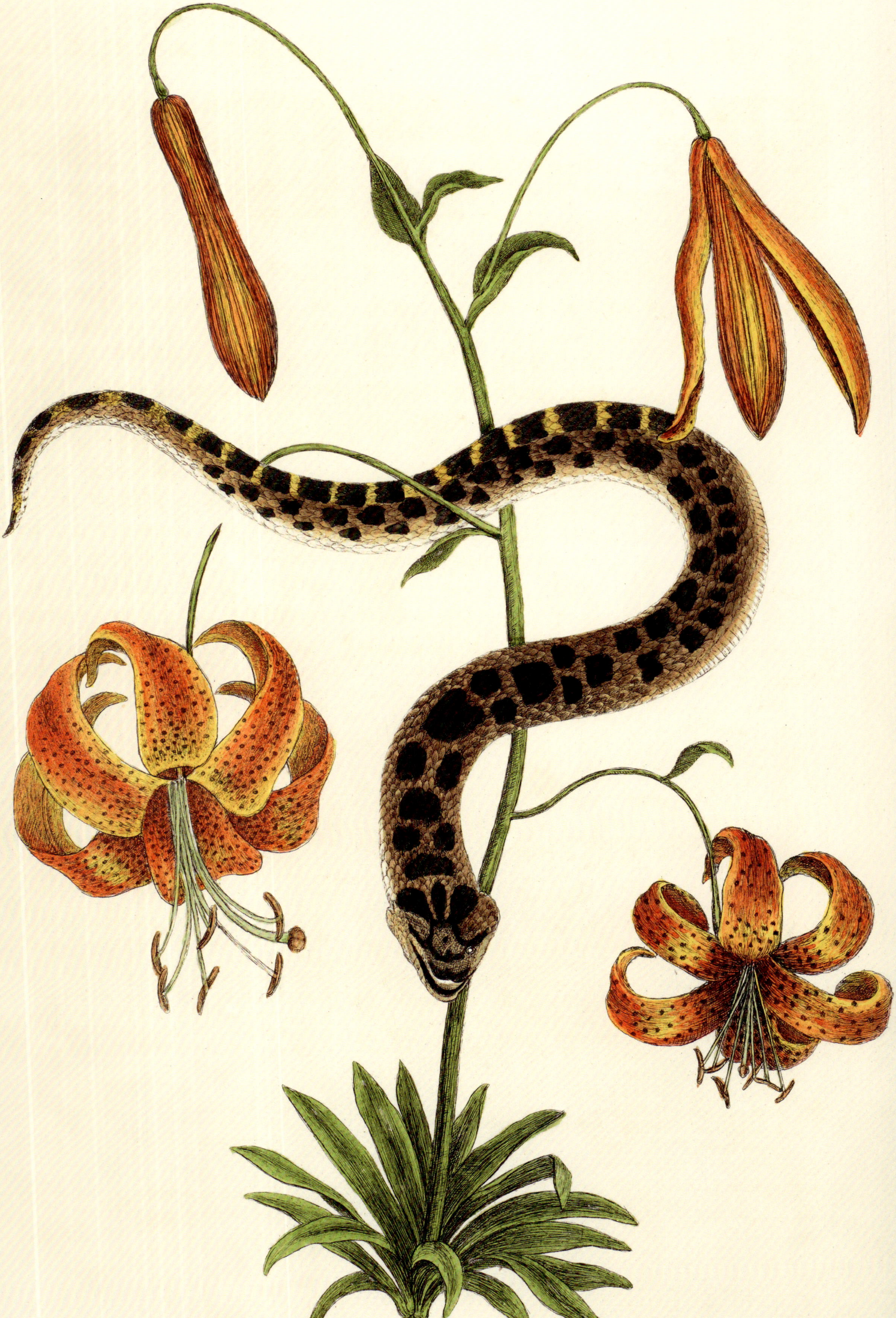

Anguis &c. Lilium &c.

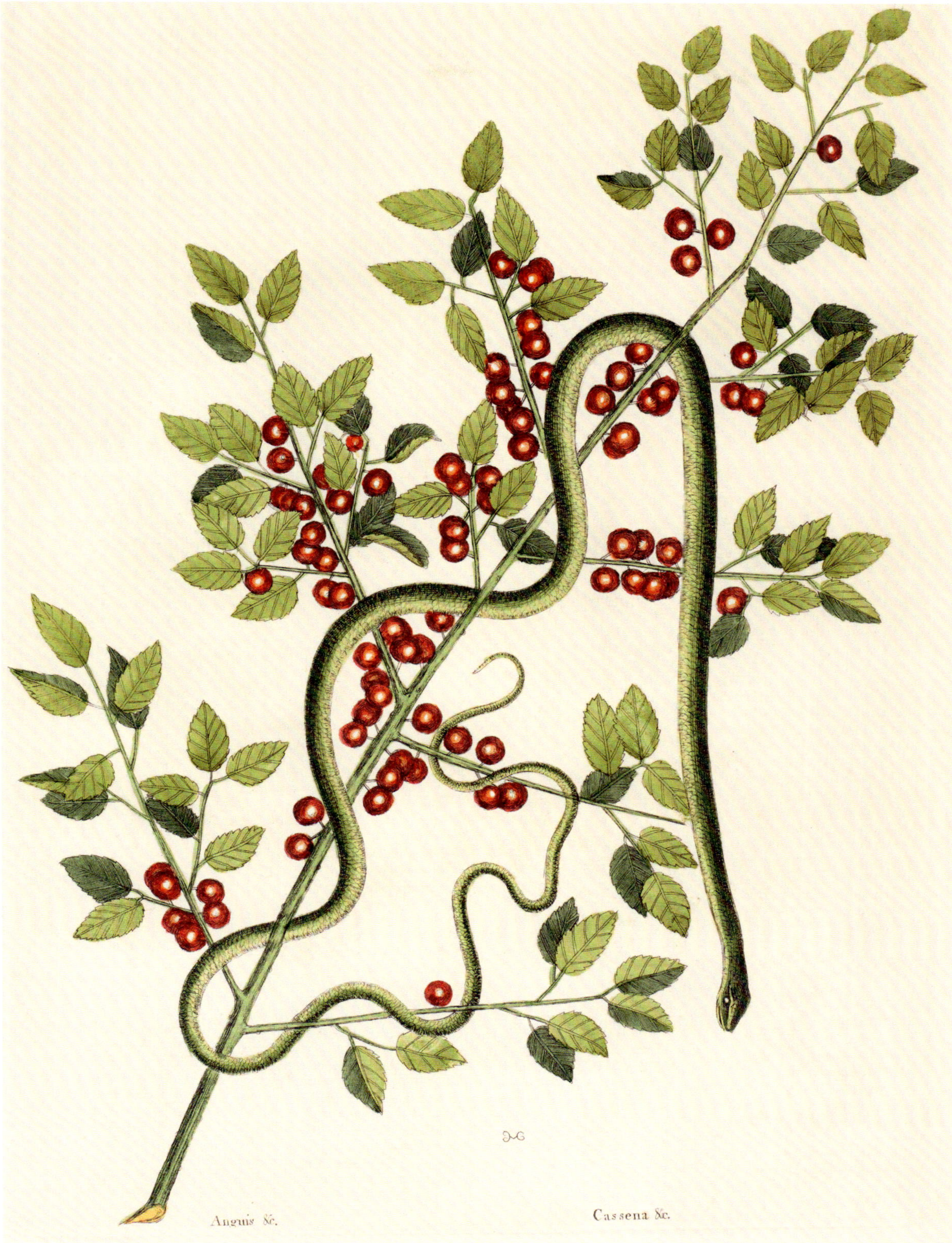

156 *left*

Eastern hog-nosed snake (*Heterodon platirhinos*); chimaera of two lilies, turk's-cap lily (*Lilium superbum*; buds and right flower) and Carolina lily (*Lilium michauxii*; leaves and left flower) (vol. 2, t. 56)

157 *above*

Smooth greensnake (*Opheodrys vernalis*); yaupon (*Ilex vomitoria*) (vol. 2, t. 57)

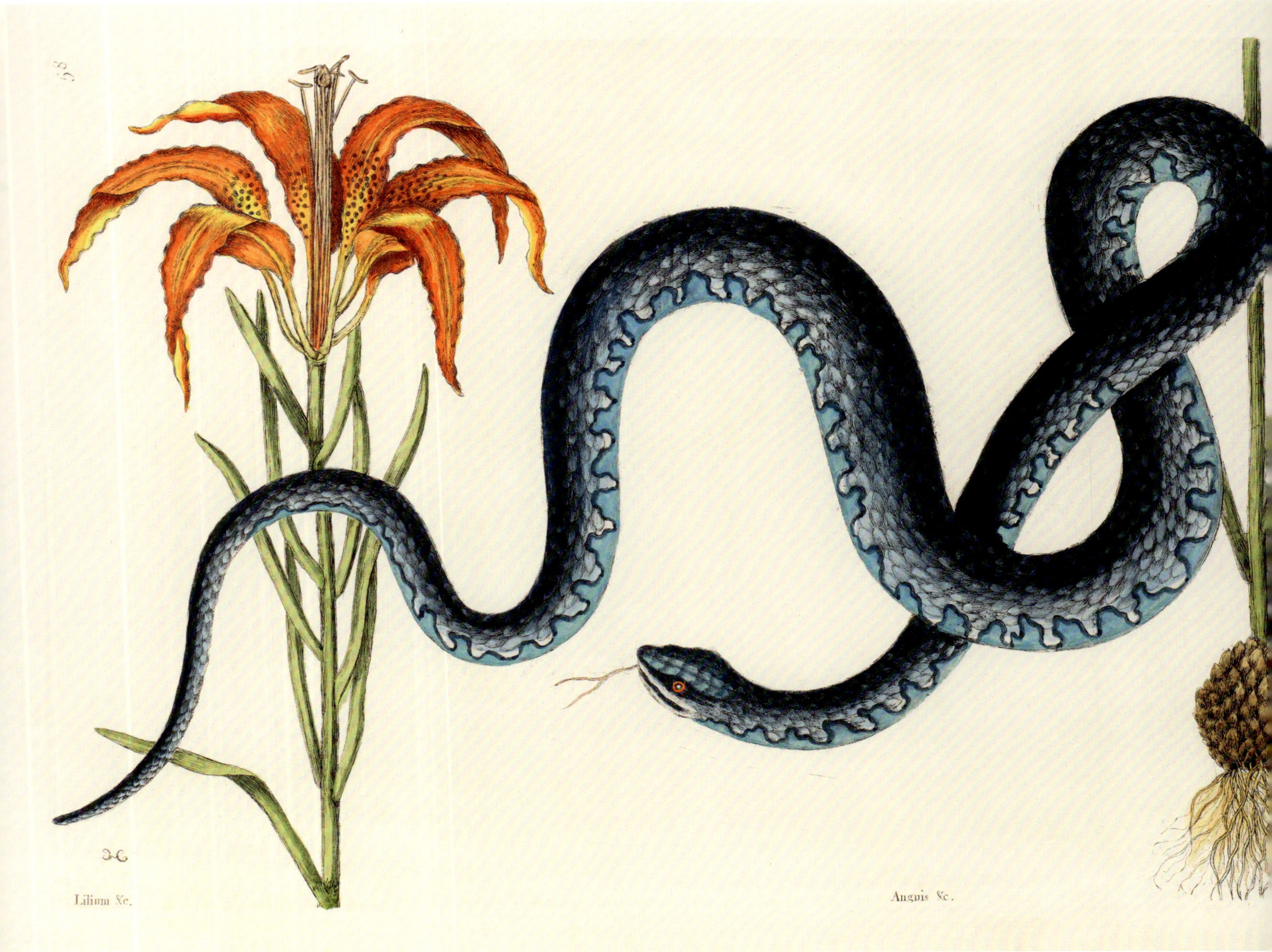

158 *above*
Eastern mudsnake (*Farancia abacura*); leopard lily (*Lilium catesbaei*) (vol. 2, t. 58)

159 *top right*
Eastern glass lizard (*Ophisaurus ventralis*); smooth purple coneflower (*Echinacea laevigata*) (vol. 2, t. 59)

160 *bottom right*
Scarletsnake (*Cemophora coccinea*); sweet potato (*Ipomoea batatas*) (vol. 2, t. 60)

69

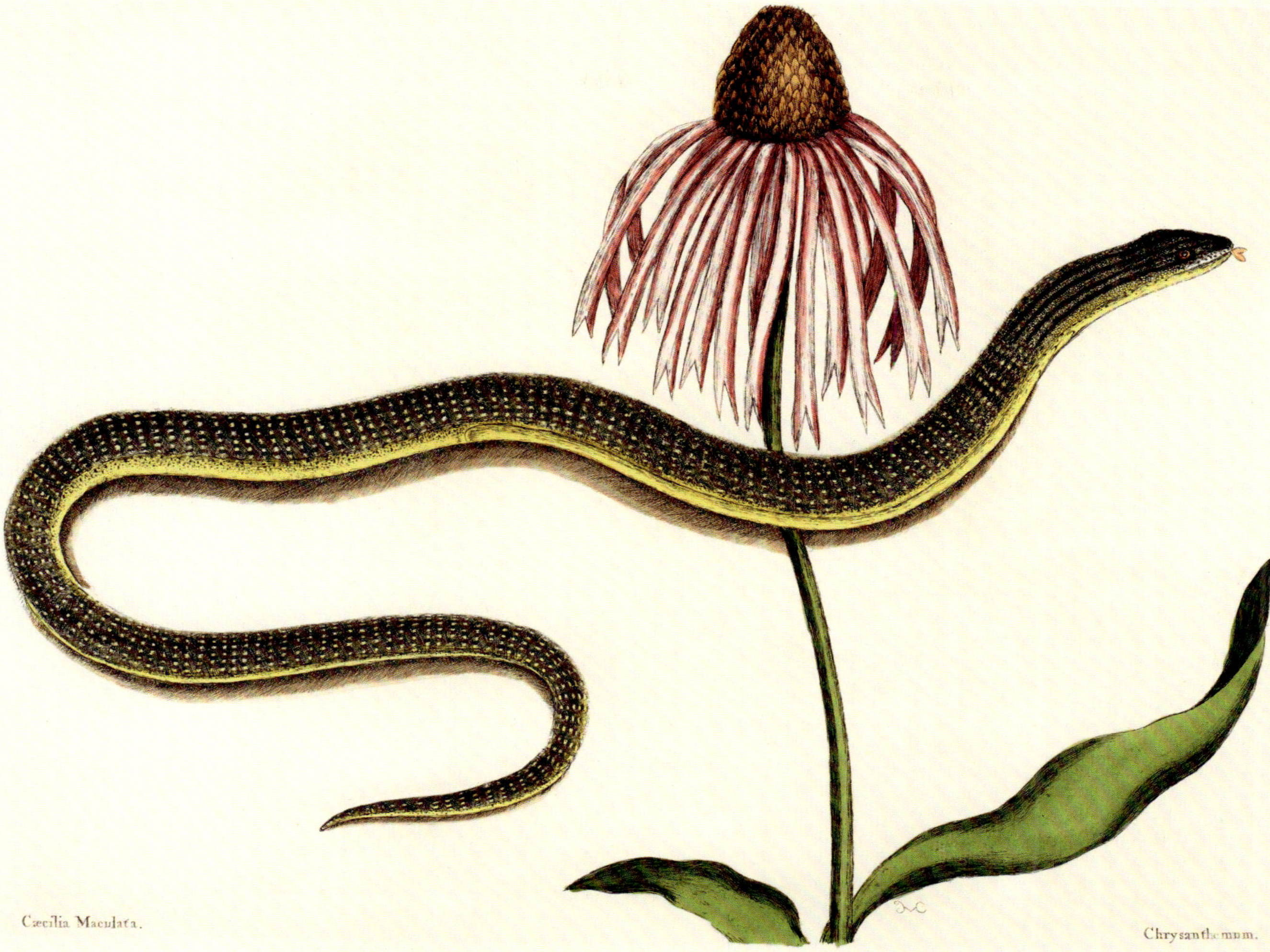

Cæcilia Maculata. Chrysanthæmum.

68

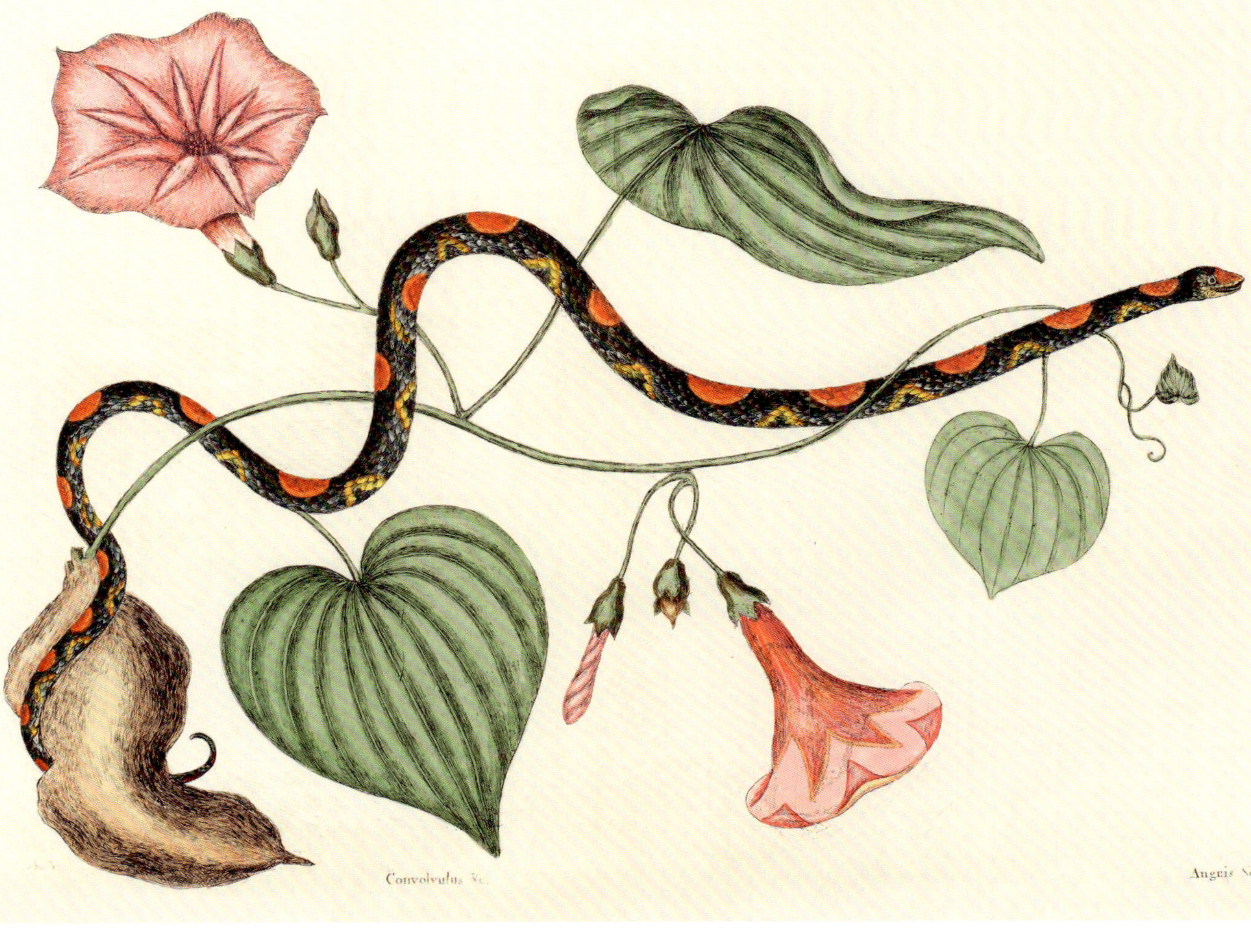

Convolvulus &c. Anguis &c.

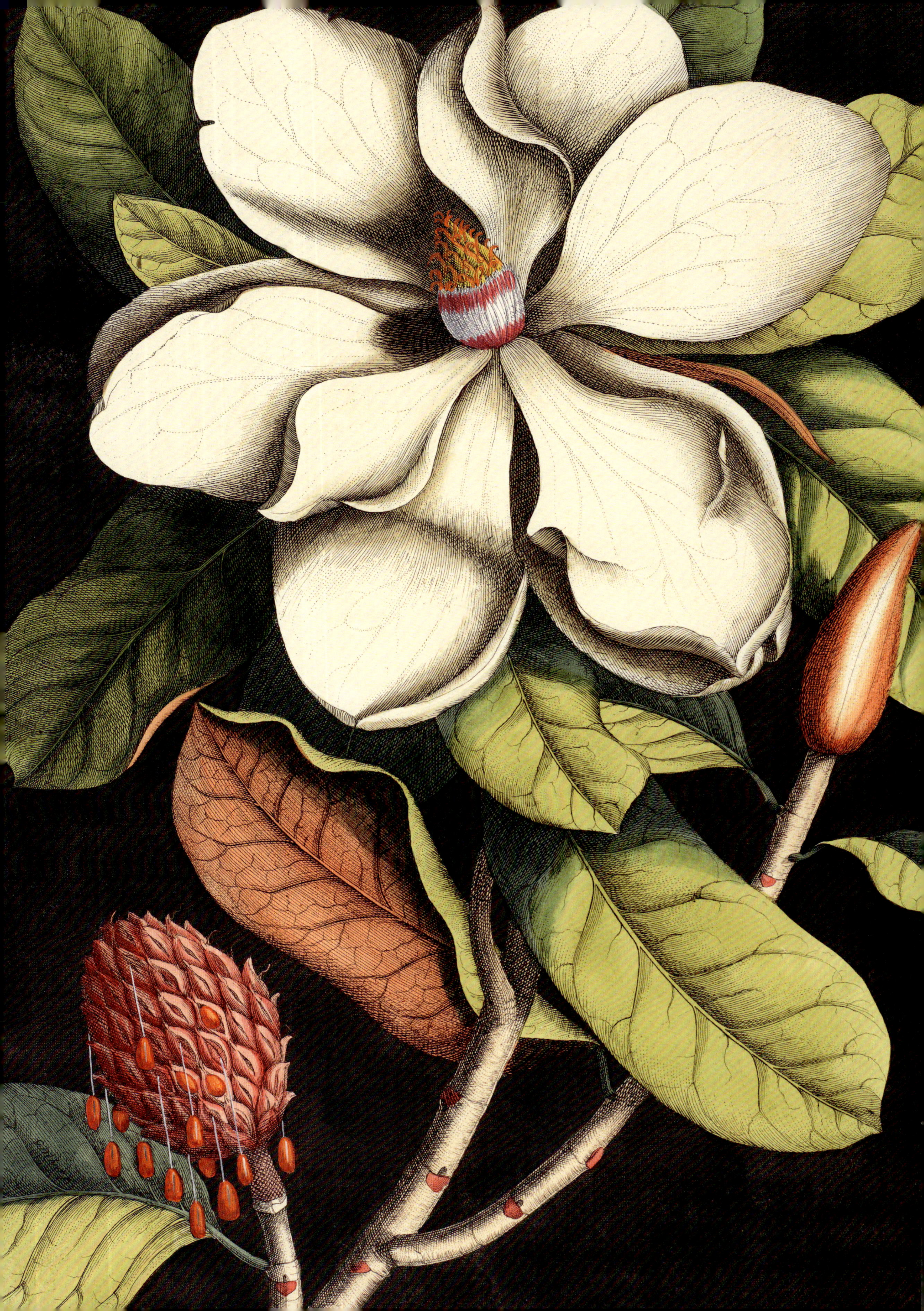

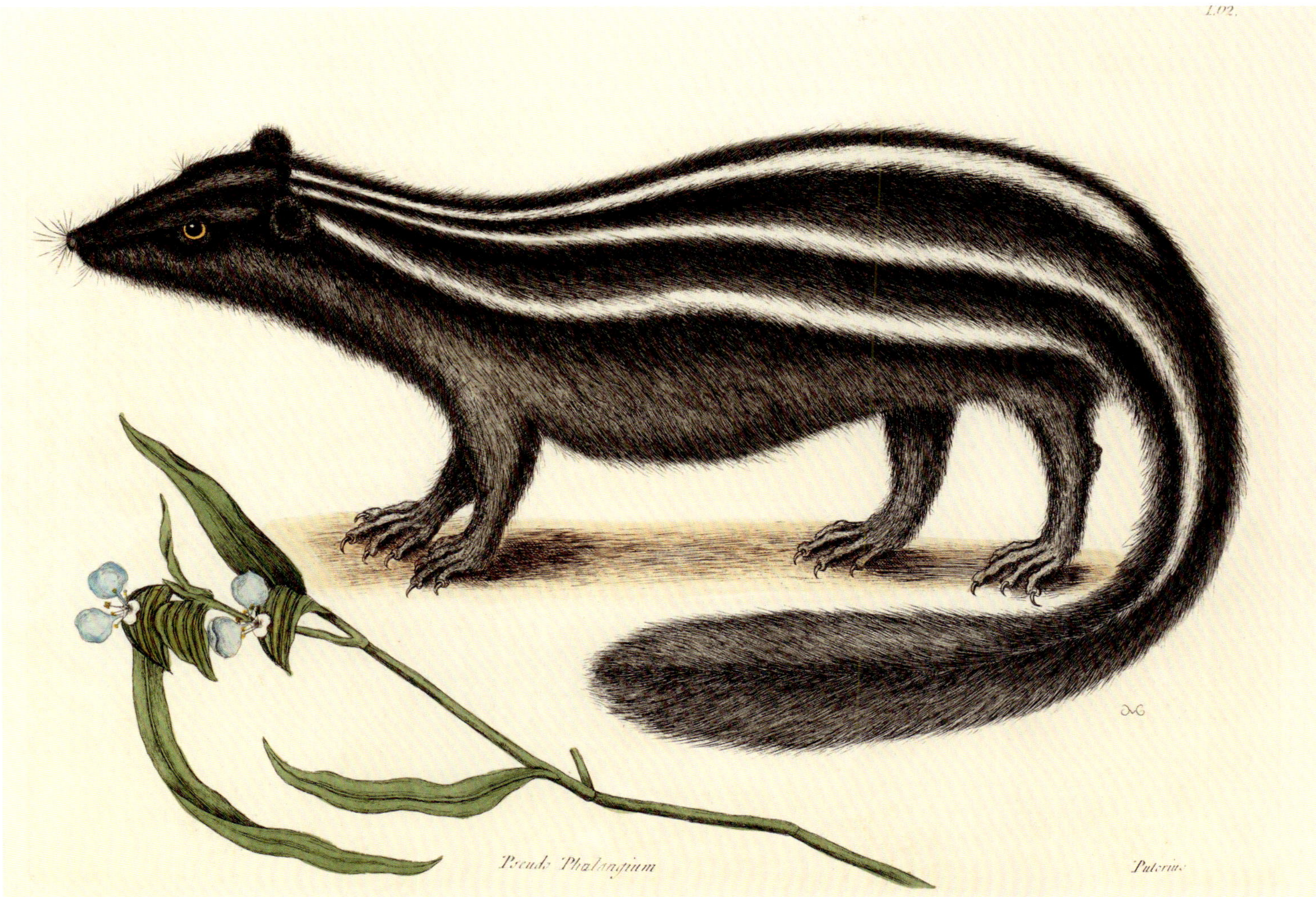

161 *left*

Southern magnolia (*Magnolia grandiflora*) (vol. 2, t. 61)

162 *above*

Eastern spotted skunk (*Spilogale putorius*); erect dayflower (*Commelina erecta*) (vol. 2, t. 62)

T. 63.

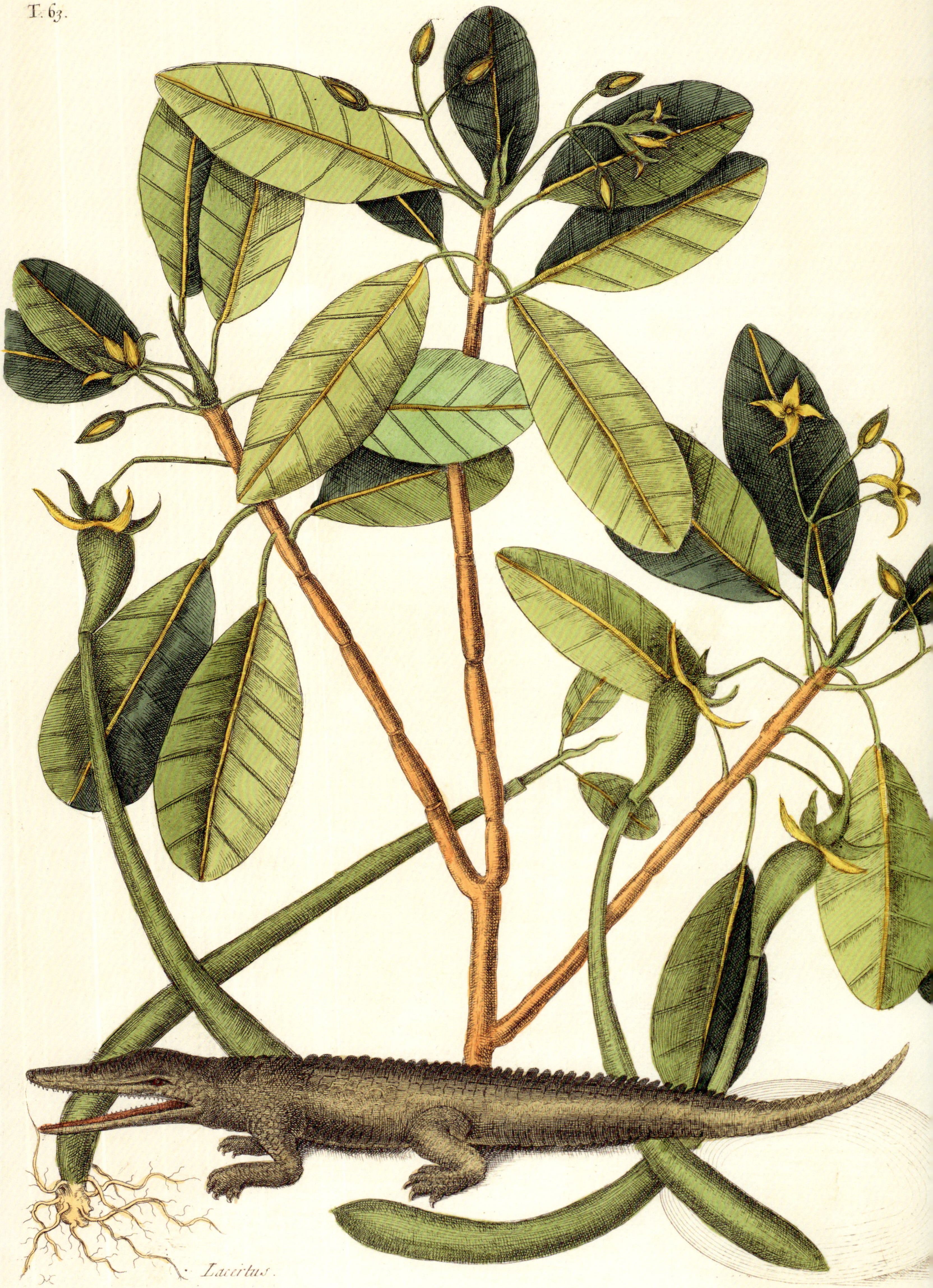

Lacertus.

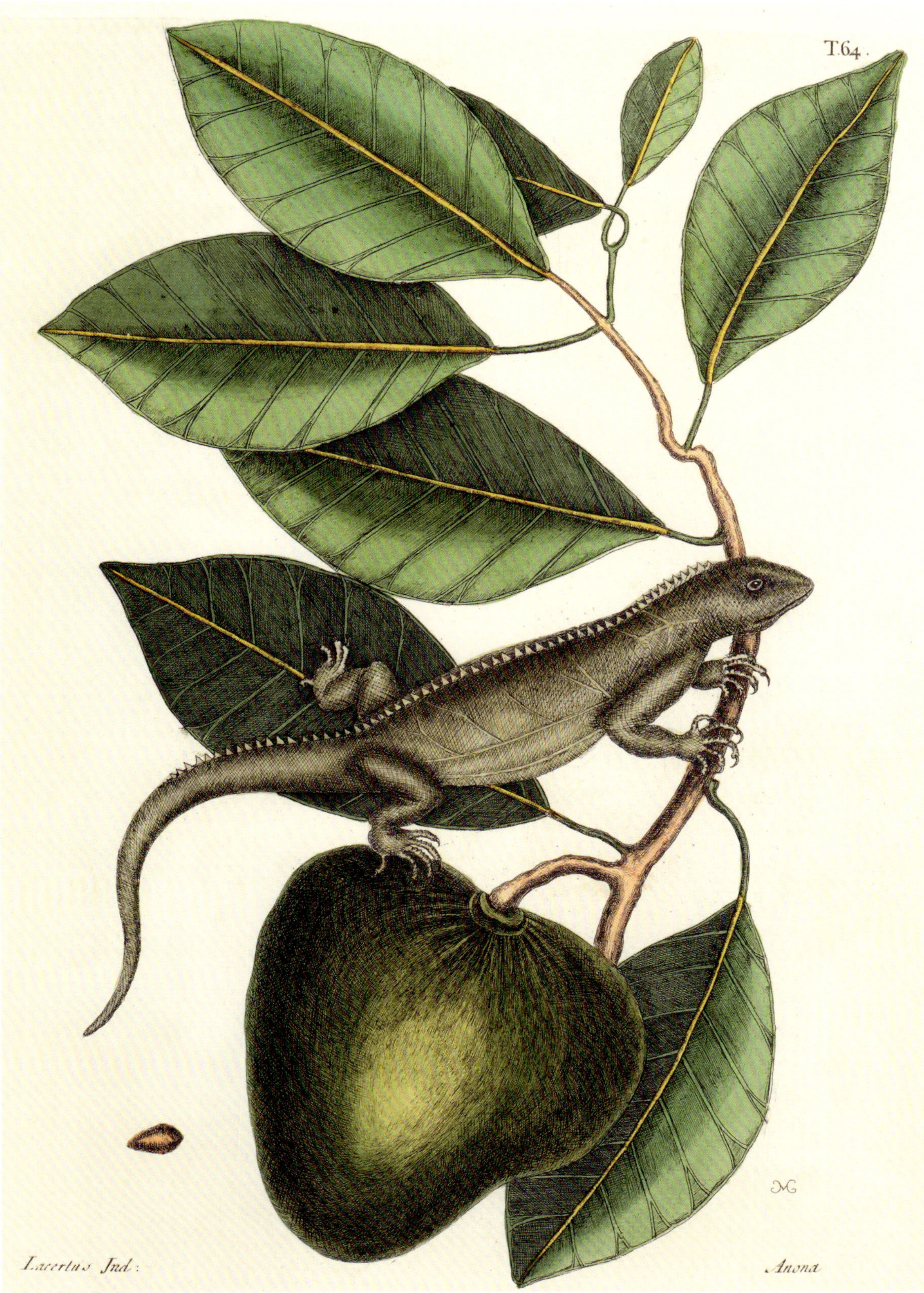

163 *left*

American alligator (*Alligator mississippiensis*); red mangrove (*Rhizophora mangle*) (vol. 2, t. 63)

164 *above*

Andros Island ground iguana (*Cyclura cychlura*); alligator apple (*Annona glabra*) (vol. 2, t. 64)

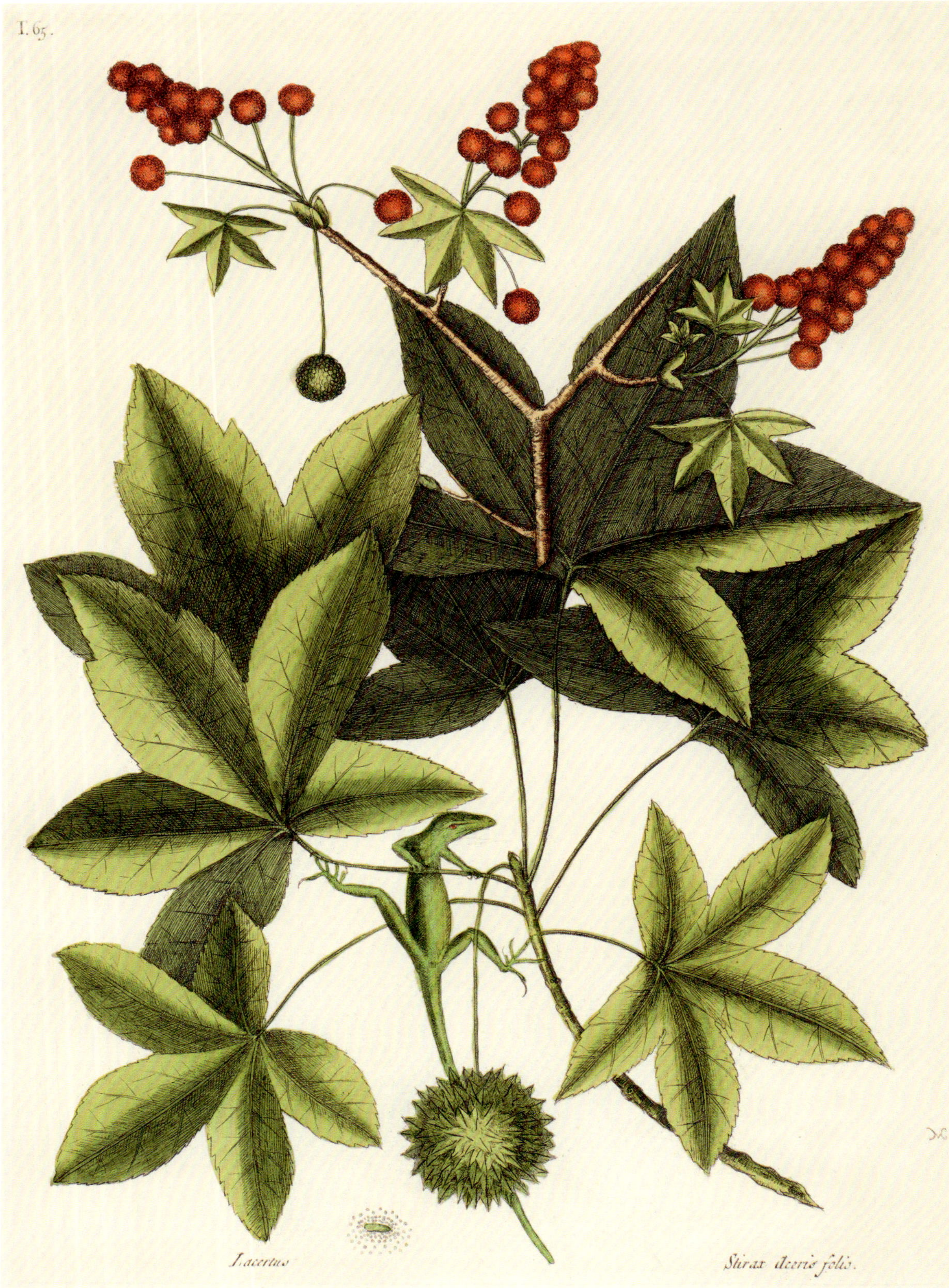

165 *above*
Green anole (*Anolis carolinensis*); sweetgum (*Liquidambar styraciflua*) (vol. 2, t. 65)

166 *right*
Graham's anole (*Anolis grahami*); logwood (*Haematoxylum campechianum*) (vol. 2, t. 66)

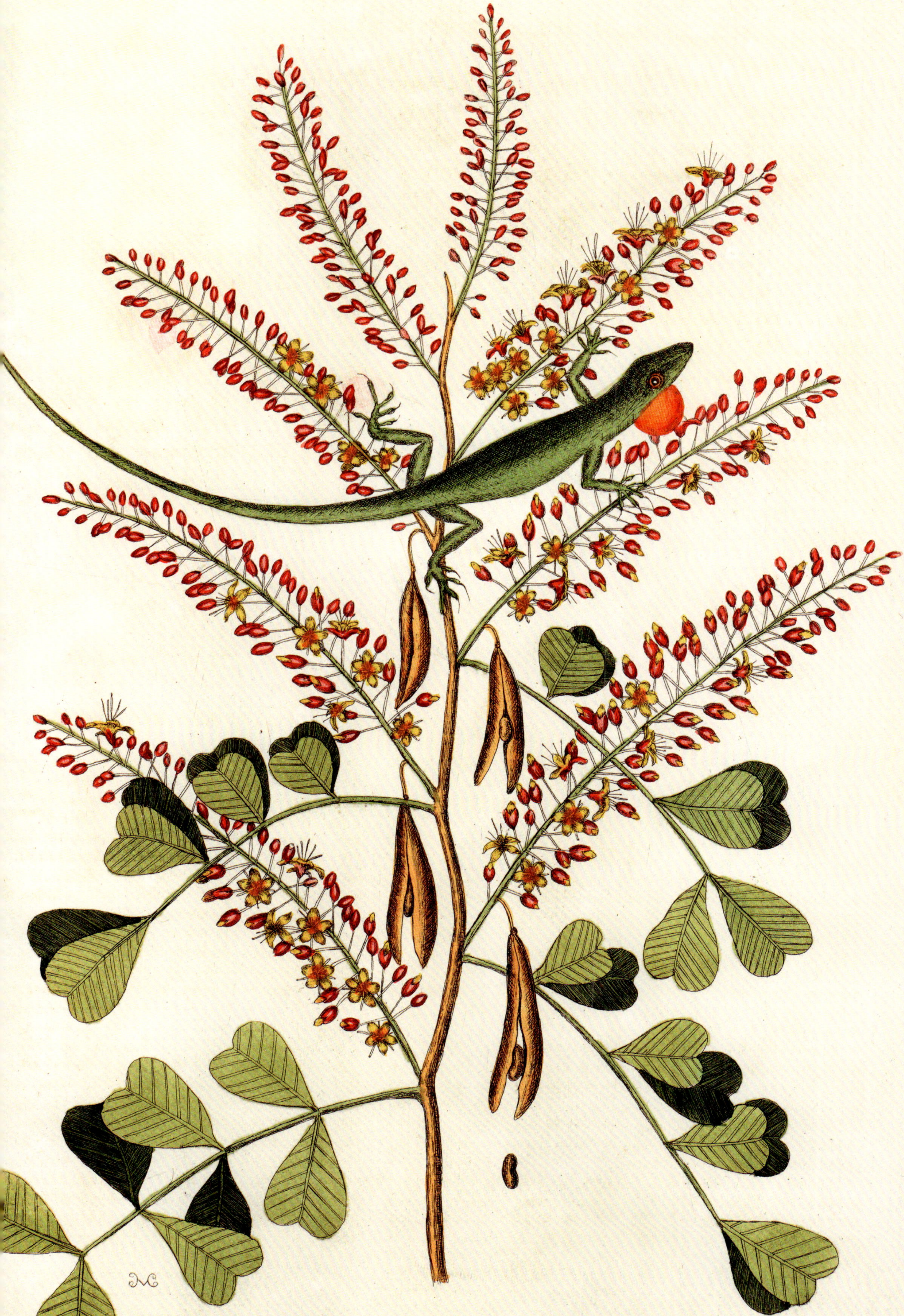

Lacertus. *Lignum Camp.*

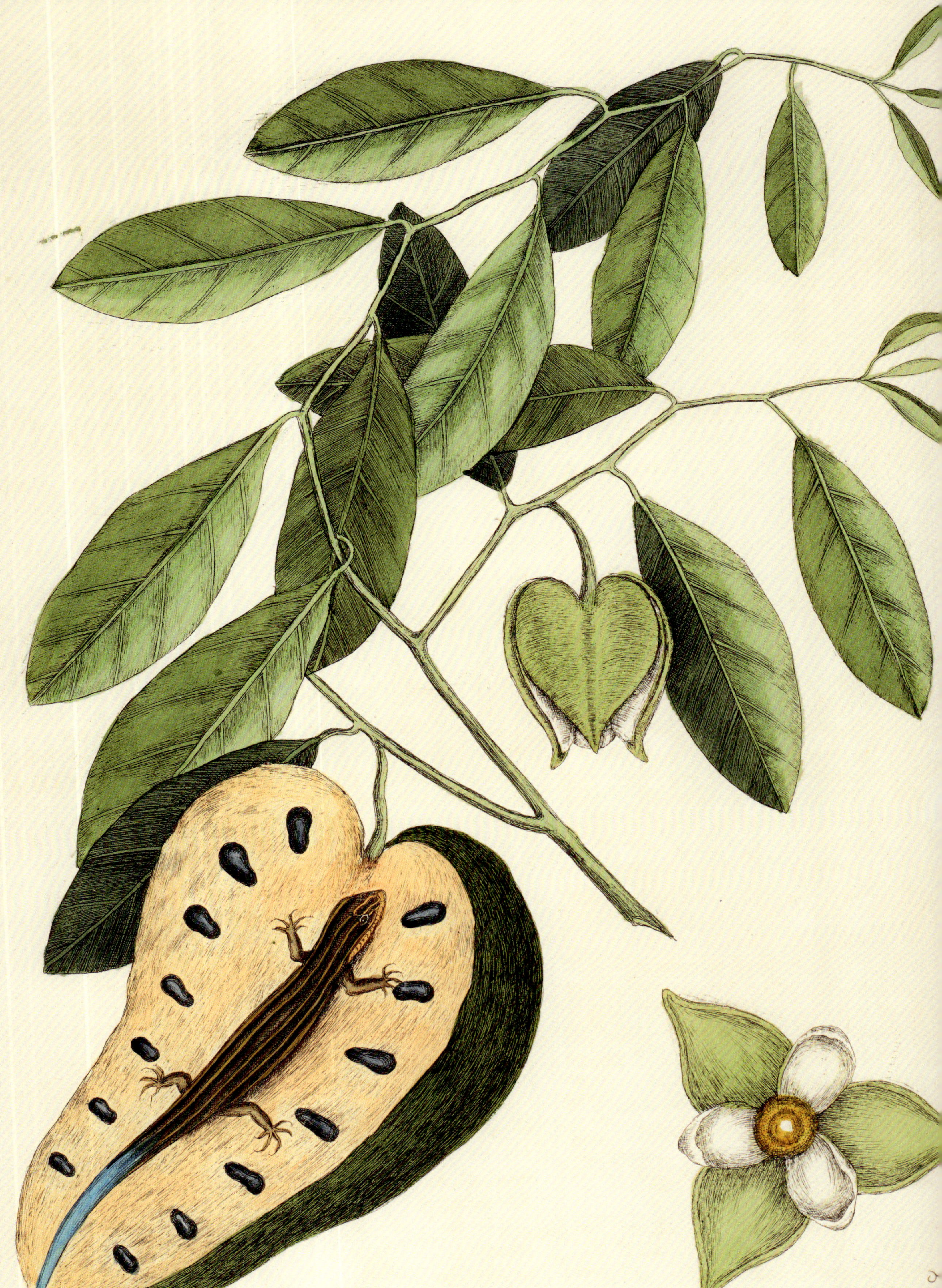

Lacertus cauda cærulea. *A*

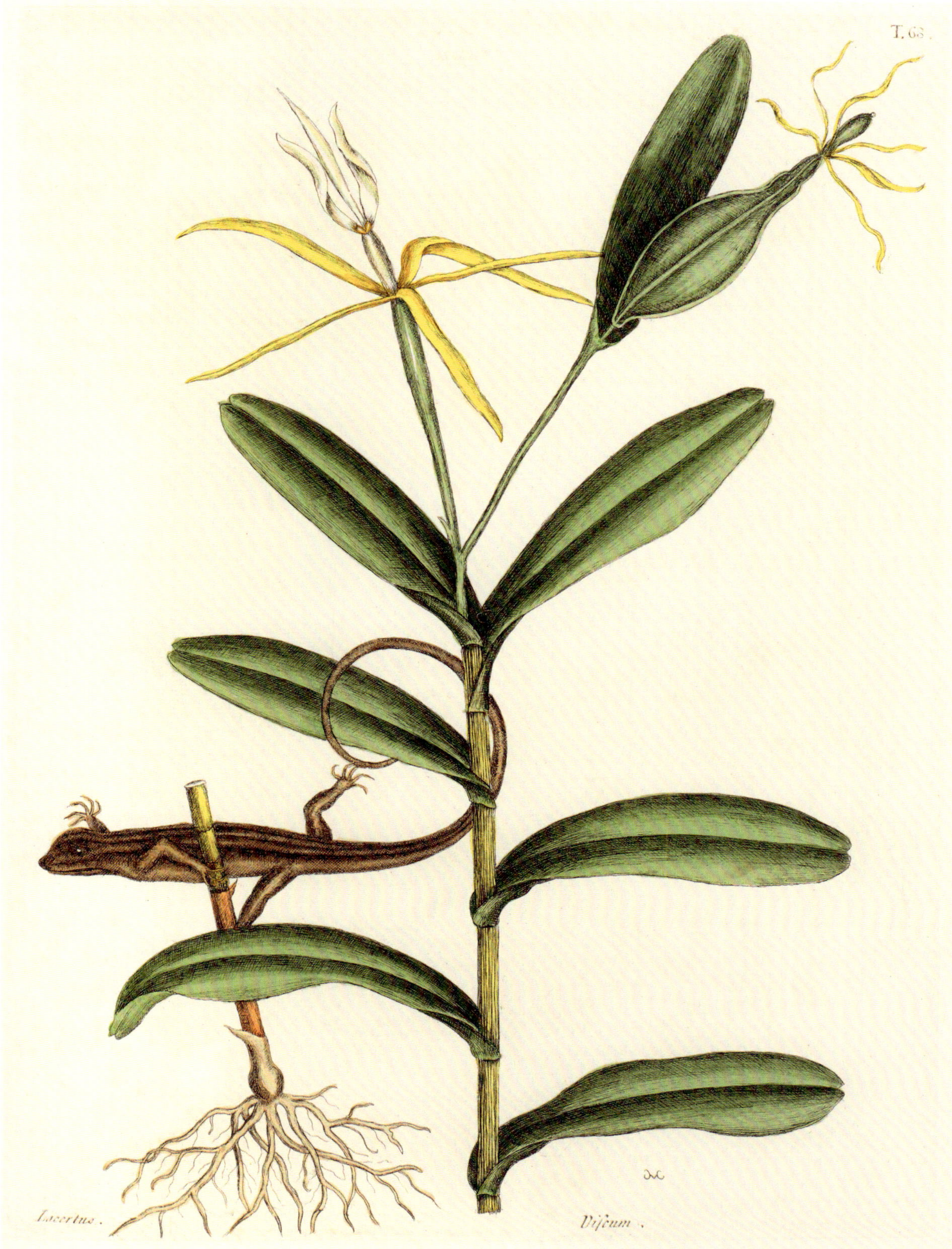

167 *left*

Common five-lined skink (*Plestioden fasciatus*); alligator apple (*Annona glabra*) (vol. 2, t. 67)

168 *above*

Northern curly-tailed lizard (*Leiocephalus carinatus*); night scented orchid (*Epidendrum nocturnum*) (vol. 2, t. 68)

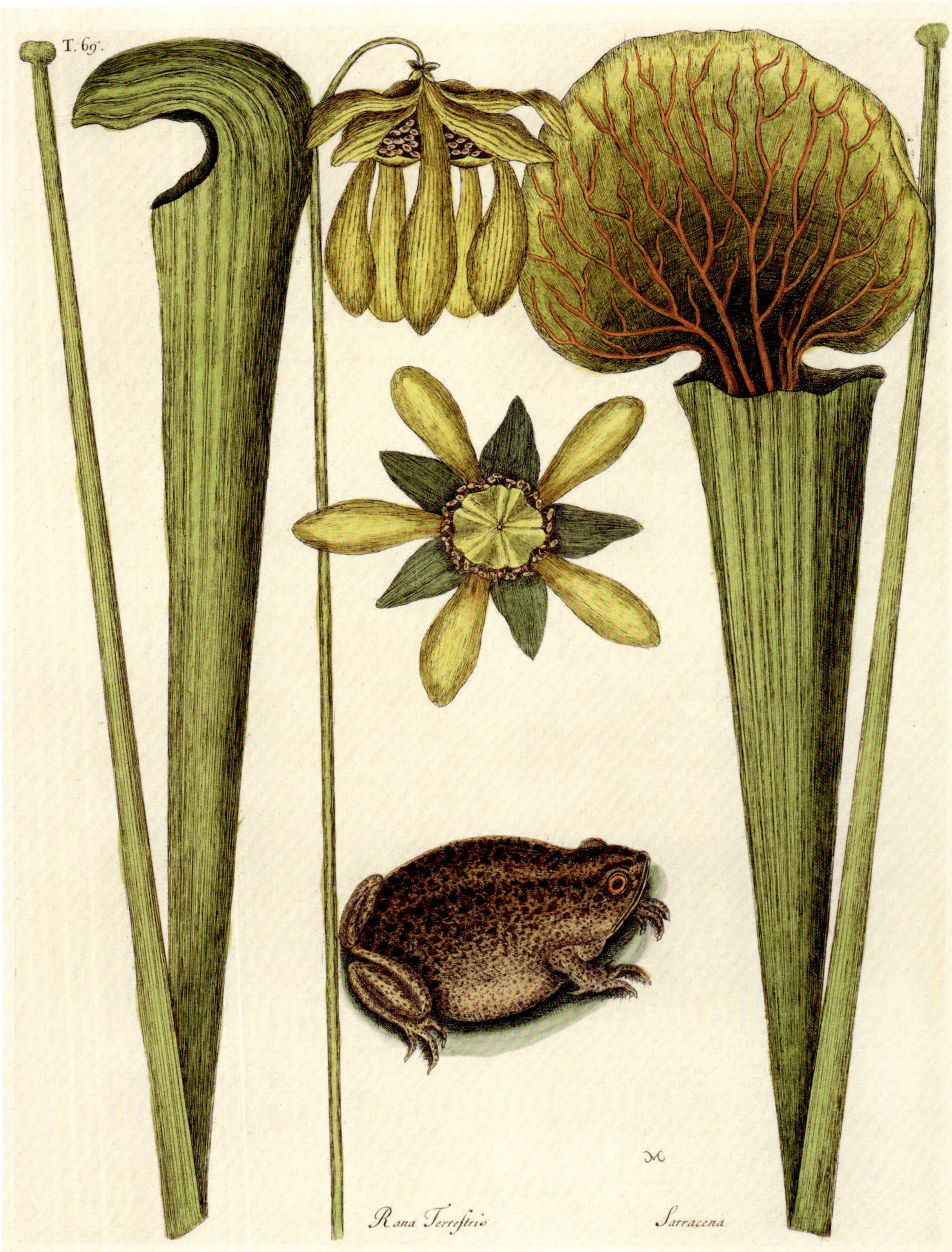

169 *above*
Southern toad (*Anaxyrus terrestris*); yellow pitcher plant (*Sarracenia flava*; left); nodding pitcher plant (*Sarracenia minor*; middle and right) (vol. 2, t. 69)

170 *right*
Southern leopard frog (*Lithobates sphenocephalus*); purple pitcher plant (*Sarracenia purpurea*) (vol. 2, t. 70)

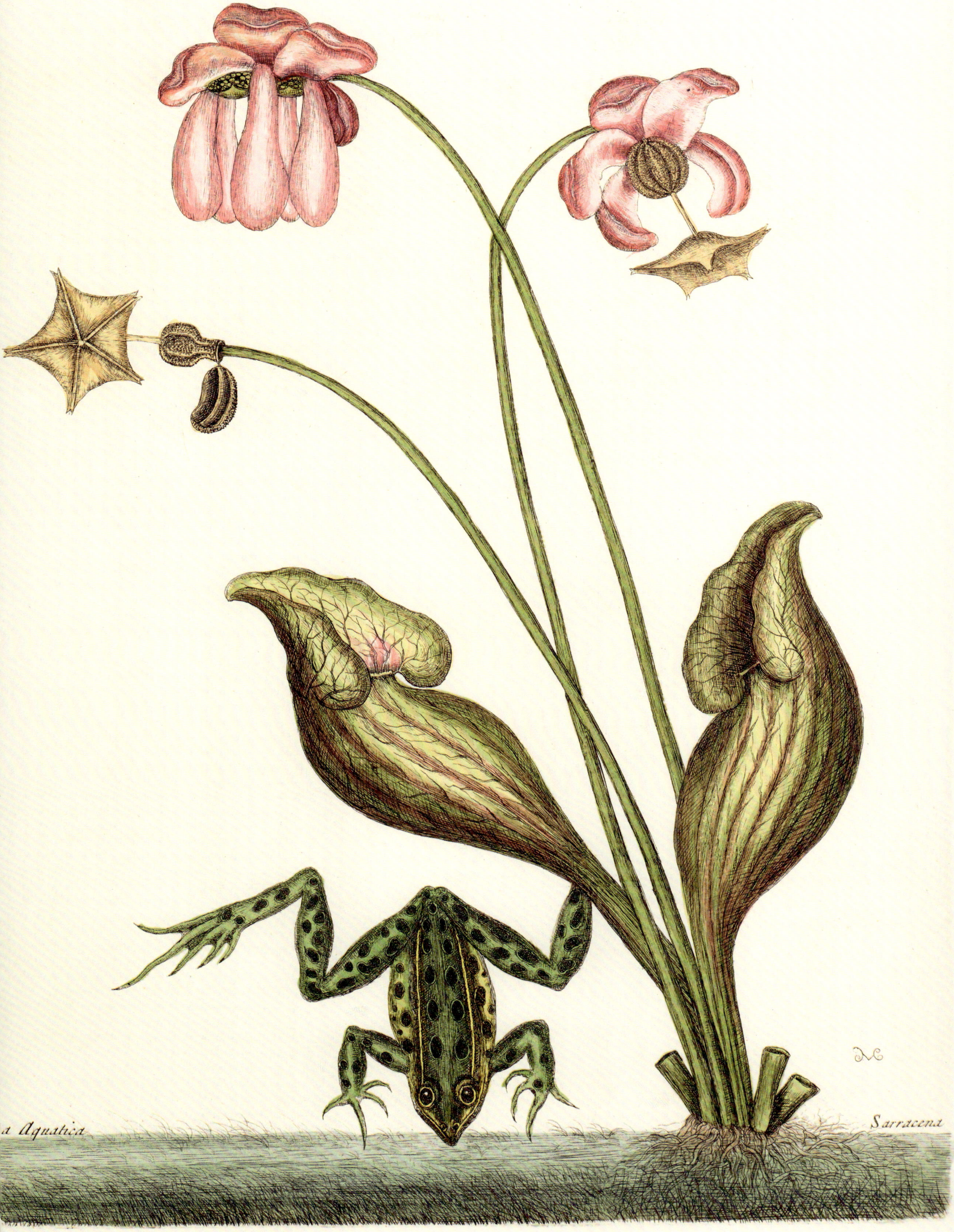
a Aquatica
Sarracena

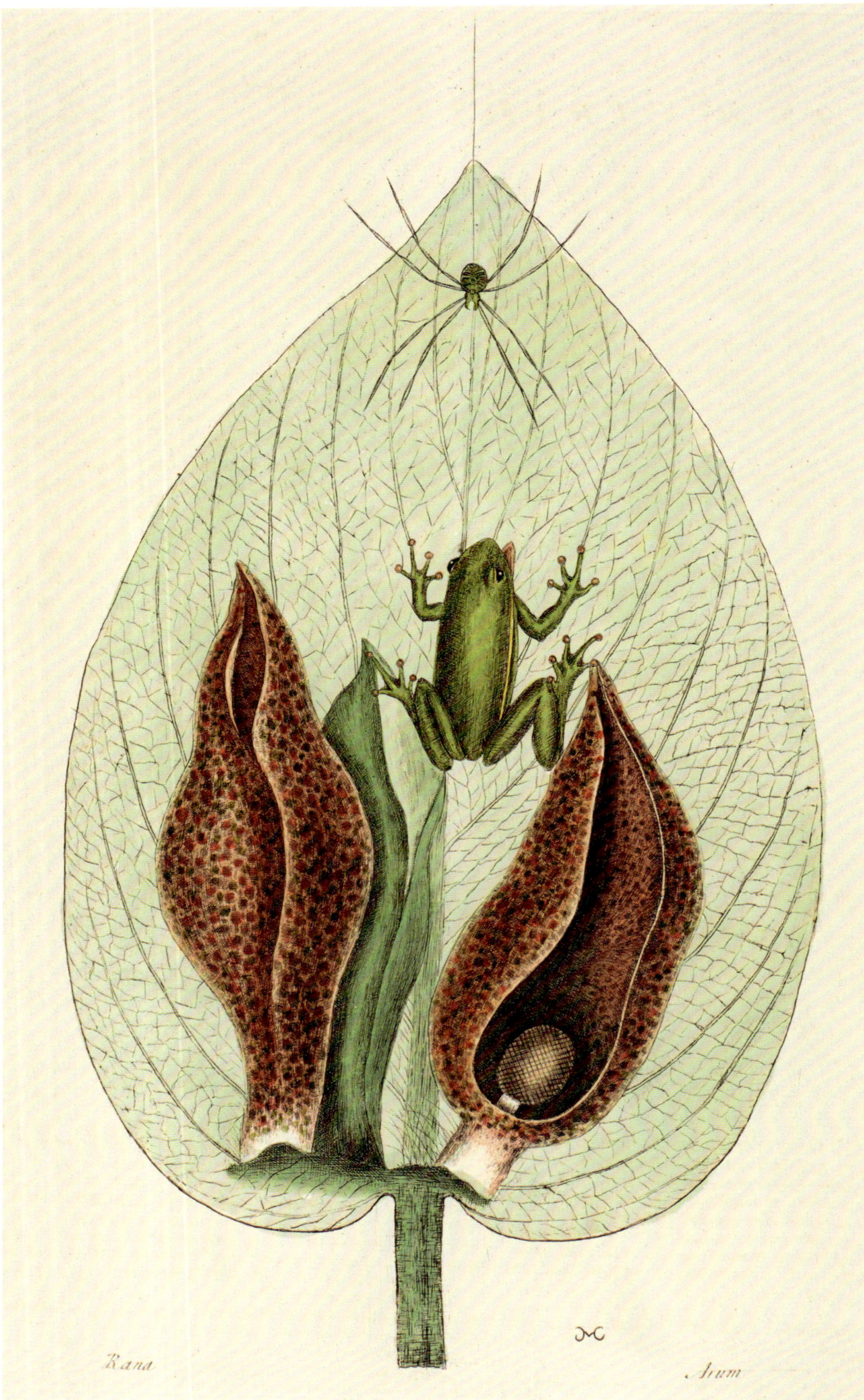

171 *above*
American green tree frog (*Dryophytes cinereus*); green lynx spider (*Peucetia viridans*); skunk cabbage (*Symplocarpus foetidus*) (vol. 2, t. 71)

172 *right*
American bullfrog (*Lithobates catesbeianus*); pink lady's slipper (*Cypripedium acaule*) (vol. 2, t. 72)

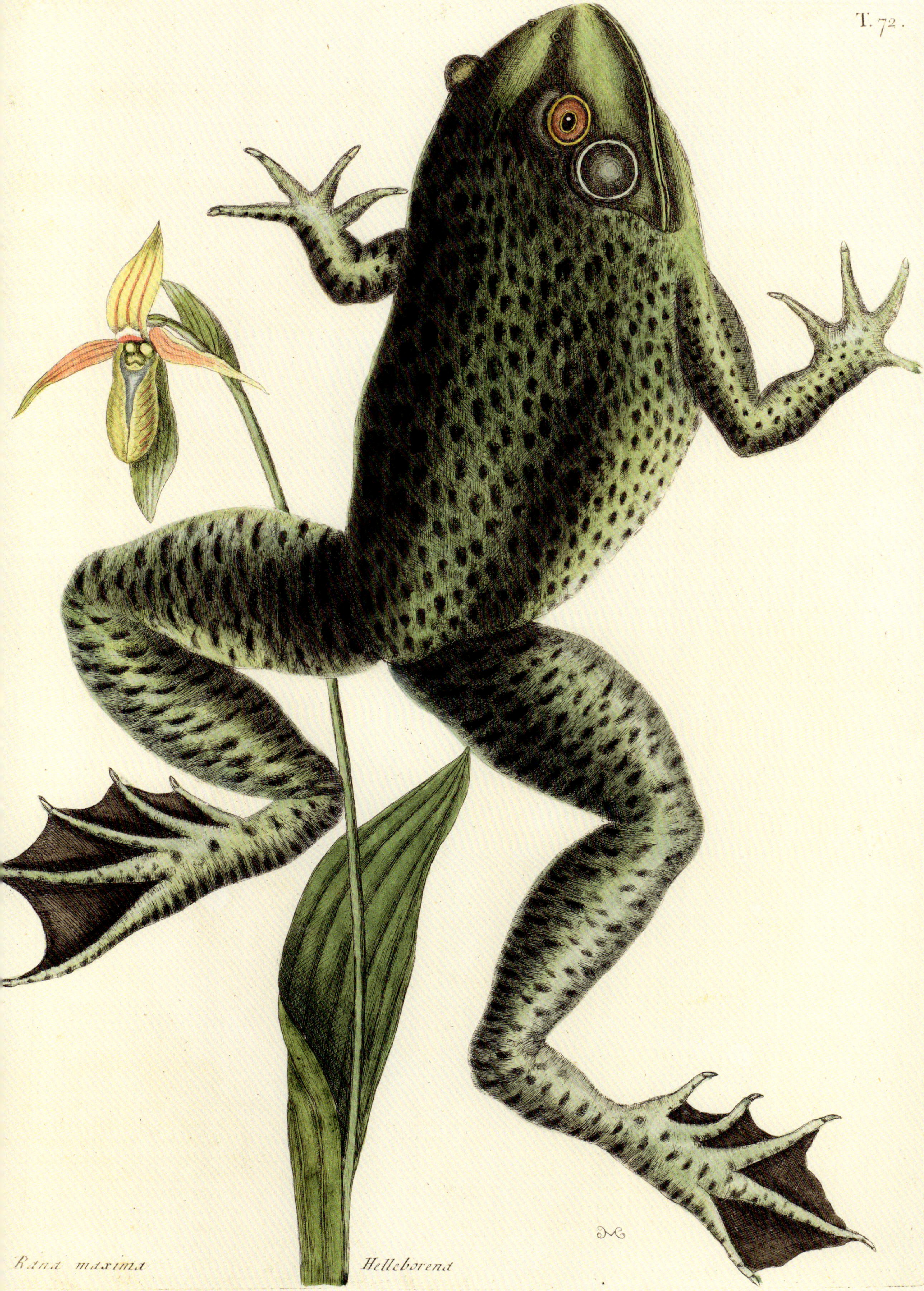

Rana maxima *Helleboreна*

T.73.

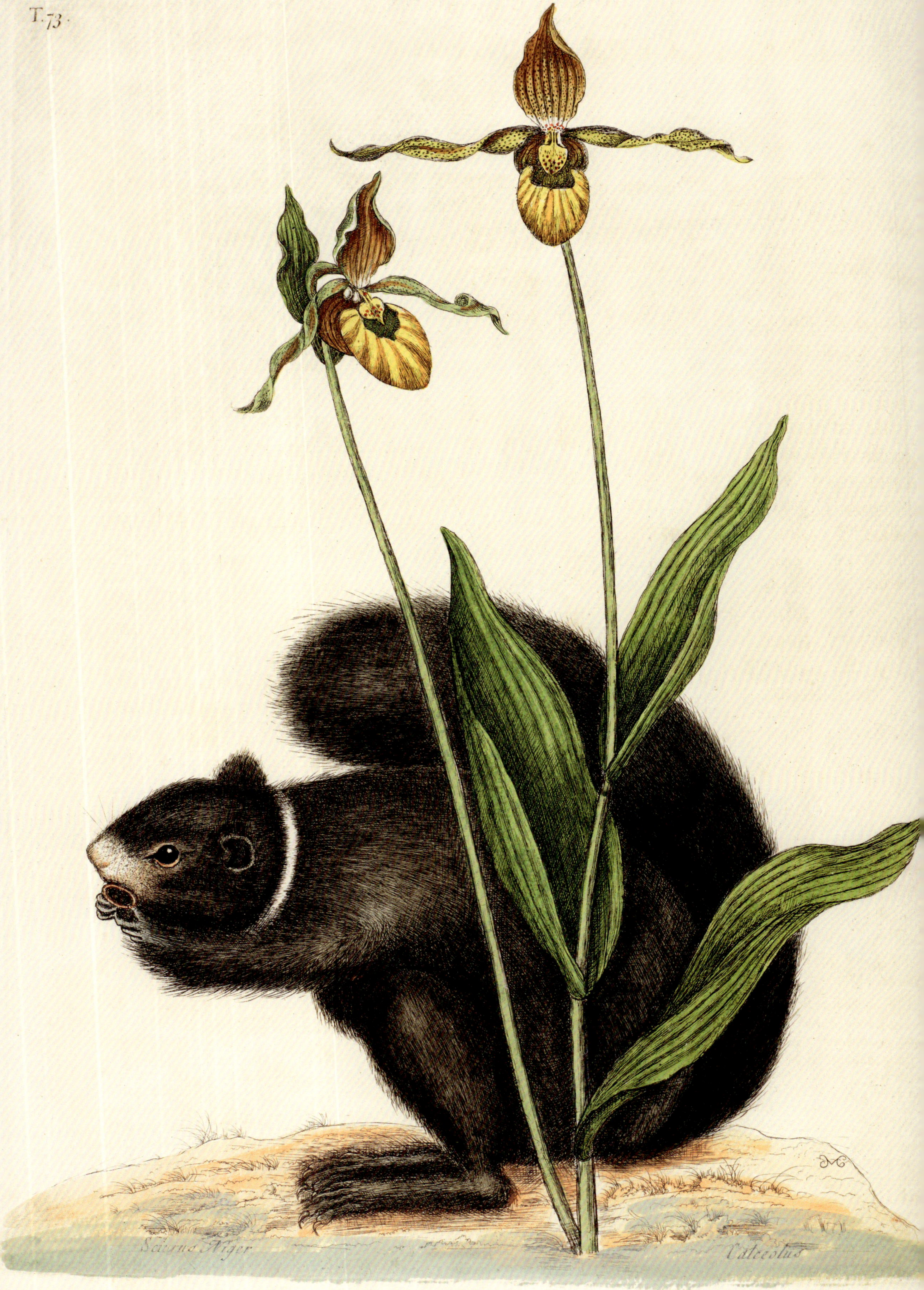

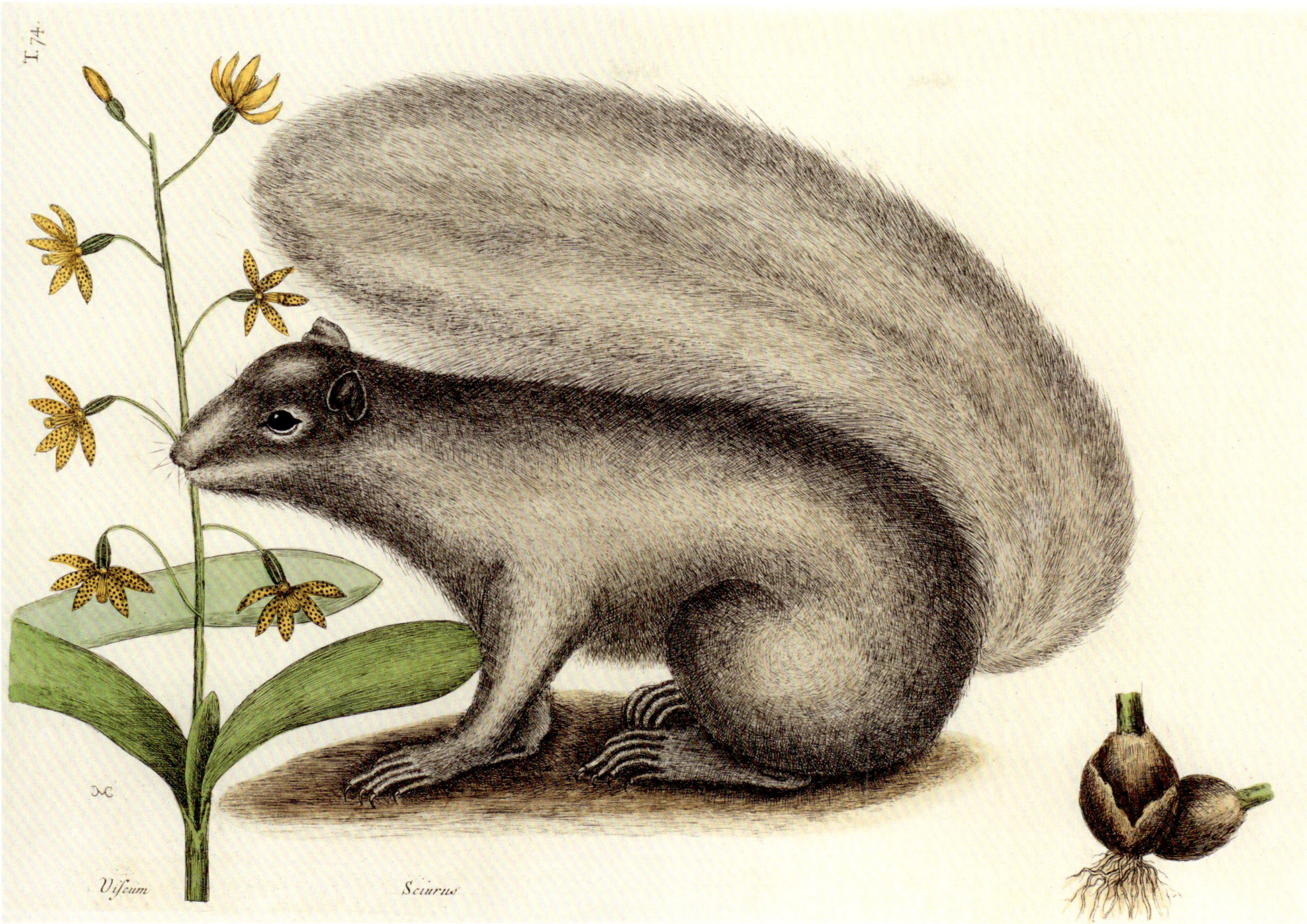

173 *left*
Eastern fox squirrel (*Sciurus niger*); greater yellow lady's slipper (*Cypripedium parviflorum*) (vol. 2, t. 73)

174 *above*
Eastern grey squirrel (*Sciurus carolinensis*); dollar orchid (*Prosthechea boothihiana*) (vol. 2, t. 74)

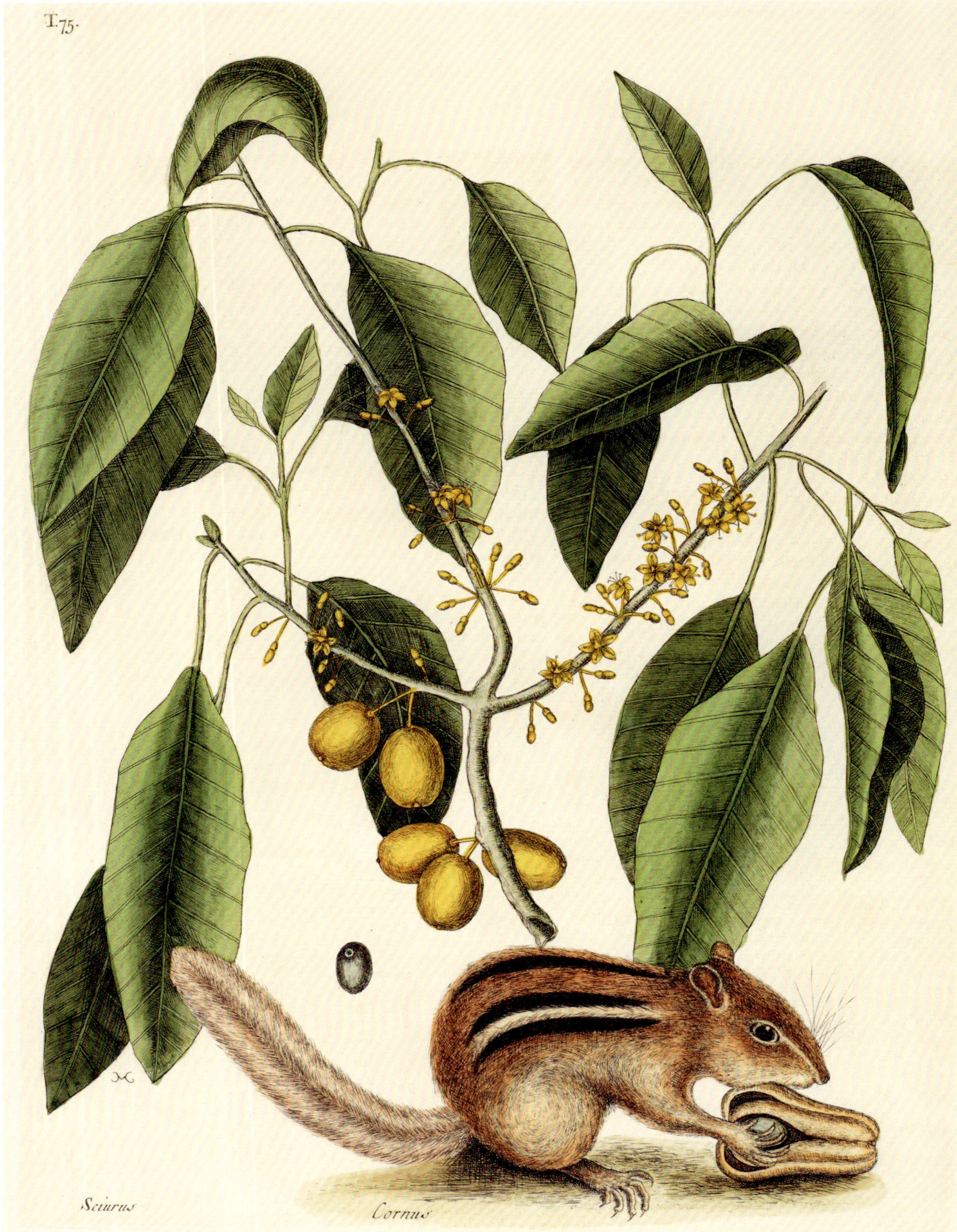

175 *above*

Eastern chipmunk (*Tamias striatus*); mastic ironwood (*Sideroxylon foetidissimum*); Carolina allspice (*Calycanthus floridus*; fruit and seed) (vol. 2, t. 75)

176 *right*

Southern flying squirrel (*Glaucomys volans*); American persimmon (*Diospyros virginiana*) (vol. 2, t. 76)

T. 76.

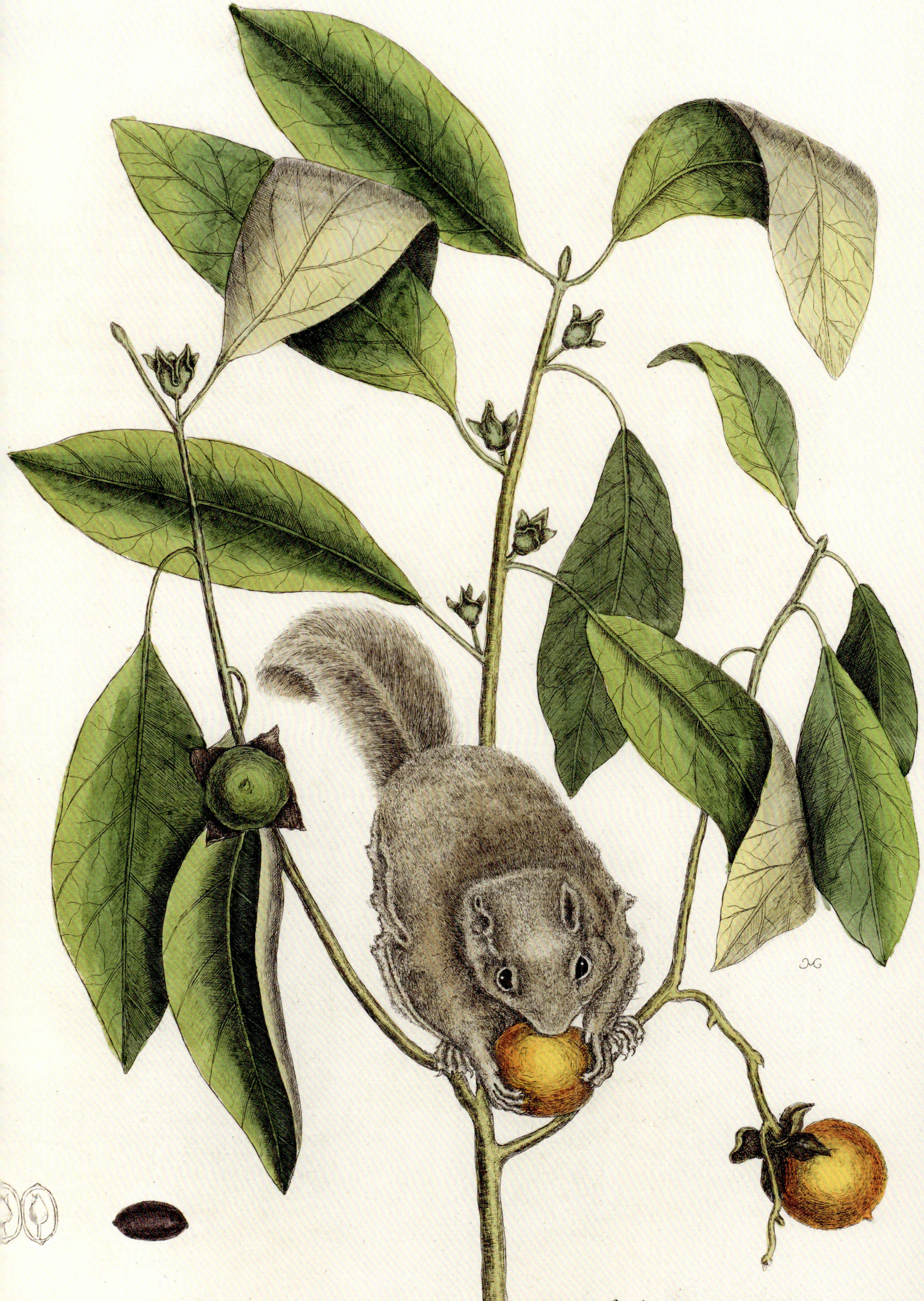

Sciurus Volans *Guajacana*

T.77.

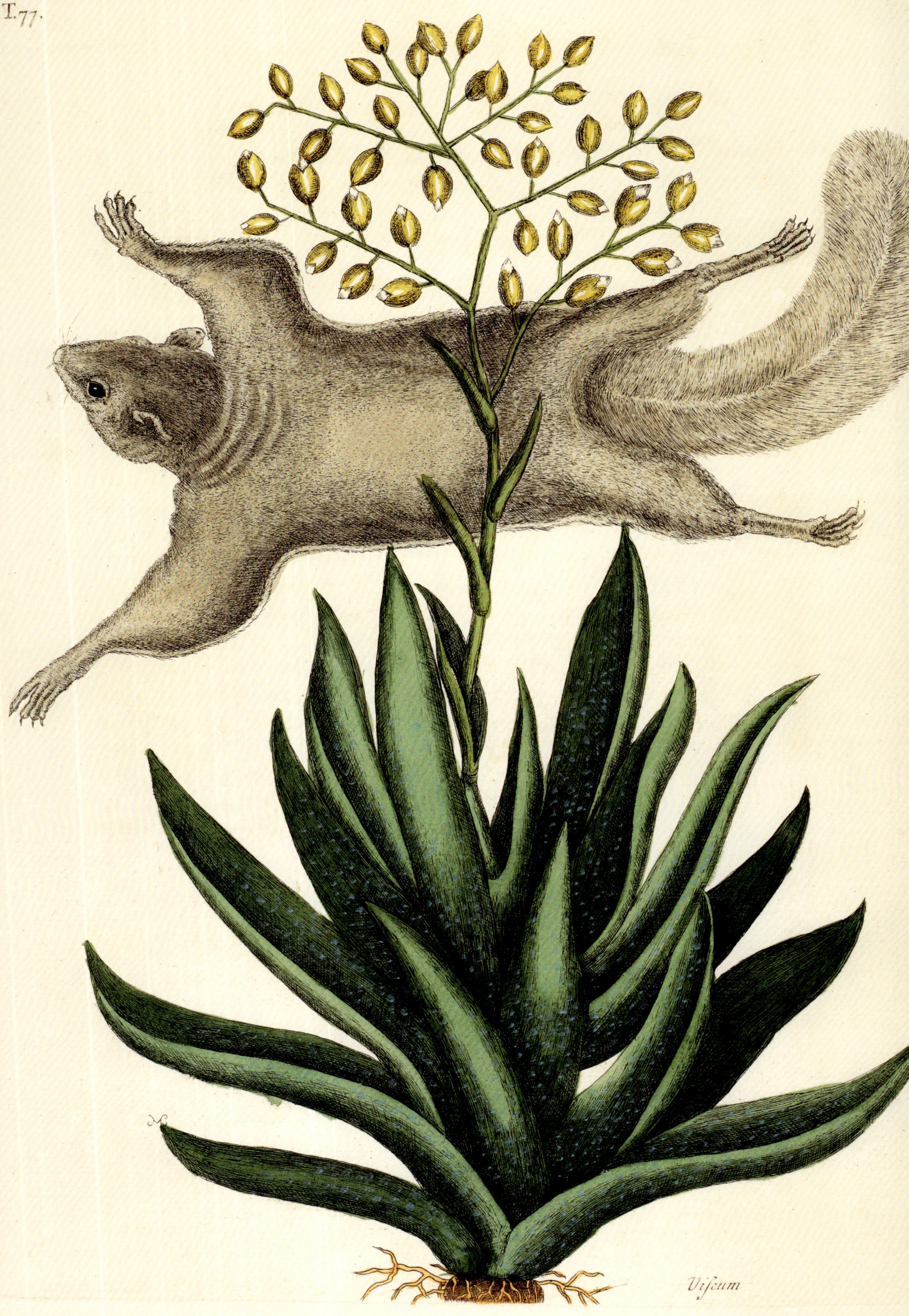

Viscum

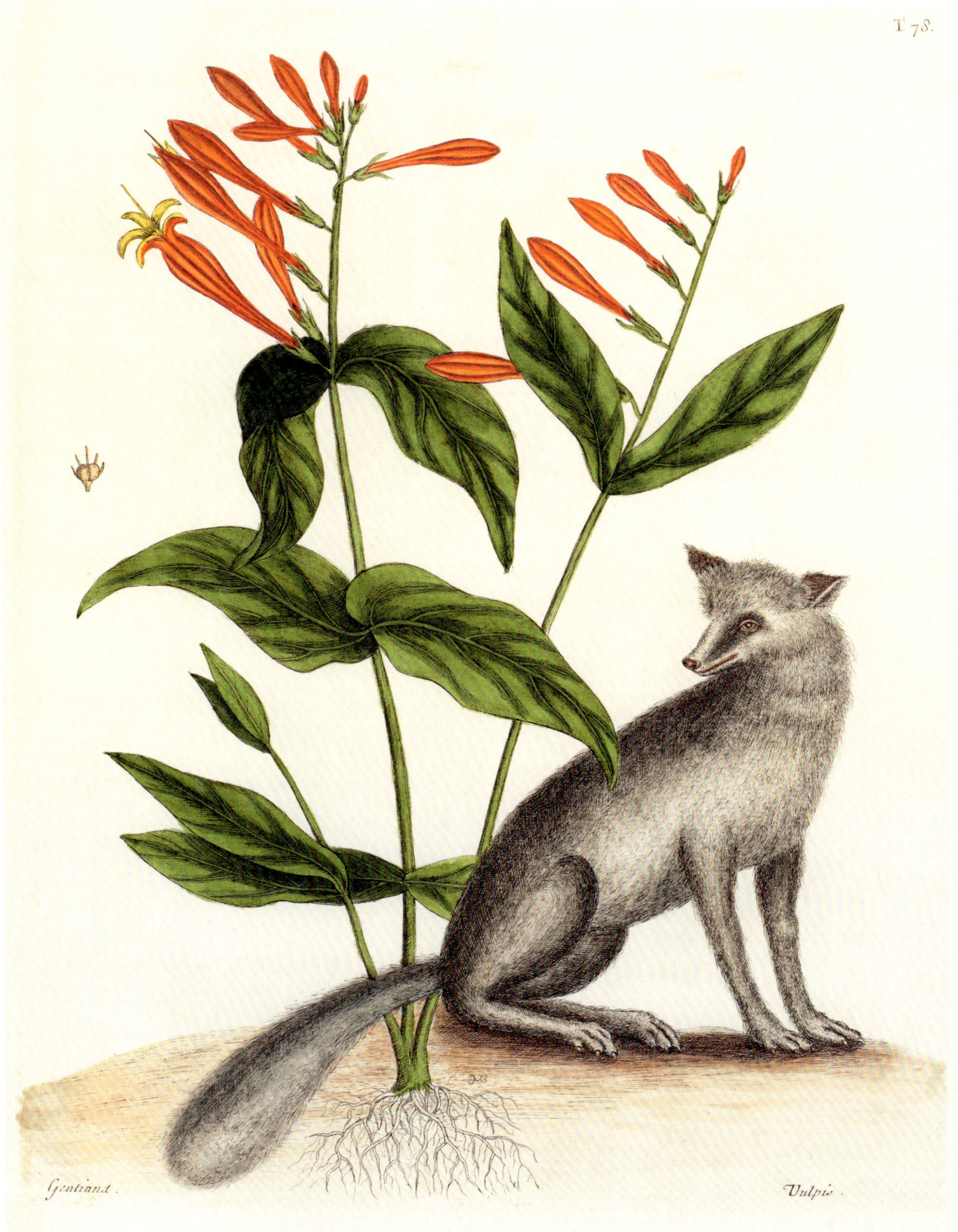

177 *left*
Southern flying squirrel (*Glaucomys volans*); powdery strap airplant (*Catopsis berteroniana*) (vol. 2, t. 77)

178 *right*
Common grey fox (*Urocyon cinereoargenteus*); indianpink (*Spigelia marilandica*) (vol. 2, t. 78)

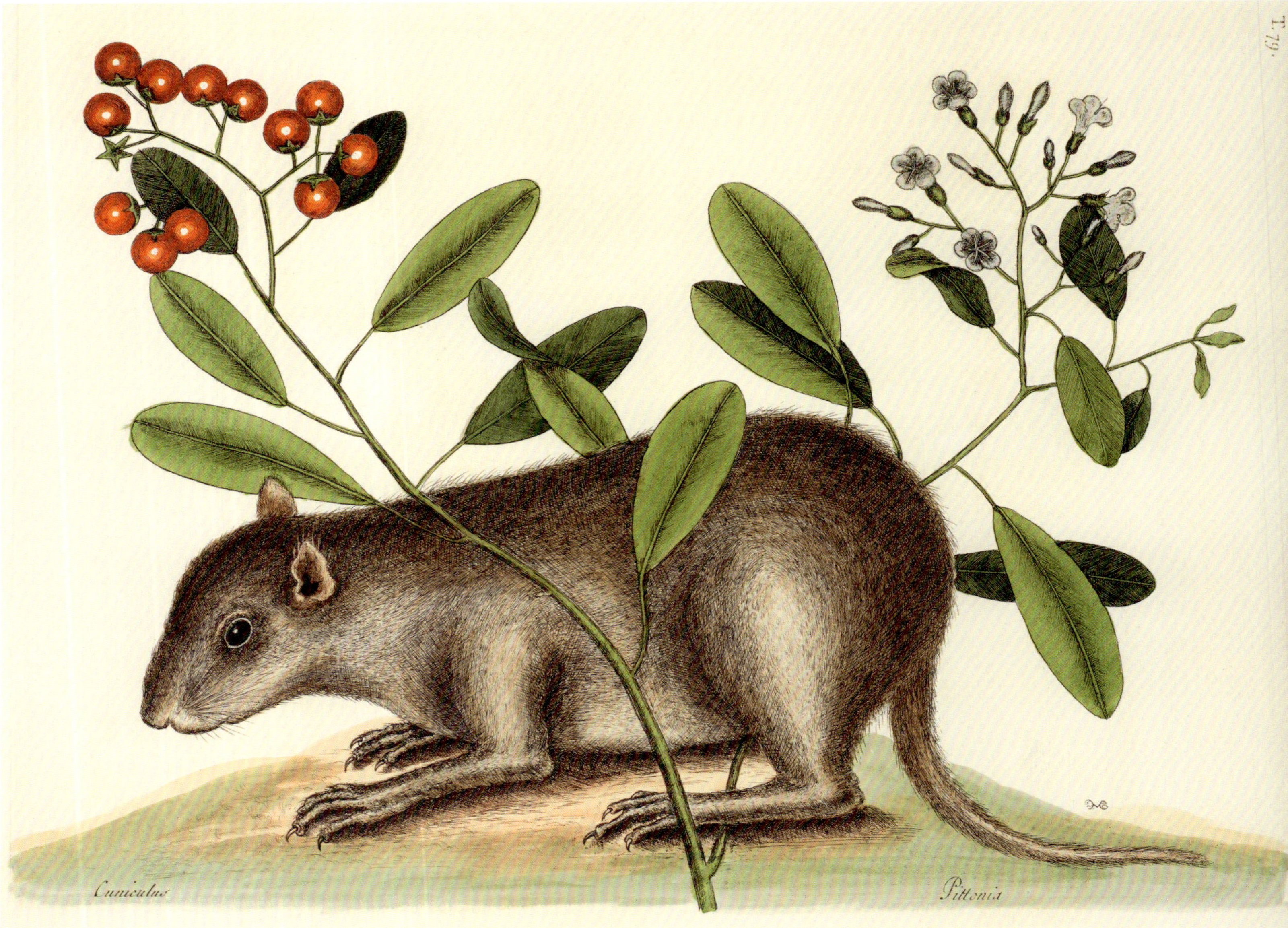

179

Bahamian hutia (*Geocapromys ingrahami*); bodywood (*Bourreria succulenta*) (vol. 2, t. 79)

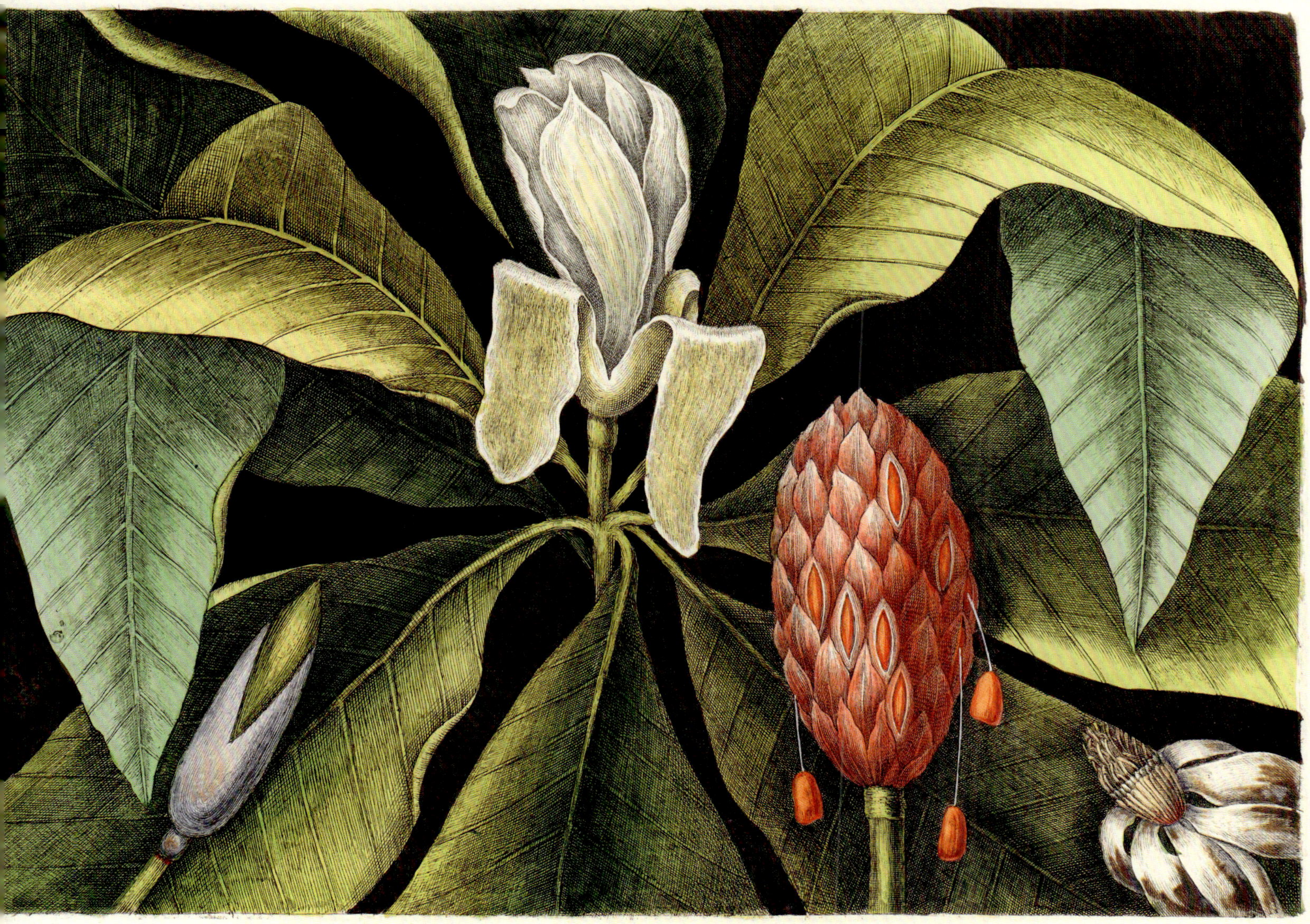

180

Umbrella magnolia (*Magnolia tripetala*) (vol. 2, t. 80)

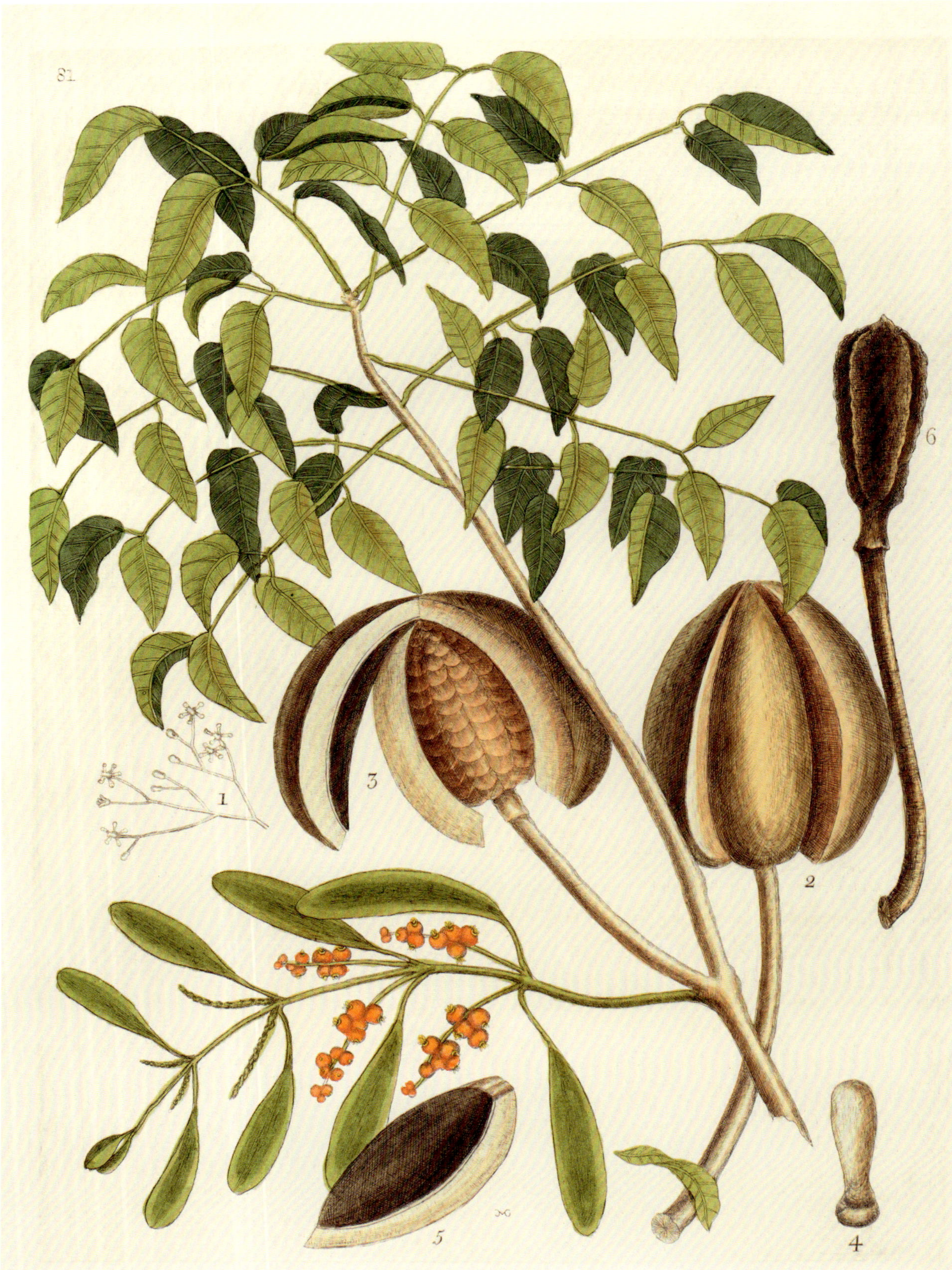

181 *above*

West Indian mahogany (*Swietenia mahogani*); mahogany mistletoe (*Phoradendron rubrum*) (vol. 2, t. 81)

182 *right*

Cross vine (*Bignonia capreolata*) (vol. 2, t. 82)

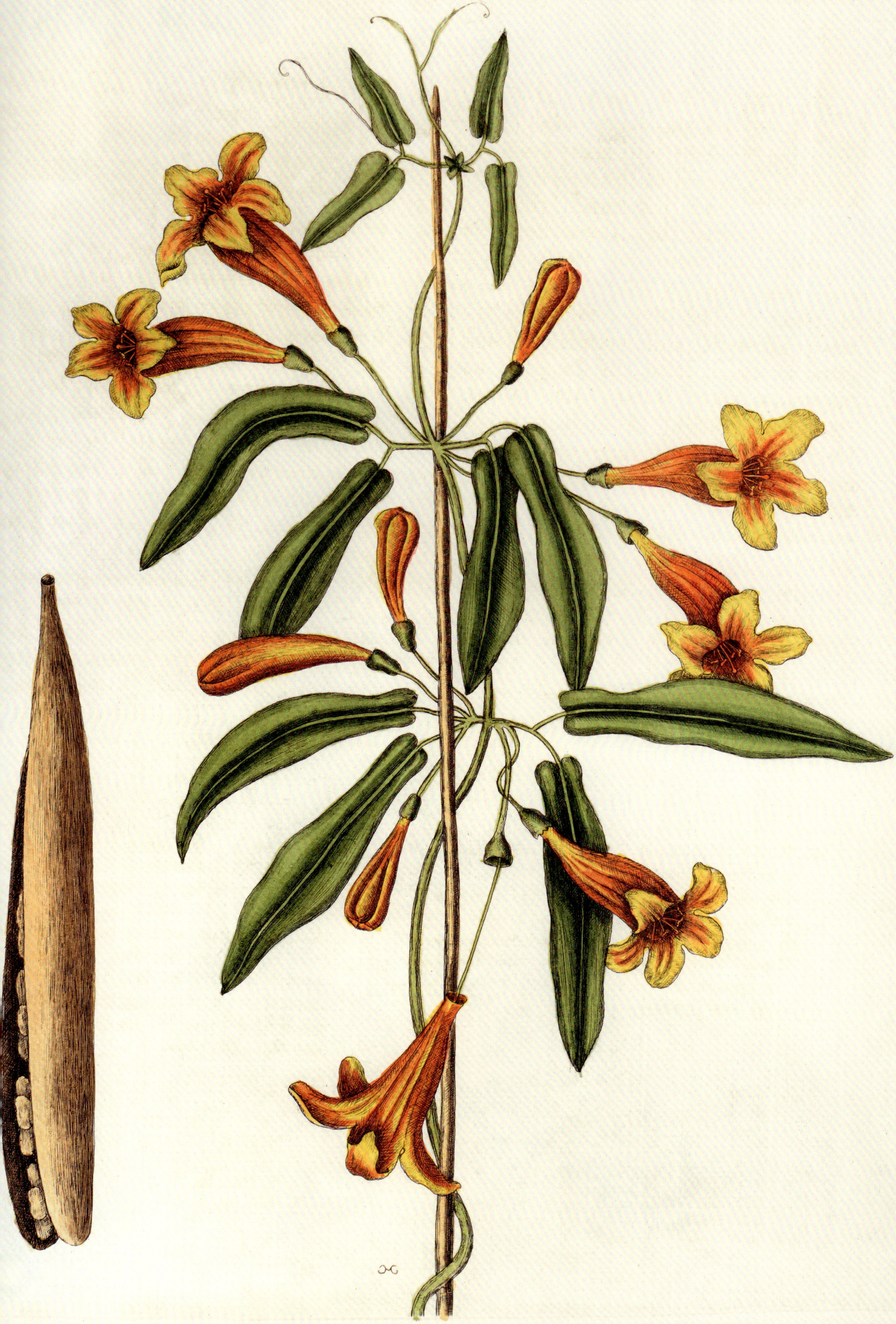

83

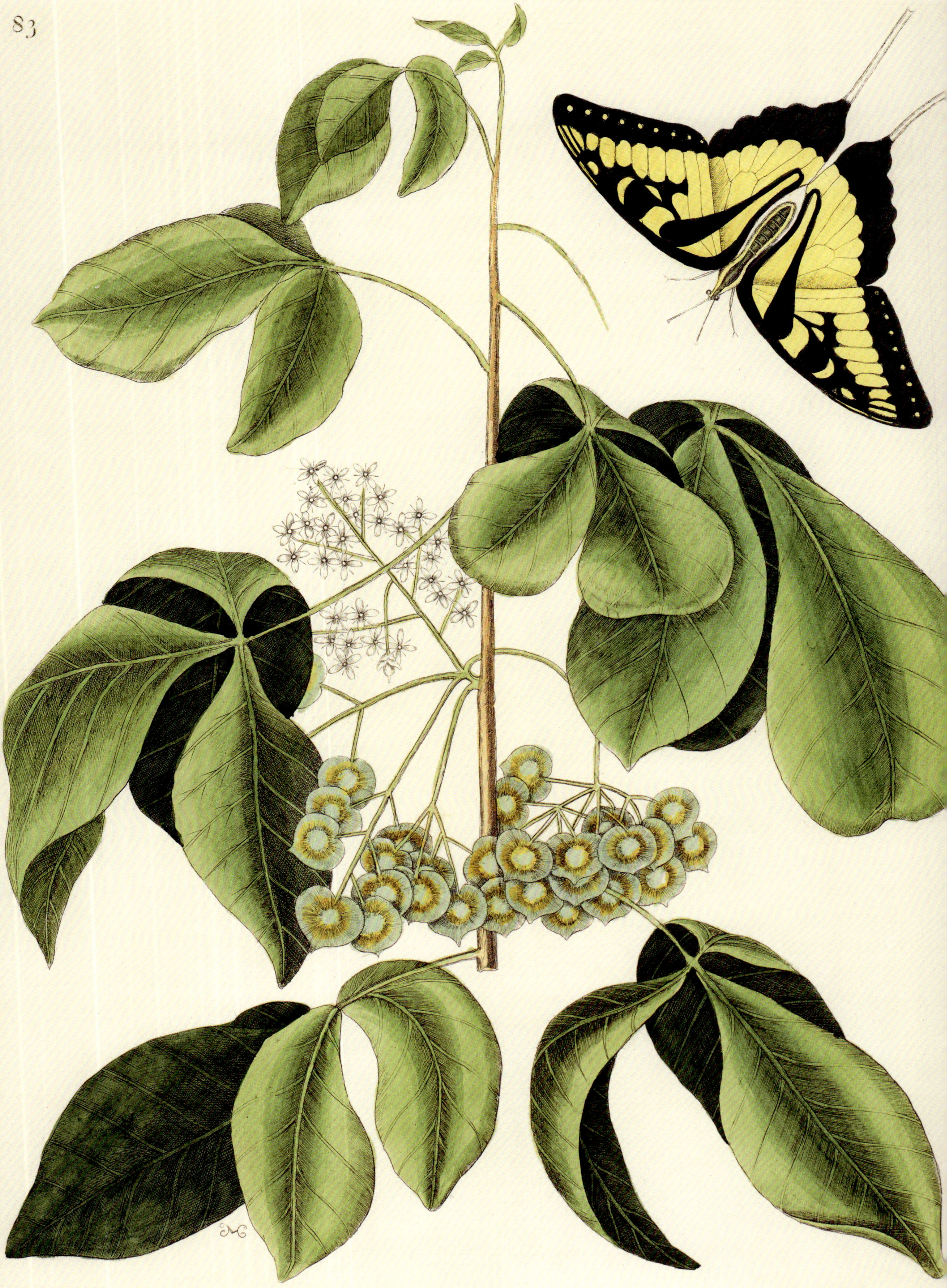

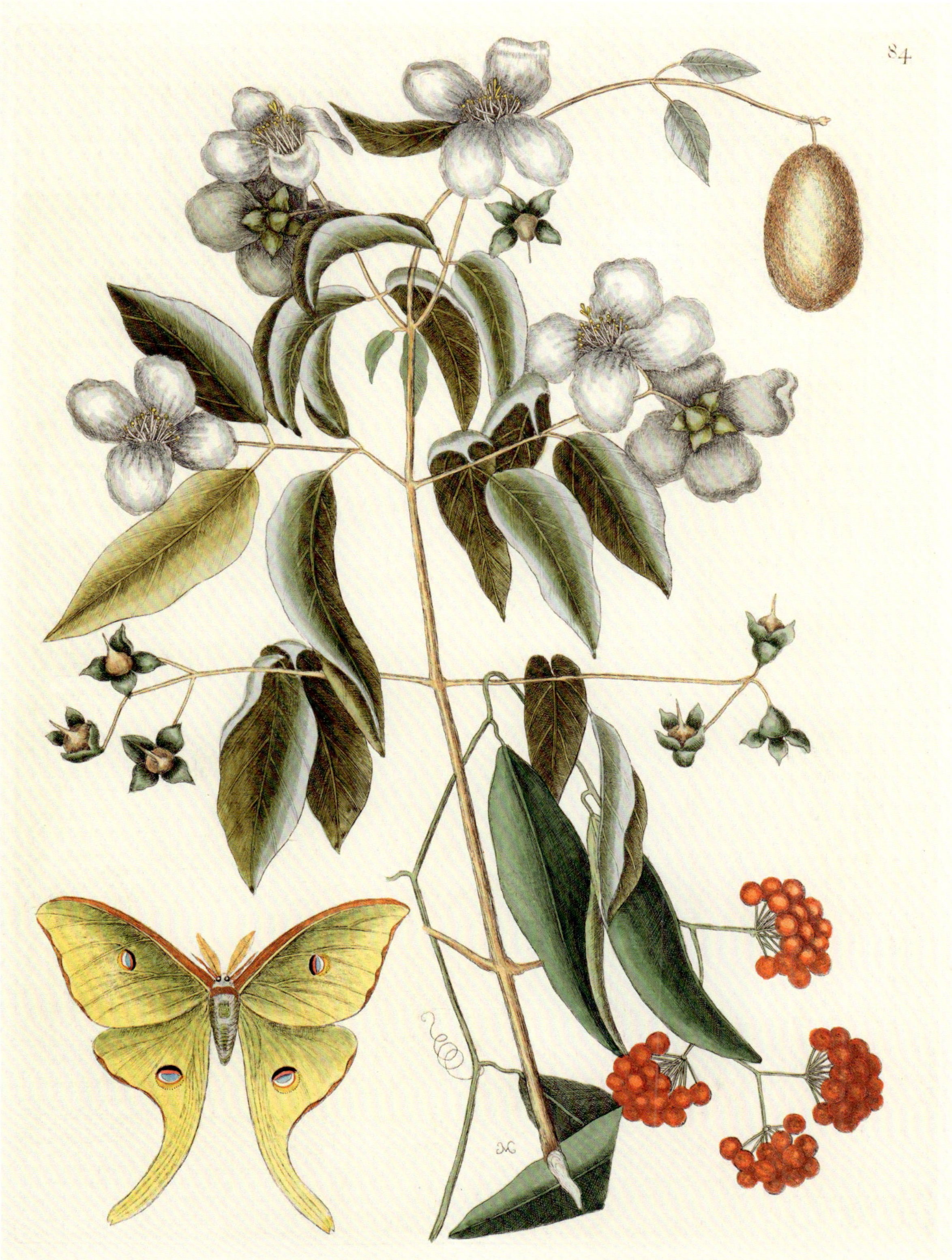

183 *left*
Hoptree (*Ptelea trifoliata*); eastern tiger swallowtail (*Papilio glaucus*) (vol. 2, t. 83)

184 *above*
Scentless mock orange (*Philadelphus inodorus*); lanceleaf greenbrier (*Smilax maritima*; climber); luna moth (*Actias luna*) (vol. 2, t. 84)

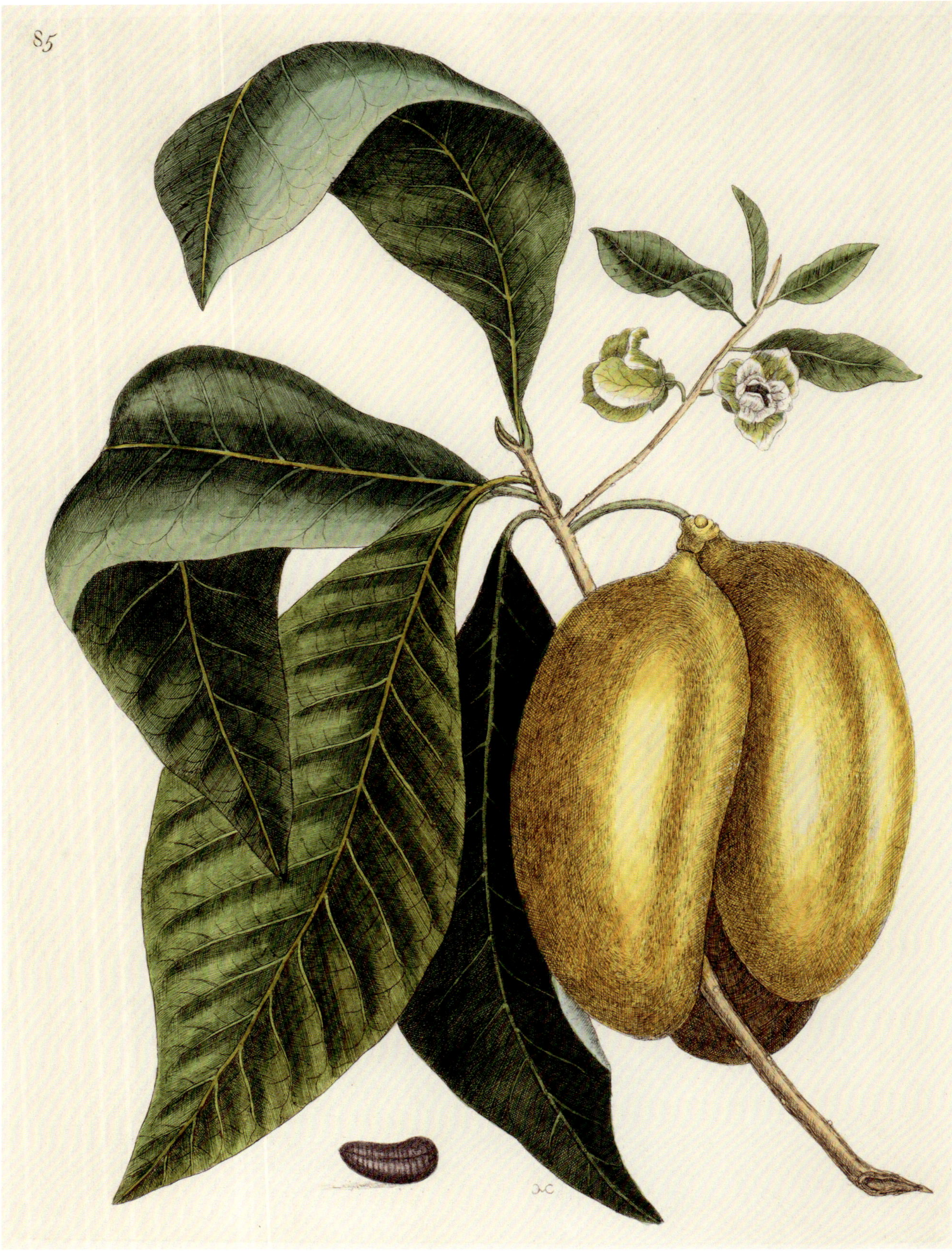

185 *above*

Pawpaw (*Asimina triloba*) (vol. 2, t. 85)

186 *right*

Bullock's heart (*Annona reticulata*); cecropia moth (*Hyalophora cecropia*) (vol. 2, t. 86)

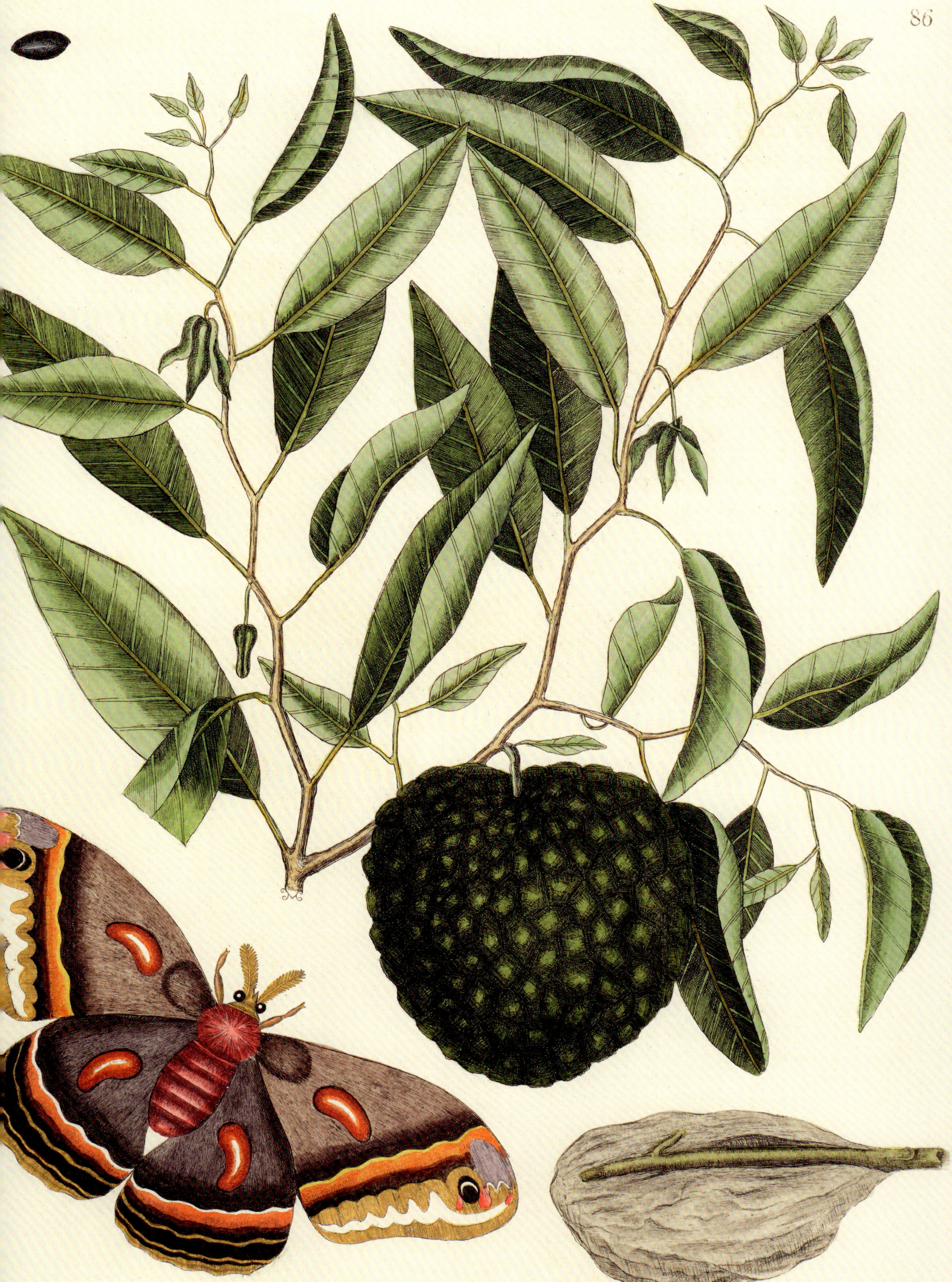

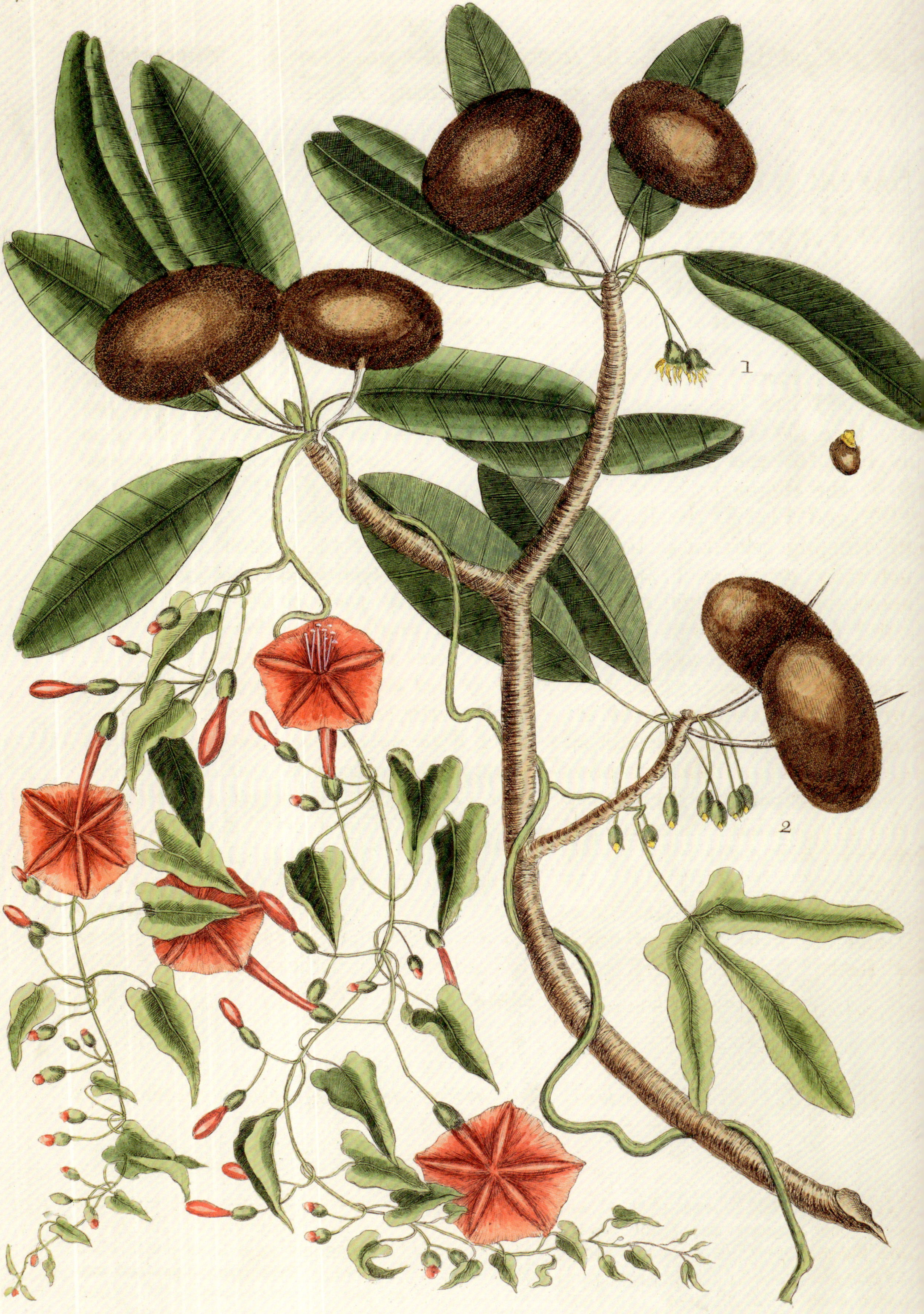
1
2

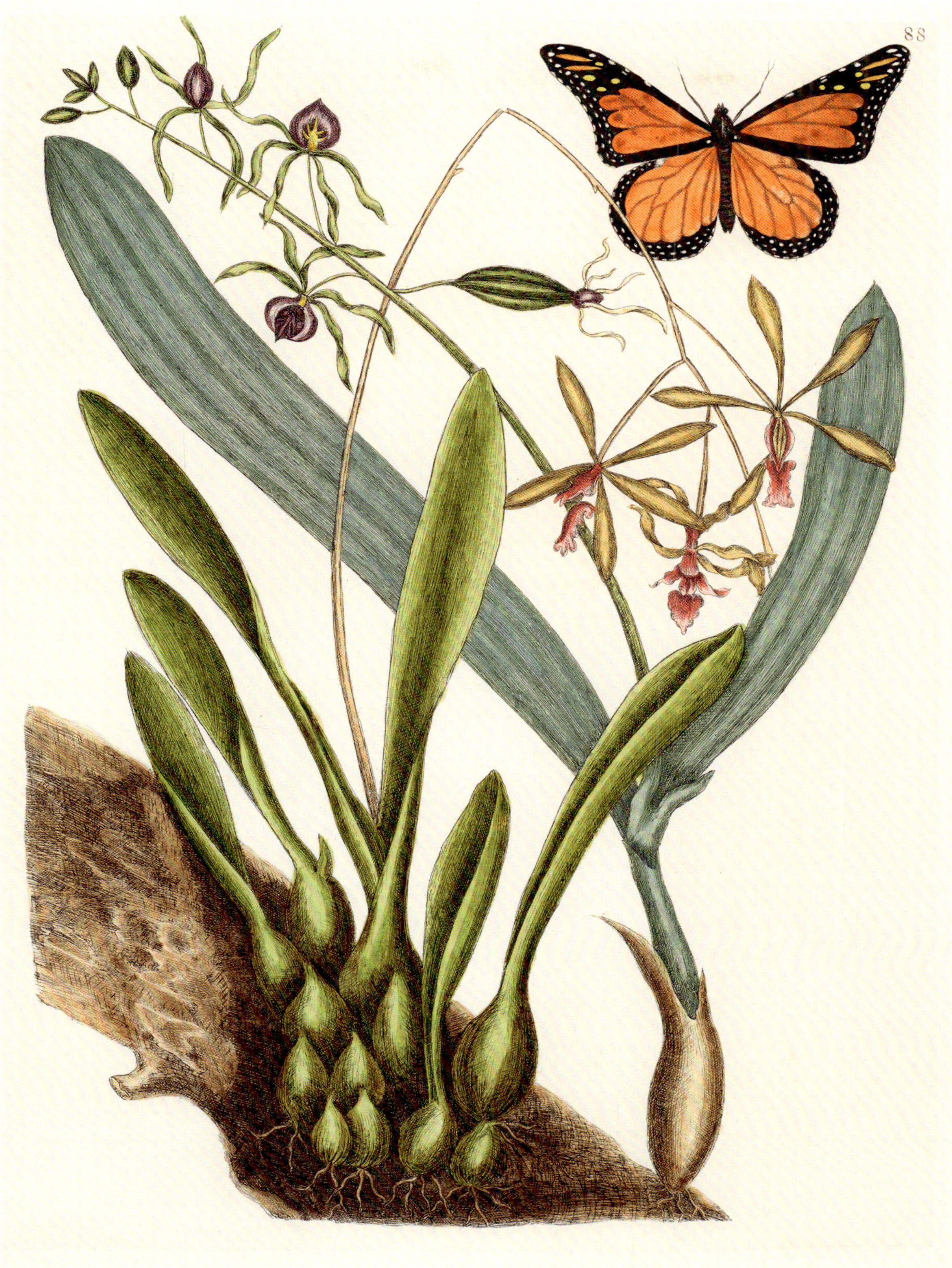

187 *left*
Calcareous morning glory (*Ipomoea microdactyla*; climber); wild dilly (*Manilkara jaimiqui*; upper) (vol. 2, t. 87)

188 *above*
Clamshell orchid (*Prosthechea cochleata*; flowers left); pleated encyclia (*Encyclia plicata*; flowers right); monarch butterfly (*Danaus plexippus*) (vol. 2, t. 88)

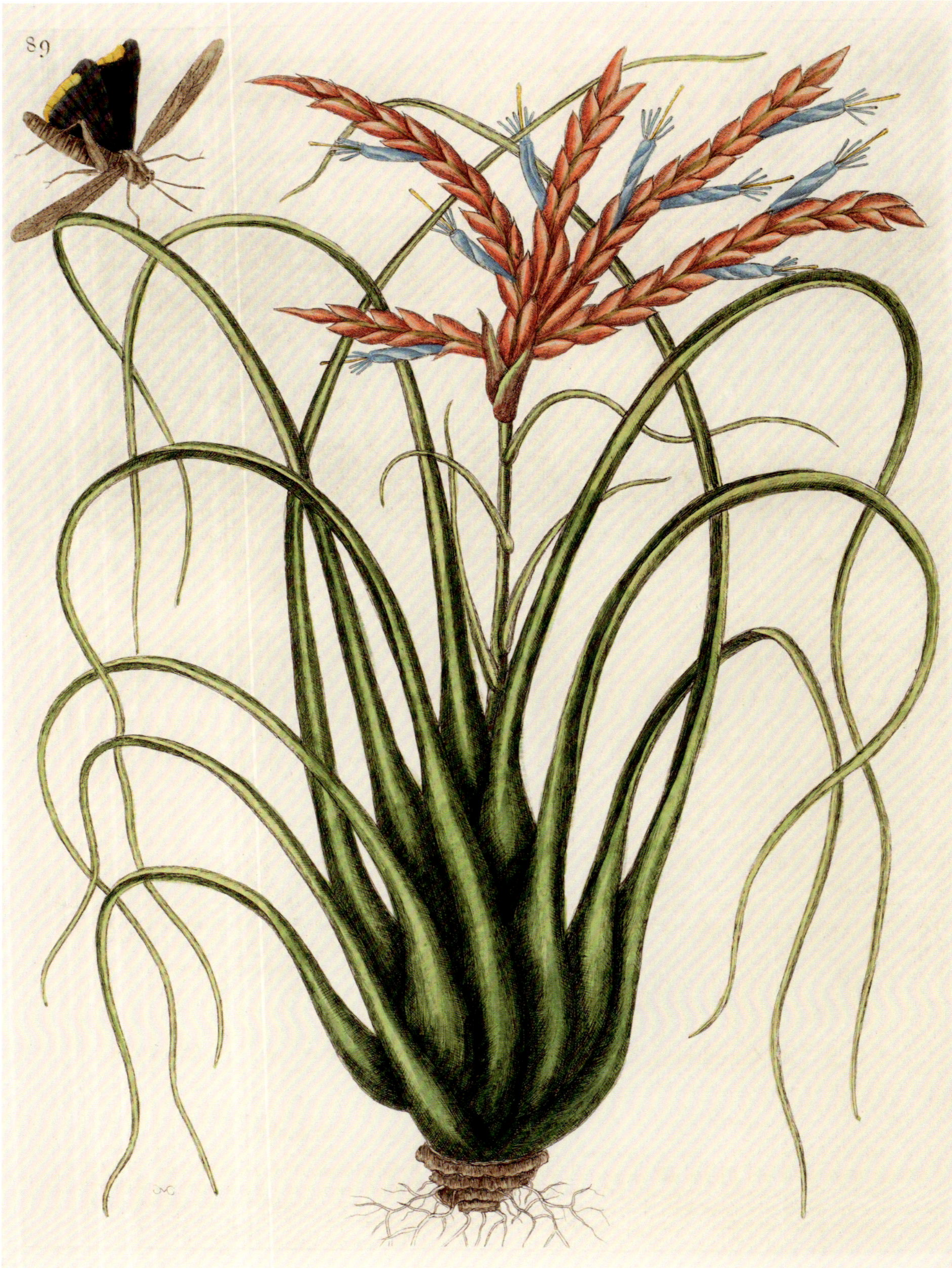

189 *above*
Northern needleleaf (*Tillandsia balbisiana*); Carolina grasshopper (*Dissosteira carolina*) (vol. 2, t. 89)

190 *right*
Tree hibiscus (*Hibiscus tiliaceus*); polyphemus moth (*Antheraea polyphemus*) (vol. 2, t. 90)

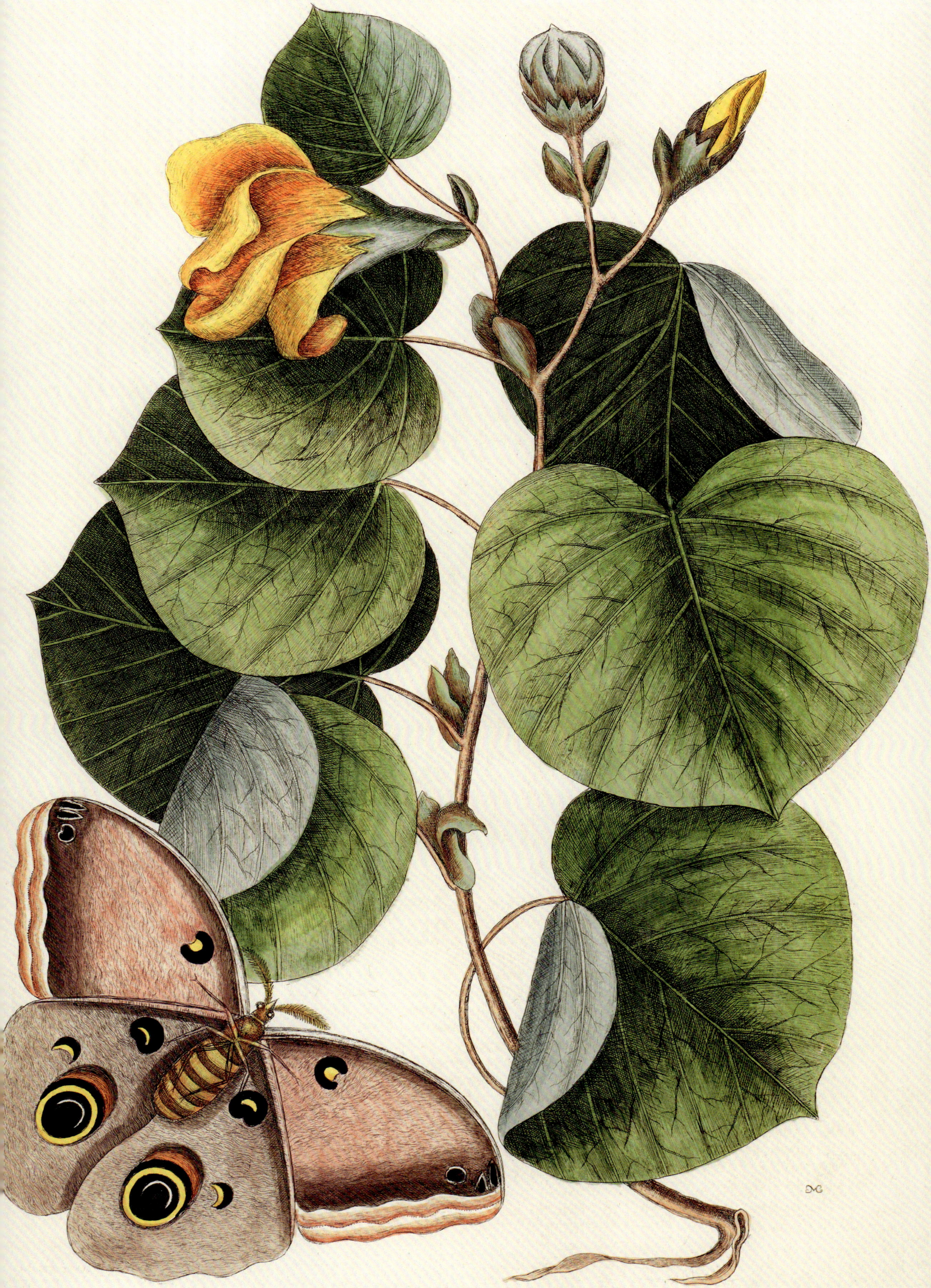

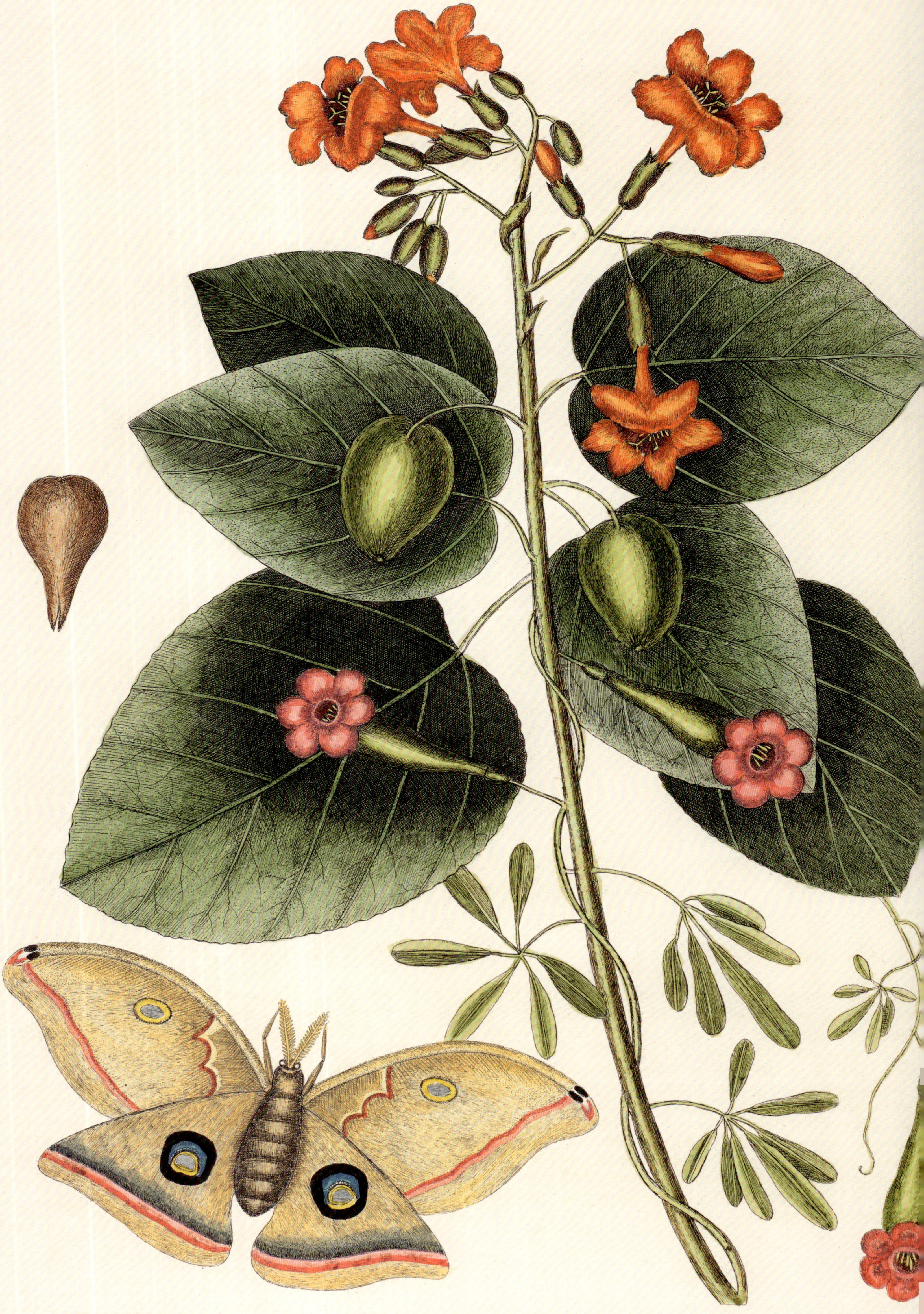

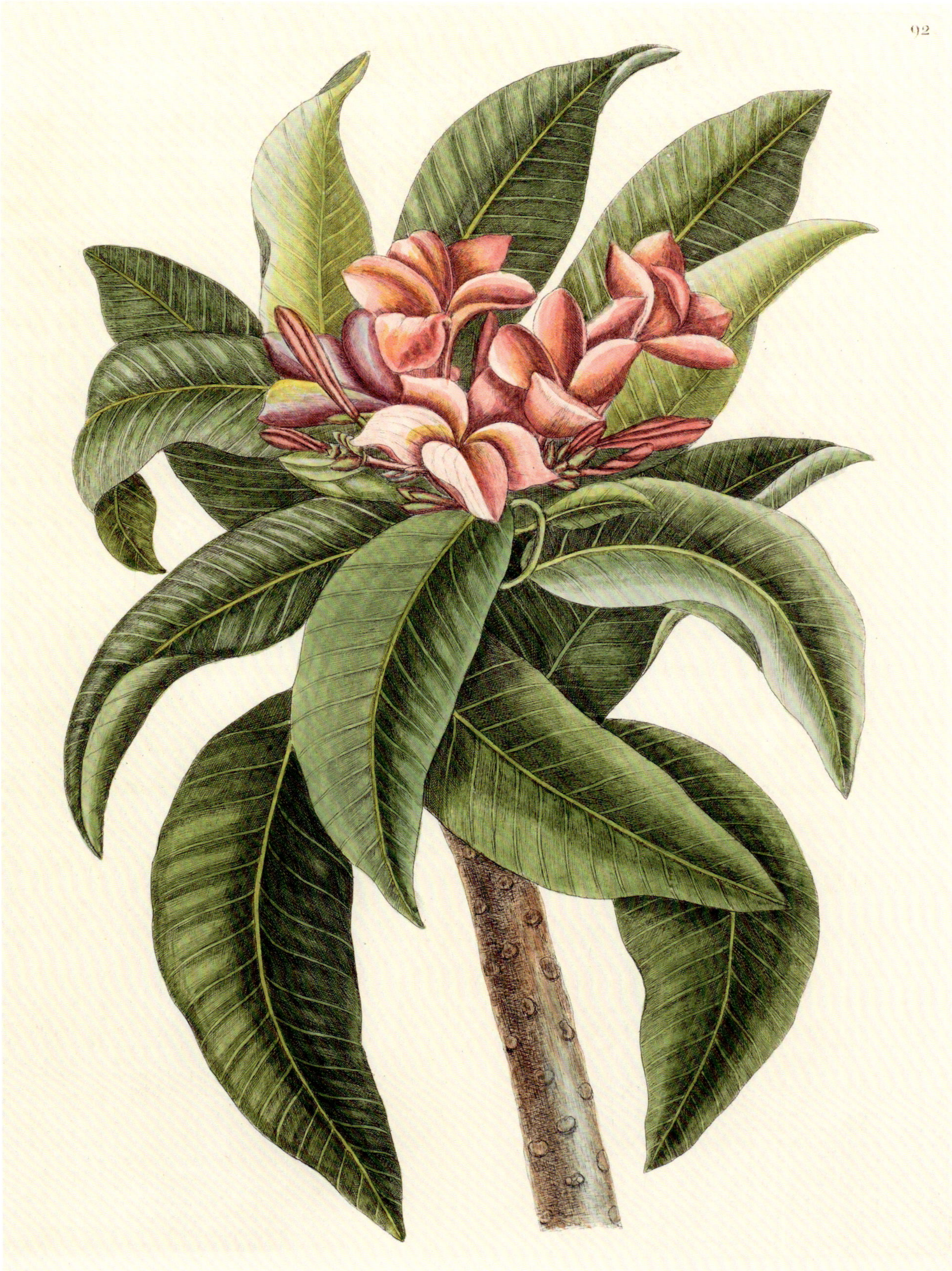

191 *left*

Geiger tree (*Cordia sebestena*); tievine (*Ipomoea carolina*); polyphemus moth (*Antheraea polyphemus*) (vol. 2, t. 91)

192 *above*

Spanish jasmine (*Plumeria rubra*) (vol. 2, t. 92)

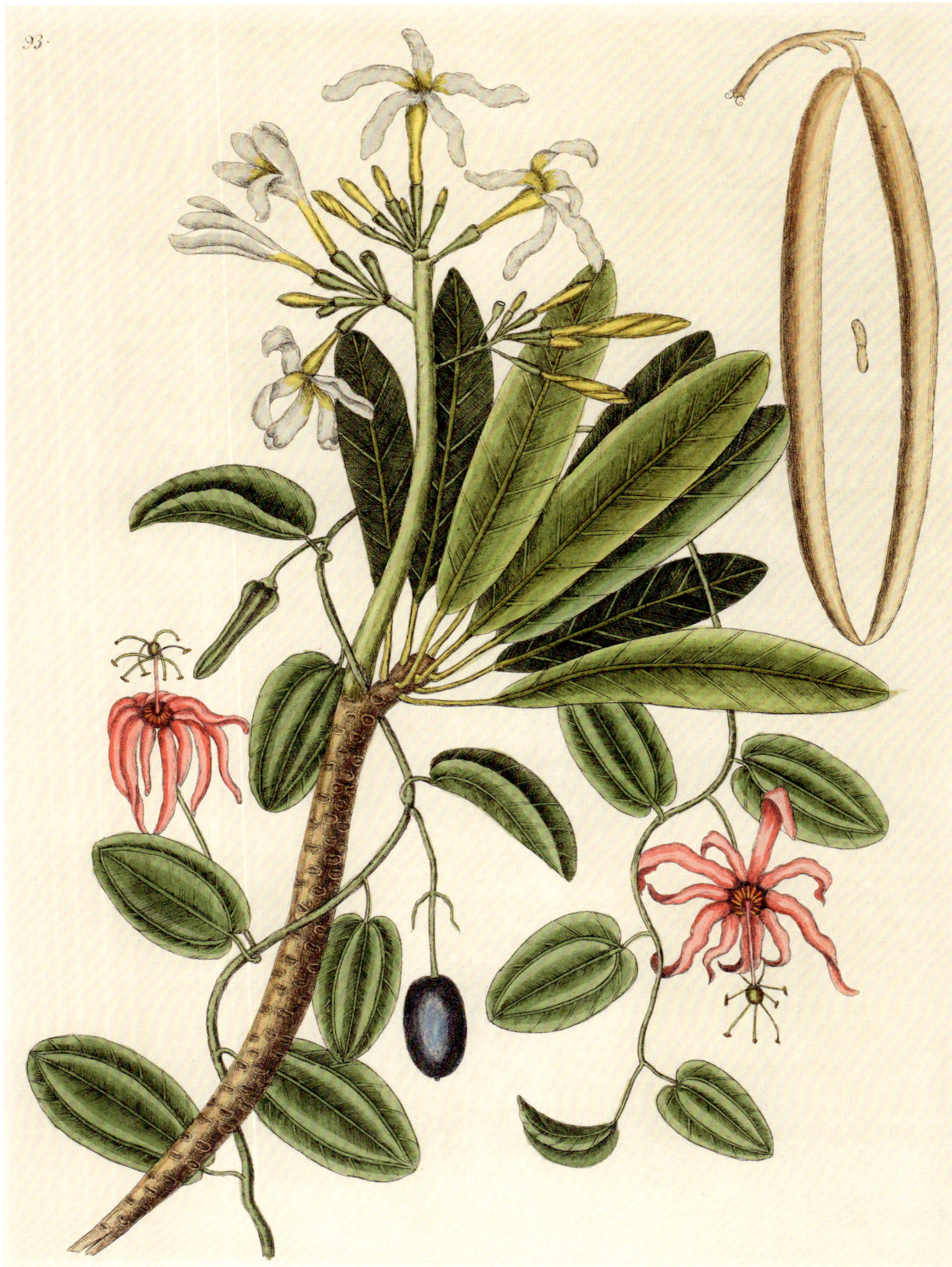

193 *above*
Frangipani (*Plumeria obtusa*); devil's pumpkin (*Passiflora cupraea*) (vol. 2, t. 93)

194 *right*
Pigeon plum (*Coccoloba diversifolia*); horned devil's caterpillar (*Citheronia regalis*) (vol. 2, t. 94)

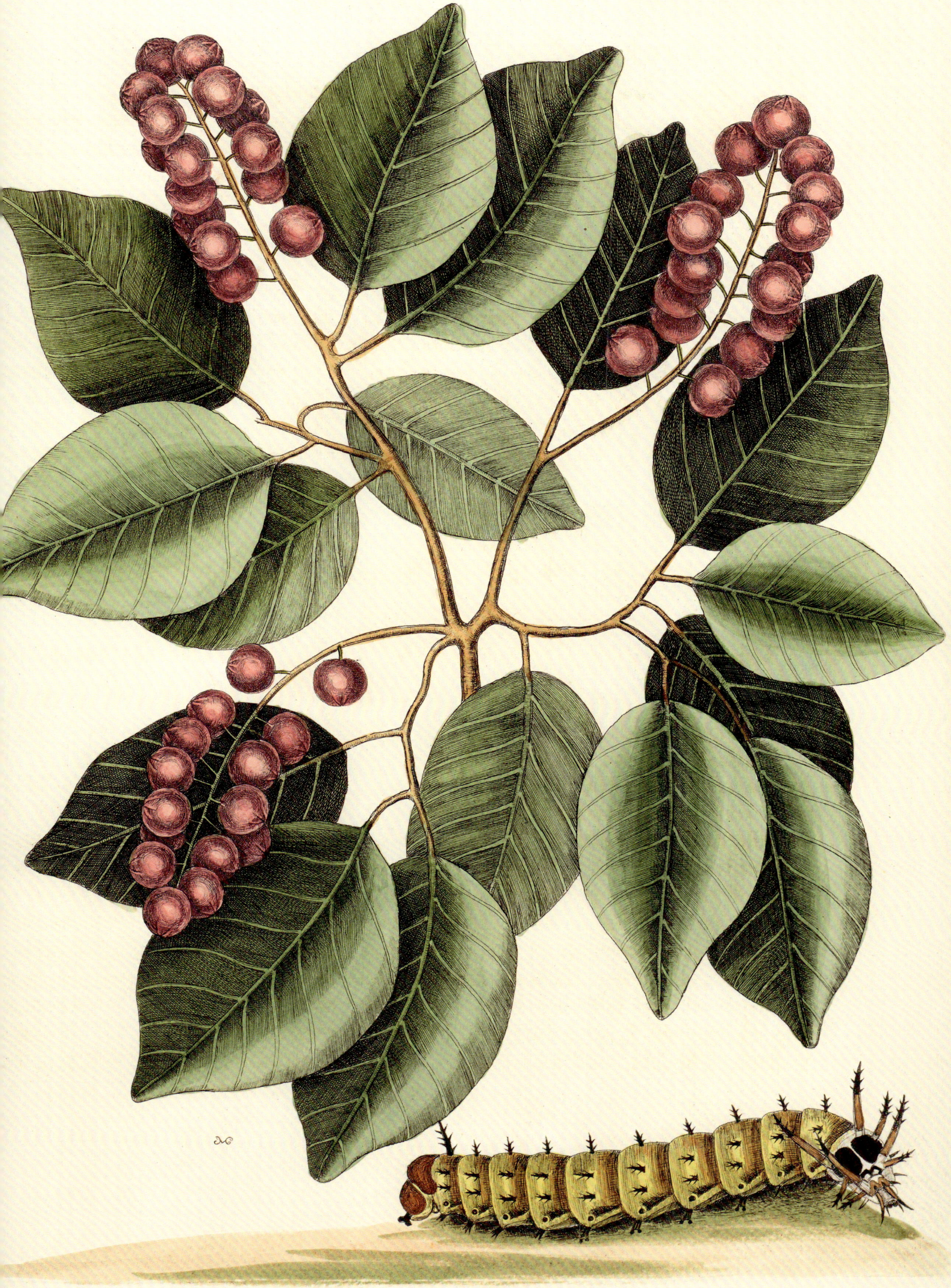

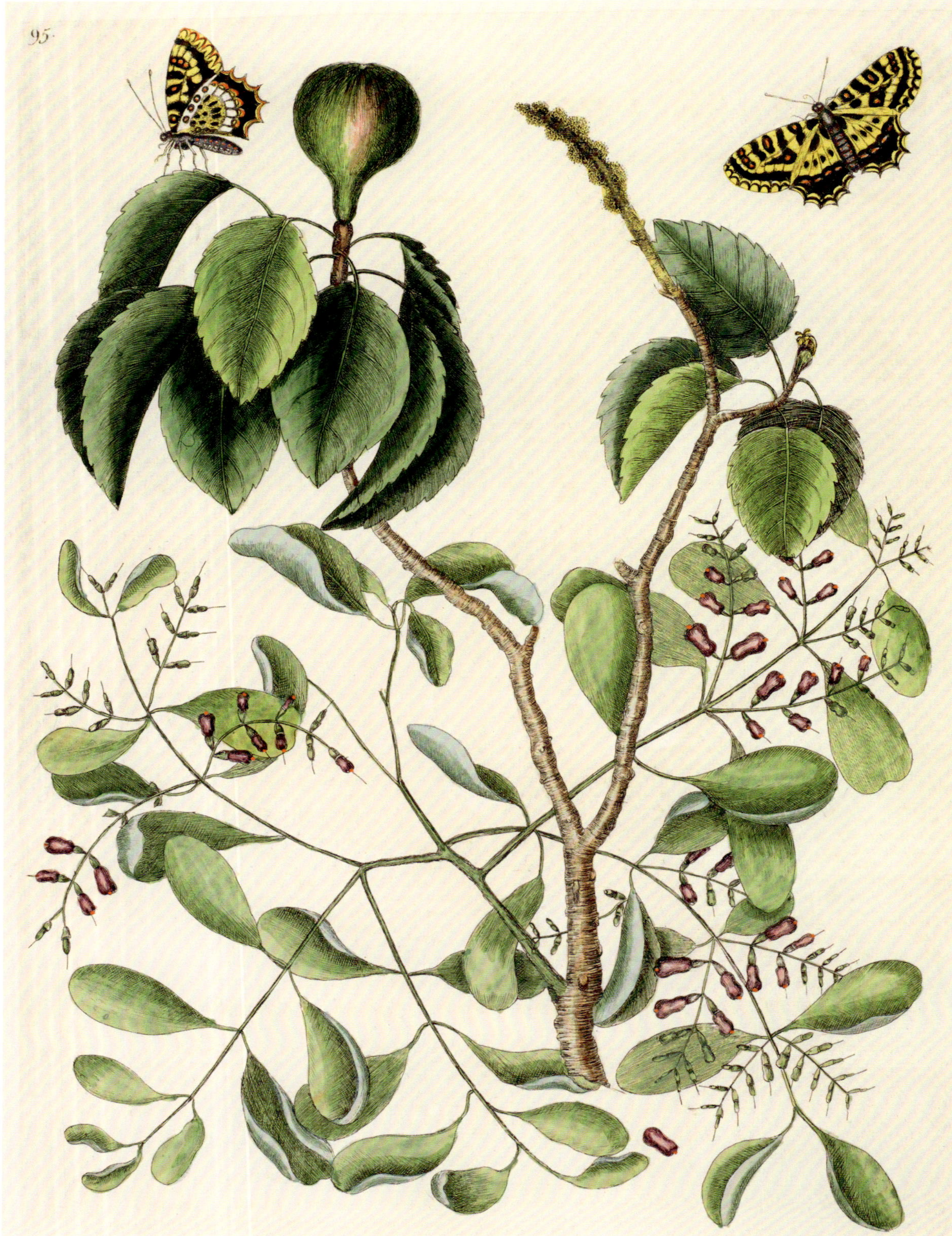

195 *above*
Manchineel (*Hippomane manicinella*; upper); smooth leachbush (*Dendropemon purpureus*; lower) Spanish festoon (*Zerynthia rumina*) (vol. 2, t. 95)

196 *right*
Sea grape (*Coccoloba uvifera*); ornate moth (*Utetheisa ornatrix*) (vol. 2, t. 96)

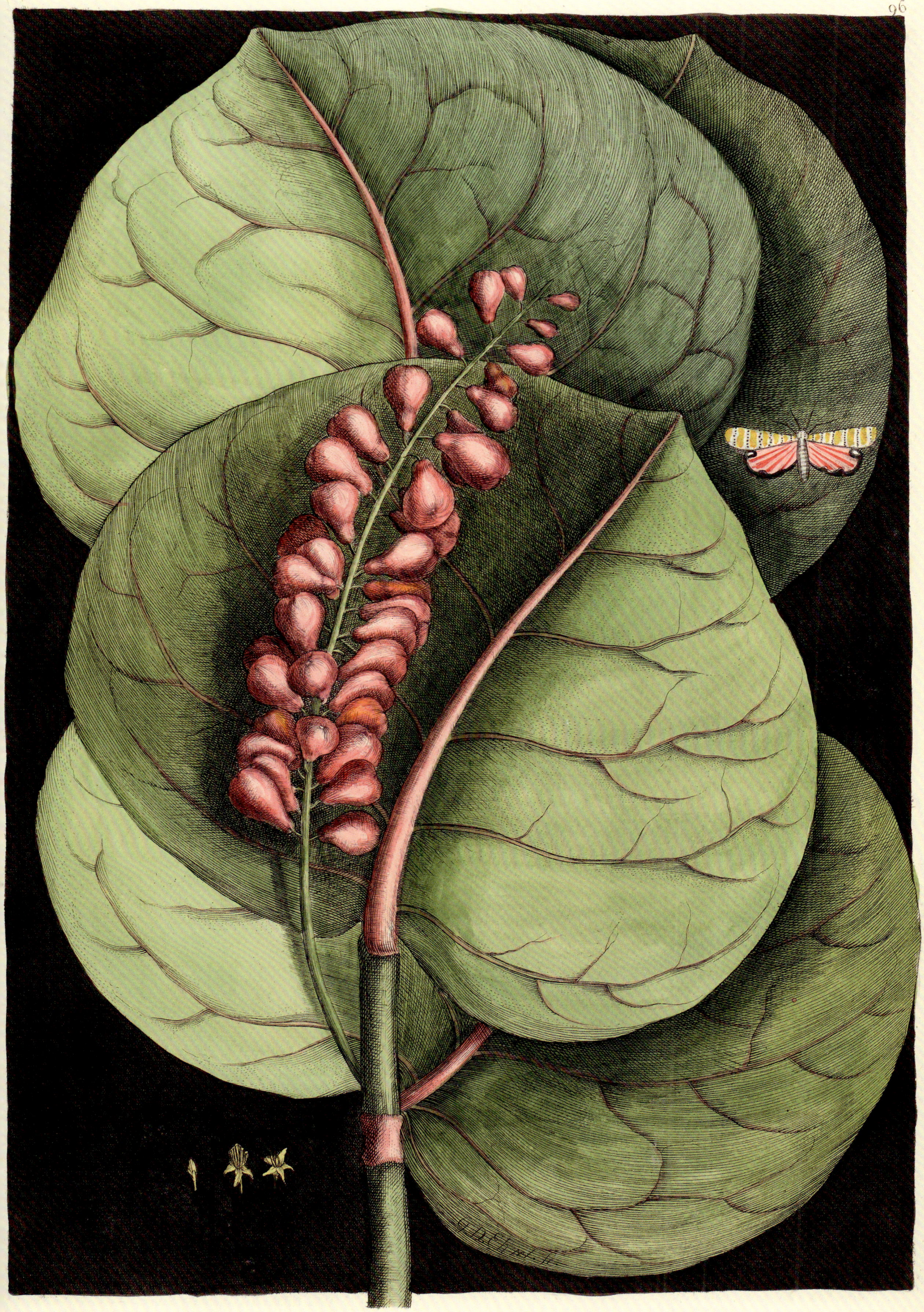
G.D.Ehret.F.

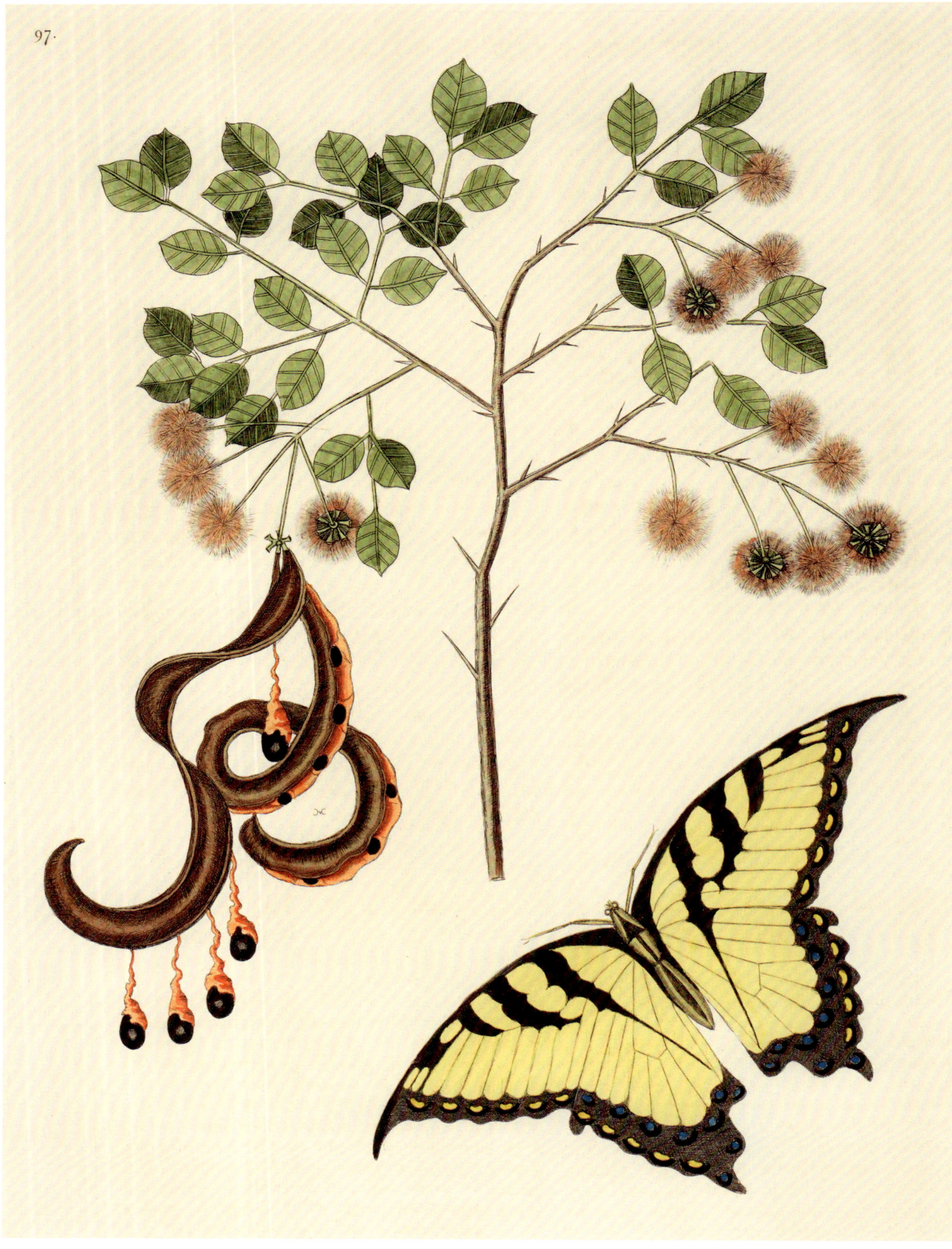

197 *above*
Bahamian cat's claw (*Pithecellobium* x *bahamense*); eastern tiger swallowtail (*Papilio glaucus*) (vol. 2, t. 97)

198 *right*
Lambkill (*Kalmia latifolia*) (vol. 2, t. 98)

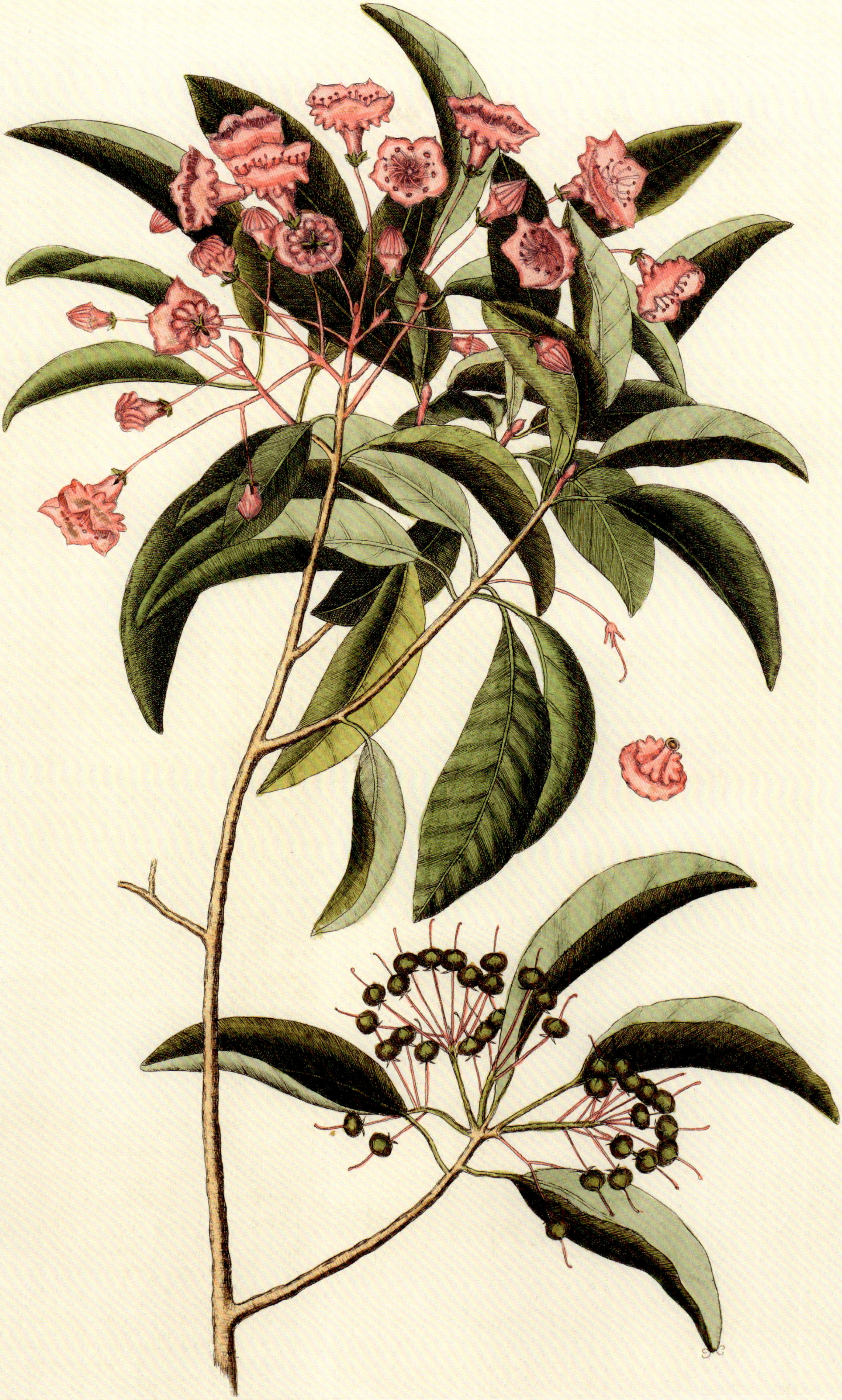

99.

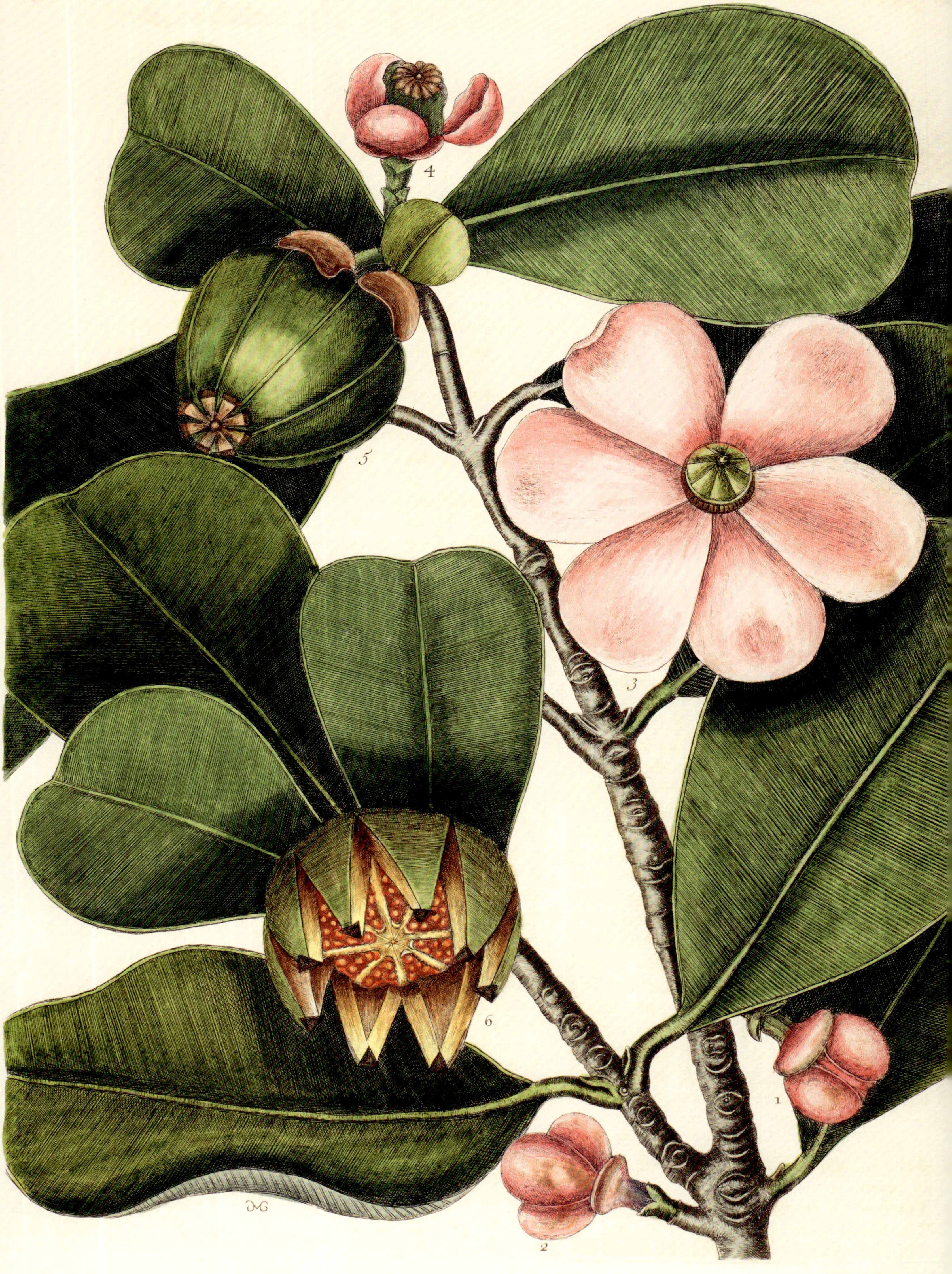

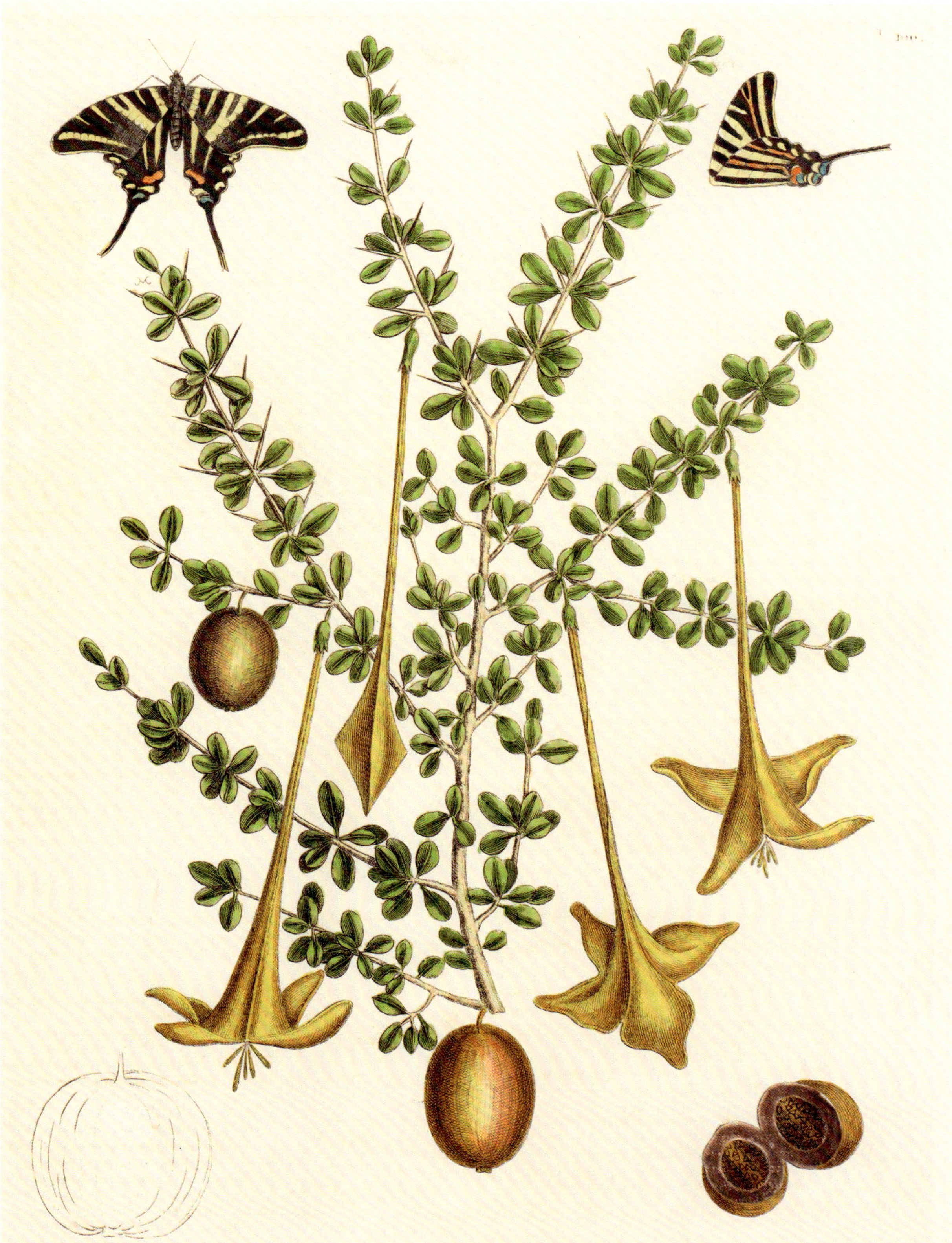

199 *left*
Pitch apple (*Clusia rosea*) (vol. 2, t. 99)

200 *above*
Lily thorn (*Catesbaea spinosa*); zebra swallowtail (*Protographium marcellus*) (vol. 2, t. 100)

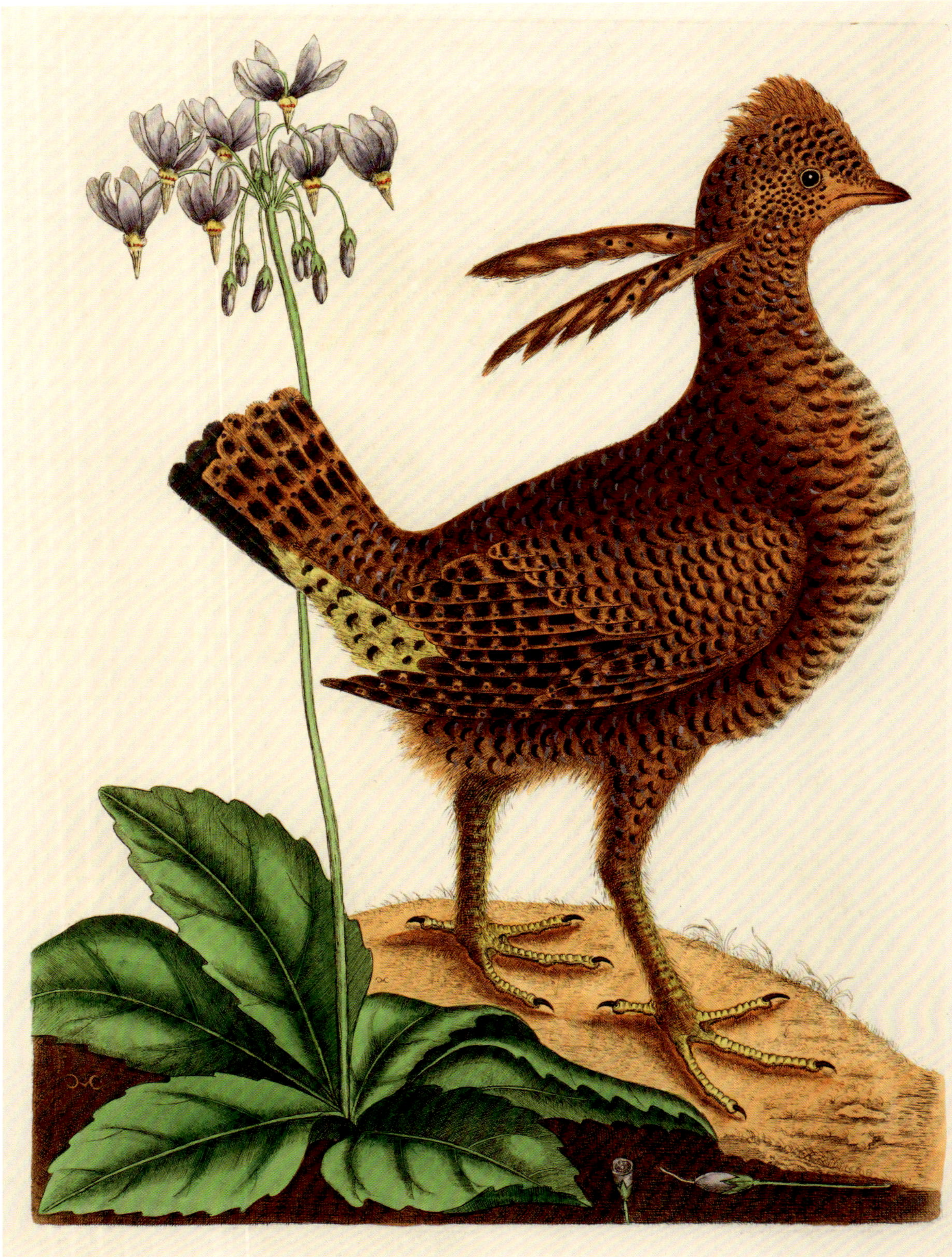

201 *above*
Greater prairie chicken (*Tympanuchus cupido*); shooting star (*Primula meadia*) (appendix, t. 1)

202 *right*
Virginian witch hazel (*Hamamelis virginiana*); probably Florida Keys centipede (*Scolopendra alternans*) (appendix, t. 2)

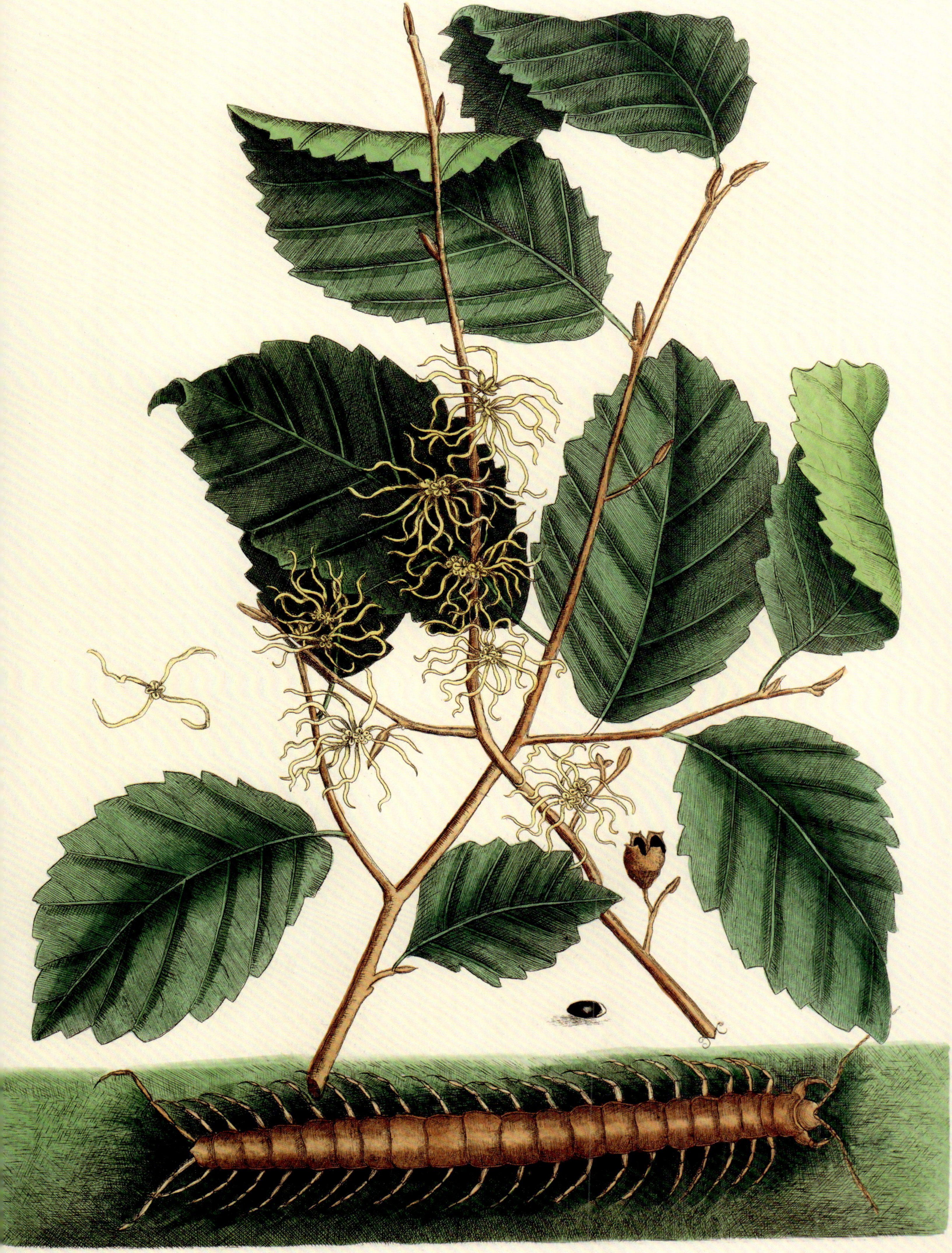

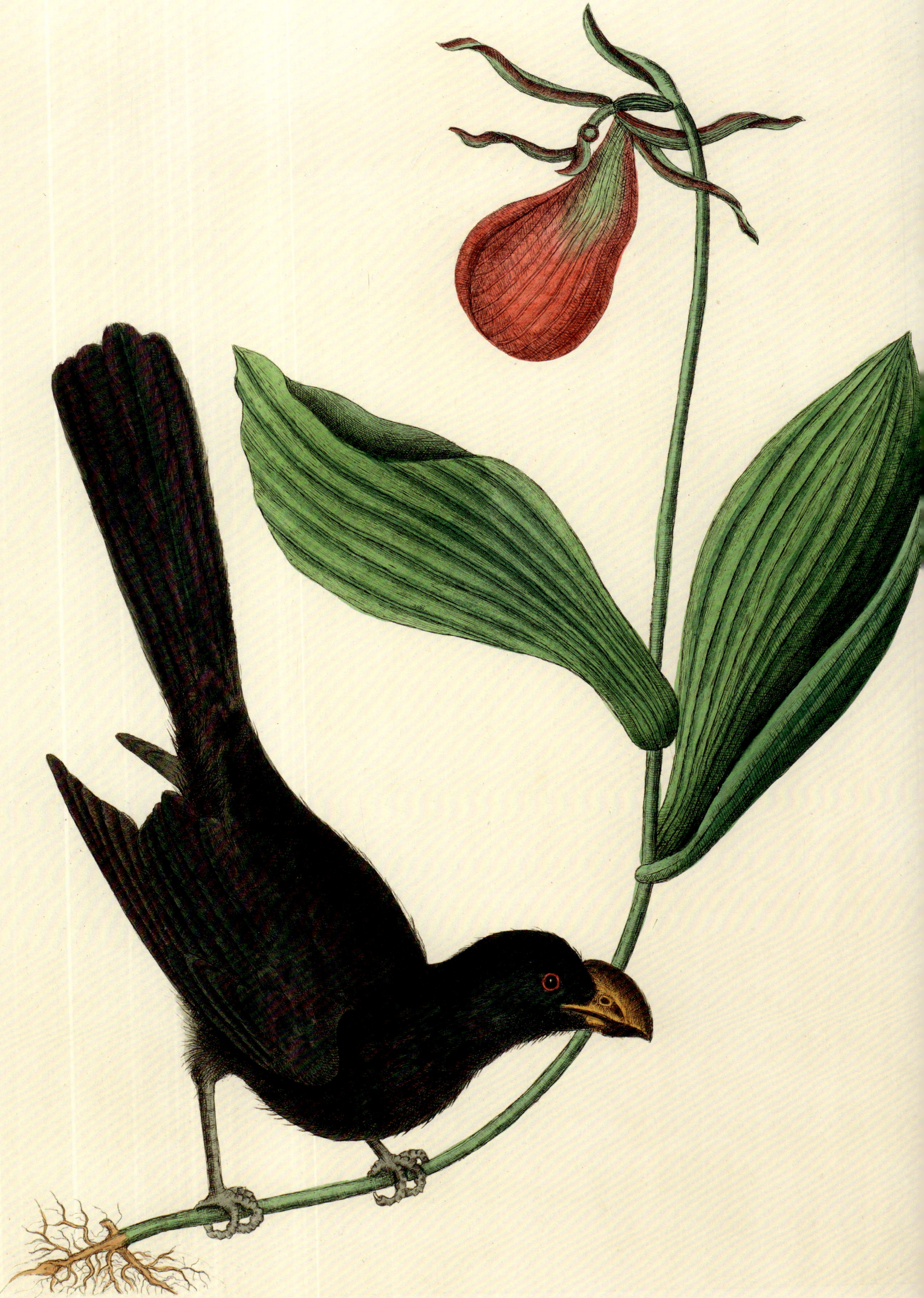

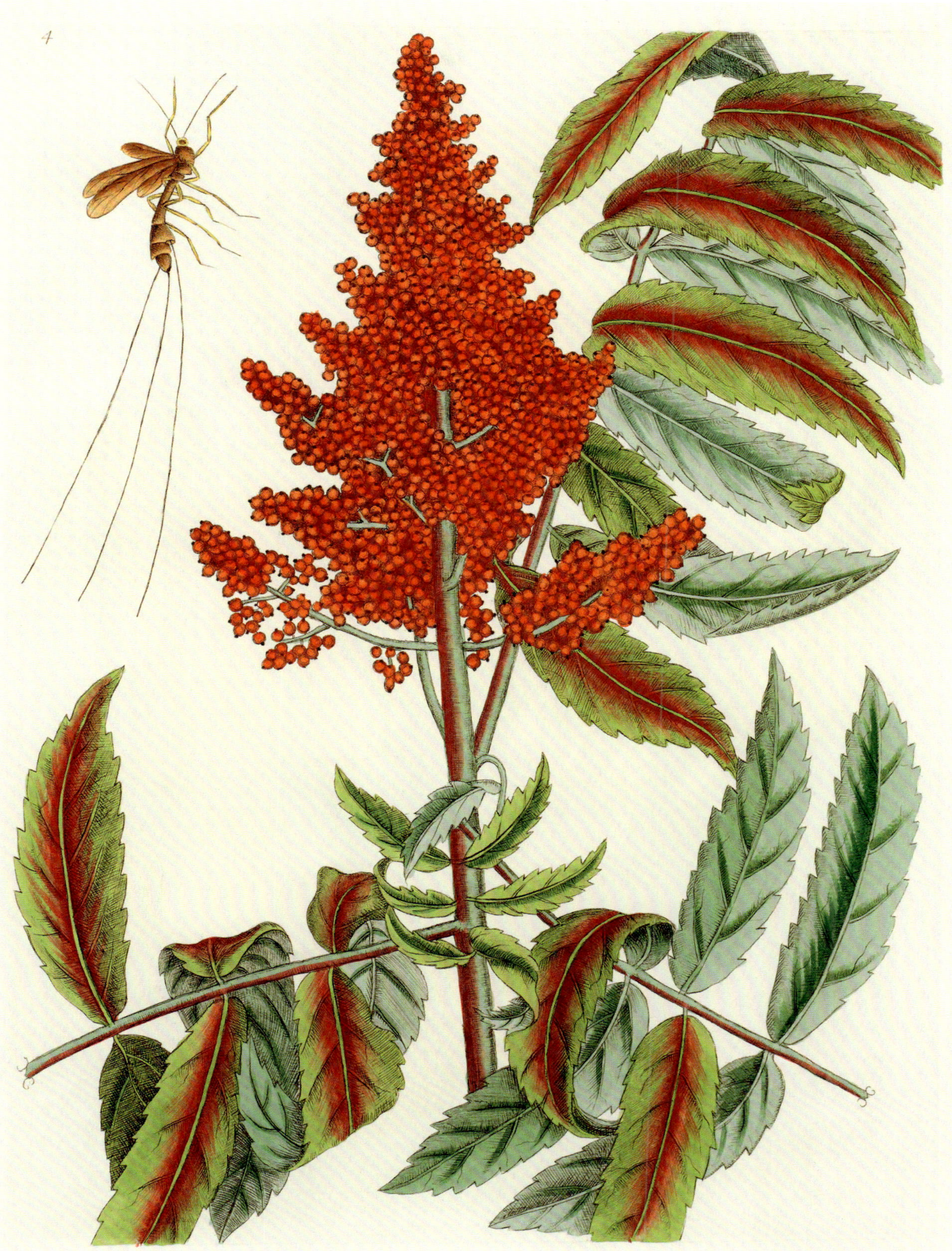

203 *left*
Smooth-billed ani (*Crotophaga ani*); pink lady's slipper (*Cypripedium acaule*) (appendix, t. 3)

204 *above*
Smooth sumac (*Rhus glabra*); giant ichneumon wasp (*Megarhyssa atrata*) (appendix, t. 4)

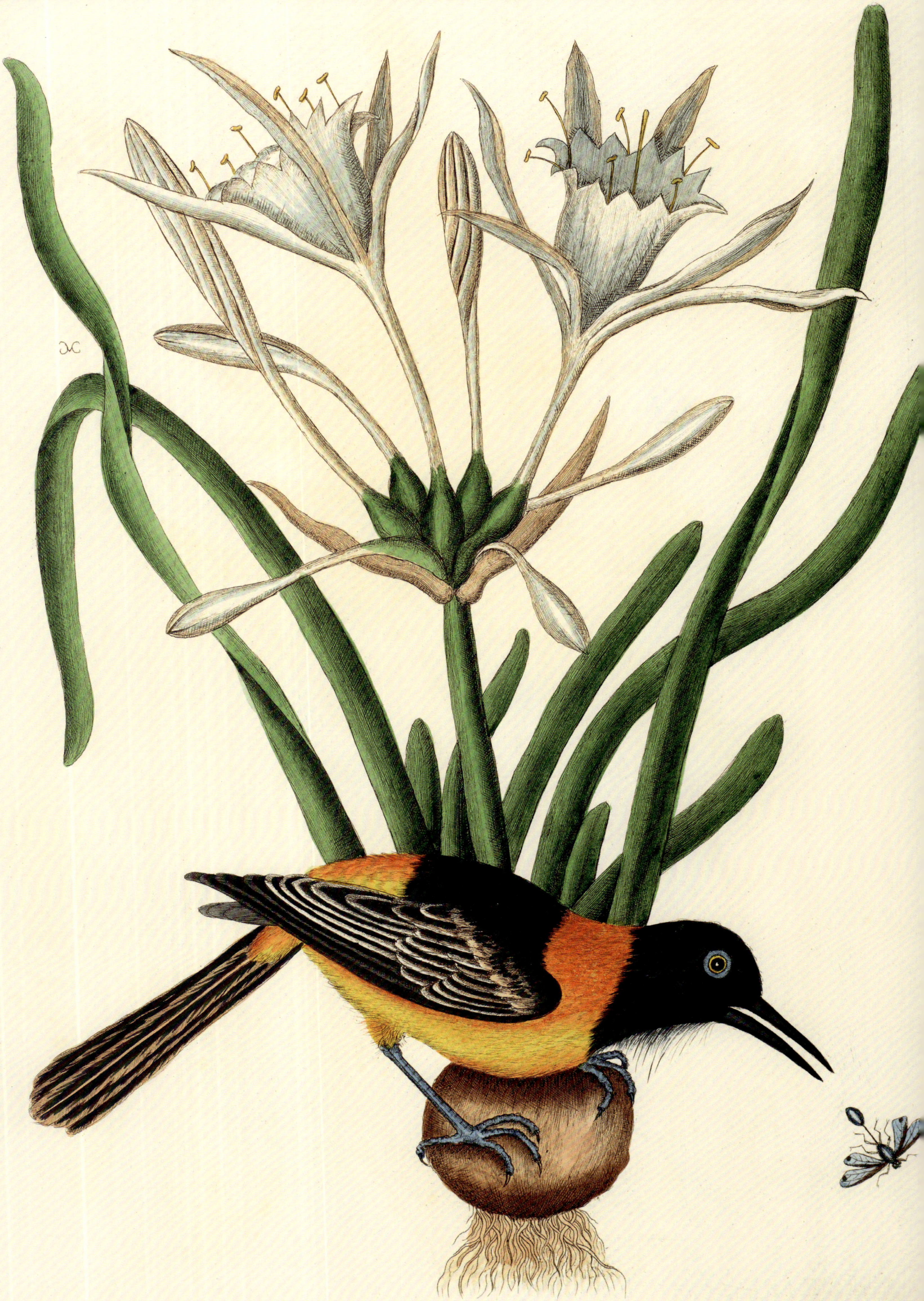
XC

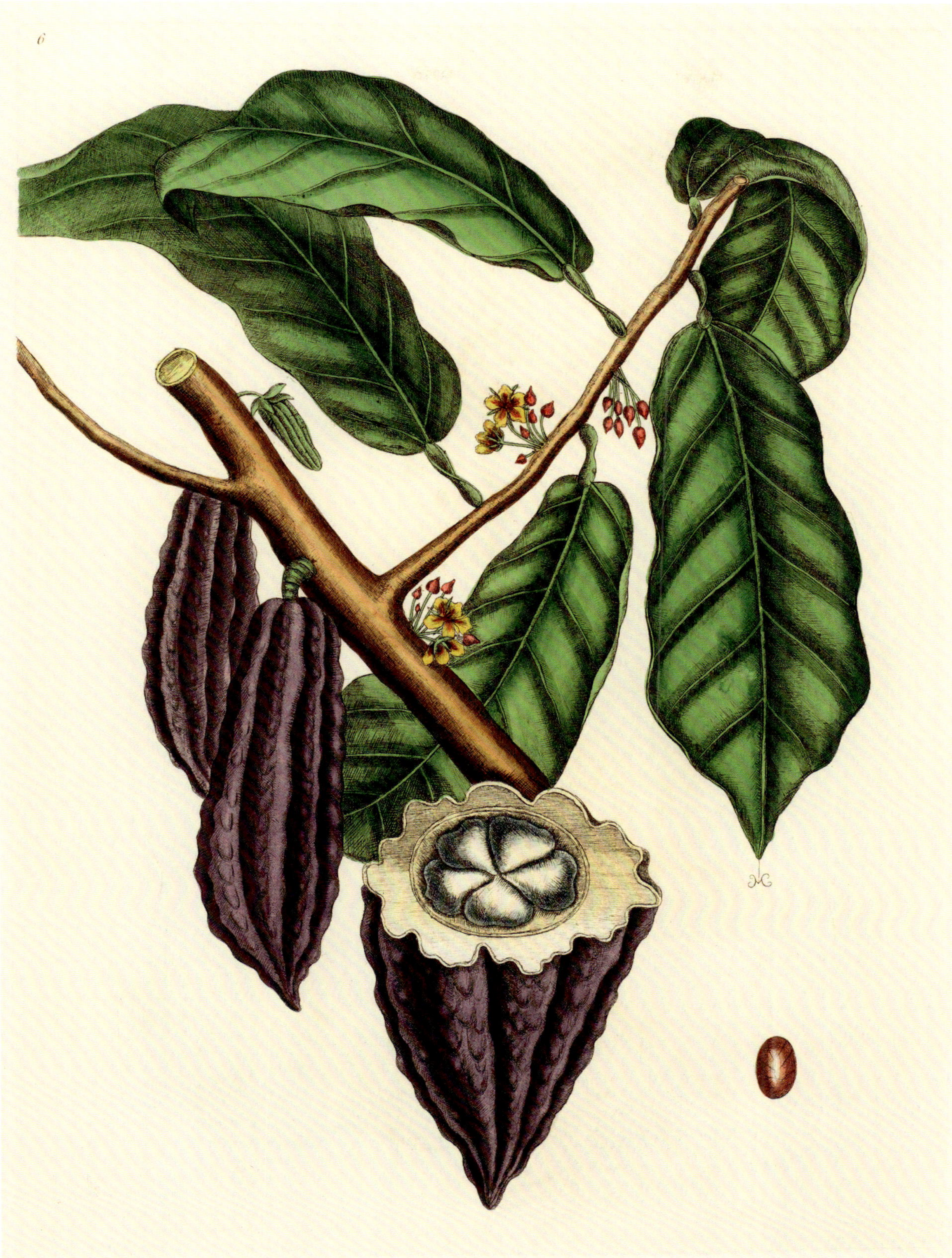

205 *left*
Venezuelan troupial (*Icterus icterus*); blue mud wasp (*Chalybion californicum*); sea daffodil (*Pancratium maritimum*) (appendix, t. 5)

206 *above*
Cacao (*Theobroma cacao*) (appendix, t. 6)

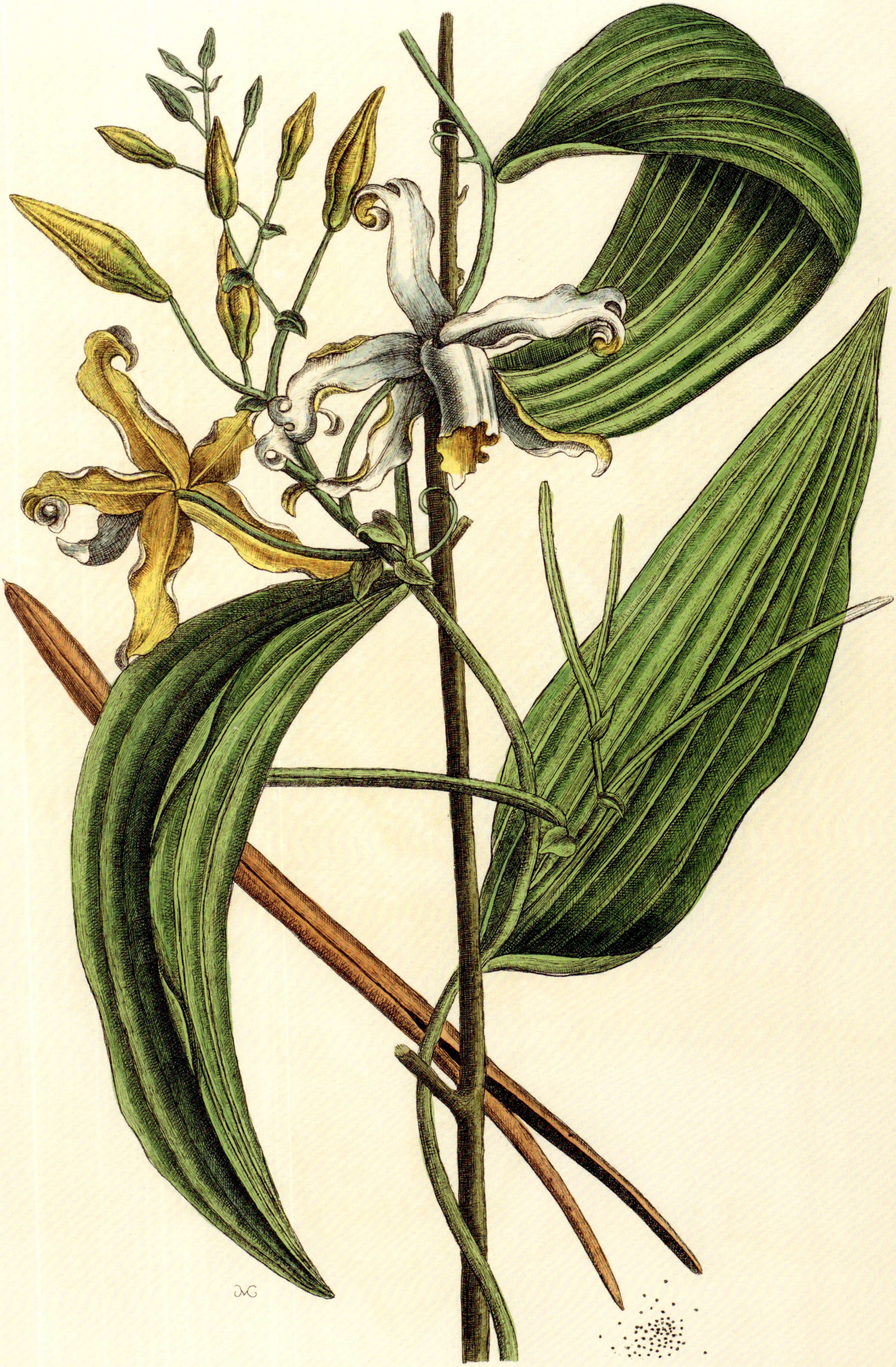

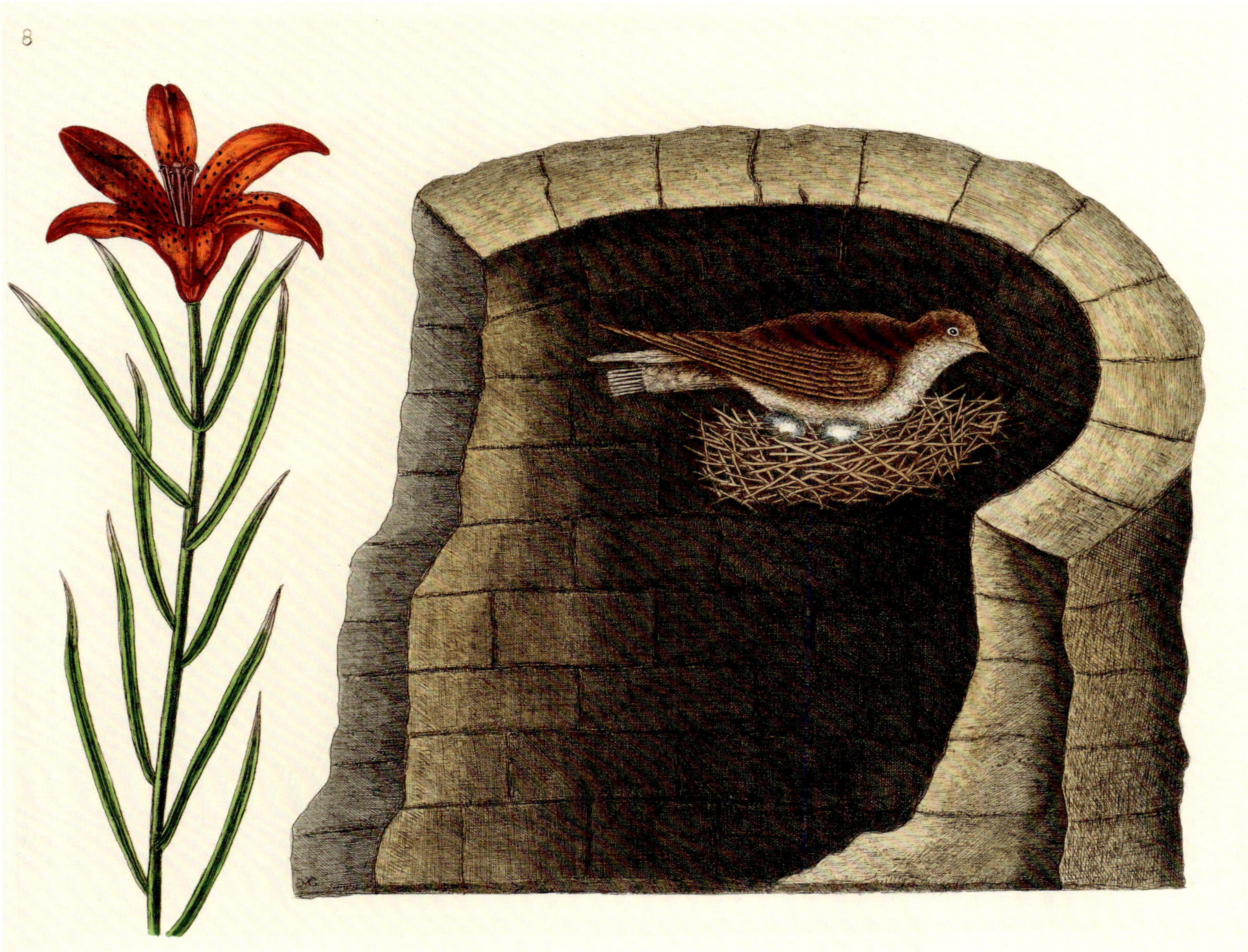

207 *left*
Probably Mexican vanilla (*Vanilla mexicana*) (appendix, t. 7)

208 *above*
Chimney swift (*Chaetura pelagica*); wood lily (*Lilium philadelphicum*) (appendix, t. 8)

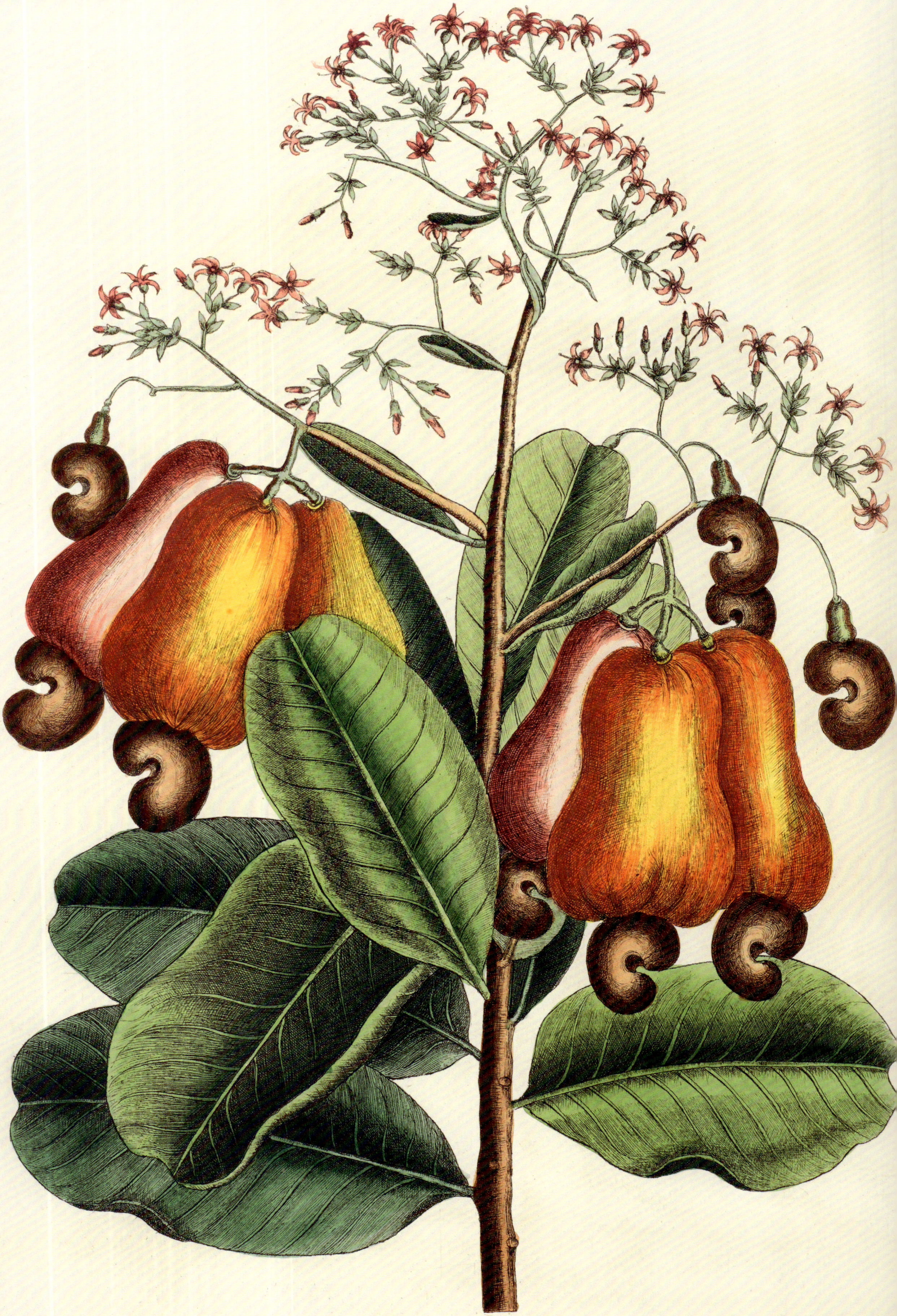

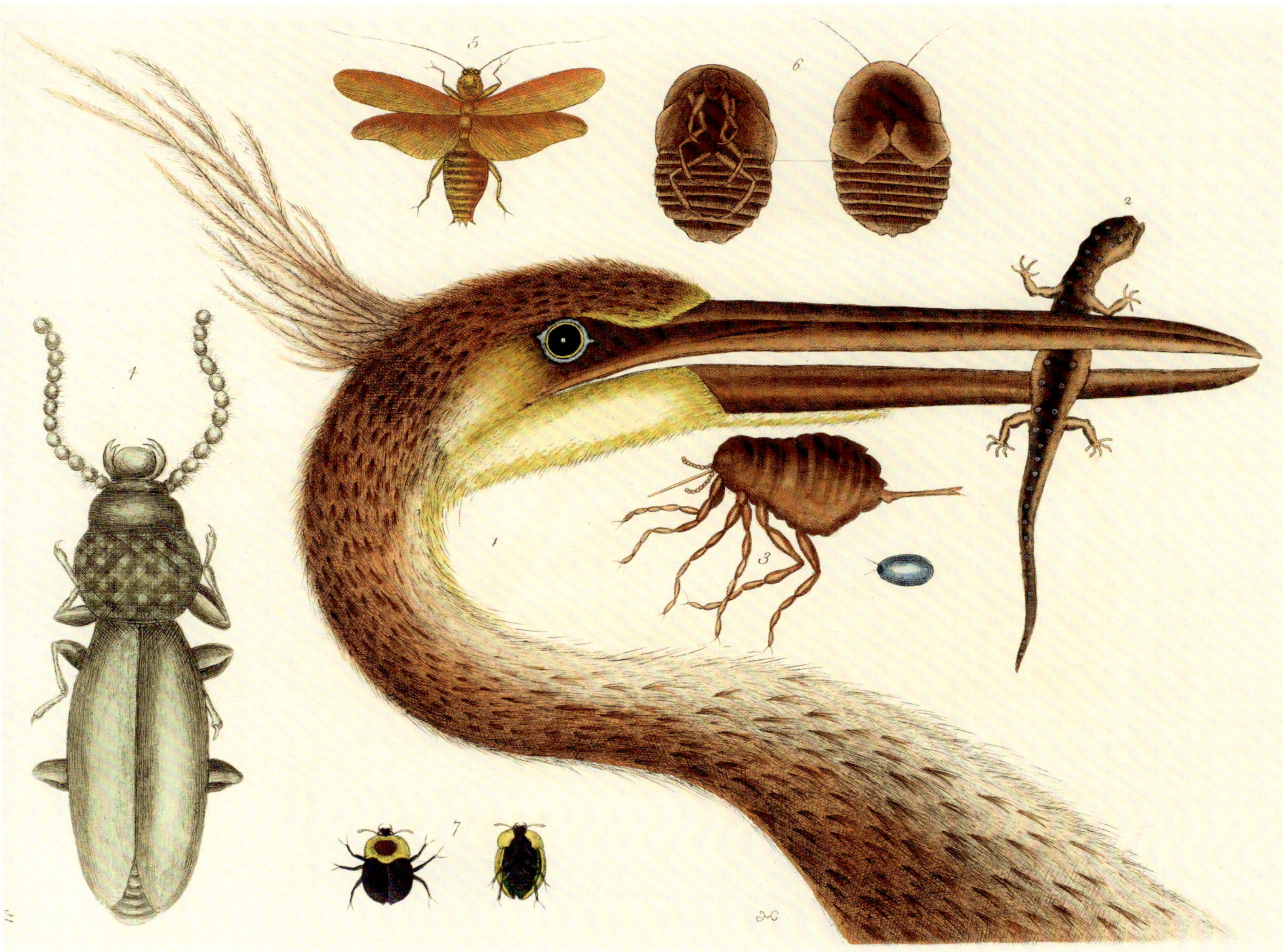

209 *left*
Cashew (*Anacardium occidentale*) (appendix, t. 9)

210 *above*
Great blue heron (*Ardea herodias*; 1); spotted salamander (*Ambystoma maculatum*; 2); chigoe flea (*Tunga penetrans*; 3); beetle, perhaps family Carabidae (4); American cockroach (*Periplaneta americana*; 5); dubia cockroach (*Blaptica dubia*; 6); American carrion beetle (*Necrophila americana*; 7) (appendix, t. 10)

II

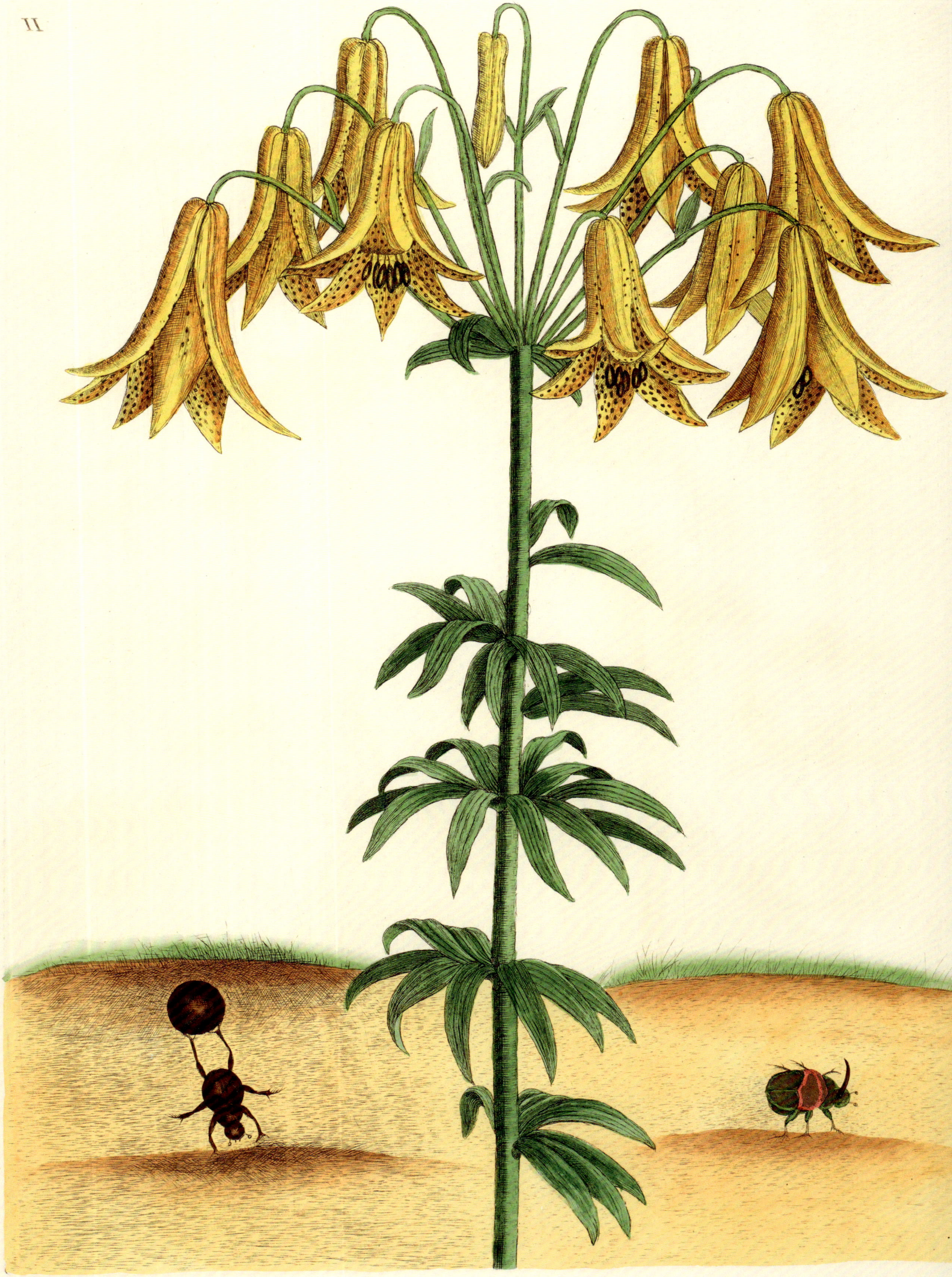

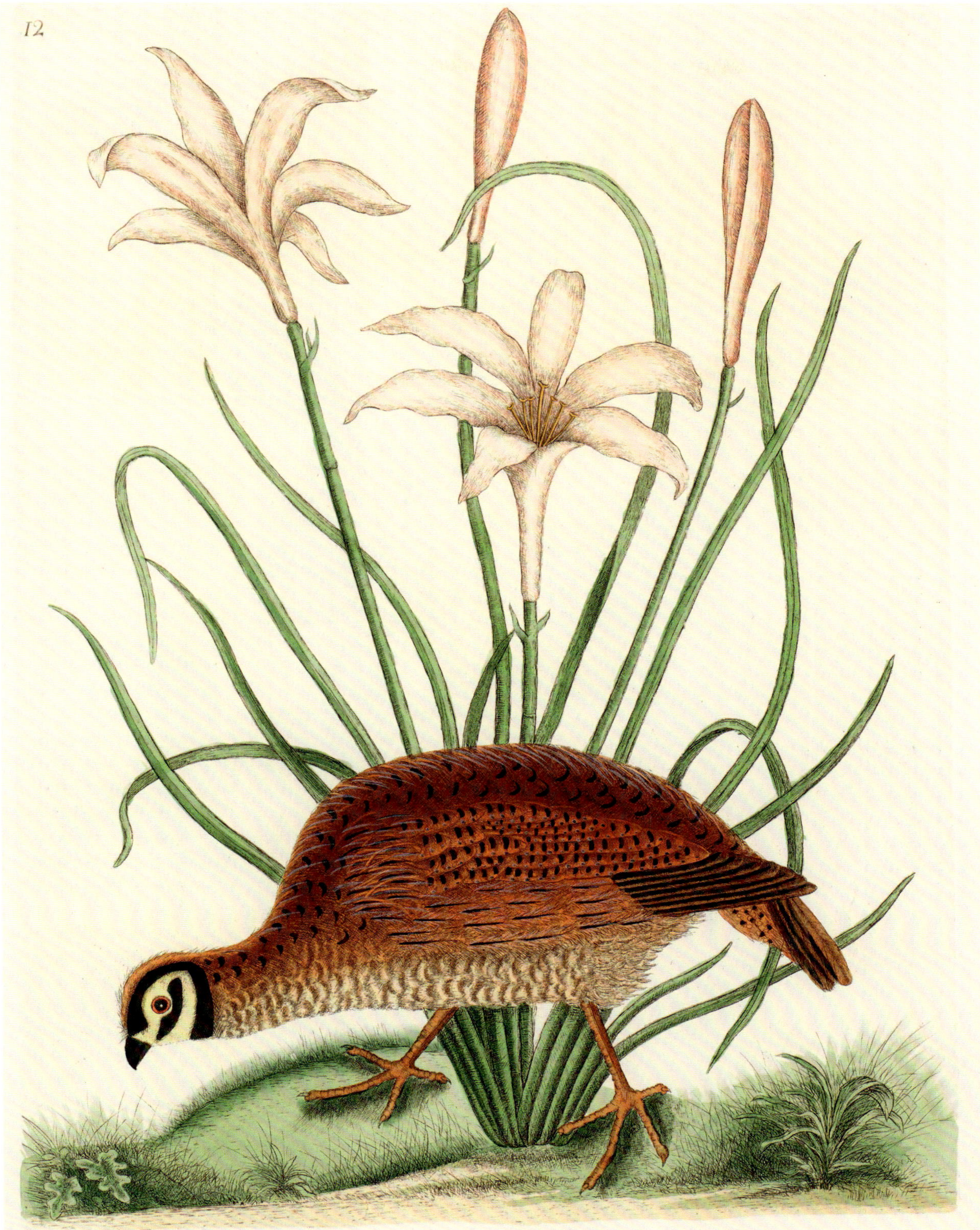

211 *left*
Canadian lily (*Lilium canadense*); dung beetle (*Canthon pilularius*, left); rainbow scarab beetle (*Phanaeus vindex*; right) (appendix, t. 11)

212 *above*
Northern bobwhite (*Colinus virginianus*); Atamasco lily (*Zaphyranthes atamasca*) (appendix, t. 12)

13

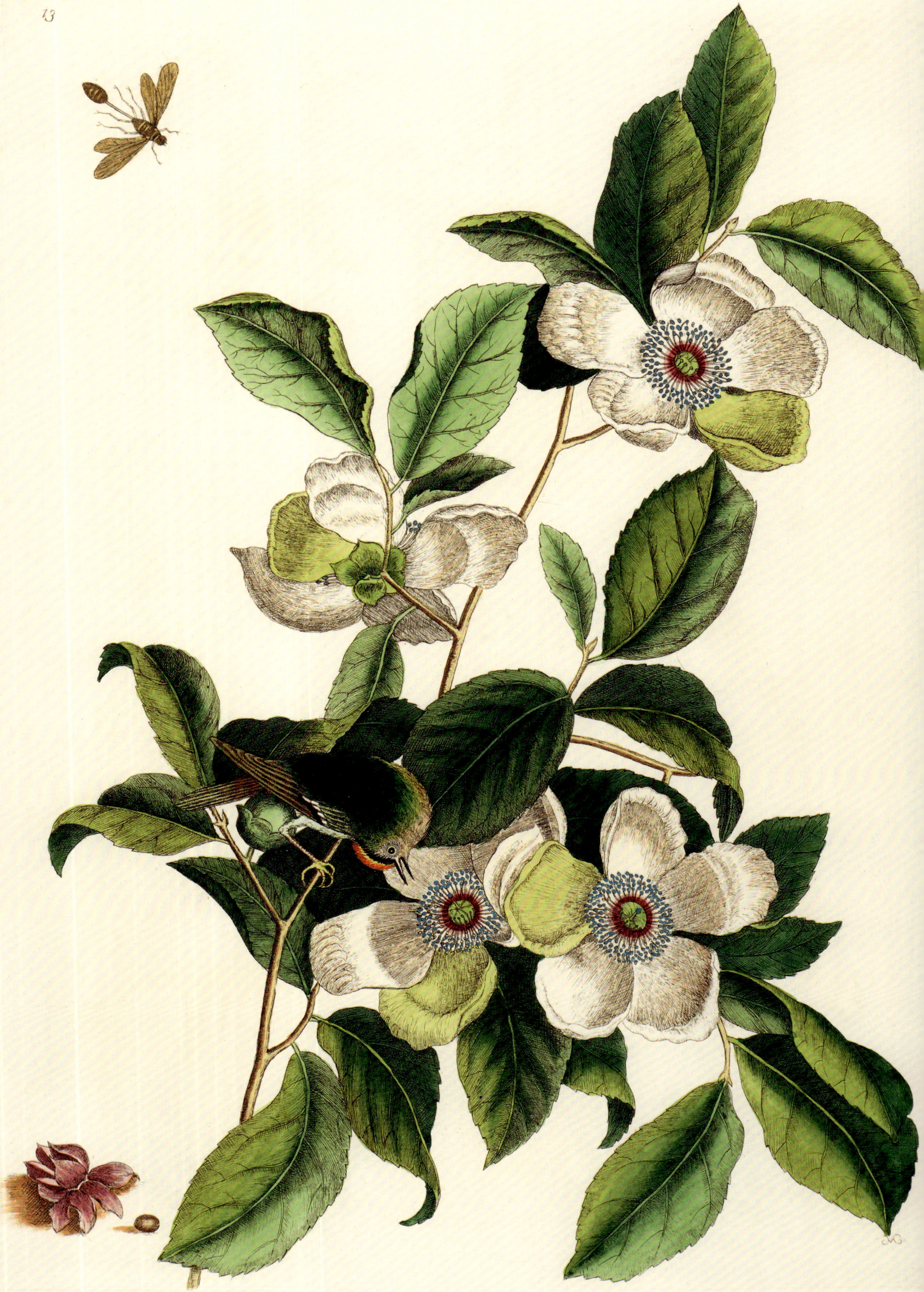

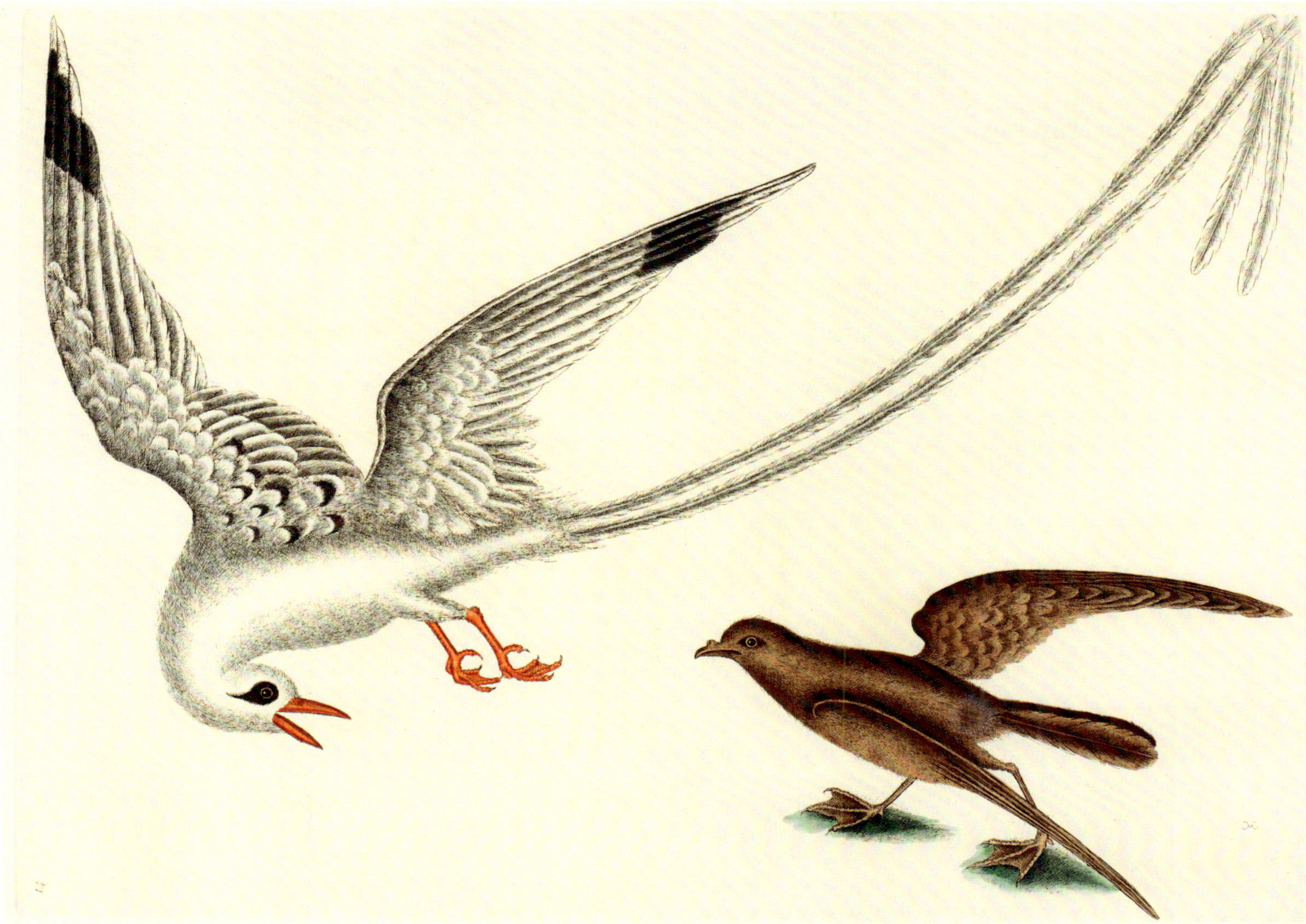

213 *left*
Ruby-crown kinglet (*Corthylio calendula*); black-and-yellow mud dauber (*Sceliphron caementarium*); silky camellia (*Stewartia malacodendron*); Carolina allspice (*Calycanthus floridus*; flower and seed, lower left) (appendix, t. 13)

214 *above*
Red-billed tropicbird (*Phaethon aethereus*; upper); European storm petrel (*Hydrobates pelagicus*; lower) (appendix, t. 14)

15

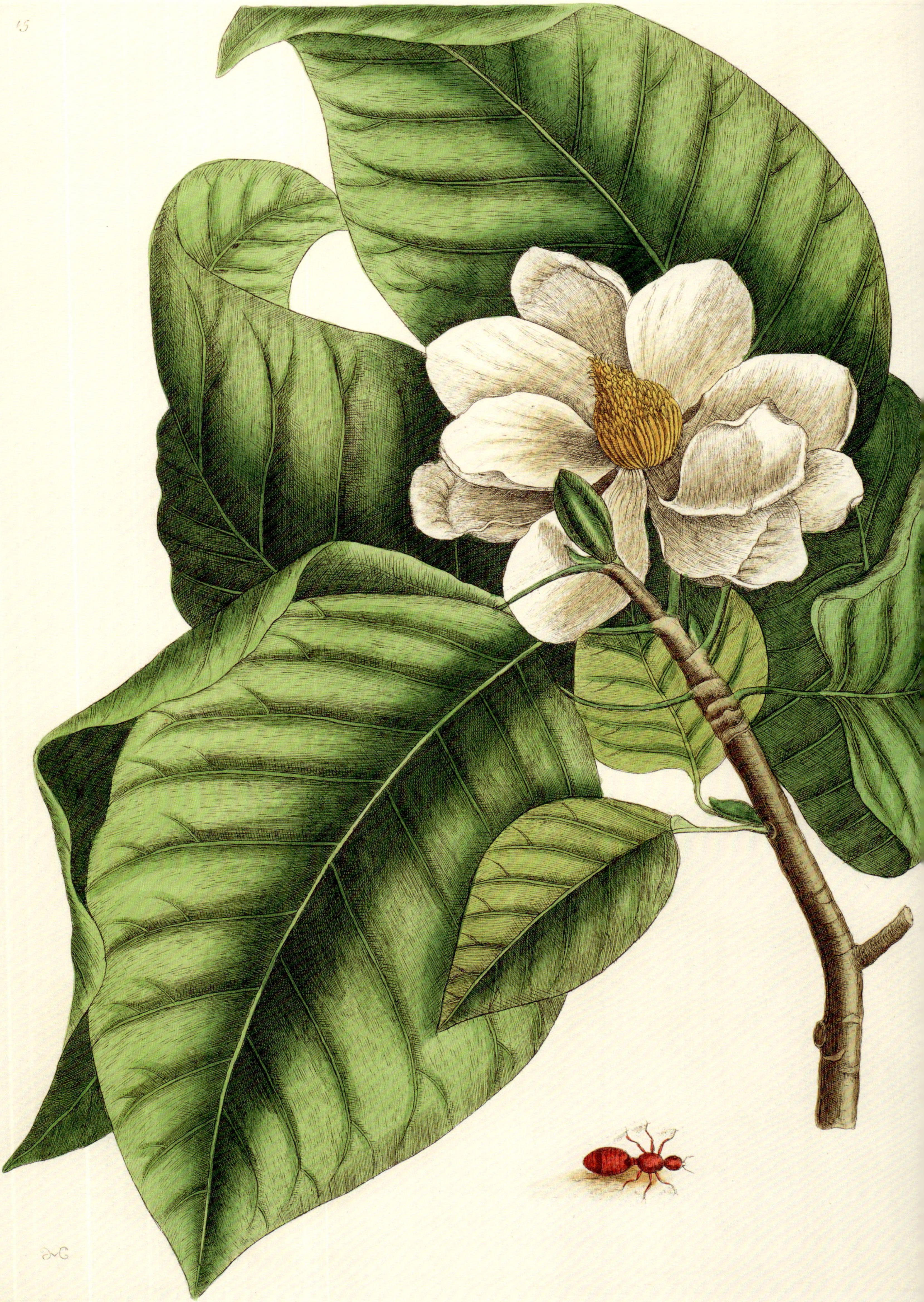

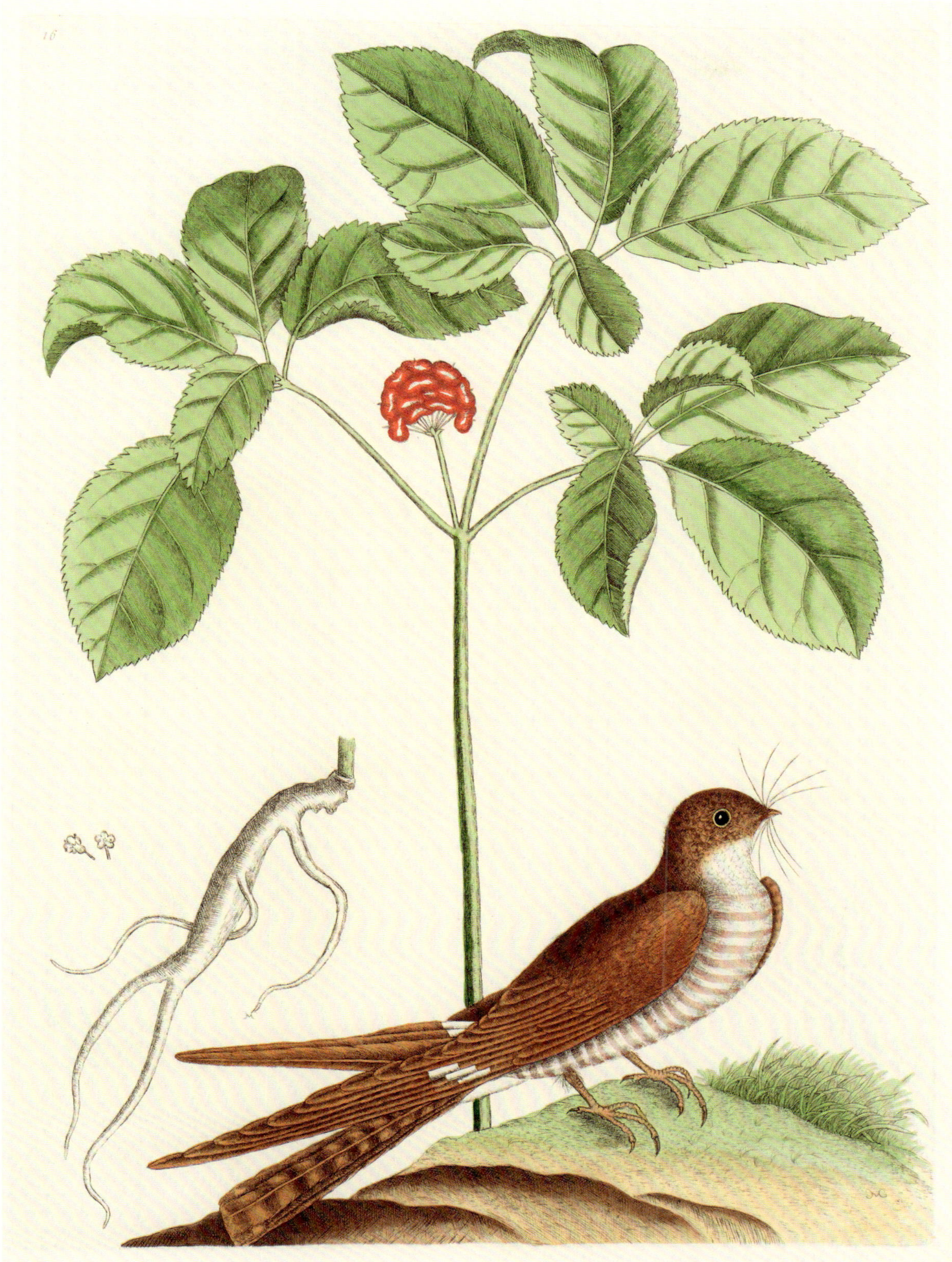

215 *left*

Sweetbay (*Magnolia virginiana*; flower); cucumber tree (*Magnolia acuminata*; leaves); velvet ant (*Dasymutilla occidentalis*) (appendix, t. 15)

216 *above*

Common nighthawk (*Chordeiles minor*); American ginseng (*Panax quinquefolius*) (appendix, t. 16)

17

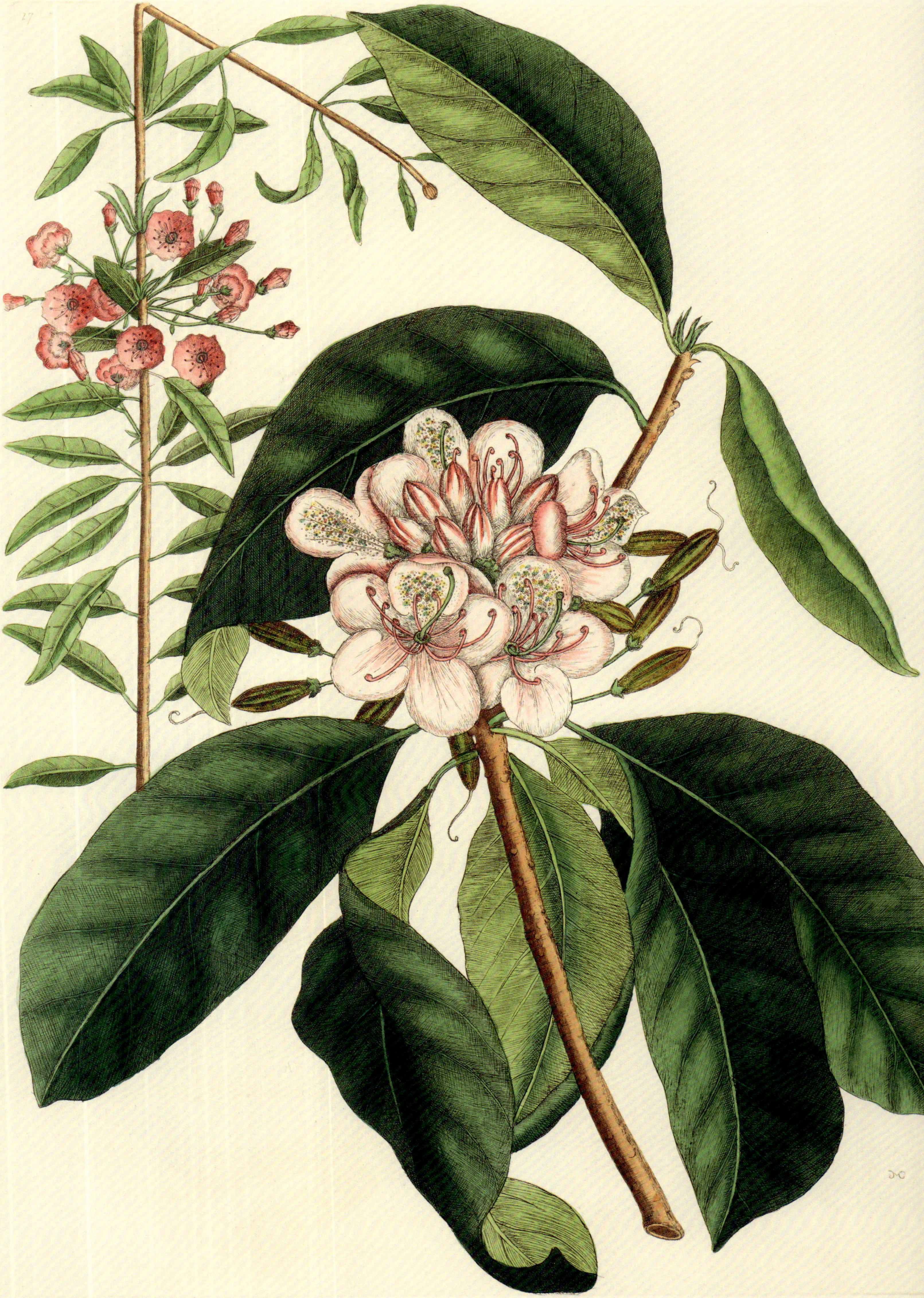

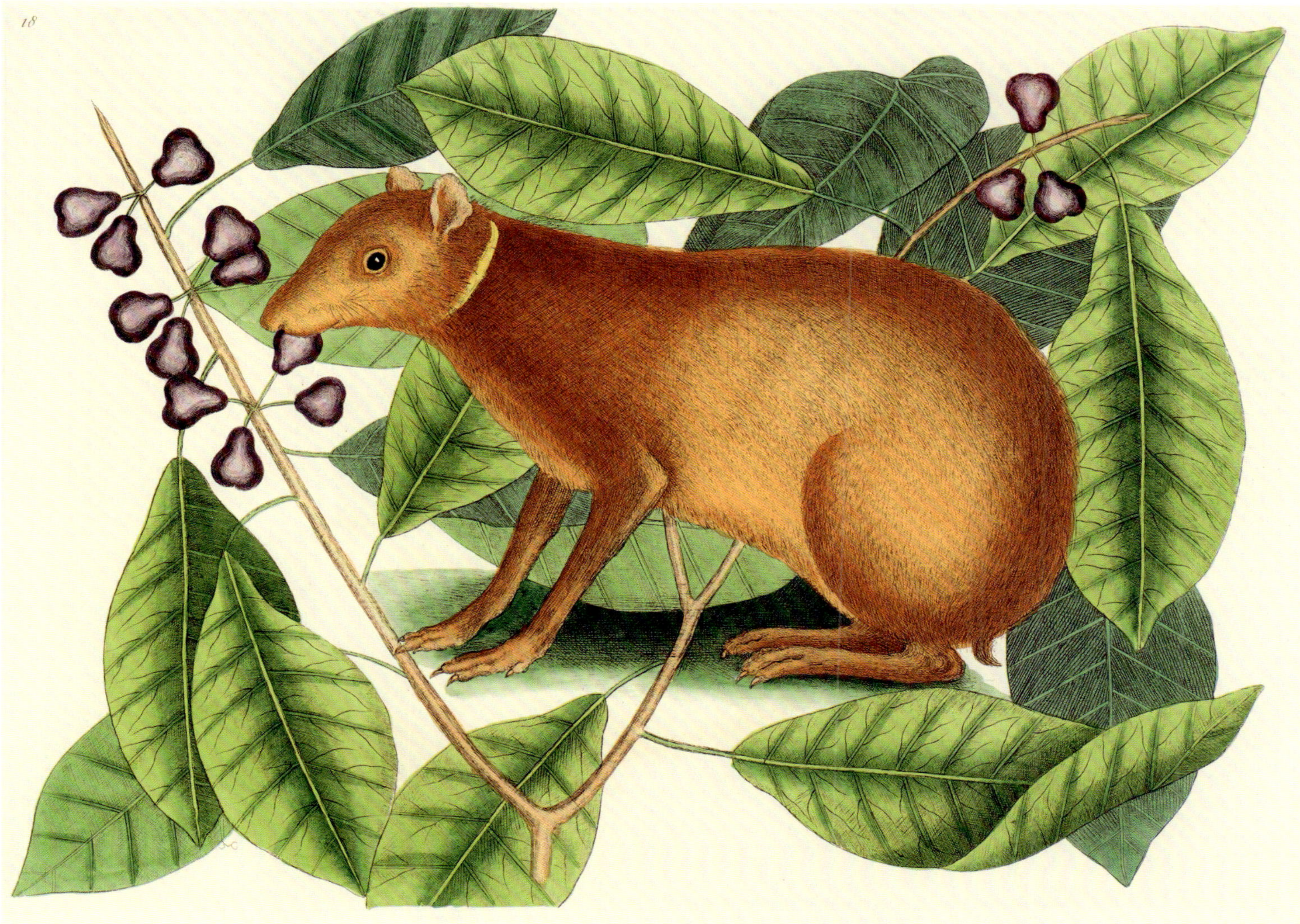

217 *left*
American rhododendron (*Rhododendron maximum*; right); sheep laurel (*Kalmia angustifolia*; left)

218 *above*
Brazilian agouti (*Dasyprocta leporina*); shortleaf fig (*Ficus citrifolia*) (appendix, t. 18)

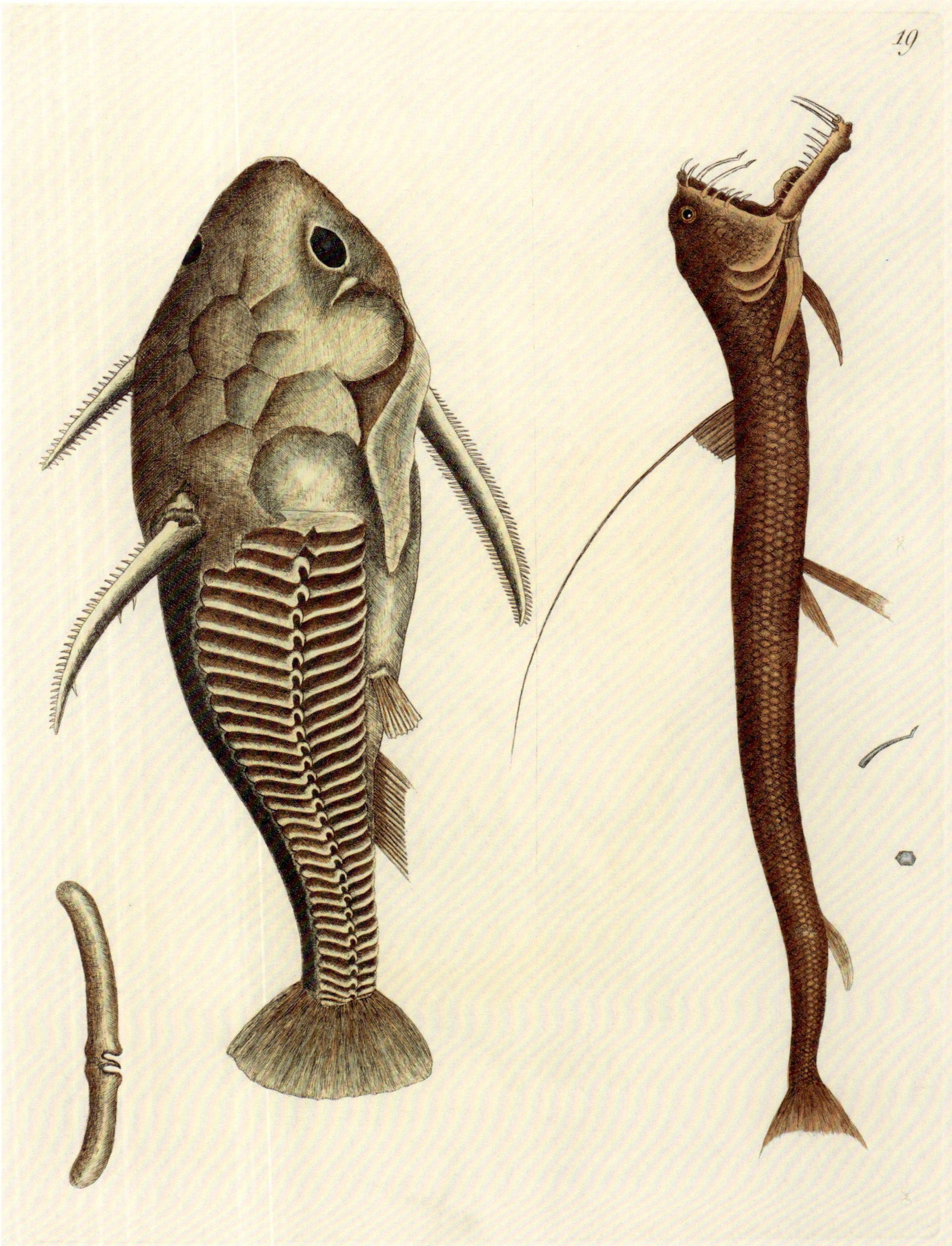

219 *above*
Spiny catfish (*Acanthodoras cataphractus*; left); viperfish (*Chauliodus sloani*; right) (appendix, t. 20)

220 *right*
American bison (*Bison bison*); rose locust (*Robinia hispida*) (appendix, t. 20)

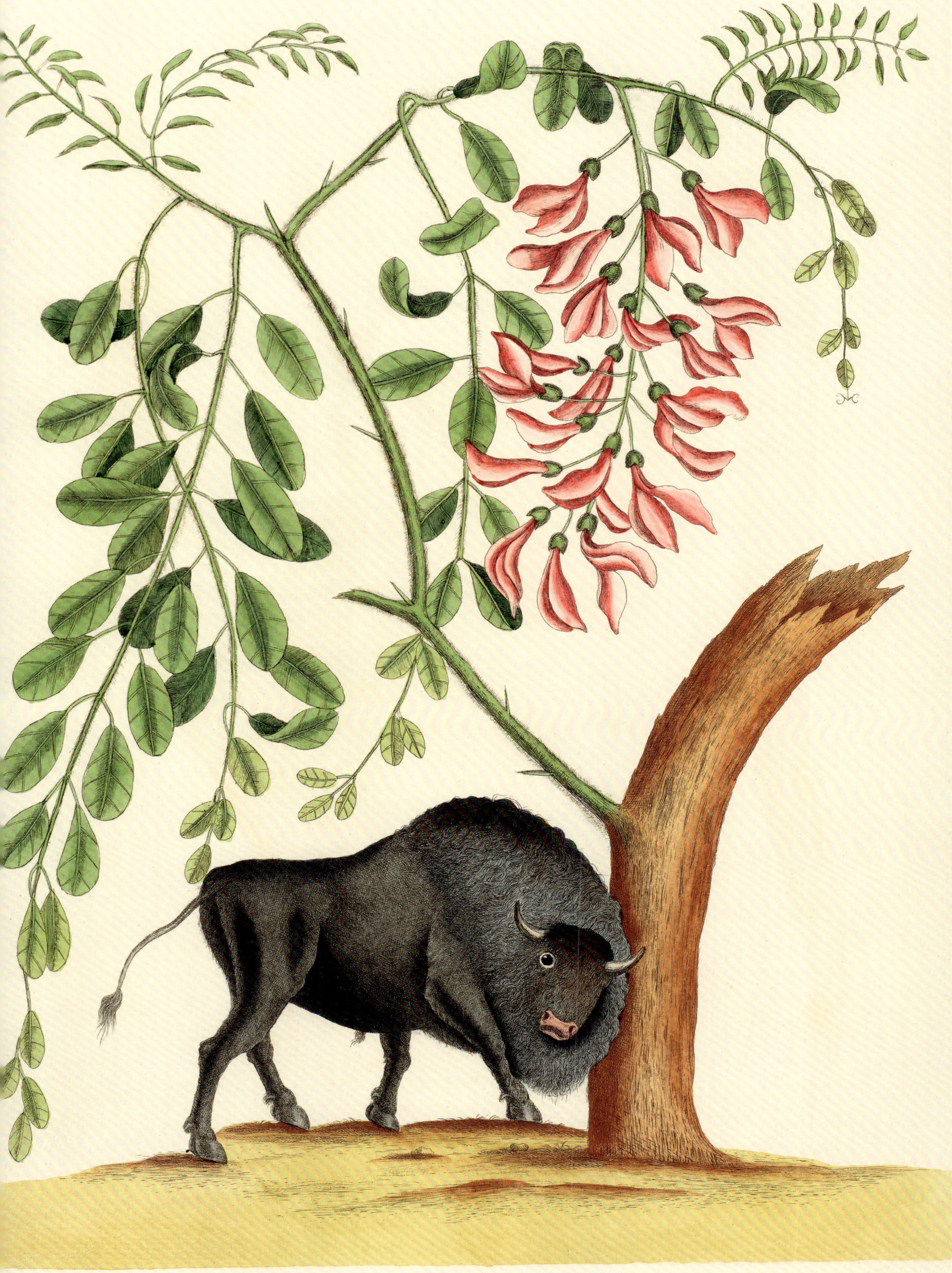

Notes

Preface

1. Miller 1730: vi.
2. McBurney 2021.
3. Livingstone 2013.
4. Oxford, Bodleian Library, Arch. Nat. hist. M 4–5.
5. Plate numbers in bold type refer to the natural history plates of Mark Catesby, which are reproduced in the present volume from a copy of the second edition of *The Natural History of Carolina, Florida, and the Bahama Islands* (Catesby 1754).
6. Edwards 1771: 1–2.
7. For example, Cates 1998; Ewan 1974; Feduccia 1985, Howard and Staples 1983; Jordan 1884; Reveal 2009, 2012, 2013, 2015; Wilbur 1990.

Chapter 1

1. Brendel 1879.
2. Elliott 2006.
3. Meltzer 2009.
4. Catesby 1754: 1.viii.
5. Hahn 2011.
6. Dethloff 1982; Riello 2013.
7. Munger 1949; Brown 2006.
8. Brockway 2002.
9. Goodman and Walsh 2001.
10. Crosby 2003; Mann 2011.
11. *Historia medicinal de las cosas que se traen de nuestras Indias Occidentales* was published in successive editions, under varying titles, with publication dates of 1565, 1569, 1571 and 1574, plus an unchanged reprint in 1580 (Frampton 1577; Boxer 1963).
12. Thorpe 1909: 53.
13. Arianrhod 2019.
14. Hariot 1590: 7.
15. Harris 2015b: 122–5.
16. Hariot 1590: 16.
17. Woodville 1790: 91–3; Cook 2011.
18. Hariot 1590: 9.
19. Conversion based on the National Archives' Currency Converter: 1270–2017 (www.nationalarchives.gov.uk/currency-converter) and appropriate adjustment for inflation between 2017 and 2023 (www.bankofengland.co.uk/monetary-policy/inflation/inflation-calculator).
20. Magnaghi 1997; Griffin 2020.
21. Tennent 1736: 48.
22. Gerard 1633: 1524. In contrast, John Ray (1693), citing Gerard, states that the plant came from the garden of William Coys in Stratford, near London.
23. Catesby 1767: 1–2; 1754: 1, pl. 55.
24. Coats 1969: 267–327.
25. Potter 2006.
26. Reveal 1992.
27. Munting 1696: 109.
28. Jocelyn 1865.
29. Lewis 1957: 41.
30. Ewan and Ewan 1970.
31. Catesby makes an indirect reference to this plant when he cites James Collins's engraving of a single leaf from Compton's garden in Plukenet (1691: t. LIV, fig. 5), when describing the southern red oak (*Quercus falcata*, Catesby 1754: pl. 21).
32. Morris 1991.
33. Frick 1987.
34. Lawson 1709: title page.
35. Bellis 2009.
36. Lawson 1709: 85–6
37. Lawson 1709: 78.
38. La Vere 2013.
39. Craton 1992.

Chapter 2

1. Lawson 1709: 85–6.
2. Biographical details of Mark Catesby's life are to be found in Nelson 2015.
3. Defoe 1724: 70.
4. Catesby 1729–47: 1, preface, v.
5. Reeds 2015: 36.
6. Nelson 2015: 5.
7. Catesby 1729–47: 1, preface, v.
8. Catesby 1754: 1.2; Brigham 1998.
9. Tinniswood 2019.
10. Catesby 1754: 1.ii.
11. Clokie 1964.
12. Harris 2020.
13. Goodricke 1885.
14. Webb 1966; Weir 1997.
15. Wood 2004.
16. Baker and Baker 1949; Johnson 1984; Stoker 2004; Meade 1974.
17. Watson 2020.
18. Handley 2002.
19. Hanham 2004.
20. Catesby 1729–47: advertisement.
21. Catesby 1729–47: 1, preface, x.
22. Delbourgo 2017.
23. Letter from Catesby to William Sherard, 16 January 1724 (Royal Society of London, Sherard Letter Book, MS 253, no. 174). For a detailed discussion of Catesby's field practices, see Harris 2015a.
24. For a detailed history of the printing ofCatesby's *Natural History*, see Overstreet 2015.
25. Royal Society of London Archive EC/1733/01.
26. Henrey 1986: 209.

Chapter 3

1. Brendel 1879.
2. Thiers 2020; Ogilvie 2006. Part of this chapter has been drawn from Harris 2015a.
3. Catesby 1754: 1.vii, v.
4. Harris 2015a.
5. Letter from Catesby to William Sherard, 5 May 1722 (Royal Society of London, Sherard Letter Book, MS 253, no. 163).
6. Letter from Catesby to William Sherard, 16 January 1724 (Royal Society of London, Sherard Letter Book, MS 253, no. 173).
7. Letter from Catesby to William Sherard, 16 January 1724 (Royal Society of London, Sherard Letter Book, MS 253, no. 173).
8. Letter from Catesby to William Sherard, 10 May 1723 (Royal Society of London, Sherard Letter Book, MS 253, no. 171).
9. Catesby 1754: 1.iii.
10. Catesby 1754: 1, preface.
11. Catesby 1754: 1.i.
12. Catesby 1754: 1.iii.
13. Catesby 1754: 1.iv.
14. Kreiner 2021.
15. Evelyn 1664: 15.
16. Evelyn 1664: 15.
17. Catesby 1763: i.
18. Catesby 1754: 1, pl. 21.
19. Catesby 1754: 1, pl. 18.
20. Catesby 1754: 1, pl. 16.
21. Catesby 1754: 1, pl. 20.
22. Catesby 1754: 1, pl. 19.
23. Catesby 1754: 1, pl. 17.
24. Catesby 1754: 1, pl. 23.
25. Catesby 1754: 1, pl. 22.
26. Catesby 1754: 1, pl. 16.
27. Hasbrouck 1891.
28. Fuller 2000.
29. Catesby 1754: 1, pl. 16.
30. Catesby 1754: 1.iii.
31. Catesby 1754: 1.xxiii–xxiv.
32. Hall 1731: 63.
33. Dethloff 1982: table 1.
34. Dethloff 1982: 238.
35. Catesby 1754: 1, pl. 14.
36. Catesby 1754: 1, pl. 14.
37. Catesby 1754: 1.iii.
38. Catesby 1754: 1.xvii.
39. Carney 2001.

40. Catesby 1754: 1.xvii.
41. Collinson 1766: 279.
42. Salley 1919; Clifton 1981.
43. Catesby 1754: 1.xxxviii.
44. Catesby 1754: 1.xxxix.
45. Hogarth 2015.
46. This identification is based on Catesby's description of the plant's habitat and flower form rather than the leaf arrangement, which is depicted as alternate rather than opposite (Catesby 1754: 1, pl. 86).
47. Catesby 1754: 2, pl. 33. Catesby's reference is incorrect; it should be Sloane 1725: 18, table 161, fig. 2.
48. Catesby 1754: 2, pl. 33.
49. Catesby 1754: 2, pl. 63.
50. Catesby 1754: 2, pl. 63.
51. W. Shakespeare, *The Tempest*, act 2, scene i.
52. Hartnoll et al. 2007.
53. Catesby 1754: 2, pl. 32.
54. Catesby 1754: 1.xxxviii.
55. Catesby 1754: 1.xxxix.
56. Catesby 1754: 1.xl.
57. Catesby 1754: 1.xl.
58. Catesby 1754: 2, pl. 81.
59. Budowski 1982.
60. Catesby 1754: 2, pl. 51.
61. Titford 2005; Ortiz-Hidalgo and Pina-Oviedo 2019; Defilipps 1998.
62. Dapson and Bain 2015.
63. Catesby 1754: 1, pl. 29.
64. It was used in North America by Native Americans (Moerman 1998: 91–2) and colonists (Millspaugh 1974: 552–6); and in Britain and Europe (Grieve 1934: 744–6).
65. Catesby 1754: 1, pl. 29.
66. Catesby 1754: 1.xxxv.
67. Feduccia 1985; Krech 2015.
68. Catesby 1754: 1.xxxv.
69. Catesby 1754: 1.xxxvi.
70. Renfrew et al. 2015.
71. Catesby 1754: 1, pl. 14.
72. Catesby 1754: 1.iii.
73. Catesby 1754: 1.xvi.
74. Catesby 1754: 1.ii.
75. Catesby 1754: 1, pl. 23.
76. Catesby 1754: 1.x.
77. Fuller 2014; Greenberg 2014.
78. Catesby 1754: 1, pl. 11.
79. Lawson 1709: 142.
80. Catesby 1754: 1, pl. 11.
81. Burgio et al. 2022.

Chapter 4

1. Catesby 1754: 1, preface, iv.
2. Catesby 1754: 1, preface, iv.
3. Catesby 1754: 1.xvi.
4. Catesby 1754: 1.viii.
5. Catesby 1754: 1.viii, xiv.
6. Catesby 1754: 1.viii.
7. Catesby 1754: 1.ix.
8. Catesby 1754: 1.x.
9. Catesby 1754: 1, pl. 83.
10. Catesby 1754: 2, pl. 107.
11. Miller 1768: entry 'Vanilla'.
12. Catesby 1754: 1, pl. 73.
13. Catesby 1754: 1.viii.
14. Catesby 1754: 1, pl. 50.
15. Catesby 1754: 1, pl. 70.
16. Catesby 1754: 1, pl. 75.
17. Moerman 1998: 91–2.
18. Catesby 1754: 2, pl. 57.
19. Catesby 1754: 1.xiv.
20. Catesby 1754: 2, pl. 65.
21. Catesby 1754: 1, pl. 35.
22. Catesby 1754: 2, pl. 54.
23. Wood 1974.
24. Nash 1992.
25. Catesby 1754: 1.xxxviii.
26. Letter from Catesby to William Sherard, 4 January 1723 (Royal Society of London, Sherard Letter Book, MS 253, no. 168).
27. Harris 2020.
28. Harris 2015b: 155–9.
29. Catesby 1754: 1.xvii.
30. Catesby 1754: 1.xviii.
31. Muñoz-Rodríguez et al. 2018.
32. Catesby 1754: 2, pl. 60.
33. Catesby 1754: 2, pl. 45. Catesby cites Hans Sloane (1707: 166) but the plant referred to there is the widespread Old World cocoyam (*Colocasia esculenta*). Cocoyam may be the more palatable species referred to by Catesby and may be the source of his confusion regarding origins.
34. Moerman 1998: 91–2.
35. Catesby 1754: 2, pl. 85.
36. Catesby 1754: 2, pl. 32.
37. Catesby 1754: 2, appendix, pl. 10.
38. Catesby 1754: 2, appendix, pl. 10. Catesby's illustration was, presumably, based on specimens examined in London.
39. Catesby 1754: 2, appendix, pl. 15. For a discussion of the defence strategies of the velvet ant see Schmidt and Blum 1977.
40. Catesby 1754: 2, appendix, pl. 15.
41. Catesby 1754: 1.vii.
42. Catesby 1754: 1.xxxv.

Chapter 5

1. Catesby 1763: i.
2. It was the relationship between Peter Collinson and John Bartram, in the middle of the eighteenth century, that solidified the place of the North American plants in British gardens (see Wulf 2008).
3. Harris et al. 2017; Harris 2023.
4. Thiers 2020.
5. Nelson and Elliott 2015.
6. Petiver, 'Brief directions for the easie making, and preserving of all natural curiosities', single sheet issued as part of Petiver (1764).
7. Conniff 2011.
8. See Thiers (2020) for a wider discussion of the modern uses of herbarium specimens.
9. Catesby 1754: 1.iv.
10. Catesby 1754: 1.v.
11. Catesby 1754: 1.xxxvii.
12. Letter from Catesby to William Sherard, 16 January 1724 (Royal Society of London, Sherard Letter Book, MS 253, no. 174).
13. Specimen of *Ulmus rubra* (Sher-0604-12, Oxford University Herbaria). A similar label is attached to the specimen in the Sloane Herbarium (HS212 f.70, Natural History Museum, London).
14. Letter from Catesby to William Sherard, 30 October 1724 (Royal Society of London, Sherard Letter Book, MS 253, no. 79).
15. Specimens of *Nymphaea odorata* (00087305N, 00087306O, 00087301J, Sher-1089 and Sher-1089-2, Oxford University Herbaria).
16. Specimen of *Nelumbo lutea* (Sher-1090-10, Oxford University Herbaria).
17. Letter from Catesby to William Sherard, 30 October 1724 (Royal Society of London, Sherard Letter Book, MS 253, no. 179).
18. Petiver, 'Brief directions for the easie making, and preserving of all natural curiosities', single sheet issued as part of Petiver (1764).
19. Specimen of *Ceratiola ericoides* (Sher-0020-3, Oxford University Herbaria).
20. Specimen of *Robinia hartwigii* (Sher-1516-10, Oxford University Herbaria).
21. Specimen of *Tilia americana* (Sher-1098-4, Oxford University Herbaria). A similar label is attached to the specimen in the Sloane Herbarium (HS212 f.69, Natural History Museum, London).
22. Specimen of *Digitaria sanguinalis* (Sher-0125-b, Oxford University Herbaria).
23. Letter from Catesby to William Sherard, 20 June 1722 (Royal Society of London, Sherard

Letter Book, MS 253, no. 164).
24. Specimen of *Styrax americanus* (Sher-sn-AIF, Oxford University Herbaria). A similar label, without the annotations, is attached to the specimen in the Sloane Herbarium (HS212 f.16, Natural History Museum, London).
25. Petiver, 'Brief directions' (Petiver 1764).
26. Letter from Catesby to William Sherard, 16 August 1724 (Royal Society of London, Sherard Letter Book, MS 253, no. 178).
27. Harris 2015a; McMillan and Blackwell 2013; McMillan et al. 2013.
28. Specimen of *Trillium cuneatum* (Sher-sn-ATH, Oxford University Herbaria).
29. Catesby 1767: ii.
30. Ulmer and MacDougal 2004.
31. Aiton 1789: 306–11.
32. Petiver 1715: 359.
33. Dillenius 1732.
34. Bradley 1718: 113–16.
35. Laird 2015a.
36. Catesby 1767: i.
37. Catesby 1754: 1, preface.
38. Catesby 1754: 1, preface, iv.
39. Laird 2015b.
40. Catesby 1767: 1.
41. Catesby 1754: 2, pl. 61.
42. Calmann 1977.
43. Gardiner 2000; Nelson 2014b.
44. Nelson 2018; Keogh 2020.
45. Catesby 1767: 1–2.
46. Catesby 1767: 1–2.
47. Evelyn 1670: 81.
48. Catesby 1767: 10.
49. Catesby 1767: 26.
50. Parkinson 1640: 386.
51. Cornut 1635: t. 103.
52. Catesby 1767: 26.
53. Catesby 1754: 1, pl. 65.
54. Bertin 1982; Van Nest et al. 2021.
55. Catesby 1767: 9.
56. Catesby 1767: title page.
57. Catesby 1767: preface, ii.
58. Catesby 1767: preface, ii.
59. Furber 1727.
60. Compton 2015.
61. Nelson 2014a.
62. Catesby 1767: 24.
63. Catesby 1754: 1, pl. 49.
64. Miller 1768: entry 'Bignonia'.
65. Miller 1768: entry 'Basteria'.
66. Miller 1755: t. LX.
67. Duhamel du Monceau 1755: t. 114.
68. Miller 1768: entry 'Basteria'.
69. Linnaeus 1759: 1066.
70. Catesby 1767: 31.
71. Letter from Peter Collinson to Johann Dillenius, 11 August 1745, attached to a specimen of *Calycanthus floridus* (Sher-1066-10, Oxford University Herbaria).
72. Catesby 1754: 2, pl. 83.
73. Miller 1768: entry 'Ptelea'.
74. Catesby 1767: 10.
75. Miller 1768: entry 'Bignonia'.
76. Catesby 1767: 31.
77. Tubby and Pérez Sierra 2015.

Chapter 6

1. Catesby 1754: 1, pl. 10.
2. Harris 2023.
3. Burchell 1824: *214*.
4. Fitch 1869: 7.
5. Quoted in Druce and Vines 1907: 158. The associated specimen appears to be missing from Oxford University Herbaria.
6. Catesby 1754: 1, preface, xi.
7. Quoted in Freer 2005: 283.
8. Harris 2015a.
9. McBurney 2021.
10. Gibson 2012.
11. Letter from Linnaeus to Ellis, 16 September 1761, in Smith 1821: 151.
12. Catesby 1754: 2, pl. 21.
13. Catesby 1754: 1.iv.
14. Catesby 1754: 1.v.
15. McBurney 2021.
16. Johnson 1973; Mitchell and Mori 1987.
17. Harris et al. 2017.
18. Catesby 1754: 2, pl. 109.
19. Catesby 1754: 2, pl. 109; Merian 1705: t. 16.
20. Freer 2005: 17.
21. Catesby 1754: 1, pl. 10.
22. Catesby 1754: 2, pl. 45.
23. Specimen of *Cocculus carolinus* collected by Catesby in 1723 (Hort-178-219b, Oxford University Herbaria).
24. Catesby 1754: 2, pl. 120.
25. Catesby reports that his herbarium specimens went to Hans Sloane in London and Johann Dillenius in Oxford, and that the rose locust was introduced by John Colleton, formerly one of the Lord Proprietors of Carolina, to his garden in Exmouth, from North America (Catesby 1754: 2, pl. 120; Smith 1912).
26. Catesby 1754: 2, pl. 63.
27. Catesby 1754: 1, pl. 73. Using Catesby's own measurements of the size of the gorgoniid stem, the specimen in his illustration is approximately 50 cm tall.
28. Martyn 1728: t. 37.
29. Dillenius 1732: 165, t. 138.
30. Dillenius 1732: 165 (author's translation). The herbarium specimen, collected by either Mark Catesby or William Houston, is preserved in William Sherard's herbarium (Sher-0007, Oxford University Herbaria).
31. Catesby 1754: 2, pl. 93.
32. Harris 2021.
33. Linnaeus 1753: 955.
34. Wilbur 1990.
35. Reveal 2012.

Chapter 7

1. Catesby *c.*1729.
2. Gascoigne 2004.
3. Overstreet 2015.
4. Gaskell 1972.
5. Catesby 1754: 1, preface.
6. Wulf 2008. Despite lending Catesby money and being identified as one of his 'Encouragers', Collinson claimed not to have a copy of the *Natural History* (Letter from Peter Collinson to Johann Dillenius, 11 August 1745, attached to a specimen of *Calycanthus floridus*: Sher-1066-10, Oxford University Herbaria).
7. Henrey 1975; Harris 2023.
8. Catesby *c.*1729.
9. Catesby *c.*1729.
10. Catesby 1754: 1.xi–xii.
11. Catesby 1754: 1.xii.
12. Catesby 1754: 1.xi.
13. Overstreet 2015.
14. Published in 1743 and bound after the preface of volume 1 of the *Natural History*.
15. Overstreet 2015: 160.
16. See Overstreet 2015: 155–72.
17. Catesby 1754: 2, pl. 1.

Chapter 8

1. Catesby 1767: preface, i.
2. Frick and Edwards 1960; Frick and Stearns 1961; McBurney and Myers 1997; Myers and Pritchard 1998; Nelson and Elliott 2015; McBurney 2021.
3. Jarvis 2007: 123–6.
4. Linnaeus 1758: 626 (author's translation).
5. Catesby 1754: 100.
6. Feduccia 1985; Krech 2015.
7. Overstreet 2015.
8. Berkeley and Berkeley 1963.

Bibliography

Aiton, W. (1789), *Hortus Kewensis, or, A Catalogue of the Plants Cultivated in the Royal Botanic Garden at Kew*, vol. III, *Diadelphia-Cryptogamia*, George Nicol, London.

Arianrhod, R. (2019), *Thomas Harriot: A Life in Science*, Oxford University Press, New York.

Baker, C.H.C., and M.I. Baker (1949), *The Life and Circumstances of James Brydges, First Duke of Chandos: Patron of the Liberal Arts*, Clarendon Press, Oxford.

Batsaki, Y., S.B. Cahalan and A. Tchikine (2016), *The Botany of Empire in the Long Eighteenth Century*, Dumbarton Oaks Research Library and Collection, Washington, DC.

Bellis, V. (2009), 'John Lawson's Plant Collections from North Carolina 1710–1711', *Castanea*, vol. 74, pp. 376–89.

Berkeley, E., and D.S. Berkeley (1963), *John Clayton: Pioneer of American Botany*, Chapel Hill, University of North Carolina Press.

Bertin, R.I. (1982), 'Floral Biology, Hummingbird Pollination and Fruit Production of Trumpet Creeper (*Campsis radicans*, Bignoniaceae)', *American Journal of Botany*, vol. 69, pp. 122–34.

Boxer, C.R. (1963), *Two Pioneers of Tropical Medicine: Garcia D'Orta and Nicolás Monardes*, Wellcome Historical Medical Library, London.

Bradley, R. (1718), *New Improvements of Planting and Gardening, Both Philosophical and Practical; Explaining the Motion of the Sapp and Generation of Plants*, W. Mears, London.

Brendel, F. (1879), 'Historical Sketch of the Science of Botany in North America from 1635 to 1840', *American Naturalist*, vol. 13, pp. 754–71.

Brigham, D.R. (1998), 'Mark Catesby and the Patronage of Natural History on the First Half of the Eighteenth Century', in A.R.W. Meyers and M.B. Pritchard (eds), *Empire's Nature: Mark Catesby's New World Vision*, University of North Carolina Press, Chapel Hill and London, pp. 91–146.

Brockway, L.H. (2002), *Science and Colonial Expansion: The Role of the British Royal Botanic Gardens*, Yale University Press, New Haven, CT, and London.

Brown, K. (2006), *The Pox: The Life and Death of a Very Social Disease*, Sutton, Stroud.

Budowski, G. (1982), 'The Socio-economic Effects of Forest Management on the Lives of People Living in the Area: The Case of Central America and Some Caribbean Countries', in E.G. Hallsworth (ed.), *Socio-economic Effects and Constrains in Tropical Management*, John Wiley, New York, pp. 87–102.

Burchell, W.J. (1824), *Travels in the Interior of Southern Africa*, vol. 2, Longman, Hurst, Rees, Orme, Brown, and Green, London.

Burgio, K.R., C.J. Carlson , A.L. Bond et al. (2022), 'The Two Extinctions of the Carolina Parakeet *Conuropsis carolinensis*', *Bird Conservation International*, vol. 32, pp. 498–505.

Calmann, G. (1977), *Ehret: Flower Painter Extraordinary: An Illustrated Biography*, Phaidon, Oxford.

Carine, M. (2020), *The Collectors: Creating Hans Sloane's Extraordinary Herbarium*, Natural History Museum, London.

Carney, J.A. (2001), *Black Rice: The African Origins of Rice Cultivation in the Americas*, Harvard University Press, Cambridge, MA.

Cates, D.L. (1998), 'Mark Catesby's Bahamian Plants', *Bahamas Journal of Science*, vol. 5, pp. 16–21.

Catesby, M. (*c*.1729), *Proposals, for Printing an Essay Towards a Natural History of Florida, Carolina and the Bahama Islands*, London.

Catesby, M. (1729–47), *The Natural History of Carolina, Florida and the Bahama Islands: Containing the Figures of Birds, Beasts, Fishes, Serpents, Insects, and Plants: Particularly, the Forest-trees, Shrubs, and Other Plants, not Hitherto Described, or Very Incorrectly Figured by Authors: Together with their Descriptions in English and French: to Which, Are Added Observations on the Air, Soil, and Waters: with Remarks Upon Agriculture, Grain, Pulse, Roots, &c*, Printed at the expence of the author, London.

Catesby, M. (1754), *The Natural History of Carolina, Florida, and the Bahama Islands*, vols 1 and 2, C. Marsh and T. Wilcox, London, p. viii.

Catesby, M. (1763), *Hortus Britanno-Americanus, or, A Curious Collection of Trees and Shrubs, the Produce of the British Colonies in North America*, W. Richardson and S. Clark for J. Ryall, London.

Catesby, M. (1767), *Hortus Europae Americanus, or, A Collection of 85 Curious Trees and Shrubs, the Produce of North America; Adapted to the Climates and Soils of Great-Britain, Ireland, and Most Parts of Europe, &c*, J. Millan, London.

Clifton, J.M. (1981), 'The Rice Industry in Colonial America', *Agricultural History*, vol. 55, pp. 266–83.

Clokie, H.N. (1964), *An Account of the Herbaria of the Department of Botany in the University of Oxford*, Oxford University Press, Oxford.

Coats, A.M. (1969), *The Quest for Plants: A History of Horticultural Explorers*, Studio Vista, London.

Collinson, P. (1766), 'An Account of the Introduction of Rice and Tar into our Colonies', *Gentleman's Magazine*, vol. 36, pp. 278–9.

Compton, J.A. (2015), 'A Vignette of Botany in the Age of Enlightenment: The Story of Catesby's Climber or the Carolina Kidney Bean Tree, *Wisteria frutescens*', *Curtis's Botanical Magazine*, vol. 32, pp. 193–247.

Conniff, R. (2011), *The Species Seekers: Heroes, Fools, and the Mad Pursuit of Life on Earth*, W.W. Norton, London.

Cook, H.J. (2011), 'Markets and Cultures: Medical Specifics and the Reconfiguration of the Body in Early Modern Europe', *Transactions of the Royal Historical Society*, vol. 21, pp. 123–45.

Cornut, J.P. (1635), *Canadensium plantarum*, Simonem Le Moyne, Paris.

Craton, M. (1992), *Islanders in the Stream: A History of the Bahamian People*, vol. 1, *From Aboriginal Times to the End of Slavery*, University of Georgia Press, Athens, GA.

Crosby, A.W. (2003), *The Columbian Exchange: Biological and Cultural Consequences of 1492*, Praeger, Westport, CT.

Dapson, R.W., and C.I. Bain (2015), 'Brazilwood, Sappanwood, Brazilin and the Red Dye Brazilein: From Textile Dyeing and Folk Medicine to Biological Staining and Musical Instruments', *Biotechnic & Histochemistry*, vol. 90, pp. 401–23.

Defilipps, R.A. (1998), 'Historical Connections between the Discovery of Brazil and the Neotropical Brazilwood, *Caesalpinia echinata* Lam.', *Archives of Natural History*, vol. 25, pp. 103–8.

Defoe, D. (1724), *A Tour Thro' the Whole Island of Great Britain, Divided Into Circuits or Journeys*, G. Strahan, London.

Delbourgo, J. (2017), *Collecting the World: The Life and Curiosity of Hans Sloane*, Penguin Books, London.

Dethloff, H.C. (1982), 'The Colonial Rice Trade', *Agricultural History*, vol. 56, pp. 231–243.

Dillenius, J.J. (1732), *Hortus Elthamensis seu plantarum rariorum quas in horto suo Elthami in cantio coluit vir ornatissimus et praestantissimus Jacobus Sherard, M.D. Soc. Reg. et Coll. Med. Lond. Soc. Sumptibus Auctoris*, London.

Druce, G.C., and S.H. Vines (1907), *The Dillenian Herbarium: An Account of the Dillenian Collection in the Herbarium of the University of*

Oxford, Clarendon Press, Oxford.
Duhamel du Monceau, H.-L. (1755), *Traité des arbres et arbustes qui se cultivent en France en pleine terre*, Chez H.L. Guerin & L.F. Delatour, Paris.
Edwards, G. (1771), 'A Catalog of the Animals and Plants Represented in Catesby's *Natural History of Carolina* with the Linnaean names', in M. Catesby, *The Natural History of Carolina, Florida, and the Bahama Islands*, Benjamin White, London.
Elliott, J.H. (2006), *Empires of the Atlantic World: Britain and Spain in America 1492–1830*, Yale University Press, New Haven, CT, and London.
Evelyn, J. (1664), *Sylva, or, A Discourse of Forest-Trees, and the Propagation of Timber in His Majesties Dominions*, J. Martyn and J. Allestry, London.
Ewan, J. (1974), 'Notes', in M. Catesby, *The Natural History of Carolina, Florida, and the Bahama Islands, Containing Two-hundred and Twenty Figures of Birds, Beasts, Fishes, Serpents, Insects, and Plants*, with an introduction by George Frick and notes by Joseph Ewan, Beehive Press, Savannah, GA, pp. 89–100.
Ewan, J., and N. Ewan (1970), *John Banister and his Natural History of Virginia 1678–1692*, University of Illinois Press, Urbana, Chicago and London.
Feduccia, A. (1985), *Catesby's Birds of Colonial America*, University of North Carolina Press, Chapel Hill.
Fitch, W.H. (1869), 'Botanical Drawing. No. I', *Gardeners' Chronicle and Agricultural Gazette*, vol. 29, p. 7.
Frampton, J. (1577), *Ioyfull Newes out of the Newe Founde Worlde, Wherein is Declared the Rare and Singular Vertues of Diuerse and Sundrie Hearbes, Trees, Oyles, Plantes, and Stones, with their Applications, as well for Phisicke as Chirurgerie*, W. Norton, London.
Freer, S. (2005), *Linnaeus' Philosophia Botanica*, Oxford University Press, Oxford.
Frick, G.F. (1987), 'Botanical Explorations and Discoveries in Colonial Maryland, 1688–1753', *Huntia*, vol. 7, pp. 5–59.
Frick, G.F., and G. Edwards (1960), 'Mark Catesby: The Discovery of a Naturalist', *Papers of the Bibliographical Society of America*, vol. 54, pp. 163–75.
Frick, G.F., and R.P. Stearns (1961), *Mark Catesby: The Colonial Audubon*, University of Illinois Press, Urbana.
Fuller, E. (2000), *Extinct Birds*, Oxford University Press, Oxford.
Fuller, E. (2014), *The Passenger Pigeon*, Princeton University Press, Princeton, NJ.
Furber, R. (1727), *A Catalogue of English and Foreign Trees: Which Will Thrive in the Natural Ground in England*, Henry Woodfall, London.
Gardiner, J. (2000), *Magnolias: A Gardener's Guide*, Timber Press, Portland, OR.
Gascoigne, B. (2004), *How to Identify Prints. A complete Guide to Manual and Mechanical Processes from Woodcut to Inkjet*, Thames & Hudson, London.
Gaskell, P. (1972), *A New Introduction to Bibliography*, Oxford University Press, Oxford.
Gerard, J. (1633), *The Herball, or, General Historie of Plantes*, Adam Islip, Joice Norton and Richard Whitakers, London.
Gibson, S. (2012), 'On Being an Animal, or, The Eighteenth-century Zoophyte Controversy in Britain', *History of Science*, vol. 50, pp. 453–76.
Goodman, J., and V. Walsh (2001), *The Story of Taxol: Nature and Politics in the Pursuit of an Anti-Cancer Drug*, Cambridge University Press, Cambridge.
Goodricke, C.A. (1885), *History of the Goodricke family*, Hazell, Watson & Viney, London.
Greenberg, J. (2014), *A Feathered River across the Sky: The Passenger Pigeon's Flight to Extinction*, Bloomsbury, New York, Princeton, NJ, and Oxford.
Grieve, M. (1934), *A Modern Herbal: The Medicinal, Culinary, Cosmetic and Economic Properties, Cultivation and Folklore of Herbs, Grasses, Fungi, Shrubs and Trees with All Their Modern Scientific Uses*, Jonathan Cape, London.
Griffin, C. (2020), 'Disentangling Commodity Histories: *Pauame* and Sassafras in the Early Modern Global World', *Journal of Global History*, vol. 15, pp. 1–18.
Griffith, L.D. (2008), *Flowers and Herbs of Early America*, Yale University Press, New Haven, CT.
Hahn, B.M. (2011), *Making Tobacco Bright: Creating an American Commodity, 1617–1937*, Johns Hopkins University Press, Baltimore, MD.
Hall, F. (1731), *The Importance of the British Plantations in America to this Kingdom; with the State of their Trade, and Methods of Improving It*, J. Peele, London.
Handley, S. (2002), 'Markham, Sir George, 3rd Bt. (1666–1736), of Sedgebrook, Lincs', The History of Parliament, https://www.historyofparliamentonline.org/volume/1690-1715/member/markham-sir-george-1666-1736.
Hanham, A.A. (2004), 'Parker, Thomas, First Earl of Macclesfield (1667–1732)', *Oxford Dictionary of National Biography*, https://doi.org/10.1093/ref:odnb/21341.
Hariot, T. (1590), *A Briefe and True Report of the New Found Land of Virginia, of the Commodities and of the Nature and Manners of the Naturall Inhabitants. Discouered by the English Colony There Seated by Sir Richard Greinuile Knight in the Yeere 1585. Which Remained Under the Gouernment of Twelue Monethes, at the Speciall Charge and Direction of the Honourable Sir Walter Raleigh Knight Lord Warden of the Stanneries who Therein Hath Beene Fauoured and Authorised by Her Maiestie and Her Letters Patents*, John Wechel / Theodor de Bry, Frankfurt.
Harris, S.A. (2015a), 'The Plant Collections of Mark Catesby in Oxford', in E.C. Nelson and D.J. Elliott (eds), *The Curious Mister Catesby: A Truly Ingenious' Naturalist Explores New Worlds*, University of Georgia Press, Athens, GA, pp. 173–88.
Harris, S.A. (2015b), *What Have Plants Ever Done for Us? Western Civilization in Fifty Plants*, Bodleian Library, Oxford.
Harris, S.A. (2020), 'William Sherard', in M. Carine (ed.), *The Collectors: Creating Hans Sloane's Extraordinary Herbarium*, Natural History Museum, London, pp. 38–43.
Harris, S.A. (2021), *Roots to Seeds: 400 Years of Oxford Botany*, Bodleian Library, Oxford.
Harris, S.A. (2023), *The Beauty of the Flower: The Art and Science of Botanical Illustration*, Reaktion Books, London.
Harris, S.A., S.K. Marner and C.E.B. Proença (2017), 'William Dampier's Brazilian Botanical Observations in 1699', *Journal of the History of Collections*, vol. 29, pp. 227–35.
Hartnoll, R.G., Baine, M.S.P., Britton, A. et al. (2007), 'Reproduction of the Black Land Crab, *Gecarcinus ruricola*, in the San Andres Archipelago, Western Caribbean', *Journal of Crustacean Biology*, vol. 27, pp. 425–36.
Hasbrouck, E.M. (1891), 'The Present Status of the Ivory-Billed Woodpecker (*Campephilus principalis*)', *The Auk*, vol. 8, pp. 174–86.
Henrey, B. (1975), *British Botanical and Horticultural Literature before 1800 comprising a History and Bibliography of Botanical and Horticultural Books Printed in England, Scotland, and Ireland from the Earliest Times until 1800*, vol. II, *The Eighteenth Century: History*, Oxford University Press, Oxford.
Henrey, B. (1986), *No Ordinary Gardener: Thomas Knowlton, 1691–1781*, Natural History Museum, London.
Hogarth, P.J. (2015), *The Biology of Mangroves and Seagrasses*, Oxford University Press, Oxford.
Howard, R.A., and G.W. Staples (1983), 'The Modern Names for Catesby's Plants', *Journal of the Arnold Arboretum*, vol. 64, pp. 511–46.
Jarvis, C. (2007), *Order Out of Chaos: Linnaean Plant Names and their Types*, Linnean Society of London, London.
Jocelyn, J. (1865), *New-Englands Rarities Discovered: in Birds, Beasts, Fishes, Serpents, and Plants of that Country. With An Introduction and Notes, by Edward Tuckerman*, William Veazia, Boston, MA.
Johnson, D. (1973), 'The Botany, Origin, and Spread of the Cashew, *Anacardium occidentale* L.', *Journal of Plantation Crops*, vol. 1, pp. 1–7.
Johnson, J. (1984), *Princely Chandos: James Brydges 1674–1744*, A. Sutton, Gloucester.
Jordan, D.S. (1884), 'An Identification of the Figures of Fishes in Catesby's *Natural History of Carolina, Florida and the Bahama Islands*', *Proceedings of the United States National Museum*, vol. 7, pp. 190–99.
Keogh, L. (2020), *The Wardian Case: How a Simple*

Box Moved Plants and Changed the World, University of Chicago Press, Chicago, IL.

Krech, S. III (2015), '"Of Birds of Passage": Mark Catesby and Contemporary Theories on Bird Migration and Torpor', in E.C. Nelson and D.J. Elliott (eds), *The Curious Mister Catesby: A 'Truly Ingenious' Naturalist Explores New Worlds*, University of Georgia Press, Athens, GA, pp. 219–30.

Kreiner, J. (2021), *Legions of Pigs in the Early Medieval West*, Yale University Press, New Haven, CT.

La Vere, D. (2013), *The Tuscarora War: Indians, Settlers, and the Fight for the Carolina Colonies*, University of North Carolina Press, Chapel Hill.

Laird, M. (2015a), 'Mark Catesby's Plant Introductions and English Gardens of the Eighteenth Century', in E.C. Nelson and D.J. Elliott (eds), *The Curious Mister Catesby: A 'Truly Ingenious' Naturalist Explores New Worlds*, University of Georgia Press, Athens, GA, pp. 265–80.

Laird, M. (2015b), *A Natural History of English Gardening, 1650–1800*, Yale University Press, New Haven, CT.

Lawson, J. (1709), *A New Voyage to Carolina; Containing the Exact Description and Natural History of That Country: Together with the Present State Thereof, and a Journal of a Thousand Miles, Travel'd Thro' Several Nations of Indians. Giving a Particular Account of Their Customs, Manners, &c*, London.

Lewis, I.F. (1957), 'Seventeenth Century Science in Old Virginia', *Virginia Journal of Science*, vol. 8, pp. 35–41.

Linnaeus, C. (1753), *Species plantarum*, vol. I, Laurence Salvius, Stockholm.

Linnaeus, C. (1758), *Systema naturae per regna tria naturae*, vol. 1, Laurence Salvius, Stockholm.

Linnaeus, C. (1759), *Systema naturae per regna tria naturae*, vol. 2, Laurence Salvius, Stockholm.

Livingstone, D.N. (2013), *Putting Science in its Place: Geographies of Scientific Knowledge*, University of Chicago Press, Chicago, IL.

MacGregor, A. (2018) *Naturalists in the Field: Collecting, Recording and Preserving the Natural World from the Fifteenth to the Twenty-First Century*, Brill, Leiden.

Magnaghi, R.M. (1997), 'Sassafras and Its Role in Early America, 1562–1662', *Terrae Incognitae: The Journal of the Society for the History of Discoveries*, vol. 29, pp. 10–21.

Mann, C.G. (2011), *1493: How Europe's Discovery of the Americas Revolutionised Trade, Ecology and Life on Earth*, Granta, London.

Martyn, J. (1728), *Historia plantarum rariorum*, Richard Reily, London.

McBurney, H. (2021), *Illuminating Natural History: The Art and Science of Mark Catesby*, Yale University Press, New Haven, CT.

McBurney, H., and A.R.W. Myers (1997), *Mark Catesby's Natural History of America: The Watercolours from the Royal Library Windsor Castle*, Merrell Holberton, London.

McMillan, P.D., and A.H. Blackwell (2013), 'The Vascular Plants Collected by Mark Catesby in South Carolina: Combining the Sloane and Oxford Herbaria', *Phytoneuron*, 2013-73, pp. 1–32.

McMillan, P.D., A.H. Blackwell, C. Blackwell and M.A. Spencer (2013), 'The Vascular Plants in the Mark Catesby Collection at the Sloane Herbarium, with Notes on Their Taxonomic and Ecological Significance', *Phytoneuron*, 2013-7, pp. 1–37.

Meade, R.H. (1974), *In the Sunshine of Life: A Biography of Dr. Richard Mead, 1673–1754*, Dorrance, Pittsburgh, PA.

Meltzer, D.J. (2009), *First Peoples in a New World: Colonizing Ice Age America*, University of California Press, Berkeley.

Merian, M.S. (1705), *Metamorphosis insectorum surinamensium*, G. Valck, Amsterdam.

Miller, P. (1730), *A Catalogue of Trees, Shrubs, Plants, and Flowers, Both Exotic and Domestic, Which Are Propagated For Sale, in the Gardens Near London* [Society of Gardeners], London.

Miller, P. (1755), *Figures of the Most Beautiful, Useful, and Uncommon Plants Described in the Gardeners Dictionary*, vol. 1, Printed for the Author, London.

Miller, P. (1768), *The Gardeners Dictionary*, Printed for the Author, London.

Millspaugh, C.F. (1974), *American Medicinal Plants: An Illustrated and Descriptive Guide to Plants Indigenous to and Naturalized in the United States which are Used in Medicine*, Dover, New York.

Mitchell, J.D., and S.A. Mori (1987), 'The Cashew and its Relatives (*Anacardium*: Anacardiaceae)', *Memoirs of the New York Botanical Garden*, vol. 42, pp. 1–76.

Moerman, D.E. (1998), *Native American Ethnobotany*, Timber Press, Portland, OR.

Morris, S. (1991), 'Legacy of a Bishop: The Trees and Shrubs of Fulham Palace Gardens Introduced 1675–1713', *Garden History*, vol. 19, pp. 47–59.

Munger, R.S. (1949), '*Guaiacum*, the Holy Wood from the New World', *Journal of the History of Medicine and Allied Sciences*, vol. 4, pp. 186–229.

Muñoz-Rodríguez, P., T. Carruthers, J.R.I. Wood et al. (2018), 'Reconciling Conflicting Phylogenies in the Origin of Sweet Potato and Dispersal to Polynesia', *Current Biology*, vol. 28, pp. 1246–56.

Munting, A. (1696), *Naauwkeurige beschryving der aardgewassen*, Pieter vander Aa, Leiden.

Myers, A.R.W., and M.B. Pritchard (1998), *Empire's Nature: Mark Catesby's New World Vision*, University of North Carolina Press, Chapel Hill and London.

Nash, R.C. (1992), 'Urbanization in the Colonial South: Charleston, South Carolina, as a Case Study', *Journal of Urban History*, vol. 19, pp. 3–130.

Nelson, E.C. (2014a), 'Catalpah – Called So by the Indians', *Hortus*, vol. 12, pp. 78–85.

Nelson, E.C. (2014b), 'Georg Dionysius Ehret, Mark Catesby and Sir Charles Wager's *Magnolia grandiflora*: An Early Eighteenth-Century Picture Puzzle Resolved', *Rhododendrons, Camellias and Magnolias*, vol. 65, pp. 36–51.

Nelson, E.C. (2015), '"The Truly Honest, Ingenious, and Modest Mr Mark Catesby, F.R.S.": Documenting his Life (1682/83–1749)', in E.C. Nelson and D.J. Elliott (eds), *The Curious Mister Catesby: A 'Truly Ingenious' Naturalist Explores New Worlds*, University of Georgia Press, Athens, GA.

Nelson, E.C. (2018), 'From Tubs to Flying Boats: Episodes in Transporting Living Plants', in A. MacGregor (ed.), *Naturalists in the Field: Collecting, Recording and Preserving the Natural World from the Fifteenth to the Twenty-First Century*, Brill, Leiden, pp. 578–606.

Nelson, E.C., and D.J. Elliott (eds) (2015), *The Curious Mister Catesby: A 'Truly Ingenious' Naturalist Explores New Worlds*, University of Georgia Press, Athens, GA.

Nicholls, S. (2009), *Paradise Found: Nature in America at the Time of Discovery*, University of Chicago Press, Chicago.

Ogilvie, B.W. (2006), *The Science of Describing: Natural History in Renaissance Europe*, University of Chicago Press, Chicago, IL, and London.

Ortiz-Hidalgo, C., and S. Pina-Oviedo (2019), 'Hematoxylin: Mesoamerica's Gift to Histopathology. *Palo de Campeche* (Logwood Tree), Pirates' Most Desired Treasure, and Irreplaceable Tissue Stain', *International Journal of Surgical Pathology*, vol. 27, pp. 4–14.

Overstreet, L.K. (2015), 'The Publication of Mark Catesby's *The Natural History of Carolina, Florida and the Bahama Islands*', in E.C. Nelson and D.J. Elliott (eds), *The Curious Mister Catesby: A Truly Ingenious' Naturalist Explores New Worlds*, University of Georgia Press, Athens, GA. pp. 155–72.

Parkinson, J. (1640), *Theatrum Botanicum: The Theater of Plants*, Thomas Cotes, London.

Petiver, J. (1715), 'Botanicaum Hortenese IV. II. Continued from No. 345', *Philosophical Transactions*, vol. 345, pp. 353–64.

Petiver, J. (1764), *Jacobi Petiveri Opera, historiam naturalem spectantia, or, Gazophylacium*, vol. 2, John Millan, London.

Plukenet, L. (1691), *Phytographia … Pars altera*, London.

Potter, J. (2006), *Strange Blooms: The Curious Lives and Adventures of the John Tradescants*, Atlantic Books, London.

Ray, J. (1693), *Historia plantarum generalis*, vol. 2, Samuel Smith & Benjamin Walford, London.

Reeds, K. (2015), 'Mark Catesby's Botanical Forerunners in Virginia', in E.C. Nelson and D.J. Elliott (eds), *The Curious Mister Catesby:*

A 'Truly Ingenious' Naturalist Explores New Worlds, University of Georgia Press, Athens, GA, pp. 27–38.
Renfrew, R., A.M. Strong, N.G. Perlut et al. (2015), 'Bobolink (*Dolichonyx oryzivorus*)', in P.G. Rodewald (ed.), *The Birds of North America*, Cornell Lab of Ornithology, Ithaca, NY, https://birdsoftheworld.org/bow/species/boboli/cur/introduction.
Reveal, J.L. (1992), *Gentle Conquest: The Botanical Discovery of North America*, Starwood, Washington, DC.
Reveal, J.L. (2009), 'Identification of the Plant and Associated Animal Images in Catesby's *Natural History*, with nomenclatural notes and comments', *Rhodora*, vol. 111, pp. 273–388.
Reveal, J.L. (2012), 'A Nomenclatural Summary of the Plant and Animal Names Based on Images in Mark Catesby's *Natural History* (1729–1747)', *Phytoneuron*, 2012-11, pp. 1–32.
Reveal, J.L. (2013), 'Identification of the Plants and Animals Illustrated by Mark Catesby for his *Natural History of Carolina, Florida and the Bahama Islands*', *Phytoneuron*, 2013-6, pp. 1–55.
Reveal, J.L. (2015), 'Identifications of the Plants and Animals Illustrated by Mark Catesby for *The Natural History of Carolina, Florida and the Bahama Islands*', in E.C. Nelson and D.J. Elliott (eds), *The Curious Mister Catesby: A 'Truly Ingenious' Naturalist Explores New Worlds*, University of Georgia Press, Athens, GA, pp. 331–50.
Riello, G. (2013), *Cotton: The Fabric that Made the Modern World*, Cambridge University Press, Cambridge.
Salley, A.S. (1919), *The Introduction of Rice Culture into South Carolina*, printed for the Commission by the State Company, Columbia, SC.
Schmidt, J.O., and M.S. Blum (1977), 'Adaptations and Responses of *Dasymutilla occidentalis* (Hymenoptera: Mutillidae) to Predators', *Entomologia Experimentalis et Applicata*, vol. 21, pp. 99–111.
Sloane, H. (1707), *A Voyage to the Islands Madera, Barbados, Nieves, S. Christophers and Jamaica; with the Natural History of the Herbs and Trees, Four-footed Beasts, Fishes, Birds, Insects, Reptiles, &c. of the Last of Those Islands*, vol. 1, London.
Sloane, H. (1725), *A Voyage to the Islands Madera, Barbados, Nieves, S. Christophers and Jamaica; with the Natural History of the Herbs and Trees, Four-footed Beasts, Fishes, Birds, Insects, Reptiles, &c. of the Last of Those Islands*, vol. 2, London.
Smith, H.A.M. (1912), 'The Baronies of South Carolina', *South Carolina Historical and Genealogical Magazine*, vol. 13, pp. 119–25.
Smith, J.E. (1821), *A Selection of the Correspondence of Linnaeus, and Other Naturalists, from the Original Manuscripts*, vol. 1, Longman, Hurst, Rees, Orme, and Brown, London.
Stoker, D. (2004), 'Harley, Edward, Second Earl of Oxford and Mortimer (1689–1741)', *Oxford Dictionary of National Biography*, https://doi.org/10.1093/ref:odnb/12337.
Tennent, J. (1736), *Every Man His Own Doctor, or, The Poor Planter's Physician: Prescribing, Plain and Easy Means For Persons To Cure Themselves of All, or Most of the Distempers, Incident to This Climate, and With Very Little Charge, the Medicines Being Chiefly of the Growth and Production of This Country*, B. Franklin, Philadelphia, PA.
Thiers, B.M. (2020), *Herbarium: The Quest to Preserve and Classify the World's Plants*, Timber Press, Portland, OR.
Thorpe, F.N. (1909), *The Federal and State Constitutions: Colonial Charters, and Other Organic Laws of the States, Territories, and Colonies, Now or Heretofore Forming the United States of America*, Government Printing Office, Washington, DC.
Tinniswood, A. (2019), *The Royal Society & the Invention of Modern Science*, Head of Zeus, London.
Titford, M. (2005), 'The Long History of Hematoxylin', *Biotechnic & Histochemistry*, vol. 80, pp. 73–8.
Tubby, K.V., and A. Pérez Sierra (2015), 'Pests and Pathogen Threats to Plane (*Platanus*) in Britain', *Arboricultural Journal*, vol. 37, pp. 85–98.
Ulmer, T., and J.M. MacDougal (2004), *Passiflora: Passionflowers of the World*, Timber Press, Portland, OR.
Van Nest, B.N., A.A. Edge, M.V. Feathers et al. (2021), 'Bees Provide Pollination Service to *Campsis radicans* (Bignoniaceae), a Primarily Ornithophilous Trumpet Flowering Vine', *Ecological Entomology*, vol. 46, pp. 117–27.
Watson, P. (2020), 'Knight, John II (c.1686–1733), of Gosfield Hall, Essex', The History of Parliament, http://www.historyofparliamentonline.org/volume/1690-1715/member/knight-john-ii-1686-1733.
Webb, S.S. (1966), 'The Strange Career of Francis Nicholson', *William and Mary Quarterly*, vol. 23, pp. 513–48.
Weir, R. (1997), *Colonial South Carolina: A History*, University of South Carolina Press, Columbia.
Wilbur, R.L. (1990), 'Identification of the Plants Illustrated and Described in Catesby's *Natural History of the Carolinas, Florida and the Bahamas*', *Sida*, vol. 14, pp. 29–48.
Wood, B. (2004), 'Perceval, John, First Earl of Egmont (1683–1748)', *Oxford Dictionary of National Biography*, https://doi.org/10.1093/ref:odnb/21911.
Wood, P.H. (1974), *Black Majority: Negroes in Colonial South Carolina from 1670 through the Stono Rebellion*, Alfred A. Knopf, New York.
Woodville, W. (1790), *Medical Botany Containing Systematic and General Descriptions, with Plates, of All The Medicinal Plants, Indigenous and Exotic, Comprehended in the Catalogues of the Materia Medica, As Published by the Royal Colleges of Physicians of London and Edinburgh: Accompanied With a Circumstantial Detail of Their Medicinal Effects, and of The Diseases In Which They Have Been Most Successfully Employed*, vol. I, James Phillips, London.
Wulf, A. (2008), *The Brother Gardeners: Botany, Empire and the Birth of an Obsession*, Heinemann, London.

Index

Numbers in italics indicate illustrations in the text, whilst those in bold indicate plates.

This book is generously supported by a donation from
THE UNWIN FUND FOR BODLEIAN LIBRARY PUBLISHING

First published in 2024 by
Bodleian Library Publishing
Broad Street, Oxford OX1 3BG
www.bodleianshop.co.uk

ISBN: 978 185124 639 7

Text © Stephen Harris, 2024

All images, unless specified below, are from Mark Catesby, *The Natural History of Carolina, Florida, and the Bahama Islands*, 1754. Bodleian Library, Arch.Nat. hist. M 4-5, © Bodleian Libraries, University of Oxford, 2024. Figs 1, 7 and 11b are from the collection of the Sherardian Library of Plant Taxonomy, University of Oxford. Figs 3, 4, 5a, 5b, 6, 8, 9, 10 and 11a are from the collection of Oxford University Herbaria, Department of Biology.

This edition © Bodleian Library Publishing, University of Oxford, 2024

Stephen Harris has asserted his right to be identified as the author of this Work.

All rights reserved.

No part of this book may be reproduced, stored in a retrieval system, or transmitted in any form or by any means, electronic, mechanical, photocopying, recording, or otherwise, without the written permission of the Bodleian Library, except for the purpose of research or private study, or criticism or review.

Publisher: Samuel Fanous
Managing Editor: Susie Foster
Editor: Janet Phillips
Picture Editor: Leanda Shrimpton
Designed and typeset by Dot Little in the Bodleian Library in 11/16pt Monotype Baskerville
Printed and bound by C&C Offset Printing Co., Ltd. on 157 gsm Chinese Golden Sun matt art paper

MIX
Paper | Supporting responsible forestry
FSC® C008047
www.fsc.org

British Library Catalogue in Publishing Data
A CIP record of this publication is available from the British Library